The Theory of Lattice-Ordered Groups

Mathematics and Its Applications

Managing Editor:

Volume 307

The Theory of Lattice-Ordered Groups

by

V. M. Kopytov
Institute of Mathematics,
Russian Academy of Sciences,
Siberian Branch,
Novosibirsk, Russia

and

N. Ya. Medvedev
Altai State University,
Barnaul, Russia

Springer-Science+Business Media, B.V.

A C.I.P. Catalogue record for this book is available from the Library of Congress.

ISBN 978-90-481-4474-7 ISBN 978-94-015-8304-6 (eBook)
DOI 10.1007/978-94-015-8304-6

Printed on acid-free paper

Originally published by Kluwer Academic Publishers in 1994.
Softcover reprint of the hardcover 1st edition 1994

Contents

Preface

A partially ordered group is an algebraic object having the structure of a group and the structure of a partially ordered set which are connected in some natural way. These connections were established in the period between the end of 19th and beginning of 20th century. It was realized that ordered algebraic systems occur in various branches of mathematics bound up with its fundamentals. For example, the classification of infinitesimals resulted in discovery of non-archimedean ordered algebraic systems, the formalization of the notion of real number led to the definition of ordered groups and ordered fields, the construction of non-archimedean geometries brought about the investigation of non-archimedean ordered groups and fields. The theory of partially ordered groups was developed by: R.Dedekind, O.Hölder, D.Gilbert, B.Neumann, A.I.Mal'cev, P.Hall, G.Birkhoff.

These connections between partial order and group operations allow us to investigate the properties of partially ordered groups. For example, partially ordered groups with interpolation property were introduced in F.Riesz's fundamental paper [1] as a key to his investigations of partially ordered real vector spaces, and the study of ordered vector spaces with interpolation properties were continued by many functional analysts since. The deepest and most developed part of the theory of partially ordered groups is the theory of lattice-ordered groups. In the 40s, following the publications of the works by G.Birkhoff, H.Nakano and P.Lorenzen the theory of lattice-ordered groups became the subject of the study in its own rights. Recently, lattice-ordered groups have cropped up in many other areas of mathematics, for instance, in the theory of Bezout domains, ordered fields, and universal algebra.

In 1963 W.Ch.Holland proved that every lattice-ordered group can be represented as a group of order-preserving permutations of a totally ordered set. This result is the cornerstone in the theory of lattice-ordered groups since it is the only tool available for studying arbitrary lattice-ordered groups. This theorem also provides an application of group-theoretical methods to l-groups.

The purpose of this book is to make the deep results of theory of lattice-ordered groups accessible to mathematicians generally, to portray its structure and indicate some of its applications to the group theory. This book is self contained for anyone familiar with the basic results on group theory. Graduate students and researchers in ordered structures and group theory will find this book valuable both an introduction into the theory of partially ordered groups, and as presentation of new ideas and results in this theory.

Chapters 1-3 contain preliminary material needed for reading the main body of the book. Here a great part of concepts and their basic properties is introduced.

In Chapter 4, the theory of ordered permutation lattice-ordered groups is considered.

Chapter 5 is devoted to the theory of right-ordered groups.

In Chapter 6, the basic facts of the theory of totally ordered groups are proved. Some applications to group theory are mentioned here too.

In Chapter 7 a lot of embedding theorems of lattice-ordered groups are proved. It is established that each lattice-ordered group G can be embedded into a divisible lattice-ordered group. Free products and amalgamation property of l-groups are dwelt with.

Chapter 8 is concerned with the properties of lattice of lattice-ordered groups such as: complete distributivity or compactness.

Chapters 9 - 13 represent a systematic study of the structure of the lattice and the semigroup of varieties of lattice-ordered groups.

The final Chapter 14 provides an exposition of the theory of quasi-varieties of lattice-ordered groups.

The statements are denoted by three numbers, e.g., Theorem 3.1.2. means: Chapter 3, Section 1, Theorem 2. They are referred to in full except when they appear in the same chapter; in such case the first two numbers are omitted.

It is a pleasure to acknowledge the influence of research which stimulated our interest in and research on ordered groups: A.I.Mal'cev's,

M.I.Kargapolov's and A.I.Kokorin's works on totally ordered groups, W.Ch.Holland's and S.H.McCleary's works on ordered permutation lattice-ordered groups, P.Conrad's works on lattice-ordered groups, D.M.Smirnov's and P.Conrad's works on right-ordered groups.

We found it expedient to use A.M.W.Glass's notation from his book [6], since it is best suited for our purposes.

May 9 1993

Valeri M. Kopytov
Institute of Mathematics
Siberian Branch RAN
Novosibirsk
Russia

Nikolai Ya. Medvedev
Altai State University
Barnaul
Russia

Symbol Index

Symbol	**(Listed in order of appearance)**	**Section**
$\bar{\prod}_{\alpha\in I} M_\alpha$	cartesian product of po-sets, M_α $(\alpha \in I)$	1.1
$\vec{\prod}_{\alpha\in I} M_\alpha$	lexicographic product of po-sets $M_{\alpha\in I}$	1.1
$x \vee y$	the least upper bound of x, y	1.2
$x \wedge y$	the greatest lower bound of x, y	1.2
$A(\Omega)$	the set of all order permutations of Ω	1.2
$\mathbf{Z}$	o-group of integers	2.1
$\mathbf{Q}$	o-group of rationals	2.1
$\mathbf{R}$	o-group of reals	2.1
x^+	$x \vee e$	2.2
x^-	$x \wedge e$	2.2
$\|x\|$	$x \vee x^{-1}$	2.2
$G \cong H$	l-groups G and H are l-isomorphic	2.3
G/H	quotient group	2.3
$A \times B$	direct product of l-groups A and B	2.3
$\vec{\prod}_{\alpha\in I} A_\alpha$	lexicographic product of l-groups	2.3
$\bar{\prod}_{\alpha\in I} G_\alpha$	cartesian product of l-groups	2.3
$\prod_{\alpha\in I} G_\alpha$	direct productof l-groups	2.3
$R(G:H)$	set of right cosets of H in G	2.3
$(M)_l$	convex l-subgroup of G generated M	2.3
$\mathcal{C}(G)$	the set of convex l-subgroups of G	3.1

Symbol	(Listed in order of appearance)	Section
$\overset{\mathcal{C}}{\bigvee}, \overset{\mathcal{C}}{\bigwedge}$	lattice operations in $\mathcal{C}(G)$	3.1
$S_A(M)$	A-invariant subsemigroup	3.2
$n(A)$	normalizer of A	3.2
$c(A)$	centralizer of A	3.2
$[A, B]$	commutator subgroup A and B	3.2
$[a, b]$	commutator a and b, $a^{-1}b^{-1}ab$	3.2
$V_\alpha(g)$	value of g	3.3
$\bar{V}_\alpha(g)$	$(V_\alpha(g), g)_l$	3.3
$g^\perp$	polar of element g	3.4
$X^\perp$	polar of set X	3.4
$\mathcal{P}(G)$	boolean algebra of polars of G	3.4
$\overset{\mathcal{P}}{\bigvee}, \overset{\mathcal{P}}{\bigwedge}$	lattice operations in $\mathcal{P}(G)$	3.4
Ω	totally ordered set	4.1
$A(\Omega)$	the group of order permutations of Ω	4.1
(G, Ω)	ordered permutations group	4.1
$\equiv$	convex G-congruence on Ω	4.2
Δ	G-block	4.2
L_ϑ	the lazy subgroup	4.2
$G_{(\Delta)}$	stabilizer of Δ	4.2
G_Δ	pointwise stabilizer of Δ	4.2
G_α	stabilizer of α	4.2
$\bar{\Omega}$	Dedekind completion of Ω	4.2
$supp(g)$	support of g	4.3
(ζ_k, ζ^k)	covering pair of G-congruences	4.5
$R(\Omega)$	elements of $A(\Omega)$ that live on the right	4.6
$L(\Omega)$	elements of $A(\Omega)$ that live on the left	4.6
$B(\Omega)$	bounded elements of $A(\Omega)$	4.6
$\mathrm{sgr}(M)$	subsemigroup generated by M	5.1
$K[G]$	group ring G	5.2
$\mathrm{gp}(M)$	subgroup or group generated by M	5.5
$S(M)$	normal subsemigroup generated by M	6.1
$x \ll y$	$x^n < y$ for all $n \in \mathbf{Z}$	6.2

Symbol	(Listed in order of appearence)	Section
$\mathcal{K}$	class of l-groups	7.1
$F_{\mathcal{K}}(G,\pi)$	$\mathcal{K}$-free l-group over po-group G	7.1
$G[\mathcal{F}]$	SP-hull of G	7.2
$A\mathrm{Wr}B$	wreath product of groups A and B	7.3
$A\mathrm{wr}B$	restricted wreath product	7.3
$X\overleftarrow{\bigcup}Y$	the order union of po-sets X and Y	7.4
$\prod\limits_{\gamma\in\Gamma}^{*} G_\gamma$	free product of groups $G_\gamma(\gamma\in\Gamma)$	7.5
$\prod\limits_{\alpha\in A}^{l-*} G_\alpha$	l-free product of l-groups $G_{\alpha\in A}$	7.5
$U(X)$	set of upper bounds of X	8.1
$D(G)$	the distributive radical of G	8.2
$x^\top$	thread of G	8.3
$\mathcal{A}$	l-variety of all abelian l-groups	9.1
$\mathcal{N}_k$	l-variety of k-nilpotent l-groups	9.1
$\mathbf{L}$	the lattice of all l-varieties	9.1
$\mathcal{W}a$	l-variety of weakly abelian l-groups	9.1
$\mathcal{N}$	l-variety of normal-valued l-groups	9.1
$\mathcal{R}$	l-variety of representable l-groups	9.1
$\mathrm{var}_l G$	l-variety generated by l-group G	9.1
$\mathrm{var}_l \mathcal{K}$	l-variety generated by class $\mathcal{K}$	9.1
$\mathrm{loc}\mathcal{N}$	class of locally nilpotent l-groups	9.5
$\mathcal{W}_{p/q}$	Feil l-variety	9.6
$(G,\Omega)\mathrm{Wr}(H,T)$	wreath product of (G,Ω) and (H,T)	11.1
$(G,\Omega)\mathrm{wr}(H,\|!T)$	restricted wreath product of	11.1
$\mathrm{Wr}^n(G,\Omega)$	n^{th} Wreath product of (G,Ω)	11.2
$\mathrm{wr}^n(G,\Omega)$	n^{th} wreath product of (G,Ω)	11.2
$\mathcal{E}$	trivial l-variety	11.2
$\mathcal{L}$	l-variety of all l-groups	11.2
$\dim\mathcal{U}$	dimension of l-variety $\mathcal{U}$	11.2
$\mathcal{S}_p$	Scrimger l-variety	12.3
$\mathcal{M}_0,\mathcal{M}^+,\mathcal{M}^-$	Medvedev's l-varieties	12.4
Λ	the lattice of all l-quasivarieties	14.1

Symbol	**(Listed in order of appearance)**	**Section**
$\mathrm{q}_l G$	l-quasivariety generated by l-group G	14.1
R_m	the l-quasivariety defined by the implication $(x^n = y^n) \Rightarrow x = y$	14.2
$Th_\forall$	universal theory of G	14.3

Chapter 1

Lattices

1.1 Partially Ordered Sets

1 ^{0}. A non-empty set M is called *partially ordered* (po-*set*) if it is equipped with binary relation $\leq$ on M satisfying the axioms:

PO.1.	$x \leq x$	(reflexivity);
PO.2.	if $x \leq x$ and $y \leq x$ then $x = y$	(antisymmetric);
PO.3.	if $x \leq y$ and $y \leq z$ then $x \leq z$	(transitivity).

This binary relation $\leq$ is called the *partial order* on the set M. A po-set M is denoted as $\{M; \leq\}$. If $x \leq y$ or $y \leq x$, then x and y are *comparable*, otherwise these elements are *incomparable*; $x < y$ means that $x \leq y$ and $x \neq y$.

A po-set M is *totally ordered* or *linearly ordered* if every two elements $x, y \in M$ are comparable, i.e., either $x \leq y$ or $y \leq x$. A totally ordered set M is *well-ordered* if every non-empty subset of M has the least element. We suppose that Zermelo's theorem is true (it is equivalent to the axiom of choice), i.e., arbitrary set can be well-ordered.

Let A be a subset of a po-set $\{M; \leq\}$, then A is also a po-set under order $\leq_{|A}$, where $x \leq_{|A} y$ if and only if $x \leq y$ in the po-set $\{M; \leq\}$.

Any subset of real numbers $\mathbf{R}$ with usual order is totally ordered.

EXAMPLE 1. Let X be an arbitrary set. There is a natural order $\leq$ on the set $F(X, \mathbf{R})$ of all functions from X into $\mathbf{R}$, where $f \leq g$ if and only if $f(x) \leq g(x)$ for all $x \in X$. This partial order on $F(X, \mathbf{R})$ is called a *pointwise order* on $F(X, \mathbf{R})$.

EXAMPLE 2. There is another partial order on $F(X, \mathbf{R})$. First we order the set X under the well-order $\prec$:

$$x_1 \prec x_2 \prec \ldots x_\alpha \prec x_{\alpha+1} \prec \ldots$$

Now, for any two given $f, g \in F(X, \mathbf{R})$ we define $f \leq g$ if and only if $f(x_\alpha) \leq g(x_\alpha)$ where x_α is the least element (under well-ordering of X) of the set $\{x_\xi | f(x_\xi) \neq g(x_\xi)\}$. This order is called a *lexicographic* order.

Let $\leq$, $\leq'$ be two partial orders on the set M. If for every two elements $x, y \in M$ such that $x \leq y$ it follows that $x \leq' y$, then the order $\leq'$ is an *extension of the order* $\leq$. It is not difficult to verify that the order on $F(X, \mathbf{R})$ from Example 2 is an extension of the order on $F(X, \mathbf{R})$ from Example 1.

2^0. Example 1 is a special case of a general algebraic construction. Let $\{M_\alpha | \alpha \in I\}$ be a collection of po-sets and let $\bar{\prod\limits_{\alpha \in I}} M_\alpha$ be the set of all functions f from I into $\bigcup\limits_{\alpha \in I} M_\alpha$, $f(\alpha) \in M_\alpha$ for all $\alpha \in I$. Now, for given two functions $f, g \in \bar{\prod\limits_{\alpha \in I}} M_\alpha$ we define $f \leq g$ if $f(\alpha) \leq g(\alpha)$ for all $\alpha \in I$. The po-set $\bar{\prod\limits_{\alpha \in I}} M_\alpha$ is called a *Cartesian product* or a *cardinal product* of po-sets M_α, $\alpha \in I$. If the set I is finite, $I = \{1, 2, \ldots, n\}$ then the Cartesian product is denoted as $M_1 \times M_2 \times \ldots \times M_n$.

It is possible to define on $\bar{\prod\limits_{\alpha \in I}} M_\alpha$ another partial order which is connected with some well-ordering of I :

$$\alpha_1 \prec \alpha_2 \prec \ldots \prec \alpha_\xi \prec \alpha_{\xi+1} \prec \ldots.$$

For $f, g \in \bar{\prod\limits_{\alpha \in I}} M_\alpha$ we set $f < g$ if $f(\alpha) < g(\alpha)$, where α is the least under the well-order of I element of the set $\{\alpha_\xi \in I | f(\alpha_\xi) \neq g(\alpha_\xi)\}$. The set $\bar{\prod\limits_{\alpha \in I}} M_\alpha$ under the above order is called a *lexicographic product* of sets M_α and denoted $\overrightarrow{\prod\limits_{\alpha \in I}} M_\alpha$. We write $M_1 \vec{\times} M_2 \vec{\times} \ldots \vec{\times} M_n$ or $\overrightarrow{\prod\limits_{1 \leq i \leq n}} M_\alpha$ if I is finite, $I = \{1, 2, \ldots, n\}$.

Finally, let $\prec$ be dual to some well-ordering totally order on I, (i.e., every non-empty subset of I has a maximal element):

$$\ldots, \alpha_{\xi-1} \prec \alpha_\xi \prec \ldots, \prec \alpha_{-1} \prec \alpha_0$$

For $f, g \in \bar{\prod\limits_{\alpha\in I}} M_\alpha$ we set $f < g$ if $f(\alpha) < g(\alpha)$ where α is the greatest element under $\prec$ of the set $\{\alpha_\xi \in I | f(\alpha_\xi) \neq g(\alpha_\xi)\}$. The set $\bar{\prod\limits_{\alpha\in I}} M_\alpha$ under this order is called an *antilexicographic product* of sets M_α and denoted as $\overleftarrow{\prod\limits_{\alpha\in I}} M_\alpha$. We write it $M_1 \overleftarrow{\times} M_2 \overleftarrow{\times} \ldots \overleftarrow{\times} M_n$ or $\overleftarrow{\prod\limits_{1\leq i\leq n}} M_\alpha$ if I is finite, $I = \{1, 2, \ldots, n\}$.

1.2 Lattices

1⁰. Let x, y be elements of the po-set M. If $u, v \in M$ and $x \leq u$, $y \leq u$, $v \leq x$, $v \leq y$ then u is an *upper bound* and v is a *lower bound* of elements x, y correspondingly. A po-set M is called *directed* or *directed up* if for every two elements their upper bound exists. A po-set M is *directed down* if every pair of elements has a lower bound.

The Cartesian and lexicographic products of directed sets are directed too, therefore the set $F(X, \mathbf{R})$ is directed under the pointwise order (see Example 1) and under the lexicographic order.

If in the po-set M for elements $x, y \in M$ there exists an upper bound a, such that $a \leq z$ for every upper bound z of elements x, y, then a is called the *least upper bound* or a *join* of x, y and it is denoted by $x \vee y$. If for elements x, y there exists a lower bound b such that $b \leq t$ for every lower bound t of element x, y, then b is the *greatest lower bound* or a *meet* of x, y and it is denoted by $x \wedge y$. It is obvious that if the join $x \vee y$ exists then it is unique. This is true for $x \wedge y$, too.

A po-set L is a *lattice* if $x \vee y$ and $x \wedge y$ exist for all $x, y \in L$. Any totally ordered set is a lattice.

EXAMPLE 1. The po-set $F(X, \mathbf{R})$ of all functions $f : X \longrightarrow \mathbf{R}$ from X to the set $\mathbf{R}$ of reals is the lattice under the pointwise order on $F(X, \mathbf{R})$ and

$$(f \vee g)(x) = \begin{cases} g(x) & \text{if} \quad f(x) \leq g(x), \\ f(x) & \text{if} \quad g(x) \leq f(x), \end{cases}$$

$$(f \wedge g)(x) = \begin{cases} f(x) & \text{if} \quad f(x) \leq g(x), \\ g(x) & \text{if} \quad g(x) \leq f(x). \end{cases}$$

EXAMPLE 2. Let $C(0,1)$ be the set of all continuous real-valued functions on the segment [0,1]. Then $C(0,1) \subseteq F([0,1], \mathbf{R})$ and $C(0,1)$ are lattices under the order induced from $F([0,1], \mathbf{R})$. It is easy to see that $f \vee g$ and $f \wedge g$ (under the order of $F([0,1], \mathbf{R})$ are continuous functions for every pair $f, g \in C(0,1)$.

EXAMPLE 3. Let Ω be a totally ordered set and $A(\Omega)$ be the set of all *automorphisms* or *order permutations* of Ω. This means that $f \in A(\Omega)$ if and only if f is a one-to-one function from Ω onto Ω and $\alpha < \beta$ in Ω implies $\alpha f < \beta f$. We define $f \leq g$ for $f, g \in A(\Omega)$ if $\alpha f \leq \alpha g$ for all $\alpha \in \Omega$. It is clear that $\leq$ is a partial order on $A(\Omega)$. Moreover, $A(\Omega)$ is a lattice. Indeed, let $f, g \in A(\Omega)$. Let, now, a be a function from Ω into Ω such that $\alpha a = \max\{\alpha f, \alpha g\}$ in Ω. If $\alpha \leq \beta$ in Ω then $\alpha f \leq \beta f$, $\alpha g \leq \beta g$, and $\max\{\alpha f, \alpha g\} \leq \max\{\beta f, \beta g\}$, i.e., a is an order-preserving function from Ω into Ω. Now, if $\gamma \in \Omega$ then there are $\alpha, \beta \in \Omega$ such that $\alpha f = \gamma$, $\beta g = \gamma$. Without loss of generality we may assume, that $\alpha \leq \beta$. Then $\alpha g \leq \beta g = \gamma = \alpha f$ and $\alpha a = \gamma$. It means that a is a function from Ω onto Ω, i.e., a is an automorphism of Ω. It is easy to verify that $a = f \vee g$ in $A(\Omega)$. By similar arguments, the function $b : \Omega \longrightarrow \Omega$ such that $\alpha b = \min\{\alpha f, \alpha g\}$ is the element of $A(\Omega)$ and $b = f \wedge g$. □

2^0. If L is a lattice then it is possible to define on L two binary operations: the join $\vee$ and the meet $\wedge$. Then L is an algebraic system of the signature $\{\vee, \wedge\}$. Any lattice can be characterized by the following identities:

L.1.	$x \vee x = x,$	$x \wedge x = x$.
L.2.	$x \vee y = y \vee x,$	$x \wedge y = y \wedge x$.
L.3.	$(x \vee y) \vee z = x \vee (y \vee z),$	$(x \wedge y) \wedge z = x \wedge (y \wedge z)$
L.4.	$(x \vee y) \wedge x = x,$	$(x \wedge y) \vee x = x$.

Identities L.1.–L.4. are called the laws of idempotenty, commutativity, associativity and absorption correspondingly.

THEOREM 1. *For any lattice L, identities* L.1.–L.4. *are valid in L. Conversely, let L be an algebraic system of the signature $\{\vee, \wedge\}$ such that identities* L.1.–L.4. *are valid in L. Then L is a lattice under the partial order defined by the rule: $x \leq y$ if and only if $x \vee y = y$.*

The proof of this theorem is straightforward. □

1.3 Properties of lattices

1⁰. A subset A of lattice L is a *sublattice* of L if $x \vee y \in A$, $x \wedge y \in A$ for any pair of elements $x, y \in A$. We note here that a subset A of the lattice L can be a lattice under the induced partial order but can not be a sublattice of L. It is easy to construct the suitable examples.

A subset A of po-set M is called *convex* if $a \leq x \leq b$, $a, b \in A$ imply $x \in A$. If a convex subset A of the lattice L is directed up and directed down, then A is a sublattice of L.

An *order map* from po-set M into po-set N is a map $\varphi : M \longrightarrow N$ such that $x\varphi \leq y\varphi$ for every $x, y \in M$, $x \leq y$. An one-to-one order mapping φ from a po-set M onto a po-set N is called an *isomorphism* of po-sets if the inverse map φ^{-1} is an order map, too. The po-sets M and N are called *order isomorphic* po-sets in this case.

Generally, an order isomorphism φ from lattice L into lattice S may not preserve joins and meets of the elements. We can guarantee only that $x\varphi \vee y\varphi \leq (x \vee y)\varphi$, $x\varphi \wedge y\varphi \geq (x \wedge y)\varphi$.

Since every lattice is a po-set and an algebraic system simultaneously we must consider different types of maps for lattices. Let L and S be lattices. A map $\varphi : L \rightarrow S$ is a *homomorphism of lattices* if $(x \vee y)\varphi = (x)\varphi \vee (y)\varphi$, $(x \wedge y)\varphi = x\varphi \wedge y\varphi$ for every $x, y \in L$. It is clear that every homomorphism of lattices is an order map.

An *isomorphism of lattices* L and S is an one-to-one homomorphism from L onto S. The lattices L, S are called *isomorphic* in this case. It is easy to see that a homomorphism from lattice L into lattice S maps L onto the sublattice of S.

Let I be a set of indeces and let for any index $\alpha \in I$ corresponds some lattice L_α. We consider the set $\prod\limits_{\alpha \in I} L_\alpha = \bar{L}$ of all functions from

I into $\bigcup_{\alpha \in I} L_\alpha$ such that $\alpha f \in L_\alpha$. We define two operations $\vee, \wedge$ on $\bar{L}$: $\alpha(f \vee g) = \alpha f \vee \alpha g$, $\alpha(f \wedge g) = \alpha f \wedge \alpha g$ for $f, g \in \bar{L}$, $\alpha \in I$. The simple verification shows that these operations satisfy axioms L.1.- L.4 and so $\bar{L}$ is a lattice. It is called a *Cartesian product of lattices* L_α, $\alpha \in I$. Of course, the Cartesian product of lattices is a po-set which is a Cartesian product of po-sets L_α, $\alpha \in I$, in the sense of definition from Section 1. We emphasize this statement because of its importance.

PROPOSITION 1. *The Cartesian product of lattices is a lattice.*

2⁰. For lexicographic products (see Section 1) the situation is more complicated. It is easy to construct an example which shows that the class of all lattices is not closed under lexicographic products. Before formulating the suitable criterion, we give two definitions. If a lattice contains the least element then this element is called the *null* of the lattice and denoted $\mathbf{0}$. If a lattice contains the greatest element then it is called the *unit* of the lattice and denoted $\mathbf{1}$.

PROPOSITION 2. *The lexicographic product of the lattices L_α, $\alpha \in I$, with the well-ordered set I is a lattice if and only if for every lattice L_α which contains incomparable elements, every L_β with $\alpha < \beta$ contains $\mathbf{0}$ and $\mathbf{1}$.*

The proof of this proposition is left to the reader. □

For a subset A of po-set M the element a is called the *least upper bound* of A if $x \leq a$ for every $x \in A$ and $a \leq u$ for any upper bound u of A. The element b is called the *greatest lower bound* of A if $b \leq x$ for every $x \in A$ and $v \leq b$ for any lower bound v of A. Then we denote $a = \sup A = \vee A$, $b = \inf A = \wedge A$. The lattice L is *complete* if for any non-empty subset A there are $\vee A$, $\wedge A$ in L. If L contains $\vee A$ and $\wedge A$ for any bounded subset A then L is a *conditionally complete lattice.*

1.4 Distributive and Modular Lattices. Boolean Algebras

1⁰. The lattice L is *distributive* if for L the distributive laws are valid:

L.5. $x \vee (y \wedge z) = (x \vee y) \wedge (x \vee z)$, $x \wedge (y \vee z) = (x \wedge y) \vee (x \wedge z)$.

The lattice L is *modular* if the modular laws are fulfilled:

L.6. $x \vee (y \wedge z) = (x \vee y) \wedge (x \vee z)$ for every $x, y, z \in L$, $x \leq y$.

The lattices from Examples 1–3 from Section 2 are distributive. It is clear that any totally ordered set is a distributive lattice. The class of all distributive lattices is closed under cardinal products, homomorphisms and sublattices.

If we consider a lattice, particularly a finite lattice, then we represent it sometimes by a graph: we represent the elements of the lattice as the points and if $x \leq y$, we place the point x below y and join these points by segment.

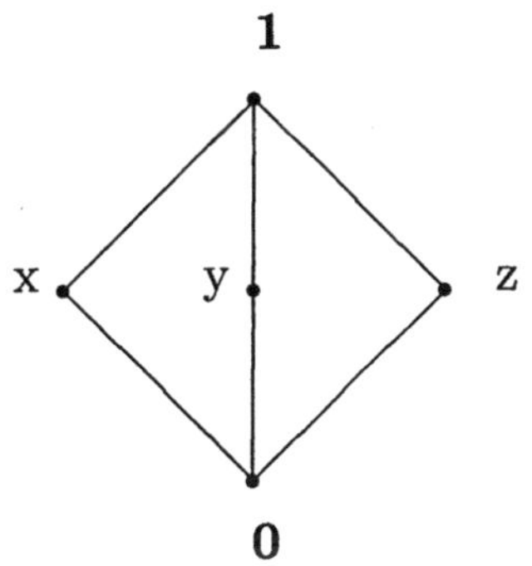

Fig. 1

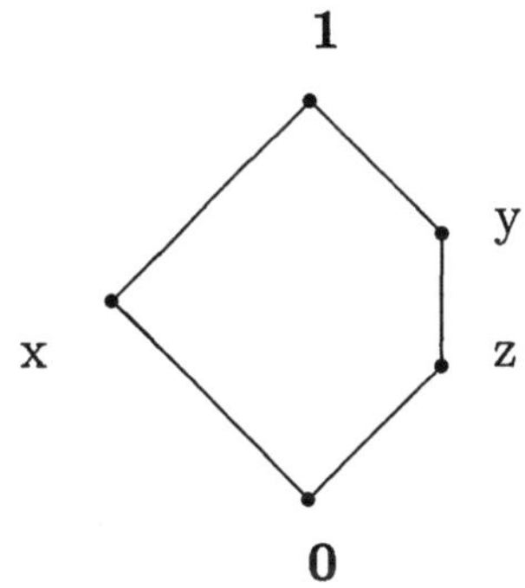

Fig. 2

EXAMPLE 1. Lattice L_1 defined by the graph of Fig.1 is nondistributive.

EXAMPLE 2. Lattice L_2 defined by the graph of Fig.2 is nondistrbutive and nonmodular.

THEOREM 1. *Lattice L is distributive if and only if $x \vee y = x \vee z$, $x \wedge y = x \wedge z$ imply $y = z$ for every elements x, y, z in L.*□

COROLLARY 1. *Lattice L is nondistributive if and only if L contains a sublattice which is isomorphic to lattice L_1 or to L_2. Lattice L is nonmodular if and only if L contains a sublattice which is isomorphic to L_2.*□

Lattice L is *Brouwerian* if L is complete and for L the following identity of the infinite distributivity is fulfilled: for all $x, y_\alpha \in L$,

$\alpha \in I$, the equality $x \wedge (\bigvee_{\alpha\in I} y_\alpha) = \bigvee_{\alpha\in I}(x \wedge y_\alpha)$ holds. If in complete lattice L the dual identity $x \vee (\bigwedge_{\alpha\in I} y_\alpha) = \bigwedge_{\alpha\in I}(x \vee y_\alpha)$ is fulfilled, then L is *dual Brouwerian.* Complete lattice L is *completely distributive* if $\bigvee_{\alpha\in A}(\bigwedge_{\beta\in B} x_{\alpha,\beta}) = \bigwedge_{\rho\in A^B}(\bigvee_{\alpha\in A} x_{\alpha,\rho})$ where A, B are sets of indeces, A^B is the set of all functions from A into B . Here $\bigvee_{\alpha\in\alpha} x_\alpha = \sup\{x_\alpha | \alpha \in A\}$, $\bigwedge_{\alpha\in A} x_\alpha = \inf\{x_\alpha | \alpha \in A\}$. It is clear that every Brouwerian lattice and every dually Brouwerian lattice are distributive.

2^0. Let L be a lattice with $\mathbf{0}$ and $\mathbf{1}$. An element x' is called a *complement* to x if $x \vee x' = \mathbf{1}$, $x \wedge x' = \mathbf{0}$. A distributive lattice B is called a *Boolean lattice* if B is a distributive lattice with $\mathbf{0}$ and $\mathbf{1}$ and for any element $x \in B$ there exists a complement x'. By Theorem 1, the complement x' for x is unique in any Boolean lattice. Then in every Boolean lattice B we can define the unary operation $'$. This operation has the following properties:

B.1. $(x')' = x,$

B.2. $(x \vee y)' = x' \wedge y', \quad (x \wedge y) = x' \vee y',$

B.3. $(x \vee x') \wedge y = y, \quad (x \wedge x') \vee y = y.$

Any Boolean lattice B can be regarded as an algebraic system with operations $\vee, \wedge, {}'$ which satisfy identities L.1.–L.5. and B.1.–B.3. This algebraic system is called a *Boolean algebra.* A subset S of a Boolean algebra B is a *subalgebra* of B if S is closed under $\vee, \wedge, {}'$. A *homomorphism* from a Boolean algebra B into a Boolean algebra C is a map $\varphi : B \to C$ such that $(x \vee y)\varphi = x\varphi \vee y\varphi$, $(x \wedge y)\varphi = x\varphi \wedge y\varphi$, $(x')\varphi = (x\varphi)'$. If φ is a one-to-one homomorphism from Boolean algebra B onto Boolean algebra C then φ is a *Boolean isomorphism.*

EXAMPLE 3. The set $\mathcal{S}(M)$ of all subsets of any set M is a Boolean algebra under set-theoretical operations: the join, the intersection and the complement. The null of $\mathcal{S}(M)$ is the empty set $\emptyset$, the unit of $\mathcal{S}(M)$ is M.

The following results show that $\mathcal{S}(M)$ is very important for the lattice theory.

THEOREM 2. *Any distributive lattice is isomorphic to some sublattice of the lattice $\mathcal{S}(M)$ of subsets for some set M.*

THEOREM 3. *Any Boolean algebra B is isomorphic to some subalgebra of the Boolean algebra $\mathcal{S}(M)$ for some set M. If B is a finite Boolean algebra then B is isomorphic $\mathcal{S}(M)$ for some finite set M.*

The proofs of these theorems can be found in many books (see for example G.Birkhoff [3] or M.Stone [1]). □

Soon we will need some information about modular lattices, i.e., about lattices that satisfy the modular law:

For any elements x, y, z of the lattice L

L.7. $$x \leq y \text{ implies } x \vee (y \wedge z) = (x \vee y) \wedge (x \wedge z).$$

It is easy to see that axiom L.6. is equivalent to the axiom

L.8. $$x \vee ((x \vee y) \wedge z) = (x \vee y) \wedge (x \vee z).$$

3⁰. The following theorem shows the connection between the lattice theory and the group theory.

THEOREM 4. *The lattice of normal subgroups of any group is modular.*

Proof. By L.8., it is enough to prove the equality $XY \cap XZ = (XY \cap Z)X$ for all normal subgroups X, Y, Z of the group G. If $a \in XY \cap XZ$, $a = uy = vz$ for $u, v \in X$, $y \in Y$, $z \in Z$ then $z = v^{-1}uy$, $z \in XY \cap Z$ and $a = vv^{-1}y \in X(XY \cap Z)$ i.e., $XY \cap XZ \subseteq (XY \cap Z)X$. Conversely, let now $a \in X(XY \cap Z)$, $a = xh$ where $x \in X, h \in XY \cap Z$. Then $h \in Z$ and $a \in XY$. As $h \in XY$, then $h = uv$ where $u \in X$, $v \in Y$ i.e., $a = xuv \in XY$. This means that $a \in XY \cap XZ$, $X(XY \cap Z) \subseteq XY \cap XZ$ and $X(XY \cap Z) = XY \cap XZ$. □

Chapter 2

Lattice-ordered groups

2.1 Definition of the l-Group

1^0. A *partially ordered group* (*po-group*) is a non-empty set G with binary operation $\cdot$ and binary relation $\leq$ such that $\{G; \cdot\}$ is a group, $\{G; \leq\}$ is a po-set and the following axioms are fulfilled:

M.1. $x \leq y$ implies $xz \leq yz$,

M.2. $x \leq y$ implies $zx \leq zy$,

for all $x, y, z \in G$.

If for the po-group G, the po-set $\{G; \leq\}$ is a directed set, or a lattice, or a totally ordered set, then G is called a *directed group*, or a *lattice-ordered group (l-group)*, or a *totally ordered group* (*o-group*), accordingly.

The additive groups $\mathbf{Z}$ – integers, $\mathbf{Q}$ – rationals, $\mathbf{R}$ – reals are the simplest examples of totally ordered groups. As usual, $\mathbf{N}$ denotes the set of positive integers.

EXAMPLE 1. Let $\{\mathbf{C}; +\}$ be the additive group of complex numbers $x + iy$, $x, y \in \mathbf{R}$. We define on $\{\mathbf{C}; +\}$ the relation of the partial order: $x + iy \leq u + iv$ if $x \leq u$, $y \leq v$ under the standard order of $\mathbf{R}$. Then $\{\mathbf{C}; +, \leq\}$ is an l-group in which $(x + iy) \vee (u + iv) = a + ib$, $(x + iy) \wedge (u + iv) = c + id$ where $a = \max(x, u)$, $b = \max(y, v)$, $c = \min(x, u)$, $d = \min(y, v)$ in $\mathbf{R}$.

EXAMPLE 2. Let $\{\mathbf{C}; +\}$ be an additive group of complex numbers. We define a totally order on $\{\mathbf{C}; +\}$: $x + iy \leq u + iv$ if $x < u$

or $x = u,\ y \le v$. Then $\{\mathbf{C};\ +, \le\ \}$ is an o-group.

EXAMPLE 3. Let $\{\mathbf{Q}^+;\ \cdot\ \}$ be the multiplicative group of positive rationals. Let us define the lattice order $\le$ on $\{\mathbf{Q}^+;\ \cdot\ \}$: $r \le s$ if $r^{-1}s \in \mathbf{Z}$. Then $\{\mathbf{Q}^+;\ \cdot, \le\ \}$ is an o-group.

EXAMPLE 4. Let Γ be a totally ordered set, $A(\Gamma)$ be a set of all one-to-one maps $f : \Gamma \longrightarrow \Gamma$ such that $\alpha \le \beta$ in Γ implies $\alpha f \le \beta f$. It is easy to see that $A(\Gamma)$ is a group under the composition: $\alpha fg = (\alpha f)g$ for any $\alpha \in \Gamma,\ f, g \in A(\Gamma)$. It was proved in Example 1.2.3. that $A(\Gamma)$ is a lattice under the partial order: $f \le g$ in $A(\Gamma)$ if $\alpha f \le \alpha g$ for all $\alpha \in \Gamma$. We claim that $A(\Gamma)$ is a l-group under this order. Let $f, g, h \in A(\Gamma),\ g \le f$. Then $\alpha g \le \alpha f$ and $\alpha gh \le \alpha fh$ for all $\alpha \in \Gamma$ i.e., $gh \le fh$. On the other hand, $\alpha h = \alpha h$ and $\alpha hg \le \alpha hf$ i.e., $hg \le hf$. So, $A(\Gamma)$ is an l-group. The elements of $A(\Gamma)$ are called *automorphisms* or *order permutations* of Γ.

EXAMPLE 5. Let $G = \mathrm{gp}\{a, b, c \mid ab = ba,\ c^{-1}ac = b,\ c^{-1}bc = a\ \}$. Then every element $x \in G$ can be written in the form: $x = c^m a^n b^k$ where $m, n, k \in \mathbf{Z}$ and this form for x is unique. We define the partial order on G : $c^m a^n b^k \le c^p a^q b^r$ if $m < p$ or $m = p,\ n \le q,\ k \le r$. Then $\{G;\ \cdot, \le\ \}$ is an l-group.

EXAMPLE 6. Let X be a set, $G = F(X, \mathbf{R})$ be a set of all functions $f : X \longrightarrow \mathbf{R}$ with the coordinate order (see Example 1.2.1): $f \le g$ for $f, g \in G$ if $f(x) \le g(x)$ for all $x \in X$. Then G is an l-group under the partial order $\le$ and the addition: $(f + g)(x) = f(x) + g(x)$.

EXAMPLE 7. The additive group $C([0, 1], \mathbf{R})$ of all continuous functions from the segment [0,1] into $\mathbf{R}$ is an l-group under the coordinate order.

EXAMPLE 8. The additive group of all differentiable functions from the segment [0,1] into $\mathbf{R}$ is a directed group but it is not an l-group under the coordinate order.

Let G be a po-group. Then the following correlations are true for all elements $x, y, z, t \in G$:

(1) $x \le y$ imply $z^{-1}xz \le z^{-1}yz$;
(2) $x \le y$ imply $y^{-1} \le x^{-1}$;
(3) $x \le y, z \le t$ imply $xz \le yt$.

2^0. The element x of po-group G is *positive* (*strongly positive*) if

$x \geq e$ ($x > e$), x is *negative* (*strongly negative*) if $x \leq e$ ($x < e$) where e is the identity element of the group G. The set of all positive elements of G is called a *positive cone.*

THEOREM 1. *The following properties are fulfilled for the positive cone of po-group* G :

O.1. $P \cdot P \subseteq P$ (P is a semigroup),
O.2. $P \cap P^{-1} = \{e\}$ (P is a pure subset),
O.3. $x^{-1}Px \subseteq P$ *for every* $x \in G$ (P is a normal subset).

If G *is a directed group then*

O.4. $$G = P \cdot P^{-1} = P^{-1} \cdot P.$$

If G *is an o-group then*

O.5. $$G = P \cup P^{-1}.$$

Conversely, if in the group G *there exists a subset* P *with properties* O.1.–O.3. *then it is possible to define a partial order* $\leq$ *on* G *such that* $\{G; \cdot, \leq\}$ *is a po-group. If* P *satisfies* O.4. *then* G *is a directed group, if* P *satisfies* O.5. *then* G *is an o-group. Moreover,* P *is a positive cone of* G.

Proof. Let G be a po-group, P be a positive cone of G. If $x, y \in P$ then $x \geq e$, $y \geq e$ and $xy \geq e$ i.e., $xy \in P$, $P \cdot P \subseteq P$. If $x \in P$, $x^{-1} \in P$ then $x \geq e$, $x^{-1} \geq e$ i.e., $x \geq e$, $x \leq e$. Therefore, $x = e$, $P \cap P^{-1} = \{e\}$.

If $a \in P$, $x \in G$ then $a \geq e$, $x^{-1}a \geq x^{-1}$, $x^{-1}ax \geq e$ i.e., $x^{-1}Px \subseteq P$. Properties O.1.-O.3. are proved.

If G is a directed group, then for every element $x \in G$ there is an element $y \in G$ such that $x \leq y$, $e \leq y$. Then $y^{-1}x \leq e$ and $x = yy^{-1}x$ where $y \in P$, $y^{-1}x \in P$, i.e. O.4. holds.

It is obvious that O.5. is true for totally ordered groups.

Conversely, let G be a group and P be a subset of G which satisfies conditions O.1-O.3. We define the relation $\leq$ on G : $x \leq y$ if $x^{-1}y \in P$. As $e \in P$ then $e = x^{-1}x$ and $x \leq x$. If $x \leq y$ and $y \leq x$ then $x^{-1}y \in P$ and $y^{-1}x = (x^{-1}y)^{-1} \in P$, i.e. $x^{-1}y \in P \cap P^{-1}$, $x^{-1}y = e$, $x = y$. Now, if $x \leq y$, $y \leq z$ then $x^{-1}y \in P$, $y^{-1}z \in P$ and, by O.1. $x^{-1}z = x^{-1}yy^{-1}z \in P$. Therefore, $\leq$ is a partial order.

We claim now that $\leq$ satisfies axioms M.1-M.2. If $x \leq y$, $x, y, z \in G$, then $x^{-1}y \in P$ and $z^{-1}x^{-1}yz \in P$, $x^{-1}z^{-1}zy \in P$ by O.3. This means that $xz \leq yz$, $zx \leq zy$ and G is a partially ordered group.

Let G satisfy condition O.4. and $x, y \in G$. Then $G = P \cdot P^{-1}$ and $x = zt^{-1}$, $y = uv^{-1}$ for some $z, t, u, v \in P$. Therefore, $x^{-1}z = t \in P$, $y^{-1}u = v \in P$, $x \leq z \leq zu$, $y \leq u \leq zu$. Consequently, $\{G; \cdot, \leq\}$ is a directed group. It is clear that the G is an o-group if it satisfies O.5.□

Group G is called an *orderable* group (*O-group*) if there exists some total order $\leq$ on G such that $\{G; \cdot, \leq\}$ is an o-group.

PROPOSITION 1. *A po-group G is an l-group if and only if G is a directed group and the positive cone P of G is a lattice.*

Proof. It is obvious that every l-group is a directed group and its positive cone is a lattice. Conversely, let G be a directed group and $P = \{x \in G \mid x \geq e\}$ be a lattice. If $x, y \in G$, then there is an element $z \in G$ such that $x^{-1} \leq z^{-1}$, $y^{-1} \leq z^{-1}$ and $e \leq z^{-1}x$, $e \leq z^{-1}y$. As P is a lattice then there are $u = z^{-1}x \vee z^{-1}y$ and $v = z^{-1}x \wedge z^{-1}y$ in P. We claim that zu is the least upper bound and zv is the greatest lower bound for x, y in G. Indeed, $z^{-1}x \leq u$, $z^{-1}y \leq u$ then $x \leq zu$, $y \leq zu$, i.e., zu is an upper bound for x, y in G. Now, if $x \leq t$, $y \leq t$ in G then $e \leq z^{-1}x \leq z^{-1}t$, $e \leq z^{-1}y \leq z^{-1}t$, i.e., $z^{-1}t$ is an upper bound for $z^{-1}x$, $z^{-1}y$ in P. As u is the least upper bound for $z^{-1}x$, $z^{-1}y$ in P then $u \leq z^{-1}t$ and $zu \leq t$. Hence, $zu = x \vee y$ in G. By similar arguments $zv = x \wedge y$ in G. □

According to Theorem 1 we can identify the partial order $\leq$ on G with the positive cone P of G. Sometimes we will note the positive cone of po-group G as $P(G)$ and the po-group G with positive cone P as (G, P).

3⁰. As it was mentioned in Section 1, it is possible to regard a lattice as a partially ordered set and as an algebraic system of the signature $\{\vee, \wedge\}$. By similar argument we can describe a lattice-ordered group as an algebraic system of the signature $l = \{\cdot, e, ^{-1}, \vee, \wedge\}$.

THEOREM 2. *Let $\{G; \cdot, e, ^{-1}, \vee, \wedge\}$ be an algebraic system such that $\{G; \cdot, e, ^{-1}\}$ is a group, $\{G; \vee, \wedge\}$ is a lattice and in G the identities*

M′ $\quad u(x \vee y)v = uxv \vee uyv, \quad u(x \wedge y)v = uxv \wedge uyv$

hold.

Then G is a lattice-ordered group under the partial order of the lattice $\{G;\ \vee, \wedge\ \}$. If G is a lattice-ordered group then in G identities M′ *hold.*

Proof. First we will prove that $x \leq y$ implies $txz \leq tyz$ for all elements $x, y, z, t \in G$. By definition of the order in lattice G, the inequality $x \leq y$ is equivalent to the equality $x \vee y = y$. Then by M' $t(x \vee y)z = txz \vee tyz = tyz$, i.e., $txz \leq tyz$.

Conversely, let G be an l-group and $x, y, z, t \in G$. Then $x \leq x \vee y$, $y \leq x \vee y$ and $txz \leq t(x \vee y)z$, $tyz \leq t(x \vee y)z$. Therefore,

(4) $$txz \vee tyz \leq t(x \vee y)z.$$

On the other hand, $txz \leq txz \vee tyz$, $tyz \leq txz \vee tyz$ i.e., $x \leq t^{-1}(txz \vee tyz)z^{-1}$, $y \leq t^{-1}(txz \vee tyz)z^{-1}$, $x \vee y \leq t^{-1}(txz \vee tyz)z^{-1}$. Therefore,

(5) $$t(x \vee y)z \leq txz \vee tyz.$$

Unequalities (4),(5) imply the fulfillment of the first axiom M′. The second axiom M′ can be proved analogously. □

PROPOSITION 2. *In any l-group G the inequality $a \wedge bc \leq (a \wedge b)(a \wedge c)$ is fulfilled for every positive elements a, b, c. If $a \wedge b = e$ then $a \wedge bc = a \wedge c$.*

Proof. If $a, b, c \geq e$ then $a \leq a^2$, $a \leq ac$, $a \leq ba$ and $(a \wedge b)(a \wedge c) = a^2 \wedge ba \wedge ac \wedge bc \geq a \wedge bc$. The second part of the proposition is a corollary of the first one. □

2.2 Calculations in l-Groups

Some properties of l-groups concerning elements and identities in l-groups were listed in Section 1. In this section we give the notions and the rules of calculations in l-groups which will be used in the book.

1^0. PROPOSITION 1. *In any l-group G the following identities are fulfilled:*

(1) $(x \vee y)^{-1} = x^{-1} \wedge y^{-1}$, $(x \wedge y)^{-1} = x^{-1} \vee y^{-1}$;

(2) $x(x \wedge y)^{-1} y = x \vee y$, $x(x \vee y)^{-1} y = x \wedge y$.

Proof. If $x \vee y \geq x, y$ then by 2.1.(2) $(x \vee y)^{-1} \leq x^{-1}, y^{-1}$, $(x \vee y)^{-1} \leq x^{-1} \wedge y^{-1}$. As $x^{-1} \wedge y^{-1} \leq x^{-1}, y^{-1}$ then $(x^{-1} \wedge y^{-1})^{-1} \geq x, y$, $(x^{-1} \wedge y^{-1})^{-1} \geq x \vee y$ and $x^{-1} \wedge y^{-1} \leq (x \vee y)^{-1}$. This inequality and $(x \vee y)^{-1} \leq x^{-1} \wedge y^{-1}$ which we obtained before, implies $(x \vee y)^{-1} = x^{-1} \wedge y^{-1}$. We can prove the dual identity analogously. Property (2) follows from identity (1) and the axioms M′1, M′2. □

PROPOSITION 2. *The lattice $\{G; \vee, \wedge\}$ is distributive for every l-group G.*

Proof. If $x \vee y = x \vee z$, $x \wedge y = x \wedge z$ for $x, y, z \in G$ then $y = (x \wedge y)x^{-1}x(x \wedge y)^{-1}y = (x \wedge y)x^{-1}(x \vee y) = (x \wedge z)x^{-1}(x \vee z) = (x \wedge z)x^{-1}x(x \wedge z)^{-1}z = z$. By Theorem 1.4.1., G is a distributive lattice.□

PROPOSITION 3. *In any l-group G the inequality $x^n \geq e$ $(n \in$ **N**$)$ implies $x \geq e$.*

Proof. If $x^n \geq e$ then by M′ we have $(x \wedge e)^n = x^n \wedge x^{n-1} \wedge \ldots \wedge x \wedge e = x^{n-1} \wedge \ldots \wedge x \wedge e = (x \wedge e)^{n-1}$, i.e., $x \wedge e = e$, $x \geq e$. □

A partial order of a po-group G is *isolated* if for every positive integer n the inequality $x^n \geq e$, $x \in G$, implies $x \geq e$. Proposition 3 asserts that the partial order of any l-group is isolated.

PROPOSITION 4. *Any l-group is a torsion-free group. If*

$$x \cdot g_1^{-1} x g_1 \cdots g_n^{-1} x g_n = e$$

in l-group G for some elements $x, g_1, \ldots, g_n \in G$, $x \neq e$, then x is incomparable with e. If G is an o-group then it is a group with unique roots.

Proof. The fact that any l-group is a torsion-free group follows from Proposition 3. In order to prove the second statement we suppose $x > e$. Then $g_i^{-1} x g_i > e$ and $x \cdot g_1^{-1} x g_1 \cdots g_n^{-1} x g_n > e$. If G is an o-group and $x^n = y^n$ for some $x, y \in G$, $n \in \mathbf{N}$, $\mathbf{x} \neq \mathbf{y}$. Let $x > y$ then $x^n > y^n$. This contradicts $x^n = y^n$. □

An element x of group G is called a Γ-*periodical element* if there are some elements $g_1, g_2, \ldots, g_n \in G$ such that $x \cdot g_1^{-1} x g_1 \cdots g_n^{-1} x g_n = e$. Proposition 3 confirms that every o-group does not contain nontrivial Γ-periodical elements. However, there are l-groups with nontrivial Γ-periodical elements. It is easy to verify that the element ab^{-1} of the group G from Example 2.1.5 is Γ-periodical: $(ab^{-1}) \cdot c^{-1}(ab^{-1})c = e$.

The following example shows that the class of all torsion-free groups does not coincide with the class of lattice-orderable groups.

EXAMPLE 1 (Gortchakov [1].)Let

$$G = \mathrm{gp}\{\ g_n, n \in \mathbf{N} |\ g_{n+1}^{-1} g_n g_{n+1} = g_n^{-1},\ g_{n+1}^{-1} g_i g_{n+1} = g_i$$

for $i < n,\ n \geq 1\}$.

It is clear that

$$g = g_1^{p_1} g_2^{p_2} \cdot \cdots \cdot g_n^{p_n}$$

for every element $g \in G$ with suitable $p_1, \ldots, p_n \in \mathbf{N}$. This form is unique for every g. Let n be a length of g if $p_n \neq 0$. We claim that every element $g \in G$ is $\Gamma-$periodical and use induction on the length for the proof. This is obvious for $n = 1$. Suppose that every element with the length less n is $\Gamma-$periodical. Let $g \in G,\ g = g_1^{p_1} \cdot \cdots \cdot g_n^{p_n},\ p_n \neq 0$. and $h = g_{n+1}^{-1} g g_{n+1} \cdot g$. Then the length of h is less then n because

$$h = g_1^{p_1} \cdots g_{n-1}^{p_{n-1}} g_n^{-p_n} \cdot g_1^{p_1} \cdot \cdots \cdot g_{n-1}^{p_{n-1}} g_n^{n} = g_1^{q_1} \cdot \cdots \cdot g_{n-1}^{q_{n-1}}\ .$$

Hence, the element h is $\Gamma-$periodical by the inductive assumption and g is $\Gamma-$periodical, too. Therefore, every partial order of G is trivial, i.e., $P(G) = \{e\}$. □

2 ^{0}. Here we obtain useful facts about decompositions of elements in l-groups. This decomposition is possible by Theorem 2.1.1: $G = P(G)P(G)^{-1}$ is true for every directed group G .

Let $x^+ = x \vee e,\ x^- = x \wedge e,\ |x| = x \vee x^{-1}$ for x$\in G$. It is obvious that $x^+, x^-, (x^{-1})^+ = (x^-)^{-1} \geq e$.

PROPOSITION 5. *In any l-group G*

$$a = a^+ a^-, \quad a^+ \wedge (a^-)^{-1} = e, \quad a^+(a^-)^{-1} = (a^-)^{-1} a^+.$$

Proof. By Proposition 1 $a(a^-)^{-1} = a(a \wedge e)^{-1} = a(a^{-1} \vee e) = e \vee a = a^+$, and $a = a^+ a^-$.

Similarly, $a^+ \wedge (a^-)^{-1} = a^+a^-(a^-)^{-1} \wedge (a^-)^{-1} = (a^+a^- \wedge e)(a^-)^{-1} = (a \wedge e)(a^-)^{-1} == a^-(a^-)^{-1} = e$.

Finally, $a^+(a^-)^{-1} = (a \vee e)(a^{-1} \vee e) = e \vee a \vee a^{-1} \vee e = e \vee a^{-1} \vee a \vee e = (a^-)^{-1}a$. □

PROPOSITION 6. *In any l-group*

(3) $(xy)^+ \leq x^+y^+, \quad (xy)^+ = x(x \wedge y^{-1})^{-1}$;

(4) $(x^n)^+ = (x^+)^n, \quad (x^n)^- = (x^-)^n$ *for* $n \in \mathbf{N}$.

Proof. (3) follows from relations: $(xy)^+ = xy \vee e \leq xy \vee x \vee y \vee e = x(y \vee e) \vee (y \vee e) = (x \vee e)(y \vee e) = x^+y^+$ and $(xy)^+ = xy \vee e = x(y \vee x^{-1}) = x(x \wedge y^{-1})^{-1}$.

It is clear that $(x^{n-k} \vee x^{-k})^n = x^{(n-k)n} \vee x^{(n-k)(n-1)-k} \vee \ldots \vee x^{(n-k)k-k(n-k)} \vee \ldots \vee x^{-kn} \geq e$ for every k, $0 \leq k \leq n$. Then, by Proposition 2.2.3 $x^{n-k} \vee x^{-k} \geq e$ and $x^n \vee e \geq x^k$. So. $(x^+)^n = (x \vee e)^n = x^n \vee x^{n-1} \vee \ldots \vee e = x^n \vee e = (x^n)^+$. □

PROPOSITION 7. *In any l-group*

(5) $$|a| = a^+(a^-)^{-1},$$

(6) $$|a^n| = |a|^n \qquad \textit{for } n \geq 0.$$

Proof. First, we note that $|a| \geq e$ for every $a \in G$. So, $|a| \geq a$, $|a| \geq a^{-1}$ then $|a| \cdot |a| \geq aa^{-1} = e$, $|a|^2 \geq e$. By Proposition 3 the order of an l-group is isolated, therefore $|a| \geq e$. Then $|a| = |a| \vee e = a \vee a^{-1} \vee e = e \vee a^{-1} \vee a \vee e = (a \vee e)(a^{-1} \vee e) = a^+(a^-)^{-1}$ and (5) is proved.

Finally, since a^+, $(a^-)^{-1}$ commute (by proposition 2.2.5), then

$$|a|^n = (a^+(a^-)^{-1})^n = (a^+)^n(a^-)^{-n} = (a^n)^+((a^n)^-)^{-1} = |a^n|. \ \square$$

PROPOSITION 8. *In any l-group*

(7) $$|x \vee y| \leq |x| \vee |y| \leq |x| \cdot |y|,$$

(8) $$|xy^{-1}| = (x \vee y)(x \wedge y)^{-1},$$

(9) $$|xy| \leq |x| \cdot |y| \cdot |x|.$$

Proof. (7) follows from relations: $|x \vee y| = (x \vee y) \vee (x \vee y)^{-1} = (x \vee y) \vee (x^{-1} \wedge y^{-1}) \leq x \vee y \vee x^{-1} \vee y^{-1} = |x| \vee |y| \leq |x| \cdot |y|$.

Now, $|xy^{-1}| = e \vee xy^{-1} \vee yx^{-1} \vee e = (x \vee y)(x^{-1} \vee y^{-1}) = (x \vee y)(x \wedge y)^{-1}$ and (8) is proved.

Finally, $|x|^{-1}|y|^{-1}|x|^{-1} \leq |x|^{-1}|y|^{-1} \leq xy \leq |x||y| \leq |x||y||x|$ and (9) is true.

Inequality (9) is similar to standard inequalities

(10) $$|xy| \leq |x||y|,$$
(10′) $$|x+y| \leq |x| + |y|,$$
which hold for reals.

It is easy to construct examples of l-groups in which relation (10) is not true.

PROPOSITION 9.(Busulini [1].) *Inequality* (10) *is fulfilled in l-group G for all elements $x, y \in G$ if and only if G is commutative.*

Proof. Let G be an l-group and x, y be commutative elements of G. Then $|xy| = xy \vee x^{-1}y^{-1} \leq xy \vee x^{-1}y \vee xy^{-1} \vee x^{-1}y^{-1} = (x \vee x^{-1})(y \vee y^{-1}) = |x||y|$. Conversely, let (10) hold in l-group G for all $x, y \in G$, and $x, y \geq e$. Then $xy = |xy| \leq |y^{-1}x^{-1}| \leq |y^{-1}||x^{-1}| = yx$. It is clear that $yx \leq xy$, too, and $xy = yx$. By Theorem 1 it follows that G is commutative. □

3⁰. The elements a, b of l-group G are *orthogonal* if $|a| \wedge |b| = e$. We write $a \perp b$ in this case. Proposition 5 shows that elements a^+, a^- are orthogonal for any element a of the l-group G.

PROPOSITION 10. *If $a \perp b$ in an l-group G, then $ab = ba$ and $|a| \vee |b| = |a||b|$.*

Proof. If $a \perp b$ in G and $a \geq e$, $b \geq e$, then $ab = a(a \wedge b)^{-1}b = a \vee b = ba$.

For $a, b \in G$ relations $e \leq a^+ \leq |a|$, $e \leq (a^-)^{-1} \leq |a|$, $e \leq b^+ \leq |b|$, $e \leq (b^-)^{-1} \leq |b|$ hold. Then $a^+ \wedge b^+ = e$, $a^+ \wedge (b^-)^{-1} = e$, $(a^-)^{-1} \wedge b^+ = e$, $(a^-)^{-1} \wedge (b^-)^{-1} = e$. Therefore, $a^+b^+ = b^+a^+$, $a^+b^- = b^-a^+$, $a^-b^+ = b^+a^-$, $a^-b^- = b^-a^-$ and $ab = ba$ because $a = a^+a^-$, $b = b^+b^-$ (see Proposition 5). □

It is easy to see that an l-group is an o-group if and only if $a \perp b$ implies $a = e$ or $b = e$.

PROPOSITION 11. *If a is a positive element of an l-group G, then the set $P_a = \{x \in G | x \wedge a = e\}$ is a subsemigroup of G and a convex sublattice of G.*

Proof. Let $x, y \in P_a$, then $a \wedge (x \wedge y) = (a \wedge x) \wedge (a \wedge y) = e$, $a \wedge (x \vee y) = (a \wedge x) \vee (a \wedge y) = e$, $a \wedge xy = a(e \wedge y) \wedge xy = a \wedge ay \wedge xy = a \wedge (a \wedge x)y = a \wedge y = e$. This means that $x \wedge y, x \vee y, xy \in P_a$ and P_a is a sublattice and a subsemigroup of G. It is obvious that P_a is convex. □

4^0. The following theorem is a cornerstone of the l-groups theory.

THEOREM 1. (Riesz decomposition. Riesz [1], G.Birkhoff [1], Lorenzen [1].) *Let G be an l-group, and $a_1 a_2 \cdots a_m = b_1 b_2 \cdots b_n$ for some elements $a_i, b_j \in G$, $a_i, b_j \geq e$ $(1 \leq i \leq m,\ 1 \leq j \leq n)$. Then there is a matrix*

$$C = \begin{bmatrix} c_{11} & c_{12} & \dots & c_{1n} \\ c_{21} & c_{22} & \dots & c_{2n} \\ \dots & \dots & \dots & \dots \\ c_{m1} & c_{m2} & \dots & c_{mn} \end{bmatrix}$$

where $c_{ij} \in G$, $c_{ij} \geq e$ such that

(11) $$a_i = c_{i1} \dots c_{in},\ b_j = c_{1j} \dots c_{mj},$$

(12) $$(c_{i+1j} \dots c_{mj}) \perp (c_{ij+1} \dots c_{in})\ (1 \leq i \leq m,\ 1 \leq j \leq n).$$

The matrix C with properties (11),(12) *is unique.*

Proof. Let us consider the case $m = 2$, $n = 2$: $a_1 a_2 = b_1 b_2$. The simple verification shows that the matrix $\begin{bmatrix} c_{11} & c_{12} \\ c_{21} & c_{22} \end{bmatrix}$ where $c_{11} = a_1 \wedge b_1$, $c_{12} = (a_1 \wedge b_1)^{-1} a_1$, $c_{21} = (a_1 \wedge b_1)^{-1} b_1$, $c_{22} = a_2 \wedge b_2$, satisfies (11),(12).

Now, we use induction on n. Suppose that the theorem is true for $p = m$, $q < n$ where $m \geq 2$, $n \geq 3$. Then $a_1 a_2 \dots a_m = b_1 b_2 \dots b_{n-2} b'_{n-1}$ where $b'_{n-1} = b_{n-1} b_n$. By the inductive assumption, there is the matrix

$$\begin{bmatrix} c_{11} & c_{12} & \dots & c_{1n-1} & d_1 \\ c_{21} & c_{22} & \dots & c_{2n-1} & d_2 \\ \dots & \dots & \dots & \dots & \dots \\ c_{m1} & c_{m2} & \dots & c_{mn-1} & d_m \end{bmatrix}$$

such that $c_{ij}, d_k \in G$, $c_{ij}, d_k \geq e$, $a_i = c_{i1} \dots c_{in-2} d_i$, $b_j = c_{1j} \dots c_{mj}$, $b_{n-1} = d_1 d_2 \dots d_m$, $c_{i+1j} \dots c_{mj} \perp c_{ij+1} \dots c_{in-2} d_i$ for all i, j, k, $1 \leq i, k \leq m$, $1 \leq j \leq n$. By the inductive assumption for the equality $d_1 d_2 \dots d_m = b_{n-1} b_n$ we have the matrix

$$\begin{bmatrix} c_{1n-1} & c_{1n} \\ c_{2n-1} & c_{2n} \\ \dots & \dots \\ c_{mn-1} & c_{mn} \end{bmatrix}$$

such that $c_{1n-1}, c_{1n}, \ldots, c_{mn-1}, c_{mn} \geq e$, $b_{n-1} = c_{1n-1} \cdots c_{mn}$, $b_n = c_{1n} \cdots c_{mn}$, $c_{in-1}c_{in} = d_i$ and $c_{i+1n-1} \cdots c_{mn-1} \perp c_{in}$ $(1 \leq i \leq m)$. Then the matrix $C = (c_{ij})$ $(1 \leq i \leq m,\ 1 \leq j \leq n)$ satisfies (11),(12).

This matrix is unique. As $e = c_{i+1j} \cdots c_{mj} \wedge c_{ij+1} \cdots c_{in}$, then $c_{ij} = c_{ij}c_{i+1j} \cdots c_{mj} \wedge c_{ij}c_{ij+1} \cdots c_{in} = (c_{1j} \cdots c_{i-1j})^{-1}b_j \wedge (c_{i1} \cdots c_{ij-1})^{-1}a_i$. Therefore, elements c_{ij} are defined consequently from c_{11} uniquely. □

COROLLARY 1. *If* $a \leq b_1 \cdots b_n$ *for* $a, b_1, \ldots, b_n \geq e$ *in any l-group* G, *then* $a = a_1 \cdots a_n$ *for some* $a_i \in G$, $e \leq a_i \leq b_i$ $(1 \leq i \leq n)$.

Proof. We apply Theorem 1 to the equality $ac = b_1 b_2 \cdots b_n$ where $c = a^{-1}b_1 b_2 \cdots b_n \geq e$.□

PROPOSITION 12. (Byrd, Lloyd, Mena, Teller [1].) *In l-group* G, $x^n = y^n$ *implies*

$$x\Big(\bigwedge_{i+j=n-1} x^i y^j\Big) = \Big(\bigwedge_{i+j=n-1} x^i y^j\Big)y$$

for any $n \in \mathbf{N}$.

Proof. (W.Ch.Holland [Private communication].) If $n = 2$ and $x^2 = y^2$ then $x(x \wedge y) = x^2 \wedge xy = y^2 \wedge xy = (y \wedge x)y$.

In general case $x(x^{n-1} \wedge x^{n-2}y \wedge \ldots \wedge xy^{n-2} \wedge y^{n-1}) = x^n \wedge x^{n-1}y \wedge \ldots \wedge x^2y^{n-2} \wedge xy^{n-1}) = y^n \wedge x^{n-1}y \wedge \ldots \wedge x^2y^{n-2} \wedge xy^{n-1}) = x^{n-1}y \wedge \ldots \wedge x^2y^{n-2} \wedge xy^{n-1} \wedge y^n = (x^{n-1} \wedge x^{n-2}y \wedge \ldots \wedge xy^{n-2} \wedge y^{n-1})y$

COROLLARY 2. *If the group* G *admits a lattice order and* $x^n = y^n$ *in* G *then* x *and* y *are conjugate elements in* G.

PROPOSITION 13. (Medvedev) *If in an l-group* G $[x, y, y] = e$ *for some elements* x, y, *then*

$$\Big[\bigvee_{j=1}^{s} \Big(y^{-lj}\Big(\bigwedge_{i=1}^{t} y^{-ki}xy^{ki}\Big)y^{lj}, y, y\Big] = e$$

for all $k, l, s, t \in \mathbf{N}$.

Here $[x, y] = x^{-1}y^{-1}xy$, $[x, y, y] = [[x, y], y]$ as usual .

Proof. Let $z = [x, y]$. If $[x, y, y] = e$, then $[z, y] = e$, $y^{-k}xy^k = xz^k$ for every $k \in \mathbf{N}$. Now,

$$[x \wedge y^{-k}xy^k, y, y] = [x \wedge xz^k, y, y] = [x(e \wedge z^k), y, y] =$$

$$= [(x \wedge xz^k)^{-1}y^{-1}(x \wedge xz^k)y, y] = [(x \wedge xz^k)^{-1}xz(e \wedge z^k), y] =$$
$$= [(e \wedge z^k)^{-1}x^{-1}xz(e \wedge z^k), y] = [z, y] = e.$$

The same arguments show that $[x \vee y^{-s}xy^s, y, y] = e$. □

2.3 Basic Facts

1⁰.It is obvious that a subgroup of po-group is a po-group too under the induced partial order. If G is an o-group then every subgroup of G is an o-group, too. This fact does not hold for l-groups: if an l-group G contains an element a incomparable with e then G contains a cyclic trivial ordered subgroup (a), i.e., in (a) every two different elements are noncomparable. Therefore, the following notion is fundamental for the l-groups theory. A subgroup H of l-group G is called an *l-subgroup* of G if H is a sublattice of the lattice G. Every l-subgroup of l-group G is a subalgebra of G in the signature $l = \{ \cdot, {}^{-1}, e, \vee, \wedge \}$. It is evident that the lattice order of l-subgroup is induced by the lattice order of l-group. The converse statement is not true. The following example illustrates this.

EXAMPLE 1. Let A be the additive group of complex numbers with the positive cone $P_1 = \{ x + iy \mid x, y \in \mathbf{R},\ x > 0 \text{ or } x = 0, y \leq 0 \}$. Let B be the additive group of complex numbers too with the positive cone $Q = \{ x + iy \mid y > 0 \text{ or } y = 0, x \leq 0 \}$. It is clear that A, B are o-groups and the direct product $G = A \times B$ is an l-group under the order $R = \{ (a, b) \mid a \in P,\ b \in Q \}$. If $H = \{ (z, z) \mid z \in \mathbf{C} \}$ then H is a subgroup of G. For $h = (z, z)$, $z = x + iy$ we have $h \leq (0, 0)$ in G iff $x \leq 0, y \leq 0$ in $\mathbf{R}$. It is clear that H is a lattice ordered group under order induced by G and H is isomorphic to the l-group of Example 2.1.1. On the other hand, H is not an l-subgroup of G since $(1 - i, 1 - i) \vee (0, 0) = (1 - i, 0) \in G$ and $(1 - i, 0) \notin H$, $(1 - i, 1 - i) \in H$.

PROPOSITION 1. *A subgroup H of l-group G is an l-subgroup of G iff $a \vee e \in H$ for any $a \in H$.*

Proof. The first statement of the proposition is evident. Conversely, let H be a subgroup of G and $a \vee e \in H$ for any $a \in H$. Then $x^{-1}y \vee e \in H$ for any $x, y \in H$ and $x(x^{-1}y \vee e) = y \vee x \in H$. Finally, $x \wedge y = (x^{-1} \vee y^{-1})^{-1} \in H$.□

The notion paramount importance in the l-group theory is convexness. We recall that a subset H of a po-set M is called *convex* if $x \leq z \leq y$, $x, y \in H$ implies $z \in H$.

PROPOSITION 2. *Any convex directed subgroup of an l-group is an l-subgroup.*

Proof. In fact, if H is a convex directed subgroup of an l-group G then for any $x \in H$ there exists $y \in H$ such that $y \leq e, y \leq x$. Then $e \leq x \vee e \leq y$ and $x \vee e \in H$. By Proposition 1, H is an l-subgroup. □

THEOREM 1. *A subgroup H of l-group G is a convex l-subgroup iff $|x| \leq |a|$ implies $x \in H$ for every $a \in H, x \in G$.*

Proof. Let H be a convex l-subgroup of l-group G, $a \in H$ and $|x| \leq |a|$. As $|x|^{-1} \leq x \leq |x|$ for every $x \in G$ then $|a|^{-1}|x|^{-1} \leq x \leq |x| \leq |a|$ and $x \in H$.

Conversely, let $x \in H$ for every $x, a \in G$, such that $a \in H$, $|x| \leq |a|$. If $x \leq y \leq z$, $x, z \in H$ then $e \leq x^{-1}y \leq x^{-1}z$, $e \leq |x^{-1}y| \leq |x^{-1}z|$. Since $|x^{-1}z| \in H$ then $x^{-1}y \in H$ and $y \in H$. Therefore, H is convex.

Further, let $a \in H$. Then $e \leq a \vee e \leq |a|$ and $a \vee e \in H$ by the assumption. By Proposition 1 we conclude that H is an l-subgroup.□

A normal convex l-subgroup of l-group is called an *l-ideal* or an *ideal* for short if no confusion arises.

The mapping φ from po-group G into po-group H is called an *order homomorphism* or *o-homomorphism* if φ is a group homomorphism of G into H and φ preserves the partial orderings. In other words, the mapping $\varphi : G \to H$ is an o-homomorphism if

(1) $$(xy)\varphi = x\varphi \cdot y\varphi,$$

(2) $$x \leq y \text{ implies } x\varphi \leq y\varphi,$$

for any $x, y \in G$.

If G, H are po-groups and φ is a group isomorphism G into H and φ is an isomorphism of po-set G into po-set H then φ is called an *order isomorphism* or *o-isomorphism*.

PROPOSITION 3. *A homomorphism φ from po-group (G, P) into po-group (H, Q) is an o-homomorphism iff $P\varphi \subset Q$.*

Proof. If φ is an o-homomorphism, then it is evident that $P\varphi \subset Q$. Conversely, let φ be a homomorphism from group G into group H such that $P\varphi \subset Q$ and $x \leq y$ in G. Then $x^{-1}y \in P$, $(x^{-1}y)\varphi \in Q$ and $(x^{-1}y)\varphi = (x\varphi)^{-1}y\varphi \in Q$, $x\varphi \leq y\varphi$ in H. □

EXAMPLE 2. Let A be the additive group of integers with trivial order and B be the additive group of integers with natural total order. If $\varphi: A \to B$ is a natural isomorphism from group A onto group B then φ is an o-homomorphism but φ is not an o-isomorphism.

The o-homomorphism $\varphi: (G,P) \to (H,Q)$ is called a *strict o-homomorphism*, if $P\varphi = Q \cap G\varphi$. It is clear that any o-isomorphism is a strict o-homomorphism. The converce statement is true too: every strict o-homomorphism of po-group which is an isomorphism of groups is an o-isomorphism.

If G, H are l-groups and $\varphi: G \to H$ is a group homomorphism from G into H such that

(3) $$(x \vee y)\varphi = x\varphi \vee y\varphi, \quad (x \wedge y)\varphi = x\varphi \wedge y\varphi,$$

for all $x, y \in G$, then φ is called an *l-homomorphism* or an *homomorphism of l-groups.* An *l-isomorphism* or an *isomorphism of l-groups* is an l-homomorphism which is an isomorphism of groups. Any l-homomorphism is a strict o-homomorphism,of course. If l-groups G and H are l-isomorphic we will often denote this fact as $G \cong H$. It is possible that a strict o-homomorphism from a lattice-ordered group G into an lattice-ordered group H is not an l-homomorphism. Example 2.1.1 illustrates it.

A po-group is *order simple* if each o-homomorphism G is either an order isomorphism or an o-homomorphism onto the identity po-group E. An l-group G is called *l-simple* or *simple l-group* if every l-homomorphism G is an l-isomorphism or l-homomorphism G onto E.

3^0. Let (G,P) be a po-group, H be a convex subgroup G, $R(G:H)$ be the set of all right cosets of Hx in G. We can define the partial order on $R(G:H)$: $Hx \leq Hy$ in $R(G:H)$ if there are $a \in Hx, b \in Hy$ such that $a \leq b$. In other words, $Hx \leq Hy$ if $x \leq hy$ for some $h \in H$.

It is obvious that $Hx \leq Hx$ .If $Hx \leq Hy$, $Hy \leq Hx$, then there exists $u, v \in H$, such that $x \leq uy$, $y \leq vx$ and $v^{-1} \leq xy^{-1} \leq u$. As H is convex, then $xy^{-1} \in H$ and $Hx = Hy$.

Finally, if $Hx \leq Hy$, $Hy \leq Hz$, then $x \leq uy, y \leq vz$ for some $u, v \in H$ and $x \leq uvz$, $uv \in H$, i.e., $Hx \leq Hz$. This establishes that the relation $\leq$ on $R(G:H)$ is a partial order and the natural map $\varphi: G \to R(G:H)$, $x\varphi = Hx$ is an o-homomorphism. The relation $\leq$ on the set $L(G:H)$ of all left cosets of H is defined similarly.

If H is a convex normal subgroup of po-group G, then the quotient group G/H with the above defined partial order is a po-group. In fact, if $x, y, z \in G$, $Hx \leq Hy$, then there is $h \in H$ such that $x \leq hy$ and $xz \leq hyz$, $zx \leq zhz^{-1}zy = h_1 zy$ where $h_1 = zhz^{-1} \in H$ too. Now, $Hx \cdot Hz = Hxz \leq Hyz = Hy \cdot Hz$ and $Hz \cdot Hx = Hzx \leq Hzy = Hz \cdot Hy$.

This means that the natural homomorphism $\varphi: G \to G/H$ is a strict o-homomorphism. It is called a *natural o-homomorphism* and po-group G/H is called a *quotient group of po-group* G by a convex normal subgroup H. So, the following statement is proved .

PROPOSITION 4. *A quotient group G/H of po-group G by a convex normal subgroup H is a po-group under the natural partial order.*

Proof. If (G, P) is a po-group and H is a convex normal subgroup of G, then the set $\bar{P} = \{\bar{x} = xH \mid x \geq h \text{ for some } h \in H\}$ is a positive cone of the quotient group G/H.□

The following example shows that sometimes a quotient group of an l-group by a convex normal subgroup can not be an l-group.

EXAMPLE 3. Let $A = (a) \times (b)$ be the direct product of two infinite cyclic groups with the order: $a^p b^n \leq a^q b^m$ if $p < q$ or $p = q$, $n \leq m$. It is easy to see that A is an o-group. Let $A' = (a') \times (b')$ be isomorphic to A o-group and $\varphi: A \to A'$ be an o-isomorphism such that $a\varphi = a'$, $b\varphi = b'$. We set on the direct product $G = A \times A'$ of o-groups A, A' the order: $xy \geq e$ for $x \in A$, $y \in A'$ if $x \geq e$ in A ,and $y \geq e$ in A'. It is clear that G is an l-group. Let $H = \{(b^{-k}, b^k),\ k \in \mathbf{Z}\}$. be a subset of G. Then it easy to see that H is a convex normal subgroup of G. The naturally ordered quotient group G/H is o-isomorphic to the group $\bar{A} = (u) \times (v) \times (w)$ with the order: $u^p v^q w^n \geq e$ if $p \geq 0$, $q \geq 0$, $p+q > 0$ or $p = q = 0$, $n \geq 0$. Here $(u), (v), (w)$ are ordered infinite cyclic groups. The po-group G/H is not an l-group as elements u,v have no meet in G/H.

THEOREM 2. *A quotient group G/H of l-group G by an ideal H is an l-group and the natural homomorphism from G onto G/H is* *an l-homomorphism* .

Proof. By Proposition 4, it is enough to prove that $(x \vee y)H = xH \vee yH$, $(x \wedge y)H = xH \wedge yH$ in G/H for every $x, y \in G$. It is evident that $(x \vee y)H \geq xH, yH$. Suppose that $zH \geq xH$, yH in G/H for any $z \in G$. Then $x \leq zu$, $y \leq zv$ for some $u, v \in H$. Since H is an ideal, then there is $h \in H$ such that $u, v \leq h$, and $x \leq zh$, $y \leq zh$, $x \vee y \leq zh$ in G, $(x \vee y)H \leq zH$, i.e., $(x \vee y)H$ is the least upper bound for xH, yH in G/H. The proof of equality $(x \wedge y)H = xH \wedge yH$ and other statements of the theorem we leave for the reader.□

EXAMPLE 4. Let $G = (a) \times (b)$ be the direct product of two infinite cyclic groups with the order: $a^n b^m \geq e$ if $n \geq 0$, $m \geq 0$. Then G is an l-group and $H = \{ a^k b^{-k} \mid k \in \mathbf{Z} \}$ is a convex normal subgroup of G which is not an ideal of G. The quotient group G/H is o-isomorphic to o-group of $\mathbf{Z}$ but the natural o-homomorphism $\varphi : G \to G/H$ is not an l-homomorphism. In fact, $aH = aH \vee bH$ in G/H but $(a \vee b)H = abH \neq aH$.□

THEOREM 3. (Homomorphism theorem) *Let $\varphi : G \to H$ be an l-homomorphism from l-group G into l-group H. Then the kernel N of φ is an ideal of G and there is an l-isomorphism $\psi : G/N \to H$ such that $x\varphi = xN\psi$ for all $x \in G$.*

Proof. It is a well-known result that N is a normal subgroup of G and there exists a group isomorphism ψ from G/N onto H such that $x\varphi = xN\psi$ and it is unique. Now we prove that N is an ideal of G. If $x \leq y \leq z$ in G for $x, z \in N$, $y \in G$. Then $e = x\varphi \leq y\varphi \leq z\varphi = e$ and $y\varphi = e$, $y \in N$, i.e., N is a convex subgroup. If $x \in N$ then $(x \vee e)\varphi = x\varphi \vee e = e$ and N is an ideal of G by Proposition 1. Finally, if $x, y \in G$ then $(xN \vee yN)\psi = ((x \vee y)N)\psi = (x \vee y)\varphi = x\varphi \vee y\varphi = xN\psi \vee yN\psi$. The dual equality follows from Proposition 2.2.1.□

It is easy to prove that Homomorphism theorem is valid for po-groups, too. In this case it is necessary to note that N is a convex normal subgroup of G and ψ is an o-homomorphism from po-group G/N into po-group H with the trivial kernel. We leave the details for the reader.□

4⁰. Now, we describe the simplest method for constructing l-groups from given l-groups. Let $\{A_\alpha | \alpha \in I\}$ be a collection of po-groups. The *Cartesian product* $\bar{A} = \bar{\prod\limits_{\alpha \in I}} A_\alpha$ of po-group A_α is the po-group which is the Cartesian product of groups A_α, and the Cartesian product of po-sets A_α. This means that $\bar{A}$ is the set of all functions $f : I \to \bigcup\limits_{\alpha \in I} A_\alpha$, $f(\alpha) \in A_\alpha$, $fg(\alpha) = f(\alpha)g(\alpha)$, $f \leq g$ iff $f(\alpha) \leq g(\alpha)$ for all $\alpha \in I$. It is clear that A is an l-group if A_α are l-groups for all $\alpha \in I$. We say in this case that A is the *Cartesian product of l-group.* The *direct product* $A = \prod\limits_{\alpha \in I} A_\alpha$ of po-groups A_α is the subgroup of $\bar{A}$ such that $f \in A$ iff $f(\alpha) \neq e$ for finite set of indeces. We write $A = A_1 \times A_2 \times \cdots \times A_n$ if $I = 1, 2, \ldots, n$. If every A_α is an l-group, then A is called the *direct product* of l-group A_α and it is easy to see that A is an l-subgroup of $\bar{A}$.

PROPOSITION 5. *Let $\{N_\alpha | \alpha \in I\}$ be the set of ideals of l-group G and $\bigcap\limits_{\alpha \in I} N_\alpha = E$. Then G is isomorphic to l-subgroup of the Cartesian product $\bar{G} = \bar{\prod\limits_{\alpha \in I}} G_\alpha$ of l-groups $G_\alpha = G/N_\alpha$.*

Here E is an identity subgroup of G as usual.

Proof. We define the map $f_g : G \to G_\alpha$, $f_g(\alpha) = gN_\alpha$ for $g \in G$, $\alpha \in I$. It is evident that f is an isomorphism from group G into group $\bar{G}$. If $x, y \in G$, then $f_{x \vee y}(\alpha) = (x \vee y)N_\alpha = xN_\alpha \vee yN_\alpha = (f_x(\alpha) \vee f_y(\alpha)$, i.e., $f_{x \vee y} = f_x \vee f_y$

Analogously, $f_{x \wedge y} = f_x \wedge f_y$ and f is an l-isomorphism. □ .

We say that l-group G is *approximated* by the family $\{ G_\alpha \mid \alpha \in I \}$ of l-groups if for any $g \in G$, $g \neq e$ there are $\alpha \in I$ and homomorphism $\varphi_\alpha : G \to G_\alpha$ such that $g\varphi_\alpha \neq e$. It is easy to see that G is approximatad by $\{ G_\alpha \mid \alpha \in I \}$ iff G is isomorphic to some l-subgroup of the Cartesian product $\bar{\prod\limits_{\beta \in K}} G_\beta$ of l-groups G_β where K is a suitable set of indeces and G_β is isomorphic to some l-group G_α, $\alpha \in I$ for every $\beta \in K$. Equivalently there exists the set $\{ N_\beta \mid \beta \in K \}$ of ideals in G such that $\bigcap\limits_{\beta \in K} N_\beta = E$ and G/N_β is isomorphic to some G_α, $\alpha \in I$ for every $\beta \in K$. We say also that G is a *subdirect product* of l-groups G_β, $\beta \in K$.

PROPOSITION 6. *Let G be a Cartesian product of l-groups A_α, $\alpha \in I$. Then $\bar{A}_\alpha = \{ f \in G \mid f(\beta) = e \text{ for } \beta \neq \alpha \}$ is an ideal of G and $\bar{A}_\alpha$ is isomorphic to A_α for any $\alpha \in I$.*

Proof. It is well-known that $\bar{A}_\alpha$ is a normal subgroup of the group G and $\bar{A}_\alpha$ is isomorphic to A_α as a group. We claim that $\bar{A}_\alpha$ is convex. Let $f \leq g \leq h$ for $f, h \in \bar{A}_\alpha$, $g \in G$. Then $e = f(\beta) \leq g(\beta) \leq h(\beta) = e$, $g(\beta) = e$ for all $\alpha, \beta \in I$, $\beta \neq \alpha$, i.e., $g \in \bar{A}_\alpha$. It is clear that $g \vee e \in \bar{A}_\alpha$ for any $g \in \bar{A}_\alpha$, i.e., $\bar{A}_\alpha$ is an ideal of G and the projection: $g \to g(\alpha)$ is an isomorphism from l-group G into $\bar{A}_\alpha$. □

We now define the following partial order on the Cartesian product of groups. Let $\{ A_\alpha \mid \alpha \in I\}$ be a set of po-groups A_α, I be a well-ordered set of indeces under some order $\prec$ and $\bar{A} = \overline{\prod\limits_{\alpha \in I}} A_\alpha$ be the Cartesian product of groups A_α. We set $f > e$ in $\bar{A}$ if $f \neq e$ and $f(\alpha) > e$ in A_α where α is the least element under the well-order of I for which $f(\alpha) \neq e$. This po-group is called a *lexicographic product of po-groups* A_α by well-ordered set I. It is denoted $\vec{A} = \overrightarrow{\prod\limits_{\alpha \in I}} A_\alpha$. It is clear that the po-set $\vec{A}$ is the lexicographic product of po-sets A_α, $\alpha \in I$ (see Chapter 1, Section 1, $\mathbf{3^0}$). The lexicographic order on the direct product of po-group by well-ordered set of indeces and *antilexicographic product* of po-groups A_α can be defined analogously.

PROPOSITION 7. *Let $\{ A_\alpha \mid \alpha \in I\}$ be a set of l-groups, I be a well-ordered set of indeces. Then $\overrightarrow{\prod\limits_{\alpha \in I}} A_\alpha$ is an l-group if and only if either A_α are o-groups for all $\alpha \in I$ or there is the largest element ζ in I and A_ζ is an l-group and A_α is an o-group for every $\alpha \neq \zeta$.*

The proof follows from Proposition 1.3.2. since the lattice of every non-trivial l-group has not the null and unit.□

PROPOSITION 8. *Each torsion free-Abelian group admits a total order.*

Proof. Let A be a torsion-free Abelian group and V be the divisible hull of A. Then V is the direct product $\prod\limits_{\alpha \in I} Q_\alpha$ where Q_α are isomorphic to additive group of rationals $\mathbf{Q}$ for every $\alpha \in I$. We totally order every Q_α , define some well-order on I and consider the lexicographic product

$\vec{A} = \overrightarrow{\prod_{\alpha \in I}} Q_\alpha$. It is clear that $\vec{A}$ is an o-group and A is embedded in V.□

5⁰. If N is a convex normal subgroup of po-group G, then quotient group G/N is also a po-group. We say in this case that G is an *extension* of po-group N by po-group G/N. If $gx > e$ for all $g \in G$, $x \in N$ such that $gN > N$ in G/N, then G is a *lexicographic extension.* We use now the standard notions of the Schreier extensions theory (see, for example, the book of M.Hall [1]). Every element g in the extension G of group N by group $\bar{G}$ is a pair $(\bar{g}, x)$ where $\bar{g} \in \bar{G}$, $x \in N$. Po-group G is a lexicographic extension of po-group (N, P_N) by po-group $(\bar{G}, \bar{P})$ if group G is an extension N by G, P_N is a normal set in G and $(\bar{g}, x) > (\bar{h}, y)$ iff $\bar{g} > \bar{h}$ in $\bar{G}$ or $\bar{g} = \bar{h}$, and $x > y$ in N.

PROPOSITION 9. *Lexicographic extension G of l-group $N \neq E$ by l-group $\bar{G}$ is an l-group if and only if G is an o-group.*□

THEOREM 4. (LEVI [1]). *Let N be a normal subgroup of group G, P_N be a partial order of N such that $g^{-1}Pg = P$ for all $g \in G$, and $\bar{P}$ be a partial order of quotient group $\bar{G} = G/N$. Then there exists a partial order P on G such that (G, P) is a lexicographic extension of po-group (N, P_N) by po-group $(\bar{G}, \bar{P})$. If (N, P_N) is an l-group and $(\bar{G}, \bar{P})$ is an o-group, then (G, P) is an l-group, if (N, P_N) and $(\bar{G}, \bar{P})$ are both o-groups, then (G, P) is an o-group, too .*

Proof. Let $P = \{\ x \in G \mid xN > N$ in G/N or $x \in N,\ x \in P_N\}$. The simple verification completes the proof. □

Chapter 3

Convex l-subgroups

3.1 The Lattice of Convex l-Subgroups

Let us denote by $C(G)$ the set of all convex subgroups of the lattice-ordered group G. The basic fact for the structure theory of l-groups is the following result.

THEOREM 1. (Birkhoff [1], Lorenzen [1].) *$C(G)$ is the complete sublattice in the lattice of all subgroups for any l-group G.*

Proof. It is obvious that $\bigwedge_{\alpha \in I} A_\alpha \in C(G)$ for every set $\{ A_\alpha \mid \alpha \in I\}$ where $A_\alpha \in C(G)$ for all $\alpha \in I$. We claim that the subgroup A of G which is generated by the set $\{ A_\alpha \mid A_\alpha \in I,\ \alpha \in I\}$ in G is a convex l-subgroup.

Let $a \in A$. Then $a = a_{\alpha_1} a_{\alpha_2} \cdots a_{\alpha_s}$ where $a_{\alpha_i} \in A_{\alpha_i}$ for $i = 1, 2, \ldots, s$. Suppose that $|x| < |a|$ for $x \in G$. By Proposition 2.2.8.(9), $|x| \leq |a_{\alpha_1}| \cdots |a_{\alpha_{s-1}}| \cdot |a_{\alpha_s}| \cdot |a_{\alpha_{s-1}}| \cdots |a_{\alpha_1}|$ holds. By Corollary 2.2.1 of the decomposition property, $|x| = b_{\alpha_1} \cdots b_{\alpha_{s-1}} b_{\alpha_s} b'_{\alpha_{s-1}} \cdots b'_{\alpha_1}$ is fulfilled where $e \leq b_{\alpha_i}, b'_{\alpha_i}$ for all $i = 1, 2, \ldots, s$. Because A_{α_i} is convex, then $b_{\alpha_i}, b'_{\alpha_i} \in A_{\alpha_i}$ for all i. Therefore $|x| \in A$. We can get $x^+, (x^-)^{-1} \in A$ analogously, hence $x = x^+ x^- \in A$. This establishes the claim: $|x| \leq |a|$ implies $x \in A$. The statement of the theorem follows now from Theorem 2.3.1.$\square$

So, $C(G)$ is a lattice for each l-group G. We denote the lattice operation in $C(G)$ by $\overset{c}{\bigvee}$, $\overset{c}{\bigwedge}$ for a distinction form set-theoretical operation

$\cup$, $\cap$. It is obvious that the lattice $C_n(G)$ of ideals of G is a sublattice of $C(G)$.

If M is a subset of l-group G then we denote a convex l-subgroup of G which is generated by M as $(M)_l$. We use this notation repeatedly throughout the book. It is clear that $(M)_l$ is the intersection of all convex l-subgroups of G which contain M and $x \in (M_l)$ iff there are elements $a_1, \ldots, a_n \in M$ such that $|x| \leq |a_1| \cdots |a_n|$. The following results about convex l-subgroups we describe in accordance with the work of K.Lorenz [1]. Let $C_f(G)$ be a set of convex l-subgroups $(a)_l$ of G for $a \in G$. For each $a \in G$ $\quad (a)_l = (a^{-1})_l = (|a|)_l$.

LEMMA 1. *$C_f(G)$ is a sublattice of lattice $C(G)$ and the mapping $\gamma \ : x \to (x)_l$, $x \in P(G)$ is a homomorphism of lattice $P(G)$ onto lattice $C_f(G)$.*

Proof. It is enough to see that $(a)_l \overset{C}{\bigvee} (b)_l = (a \vee b)_l$ and $(a)_l \overset{C}{\bigwedge} (b)_l = (a \wedge b)_l$ for all $a, b \in P(G)$.

Let $|x| \leq (a \wedge b)^n$, $n \in \mathbf{N}$. Then $|x| \leq a^n$, $|x| \leq b^n$, hence $x \in (a)_l \overset{C}{\bigwedge} (b)_l$. Conversely, $|x| \leq a^n, b^m$ implies $|x| \leq |a^n| \wedge |b^m| \leq a^n \wedge a^{n-1} b^m \wedge \cdots \wedge a(b^m)^{n-1} \wedge b^{mn} = a \wedge b^m)^n \leq (b^m \wedge b^{m-1} a \cdot \wedge b a^{m-1} \wedge a^m = (b \wedge a)^{mn}$. Hence $(a)_l \overset{C}{\bigwedge} (b)_l \subseteq (a \wedge b)_l$ and $(a)_l \overset{C}{\bigwedge} (b)_l = (a \wedge b)_l$.

In order to prove the second inequality we note that $(a \vee b)_l \subseteq (a)_l \overset{C}{\bigvee} (b)_l$ follows immediately from $(a)_l \overset{C}{\bigvee} (b)_l = (\{a, b\})_l$. Conversely, if $x \in (a)_l \overset{C}{\bigvee} (b)_l$, then $|x| \leq a^{n_1} b^{m_1} \cdots a^{n_k} b^{m_k} \leq (a^n b^n)^k$ where $n = \max\limits_{i=1}^{k} \{n_i, m_i\}$ in $\mathbf{Z}$. As $a^n b^n \leq a^{2n} \vee a^{2n-1} b \vee \cdots a^n b^n \vee \cdots \vee a b^{2n-1} \vee b^{2n} = (a \vee b)^{2n}$ then $|x| \leq (a \vee b)^{2n}$ and $x \in (a \vee b)_l$, $\quad (a \vee b)_l = (a)_l \overset{C}{\bigvee} (b)_l$. □

It is clear now that lattice $C_f(G)$ is distributive because it is a homomorphical image of the distributive lattice $P(G)$.

We can get some information about $C(G)$ from lattice $C_f(G)$ as $C(G)$ is a completion of $C_f(G)$: for every convex l-subgroup $H \in C(G)$ is true $H = \sup_{C(G)} \{ \ (a)_l \mid a \in H \}$.

LEMMA 2. *If G is an l-group, $A, B \subseteq G$, then*

$$(1) \qquad (A)_l = \sup a \in A_{C(G)}(|a|)_l = (\bigcup_{a\in A}\{|a|\})_l,$$

$$(2) \qquad (A)_l \overset{C}{\bigwedge}(B)_l = (C)_l \text{ where } C = \{|a| \wedge |b| \mid a \in A,\ b \in B\}.$$

Proof. (1) is evident.

Let $x \in (C)_l$. Then $|x| \le (|a|_1 \wedge |b|_1)^{n_1} \cdots (|a_s| \wedge |b_s|)^{n_s}$ where $a_i \in A$, $b_i \in B$, $n_i \in \mathbf{Z}$ for $i = 1, 2, \ldots, s$. Hence $|x| \le |a_1|^{n_1} \cdots |a_s|^{n_s}$, $|x| \le |b_1|^{n_1} \cdots |b_s|^{n_s}$ and $x \in (A)_l \wedge (B)_l$, $(C)_l \subseteq (A)_l \overset{C}{\bigwedge}(B)_l$. Conversely, let $|x| \le |a_1|^{n_1} \cdots |a_r|^{n_r}$, $|x| \le |b_1|^{m_1} \cdots |b_s|^{m_s}$ for some $n_i, m_j \in \mathbf{N}$. Then $|x| \le |a_1|^{n_1} \cdots |a_r|^{n_r} \wedge |b_1|^{m_1} \cdots |b_s|^{m_s}$ and by Proposition 2.1.2.

$$|x| \le (|a_1|^{n_1} \cdots |a_r|^{n_r} \wedge |b_1|^{m_1}) \cdots (|a_1|^{n_1} \cdots |a_r|^{n_r} \wedge |b_s|^{m_s} \le$$
$$\le (|a_1| \wedge |b_1|)^{n_1 m_1} \cdots (|a_i| \wedge |b_j|)^{n_i m_j} \cdots (|a_r| \wedge |b_s|)^{n_r m_s}.$$

Therefore $x \in (C)_l$, $(A)_l \overset{C}{\bigwedge}(B)_l$ and (2) is proved, too. □

The following important result was due to G.Birkhoff [1] for ideals and to K.Lorenz [1] and M.Jakubikova [1] in general case.

THEOREM 2. *The lattice $C(G)$ of convex l-subgroups of l-group G is distributive and moreover it is Browerian.*

Let G be an l-group, $A, B_\alpha \in C(G)$, $\alpha \in I$ and $A^+ = A \cap P(G), B_\alpha^+ = B_\alpha \cap P(G)$. Then $A = (A^+)_l$, $B_\alpha = (B_\alpha^+)_l$, $\overset{C}{\bigvee_{\alpha\in I}} B_\alpha = (\bigcup_{\alpha\in I} B_\alpha^+)_l$ and $A \overset{C}{\bigwedge}(\overset{C}{\bigvee_{\alpha\in I}} B_\alpha) = (A^+)_l \overset{C}{\bigwedge}(\bigcup_{\alpha\in I} B_\alpha^+)_l = (M)_l$ where $M = \{\, a \wedge b \mid a \in A^+, b \in \bigcup_{\alpha\in I} B_\alpha^+ \,\}$. Then

$$A \overset{C}{\bigwedge}(\overset{C}{\bigvee_{\alpha\in I}} B_\alpha) = (M)_l = (\bigcup_{\alpha\in I} M_\alpha)_l = \overset{C}{\bigvee_{\alpha\in I}}(M_\alpha)_l = \overset{C}{\bigvee_{\alpha\in I}}(A \overset{C}{\bigwedge} B_\alpha). \square$$

The following example shows that $C(G)$ can not be dually Browerian and completely distributive.

EXAMPLE 1. Let $G = \overline{\prod_{n\in\mathbf{N}}} G_n$ be the Cartesian product of o-groups G_n, $(n \in \mathbf{N})$ which are isomorphic to the additive group $\mathbf{Z}$ of integers with natural total order. Let $A = \prod_{n\in\mathbf{N}} G_n$ be the direct product of o-groups $G_n \quad (n \in \mathbf{N}), \quad A \subseteq G$. Let $B_n = \{ f \in G \mid f(n) = 0 \}$ for every $n \in \mathbf{N}$. It is clear that A and B_n are convex l-subgroups of G and $G = A \overset{c}{\bigvee} B_n$ for every $n \in \mathbf{N}$. On the other hand, $\overset{c}{\bigwedge\limits_{n \in \mathbf{N}}} B_n = E$ and

$E = \overset{c}{\bigwedge\limits_{n\in\mathbf{N}}} (A \overset{c}{\bigvee} B_n) \neq G = \overset{c}{\bigvee\limits_{n\in\mathbf{N}}} (A \overset{c}{\bigwedge} B_n)$.□

3.2 Archimedean o-Groups. Convex Subgroups in o-Groups

Now we consider two types of l-groups which have the simple structure of l-subgroups .

1^0. The simplest structure of convex l-subgroup are possessed by l-groups G with two elements in the lattice $C(G)$, i.e., l-groups which contain no proper convex l-subgroups.

An l-group G is called *Archimedean* if for G the following law (the *Archimedean axiom*) holds.

(A) For every two strictly positive elements $a, b \in G$ there exist the natural numbers n, m such that $a^n \not< b, \quad b^m \not< a$.

PROPOSITION 1. *An l-group G contains no proper convex l-subgroups if and only if G is an Archimedean group.*

Proof. If $a, b \in G, \ a > e, \ b > e$ and G contains no proper convex l-subgroups, then $(a)_l = (b)_l = G$. Hence $b \in (a)_l, \ a \in (b)_l$. By the definition $(a)_l = \{ x \in G \mid |x| \leq a^n \}$ for suitable natural n and $b \leq a^n$ for some n. By similar arguments $a \leq b^m$ for some m and G is an Archimedean l-group. We claim now that G is an o-group. If it is not true, then G contains two orthogonal elements $a, b, \ a \neq e \neq b, \ a \wedge b = e$. Then $x \in (a)_l$ implies $|x| \wedge b = e$ (see Proposition 2.2.11 and $b \notin (a)_l$). This contradicts our assumptions about G.

Converse statement is obvious.□

The following results is a cornestone of the ordered groups theory and it is well known as Hölder's theorem.

THEOREM 1. *An Archimedean o-group is o-isomorphic to some subgroup of additive group* **R** *of reals with the natural order.*

Proof. Let G be an Archimedean o-group. We will consider two cases:

(1) there is the least positive element a in G,

(2) there is no least positive element in G.

In the case (1) for every element $b \in G$ there exsists an integer n such that $a^n \le b < a^{n+1}$, $e \le ba^{-n} < a$. As a is the least positive element of G then $ba^{-n} = e$, $b = a^n$. Hence G is isomorphic to the additive group $\mathbf{Z}$ of integers with natural order.

In the case (2) G contains no least positive element. Then for any $a > e$ there exsists $c > e$ such that $c^2 \le a$. In fact, for a there exsists $b \in G$ such that $e < b < a$. If $b^2 \le a$, then $c = b$, if $b^2 > a > b$, then $e > b^{-1}ab > b^{-1}$ and $a > ab^{-1} \cdot ab^{-1} > ab^{-1} > e$, $c = ab^{-1}$.

We will prove now the commutativity of G. Suppose that there are two elements $x, y \in G$ such that $xy \ne yx$ i.e., $z = [x, y] = x^1y^{-1}xy \ne e$. We can assume that $z > e$. As we see above there exsists an element $t \in G$, $t > e$ such tat $t^2 < z$. Because G is an Archimedean o-group, then there exists the integer n, m such that $t^n \le x < t^{n+1}$, $t^m \le y < t^{m+1}$. Hence $xy < t^{n+m+2}$, $x^{-1}y^{-1} \le t^{-n-m}$. Therefore, $z = x^{-1}y^{-1}xy < t^2$ and this contradicts the choice of element t. So, G is an Abelian group.

Finally, let A be a divisible hull of G, $G \subseteq A$. Let $P(A) = \{\, x \in A |$ there exists a natural number n such that $x^n \in G$, $x^n \ge e$ in $G\}$. It is clear that $P(A)$ is a pure linear subsemigroup of A. Hence $P(A)$ defines the total order on A which is extended to the total order of G. It is clear that $(A, P(A))$ is an Archimedean o-group. Because A is a divisible Abelian torsion-free group, then A contains subgroup Q which is o-isomorphic to the additive totally ordered group $\mathbf{Q}$ of rationals. Therefore, every element $g \in A$ defines the Dedekind section α_g in Q. We can identify the real number with α_g and denote this number by α_g, too. The simple verification shows that the map:$g \to \alpha_g$ is an o-isomorphism from G into o-group $\mathbf{R}$ of reals.□

PROPOSITION 2. *If A, B are subgroups of the additive o-group $\mathbf{R}$ of reals and $\varphi : A \to B$ is an o-isomorphism between A and B, then there exists some poositive real number r such that $s\varphi = rx$ for all $x \in G$.*

Proof. Let $(a\varphi) \cdot (b\varphi)^{-1} \ne ab^{-1}$ for some elements $a, b \in A$, $a, b > 0$ in $\mathbf{R}$. Let $(a\varphi)(b\varphi)^{-1} < ab^{-1}$. Then there is a rational number

$mn^{-1}, m, n \in \mathbf{N}$ such that $(a\varphi)(a\varphi)^{-1} < mn^{-1} < ab^{-1}$. Then $mb < na$, $m(b\varphi) > n(a\varphi)$. This is a contradicton.□

COROLLARY 1. *The group of order automorphisms of Archimedean o-group is isomorphic to some subgroup of the multiplicative group* $\mathbf{R}^+$ *of positive reals.*□

2⁰. We can use Hölder's theorem for the description of l-groups G such that the lattice $C(G)$ of convex l-subgroups of G is totally ordered.

PROPOSITION 3. *The lattice* $C(G)$ *is totally ordered for l-group* G *if and only if* G *is o-group.*

Proof. It is obvious that $C(G)$ is totally ordered if G is an o-group. The proof of a converse statement coinsides with the analogus part of Proposition 1 and we leave it for the reader.□

We can describe the structure of o-group G by the properties of $C(G)$. We give the definitions which will be usefull for us. Let G be a group and U be some system of subgroups of G which is totally ordered under the set-theoretical inclusion. The system U is called *complete* if $\bigcup_{\alpha\in I} G_\alpha \in U$, $\bigcap_{\alpha\in I} G_\alpha \in U$ for every set $\{ G_A | \ \alpha \in I \}$ of subgroups $G_\alpha \in U$. Now, U is *infrainvariant* if U is complete $E \in U, G \in U$ and $x^{-1}Ax \in U$ for every $A \in U$, $x \in G$. If U is a complete system of subgroups of G, $A, B \in U$, $A \subset B$, $A \neq B$ and for any $C \in U$ the inclusion $A \subseteq C \subseteq B$ implies $A = C$ or $C = B$, then then the pair A, B is called a *jump* in U. We denote this fact $A \prec B$. It is clear that every element $g \in G$, $g \neq e$ determines a jump $A_g \prec B_g$ in the complete system U of subgroups such that $g \in B_g \setminus A_g$. Here $A_g = \bigcup_{\gamma\in\Gamma} G_g$ where $\{ G_\gamma \mid \gamma \in \Gamma\}$ is the set of all subgroups G_γ from U such that $g \notin G_\gamma$, $B_g = \bigcap_{\delta\in\Delta} G_\delta$ where $\{ G_\delta \mid \delta \in \Delta\}$ is the set of all $G_\delta \in U$, $g \in G_\delta$.

We use the following standard group-theoretical notations. Let G be a group, H be a subgroup of G. Then

$$n(H) = N_G(H) = \{ x \in G \mid x^{-1}Hx = H\},$$

$$c(H) = c_G(H) = \{ x \in G \mid x^{-1}hx = h \text{ for all } h \in H \}.$$

It is clear that $n(H)$, $c(H)$ are subgroups of G, $c(H)$ is a normal subgroup of $n(H)$. The subgroup $n(H)$ is called a *normalizer* of H ,

$c(H)$ is called a *centralizer* of H . If $A \prec B$ is a jump in the infanvariant system U of subgroups in any group G, then $n(A) = n(B)$. It is also clear that every element $x \in n(B)$ induces an automorhism of the quotient group $B/A : bA \to x^{-1}bxA$. The group of all these automorphisms of B/A is isomorphic to the quotient group $n(A)/c(B/A)$. Here $c(B/A) = \{ x \in n(B) \mid x^{-1}bxA = bA\}$ is a *centralizer of jump* $A \prec B$.

THEOREM 2. (Iwasawa [2], Mal'cev [3].) *The system $C(G)$ of all convex subgroup of o-group G is infrainvariant. If $A \prec B$ is an jump in $C(G)$, then the quotient group B/A is order-isomorpphic to a subgroup of the additive o-group of the reals. The group of automorphisms of the quotient group B/A induced by the inner automorphisms of the normalizer $n(A)$ of the subgroup A in G is isomorphic to a subgroup of the multilicative group of positive real numbers.*

Proof. Let G be an o-group. By Proposition 3 the lattice $C(G)$ of convex subgroups of G is a complete chain. If $H \in C(G)$, then $g^{-1}Hg$ is a convex l-subgroup of G for all $g \in G$. Hence $C(G)$ is an infrainvariant system. If $A \prec B$ is a jump in $C(G)$, then A is normal in B. By Theorem 1 the quotient group B/A is Archimedean and it is order isomorphic to a subgroup of additive o-groups $\mathbf{R}$ of reals and the action of the elements of $n(A)$ on B/A is represented by the multiplication by positive numbers (see Corollary 1 of Proposition 2).□

Now we need to introduce some definitions. Let G be a group, A be a group of operators on G. The partial order $\leq$ is called *A-order* if $g \leq e$ implies $g^{\alpha} \leq e$ for all $g \in G$, $\alpha \in A$. Let $S_A(M)$ denote an invariant under A subsemigroup of G generated by subset M of G and $S'_A(M) = S_A(M) \cup \{e\}$.

LEMMA 1. (Podderyugin [1].) *Let G be an Abelian group with a commutative group of operators A. Then every partial A-order of G can be extended to total A-order of G if and only if for every $g \in G$, $\alpha_1, \dots, \alpha_n \in A$*

$$g^{\alpha_1+\alpha_2+\cdots+\alpha_n} = g^{\alpha_1}g^{\alpha_2}\cdots g^{\alpha_n} = e \textit{ implies } g = e$$

Proof. Let $P = \{ x \in G \mid x > e \}$ be a total A-order on G and $g^{\alpha_1+\cdots\alpha_n} = e$ for some $g \in G$, $\alpha_1, \dots, \alpha_n \in A$. Let $g > e$, then $g^{\alpha_i} > e$ and $g^{\alpha_1+\cdots+\alpha_n} > e$. This is a contradiction.

Conversely, let $g^{\alpha_1+\cdots+\alpha_n} = e$ imply $g = e$ for any $g \in G$, $\alpha_i \in A$ and P be a maximal A-order of G. If P is not total, then there is $g \in G$ such that $g \notin P \cup P^{-1}$.

We claim that $S_A(g) \cap P \neq \emptyset$ and $S_A(g^{-1}) \cap P \neq \emptyset$. Let, for example, $S_A(g) \cap P = \emptyset$. Then $Q = P \cdot S'_A(g^{-1})$ is an A-invariant subsemigroup of G. As P is a maximal A-order of G, then Q is not a partial A-order of G and $Q \cap Q^{-1} \neq \emptyset$. Hence $p(g^{-1})^{\alpha_1+\cdots+\alpha_r} = (q(g^{-1})^{\beta_1+\cdots+\beta_s})^{-1}$ for some $p, q \in P$, $\alpha_1, \ldots, \alpha_r, \beta_1, \ldots, \beta_s \in A$. By commutativity of G it follows that $g^{\alpha_1+\cdots+\alpha_r+\beta_1+\cdots+\beta_s} = pq$. This contradicts our assumption.

So, $S_A(g) \cap P \neq \emptyset$ and $S_A(g^{-1}) \cap P \neq \emptyset$. Then there are $\alpha_1, \ldots, \alpha_r$, $\beta_1, \ldots, \beta_s \in A$ such that $g^{\alpha_1+\cdots+\alpha_r} \geq e$ and $g^{\beta_1+\cdots\beta_s} \leq e$ under order P. Hence

$$e \leq (g^{\alpha_1+\cdots+\alpha_r})^{\beta_1} \cdots (g^{\alpha_1+\cdots+\alpha_r})^{\beta_s} = g^{\alpha_1\beta_1} \cdots g^{\alpha_i\beta_j} \cdots g^{\alpha_r\beta_s} =$$
$$g^{\beta_1\alpha_1} \cdots g^{\beta_i\alpha_i} \cdots g^{\beta_s\alpha_r} = (g^{\beta_1+\cdots+\beta_s})^{\alpha_1} \cdots (g^{\beta_1+\cdots+\beta_s})^{\alpha_r} \leq e.$$

Then $g^{\alpha_1\beta_1} \cdots g^{\alpha_i\beta_j} \cdots g^{\alpha_r\beta_s} = e$ and $g = e$, a contradiction.□

The subset H of group G is called *strongly isolated* if $x^{1+g_1+\cdots+g_n} = x \cdot x^{g_1} \cdots x^{g_n} \in H$ implies $x \in H$ for all $x, g_1, \ldots, g_n \in G$. A totally ordered complete system $C(G)$ of subgroups of G is called *soluble* if for every jump $A \prec B$ in $C(G)$ the subgroup A is normal in B and the quotient group B/A is Abelian.

Let A, B be any subgroups in group G. A *commutator subgroup of A and B* is the subgroup $[A, B]$ of G generated by all commutators $[a, b]$ where $a \in A$, $b \in B$. Since $[a, b] = [b, a]^{-1}$ we have $[A, B] = [B, A]$ (here $[a, b] = a^{-1}b^{-1}ab$ as usualy). Now we are ready to describe totally ordered groups in the terms of the subgroup systems in G.

THEOREM 3. (Mal'cev [3], Podderyugin [2], Rieger [1]) *A group G is orderable if and only if G has a totally ordered soluble infrainvariant subgroup system $C(G)$ of G such that*

(3) *all subgroups of $C(G)$ are strongly isolated in G*
(4) *for any jump $A \prec B$ in $C(G)$ holds $[[n(A), n(A)], B] \subseteq A$.*

Moreover, the subgroups of the system $C(G)$ are convex subgroups under some total order on G.

Proof. If G is a totally ordered group the Theorem 2 guarantees the existence of such a system. It is the system $C(G)$ of all convex subgroups of G. In fact, if $x \notin H$, $x > e$, then $x^{1+g_1+\cdots+g_n} > h$ for every $x, g_1, \ldots, g_n \in G$, $h \in H$ and (3) holds for $C(G)$. The property (4) is equivalent to commutativity of inner automorphisms induced by $n(A)$ on B/A.

Conversely, let $C(G)$ be a totally ordered under inclusion soluble infrainvariant subgroup system in G with properties (3)-(4). Divide the jumps in $C(G)$ into equivalence classes $\{ A^g_\alpha \prec B^g_\alpha \mid g \in G \}$, $A^g_\alpha = g^{-1}Ag$. Choose a representative $A^*_\alpha \prec B^*_\alpha$ in each class. The automorphisms of B^*_α/A^*_α induced by inner automorphisms of $n(A^*_\alpha)$ are commutative on B^*_α/A^*_α by (4). Therefore $n(A_\alpha)$ induced a commutative group of operators on B^*_α/A^*_α. By (3) for all $b \in B^*_\alpha$, $g_1, \ldots, g_n \in n(A^*_\alpha)$ the equality $(bA^*_\alpha)^{1+g_1+\cdots+g_n} = A^*_\alpha$ implies $b^{1+g_1+\cdots+g_n} \in A^*_\alpha$, $b \in A^*_\alpha$. By Lemma 1 a quotient group B^*_α/A^*_α has a total $n(A^*_\alpha)$-order. Now we fix an $n(A^*_\alpha)$-order P^*_α on B^*_α/A^*_α.

Every element $x \in G$, $x \neq e$ determines a jump $A_\alpha \prec B_\alpha$ in $C(G)$. Let $g(x) \in G$ such that $A^*_\alpha = A^{g(x)}_\alpha$. We define the set P in G by the rule:

$$P = \{ x \in G \mid x \neq e \text{ and } x^{g(x)}A^*_\alpha \in P^*_\alpha \}$$

The direct verification shows that P is independent of the choice of element $g(x)$ for $x \in G$ and P defines the total order on G such that all subgroups $H \in C(G)$ are convex in G under P.□

The proof of the following theorem is similar.

THEOREM 4. (Kokorin [6].) *Group G is orderable if and only if G has a totally ordered under inclusion soluble infrainvariant subgroups system $C(G)$ such that for any jump $A \prec B$ in $C(G)$ the quotient group B/A has a total $n(A)$-order. Moreover, the subgroups of $C(G)$ are convex subgroups under some total order of G.*□.

3.3 Prime Subgroups

1⁰. Let G be an l-group, H be a convex l-subgroup of G and $R(G:H)$ be the set of right cosets of H in G. As it was mentioned in Chapter 2 §3 the set $R(G:H)$ is a po-set under ordering : $Hx \leq Hy$ if $hx \leq y$ for some $h \in H$.

PROPOSITION 1. *Let G be an l-group and H be a convex l-subgroup og G. Then the set $R(G:H)$ is a lattice under the induced order and $Hx \vee Hy = H(x \vee y)$, $Hx \wedge Hy = H(x \wedge y)$ for all $x, y \in G$.*

Proof. By definition of the partial order on $R(G:H)$ it is clear that $Hx \leq H(x \vee y)$ and $Hy \leq H(x \vee y)$ in $R(G:H)$. Let $Hz \geq Hx$, $Hz \geq Hy$ for some $z \in G$. Then $h_1 z \geq x$, $h_2 z \geq y$ for some $h_1, h_2 \in H$. Let $h = h_1 \vee h_2$ in G. Then $h \in H$ and $hz \geq h_1 z \geq x$, $hz \geq h_2 z \geq y$, $hz \geq x \vee y$. Hence $Hz \geq H(x \vee y)$ and $H(x \vee y)$ is the least upper bound of Hx, Hy in $R(G:H)$. By similar arguments $H(x \wedge y)$ is the greatest lower bound of Hx, Hy in $R(G:H)$.□

It is evident that the map $\varphi : x \to Hx$ is a homomorphism of the lattice G onto lattice $R(G:H)$. For $g \in G$ we define the map $R_H(g) : R(G:H) \to R(G:H)$ by the rule: $(Hx)R_H(g) = Hxg$.

PROPOSITION 2. *The map $R_H(g)$ is an automorphism of lattice $R(G:H)$ for all $g \in G$.*□

2^0. A convex l-subgroup H of l-group G is called a *prime* subgroup if the set $R(G:H)$ is totally ordered under the induced order.

THEOREM 1. (Holland [2], Conrad [3].) *A convex l-subgroup H of l-group G is a prime subgroup if and only if $a \wedge b \notin H$ for all positive elements $a, b \in G \setminus H$.*

Proof. Let H be a prime subgroup of G and $a, b > e$, $a, b \in G \setminus H$. Since the set $R(G:H)$ is totally ordered, then $H(a \wedge b) = \min\{Ha, Hb\} \neq H$ in $R(G:H)$. Hence $a \wedge b \notin H$.

Conversely, let for all $a, b \in G \setminus H$, $a, b > e$ hold $a \wedge b \notin H$. Suppose on the contrary that there are incomparable elements Hx, Hy in $R(G:H)$. Then $Hx \wedge Hy = H(x \wedge y)$ and $Hx(x \wedge y)^{-1} \wedge Hy(x \wedge y)^{-1} = H$, $Hx(x \wedge y)^{-1} \neq H$, $Hy(x \wedge y)^{-1} \neq H$ So $x(x \wedge y)^{-1}, y(x \wedge y)^{-1} \notin H$. But $x(x \wedge y)^{-1} \wedge y(x \wedge y)^{-1} = e \in H$, a contradiction.□

THEOREM 2. (Holland [2], Conrad [3].) *A convex l-subgroup H of l-group G is a prime subgroup if and only if the set of all convex l-subgroups of G containing H is totally ordered under inclusion.*

Proof. Let H be a prime subgroup of l-group G. Suppose that there are convex l-subgroups A, B such that $H \subseteq A$, $H \subseteq B$ and $A \not\subseteq B$, $B \not\subseteq A$. Let $a > e$, $a \in A \setminus B$, $b > e$, $b \in B \setminus A$. Then $a' = a(a \wedge b)^{-1} \in A \setminus B$, and $b' = b(a \wedge b)^{-1} \in B \setminus A$. But $a' \wedge b' = e$, a contradiction.

Conversely, let the set of all convex l-subgroup of G containing H be totally ordered by inclusion. Assume on the contrary that H is not prime. Then by Theorem 1 there are $a, b > e,\ a, b \in G \setminus H$ such that $a \wedge b \in H$. Now we consider the subsets A, B in G defined by the rules:

$$A = \{\ x \in G \mid |x| \wedge b \in H\ \}, \quad B = \{\ x \in G \mid |x| \wedge a \in H\ \}.$$

It is obvious that $A \not\subseteq B,\ B \not\subseteq A$ and $H \subseteq A,\ H \subseteq B$. We claim that A, B are convex l-subgroups in G. Let $x \in A,\ y \in A$, then $e \leq |x| \wedge b = h_1 \in H$ and $e \leq |y| \wedge b = h_2 \in H$. By Proposition 2.2.8(9)

$$\begin{aligned}
&|xy^{-1}| \wedge b \leq |x| \cdot |y| \cdot |x| \wedge b = |x| \cdot |y| \cdot |x| \wedge |x| \cdot |y| \cdot b \wedge b = \\
&= |x| \cdot |y|(|x| \wedge b) \wedge b = |x| \cdot |y| \cdot h_1 \wedge b \leq |x| \cdot |y| h_1 \wedge bh_1 = \\
&= (|x| \cdot |y| \wedge b)h_1 = (|x| \cdot |y| \wedge |x| \cdot b \wedge b)h_1 = (|x|(|y| \wedge b) \wedge b)h_1 = \\
&= (|x|h_2 \wedge b)h_1 \leq (|x| \cdot h_2 \wedge b \cdot h_2)h_1 = (|x| \wedge b)h_2 \cdot h_1 =
\end{aligned}$$

$= h_1 h_2 h_1 \in H$. By our assumption, it follows that $xy^{-1} \in A$. Therefore A is a subgroup of G. Let now $e \leq x \leq a$ and $a \in H$. Then $e \leq x \wedge b \leq a \wedge b \in H$ and $x \wedge b \in H$, too. This means that A is a convex subgroup in G. If $x \in A$, then $|x \vee e| = x \vee e \leq |x|$ and $e \leq |x \vee e| \wedge b \leq |x| \wedge b \in H$. So A is a convex l-subgroup in G. By similar arguments B is a convex l-subgroup of G, too. By definition A and B contain H and are incomparable under inclusions, a contradiction.□

A subset X of the lattice L is called a *root system* if for all $x \in X$ the subset $U_x = \{\ y \in L \mid x \leq y\ \}$ is totally ordered and is contained in X.

COROLLARY 1. (Conrad [3].) *The set of all prime subgroups $C_0(G)$ of l-group G is a root system in the lattice $C(G)$ of all convex l-subgroups of G.*

3^0. Prime subgroups are of profound importance in the theory of l-groups. Now we will establish the existence of numerous prime subgroups in any l-group.

Let G be an l-group, $g \in G$ and $g \neq e$. Let $\{V_\alpha(g) \mid \alpha \in I\ \}$ be a set of all convex l-subgroups of G maximal with respect to not containing $g \in G$. It is evident that for every $g \in G,\ g \neq e$ the set $\{V_\alpha(g) \mid \alpha \in I\ \}$ is not empty. Each V_α is called a *value* of g and is a *regular subgroup* of G. Let $\bar{V}_\alpha(g) = (V_\alpha(g), g)_l$ be a convex l-subgoup of G generated by $V_\alpha(g)$ and g. It is clear that the pair $V_\alpha(g) \prec \bar{V}_\alpha(g)$ is a jump in the lattice $\mathcal{C}(G)$ of convex l-subgroups of G and $\bar{V}_\alpha(g)$ is a cover of $V\alpha(g)$ in $\mathcal{C}(G)$.

THEOREM 3. (Holland [2].) *Let G be an l-group and V be a regular subgroup of G. Then V is a prime subgroup of G.*

Proof. Let V be a regular subgroup of G. Then $V = V_\alpha(g)$ for some $g \in G$, $g \neq e$ and $\bar{V}_\alpha(g)$ covers $V_\alpha(g)$ in the lattice $\mathcal{C}(G)$. It is clear that if H is a convex l-subgroup of G and $V_\alpha(g) \subset H$, $V_\alpha(g) \neq H$, then $\bar{V}_\alpha(g)$ has tis property, too. Let $C = \{ x \in G \mid |x| \wedge |g| \in V_\alpha(g) \}$. By similar arguments as in the proof of Theorem 2, it follows that C is a convex l-subgroup of G containing $V_\alpha(g)$. If C contains $V_\alpha(g)$ properly, then C contains $\bar{V}_\alpha(g)$, too. Then $g \wedge |g| \in V_\alpha(g)$, a contradiction. Therefore $C = V_\alpha(g)$.

Let now $a, b \in G$, $a, b > e$, $a \wedge b \in V_\alpha(g)$ and $a \notin V_\alpha(g)$. Let $A = \{ x \in G \mid |x| \wedge b \in V_\alpha(g) \}$. By similar arguments A is a convex l-subgroup of G. It is obvious that A strictly contains $V_\alpha(g)$ as $a \in A$, $a \notin V_\alpha(g)$. Hence $\bar{V}_\alpha(g) \subseteq A$. By definition A $|g| \wedge b \in V_\alpha(g)$ and $b \in C = V_\alpha(g)$. Therefore for any two positive elements $a, b \in G$ such that $a \notin V_\alpha(g)$, $a \wedge b \in V_\alpha(g)$ it follows that $b \in V_\alpha(g)$. Thus by Theorem 1, $V = V_\alpha(g)$ is a prime subgroup of G.□

PROPOSITION 3. (Conrad [3].) *Each convex l-subgroup of an l-group G can be obtained as intersection of regular subgroups.*

Proof. Let H be a convex l-subgroup of G and $x \in G \setminus H$. By Zorn's Lemma there is a convex l-subgroup $V(x)$ containing H and maximal with the property $x \notin V(x)$. It is clear that $H = \bigcap_{x \in G \setminus H} V(x)$.□

A convex l-subgroup H of l-group G is called a minimal prime subgroup if H is a prime subgroup and each convex l-subgroup H_1 which is strictly contained in H is not prime. From Theorems 1,2, it is apparent that the intersection of the totally ordered set of prime subgroups is a prime subgroup, too. Since each prime subgroup may evidently be embedded in a maximal totally ordered set of prime subgroups, it follows that each prime subgroup contains a (not necessarry unique) minimal prime subgroup. Therefore the following statement is proved.

PROPOSITION 4. *Each prime subgroup contains a minimal*

prime subgroup. □

It is clear that if a prime subgroup H is normal in G, then the quotient group G/H is a totally ordered group. Each convex subgroup of o-group G is evidently a prime subgroup in G.

Values are useful in distinguishing the order on G.

PROPOSITION 5. (Holland [3].) *If G is an l-group, then $f \leq g$ in G if and only if $Vf \leq Vg$ for all regular subgroups V in G.*

Proof. If $f \leq g$ in G, then $Vf \leq Vg$ for all convex subgroups V. Conversely, if $f \not\leq g$ in G, then $fg^{-1} \vee e \neq e$, so there is a value $V_\alpha(fg^{-1} \vee e) = V$. Now $e < fg^{-1} \notin V$. Hence $V < V(fg^{-1} \vee e) = V(f \vee g)g^{-1}$. Thus $Vg < V(f \vee g) = \max\{Vf, Vg\} = Vf$ since V is a prime subgroup. Consequently, $Vf \not\leq Vg$.□

3.4 Polars

1⁰. Let M be a non-empty subset of l-group G. Let

$$M^{\perp} = \{ \ x \in G \mid |x| \wedge |y| = e \text{ for all } y \in M \ \},$$

be the *polar* of M.

PROPOSITION 1. *Each polar $M^{\perp}$ of l-group G is a convex l-subgroup of G.*

Proof. Let G be an l-group, $M \subseteq G$ and $x, y \in M^{\perp}$. Then $|xy^{-1}| \leq |x| \cdot |y| \cdot |x|$ and for all $m \in M$

$$\begin{aligned} |xy^{-1}| \wedge |m| \leq |x| \cdot |y| \cdot |x| \wedge |x| \cdot |y| \cdot |m| \leq \\ \leq |x| \cdot |y|(|x| \wedge |m|) \wedge |m| = |x| \cdot |y| \wedge |m| \cdot |y| \wedge |m| = \\ = (|x| \wedge |m|)|y| \wedge |m| = |y| \wedge |m| = e \end{aligned}$$

holds.

Hence $M^{\perp}$ is a subgroup of G. If $e \leq z \leq y$, $y \in N^{\perp}$, then $e \leq z \wedge |m| \leq y \wedge |m| = e$ and $z \in M^{\perp}$. Let $z \in M^{\perp}$ then $|z \vee e| \leq |z|$ and $e \leq |z \vee e| \wedge |m| \leq |z| \wedge |m| = e$. By Proposition 2.3.1, the polar is a convex l-subgroup of G.□

By Proposition 1, the set $\mathcal{P}(G)$ of all polars of G is contained in the lattice $\mathcal{C}(G)$ of all convex l-subgroups of G. Now we establish some useful properties of polars of G.

PROPOSITION 2. (Šik [2].) *For any two subsets X and Y of l-group G*

(1) *if* $X \subseteq Y$ *then* $Y^{\perp} \subseteq X^{\perp}$,

(2) $X^{\perp\perp\perp} = X^{\perp}$.

We leave the proof for the reader.□

LEMMA 1. (Šik [2].) *For every subset X of an l-group G the subset $X^{\perp\perp}$ is the least polar of G containing X.*

Proof. It is clear that $X^{\perp\perp}$ is a polar of G and $X \subseteq X^{\perp\perp}$. If $Y^{\perp} \supseteq X$, then by Proposition 1, $Y^{\perp\perp} \subseteq X^{\perp}$, $Y^{\perp} = Y^{\perp\perp\perp} \supseteq X^{\perp\perp}$.□

THEOREM 1. (Šik [8].) *The set $\mathcal{P}(G)$ of all polars of l-group G is a complete Boolean algebra under inclusion and for all $X \in \mathcal{P}(G)$,*

(3) $X \overset{\mathcal{P}}{\bigwedge} Y = X \cap Y = (X^{\perp} \cup Y^{\perp})^{\perp}$,

(4) $X \overset{\mathcal{P}}{\bigvee} Y = (X \cup Y)^{\perp\perp} = (X^{\perp} \cap Y^{\perp})^{\perp}$

(5) *a complement X' of the polar X in $\mathcal{P}(G)$is equal to $X^{\perp}$.*

Here $\overset{\mathcal{P}}{\bigwedge}, \overset{\mathcal{P}}{\bigvee}$ are operations in the Boolean algebra $\mathcal{P}(G)$.

Proof. By Lemma 1, the polar $(\bigcup_{\alpha\in A} X_{\alpha}^{\perp})^{\perp\perp}$ is the supremum of the family of polars $\{X_{\alpha} | \alpha \in I\}$. Let now $\{X_{\alpha} | \alpha \in I\}$ be a family of subsets of G. We claim that $\bigcap_{\alpha\in I} X_{\alpha}^{\perp} = (\bigcup_{\alpha\in I} X_{\alpha})^{\perp}$. The inclusion $\bigcap_{\alpha\in I} X_{\alpha}^{\perp} \subseteq (\bigcup_{\alpha\in I} X_{\alpha})^{\perp}$ is evident. Let $|z| \wedge |y| = e$ for al $z \in \bigcup_{\alpha\in I} X_{\alpha}$. Then $|x_{\alpha}| \wedge |y| = e$ for all $\alpha \in I$, $x_{\alpha} \in X_{\alpha}$. Hence $y \in X_{\alpha}$ for all $\alpha \in I$ and $y \in \bigcap_{\alpha\in I} X_{\alpha}$.

Thus $\mathcal{P}(G)$ is a complete lattice. Now we define the operations $\overset{\mathcal{P}}{\bigvee}$ and $\overset{\mathcal{P}}{\bigwedge}$ by the rules:

$$X \overset{\mathcal{P}}{\bigvee} Y = \bigcap_{Z\supseteq X\cup Y} Z, \quad X \overset{\mathcal{P}}{\bigwedge} Y = (X \cup Y)^{\perp} = (X^{\perp} \cup Y^{\perp})^{\perp}$$

for $X, Y, Z \in \mathcal{P}(G)$. By Lemma 1, $X \overset{\mathcal{P}}{\bigvee} Y = (X^\perp \cap Y^\perp)^\perp$, then

$$X \overset{\mathcal{P}}{\bigvee}(Y \overset{\mathcal{P}}{\bigwedge} Z) = X \overset{\mathcal{P}}{\bigvee}(Y^\perp \cup Z^\perp)^\perp = (X^\perp \cap (Y^\perp \cup Z^\perp))^\perp =$$
$$= ((X^\perp \cap Y^\perp) \cup (X^\perp \cap Z^\perp))^\perp = (X^\perp \cap Y^\perp)^\perp \overset{\mathcal{P}}{\bigwedge}(X^\perp \cap Z^\perp)^\perp =$$
$$= (X \overset{\mathcal{P}}{\bigvee} Y) \overset{\mathcal{P}}{\bigwedge}(X \overset{\mathcal{P}}{\bigvee} Z).$$

Thus, the lattice $\mathcal{P}(G)$ is distributive.Finally, we notice that $\mathcal{P}(G)$ is a lattice with complementation $'$ defined by the rule: $X' = X^\perp$ for all $X \in \mathcal{P}(G)$. In fact $X \overset{\mathcal{P}}{\bigwedge} X^\perp = E = G^\perp$ is a null of $\mathcal{P}(G)$ and $X \overset{\mathcal{P}}{\bigvee} X^\perp = (X^\perp \cap X)^\perp = E^\perp = G$ is a unit of $\mathcal{P}(G)$.□

Now we notice that if X is a polar of G and Y is a polar of X, then Y is a polar of G, too.

2^0. A Boolean algebra $\mathcal{P}(G)$ is not a sublattice of the lattice of convex l-subgroups of any l-group G in general.

PROPOSITION 3. *The Boolean algebra $\mathcal{P}(G)$ of polars of l-group G is a sublattice of the lattice $\mathcal{C}(G)$ of all convex l-subgroups of G if and only if $G = H \times H^\perp$ for all $H \in \mathcal{P}(G)$.*

Proof. Let $\mathcal{P}(G)$ be a sublattice of $\mathcal{C}(G)$. Then $H \cdot H^\perp = H \overset{\mathcal{P}}{\bigvee} H^\perp = G$, $H \cap H^\perp = E$ and elements of H and $H^\perp$ are commute. Therefore H and $H^\perp$ are ideals of G and $G = H \times H^\perp$. The converse statement is obvious.□

If $G = H \times H^\perp$ is the direct product of H and $H^\perp$ for every polar H of G, then G is said to be *strongly projectable* .

THEOREM 2. (Šik [2].) *The Boolean algebra $\mathcal{P}(G)$ of l-group G is a complete sublattice of the lattice $\mathcal{C}(G)$ if and only if G is a direct product of totally ordered groups.*

Proof. Let $G = \prod_{\alpha \in A} G_\alpha$ be the direct product of the family $\{G_\alpha | \alpha \in A\}$ of totally ordered groups G_α and X be any polar of G. Let $T(X) = \{\alpha \in A | x(\alpha) \neq e$ for some $x \in X\}$. Then

$$X^\perp = \{g \in G | g(\alpha) = e \text{ for all } \alpha \in T(X)\}$$

and

$$X = X^{\perp\perp} = \{g \in G | g(\beta) = e \text{ for all } \beta \in A \setminus T(X)\} =$$
$$= \text{gp}\{G_\beta | \alpha \in T(X)\}.$$

This means that every polar X is generated as a group by the family of all convex l-subgroups $\{G_{(\alpha)} | \alpha \in T(X)$. Then a convex l-subgroup Y which is generated (as a subgroup) by the set $\{X_\gamma | \gamma \in \Gamma\}$, of polars in G is a polar of G, too. Hence $\mathcal{P}(G)$ is a complete sublattice of the lattice $C(G)$.

Conversely, let $P(G)$ be a complete sublattice of $\mathcal{C}(G)$. Assume that G is not o-group. Then there is an element $a \in G$ incomparable with e. Since the elements $a^+, (a^-)^{-1}$ are orhtogonal and not e, then the polar $(a^+)^\perp$ is a non-trivial polar of G. Let K be any non-trivial polar of G, A be a polar of K and B be a complement of A in $\mathcal{P}(K)$. Then $G = A \times A^\perp$ and by distributivity of the lattice $\mathcal{C}(G)$ it follows that $K = (A^{\perp\perp} \cap K) \times (A^\perp \cap K)$. It is clear that $A^\perp \cap K = B$ and $A^{\perp\perp} \cap K = A$. Therefore $\mathcal{P}(K)$ is a complete sublattice of $\mathcal{P}(G)$.

By our assumption, G is not an o-group. Then there are orthogonal elements $a, b \in G$, $a, b > e$. Let $c = ab^{-1}$ and $M(a, b) = \{K_\alpha | \alpha \in I\}$ be the family of all polars of G not containing c. By the choice of c, the family $M(a, b)$ is non-empty. Since the join of the totally ordered set polars from $M(a, b)$ is a polar of G from $M(a, b)$, then by Zorn's Lemma, there is a maximal element M in $M(a, b)$. So $G = M \times M^\perp$ and the polar $M^\perp = M_1$ is indecomposable into the direct product since M is a maximal element in $M(a, b)$.

Thus it is proved that G is decomposed into the direct product of o-group M_1 and l-group $G_1 = M^\perp = M$. The l-group G_1 also satisfies the hypothesis of the theorem. Let us suppose that there is a system of polars $M_\xi, \xi < \alpha$ where ξ, α are ordinals such that for every $\xi < \alpha$, the polar M_ξ is a direct product of o-groups and $M_\xi = M_{\xi-1} \times C_\xi$ for some o-group C_ξ if ξ is a successor ordinal and $M_\xi = \bigcup_{\eta<\xi} M_\eta$ if ξ is a limit ordinal. Now we define a polar M_α as follows: $M_\alpha = \bigcup_{\xi<\alpha} M_\xi$ for a limit ordinal α.

If α is a successor ordinal, then $G = M_{\alpha-1} \times M_{\alpha-1}^\perp$. By above arguments, there exists a totally ordered polar C_α in $M_{\alpha-1}^\perp$. Then we set $M_\alpha = M_{\alpha-1} \times C_\alpha$. By definition, it follows that $G = M_\zeta$ for some ordinal ζ.□

In the hypothesis of Theorem 2, the Boolean algebra $\mathcal{P}(G)$ of polars of G is atomic and totally ordered direct factors of G are precisely the atoms of $\mathcal{P}(G)$. Thus if l-group G is the direct product of o-groups, then this decomposition of G in to the direct product of o-groups is unique.

3.5 Lattice-Ordered Groups with Finite Boolean Algebra of Polars

Now we are ready to obtain the description of l-groups with finite Boolean algebra of polars. This description for Abelian l-groups was due to Jaffard [1] and, in general case, to Conrad [2].

Let G be an l-group and $\mathcal{P}(G)$ be the Boolean algebra of polars of G. For every $K \in \mathcal{P}(G)$, let K^0 be the convex l-subgroup of K generated by all polars $X \in \mathcal{P}(G)$ such that $X \subseteq K$, $X \neq K$. It is clear that K^0 is an ideal in K (of course, it is possible that $K = K^0$).

LEMMA 1. *Each polar K of l-group G is a lexicographic extension of the convex l-subgroup K^0 by totally ordered group K/K^0.*

Proof. Case $K = K^0$ is obvious. Let $K \neq K^0$, $g \in K \setminus K^0$, $g > e$. On the contrary, assume that there is $a > e$, $a \in K^0$ such that $a \wedge g \neq a$, $a \wedge g \neq g$. Let $g_1 = g(g \wedge a)^{-1}$, $a_1 = a(g \wedge a)^{-1}$. Then $g_1 > e$, $g_1 \in K \setminus K^0$ and $a_1 > e$, $a_1 \in K^0$. Thus $a_1 \in (g_1)^\perp \cap K$ and $(g_1)^\perp \cap K \neq E$. We claim that $K = (g_1)^{\perp\perp}$. In fact, as $g_1 \in K$ then $(g_1)^{\perp\perp} \subseteq K^{\perp\perp} = K$. If $(g_1^{\perp\perp}) \neq K$, then (g_1) is contained in some polar X, $X \subseteq K$, $X \neq K$. Then $g_1 \in K^0$, a contradiction to $g \in K \setminus K^0$. By Proposition 2.3.9, the quotient group K/K^0 is totally ordered.□

LEMMA 2. *The subgroup S of l-group G generated by all minimal polars K_α, $\alpha \in I$ is a direct product of totally ordered groups $K_\alpha, \alpha \in I$ and is a ideal of G.*

Proof. It is clear that each minimal polar K_α is a totally ordered convex l-subgroup of G. It is also evident that the subgroup K_α and K_β are orthogonal for all $\alpha, \beta \in I$, $\alpha \neq \beta$. Therefore, K_α is orthogonal to the convex l-subgroup $(\bigcup_{\beta\neq\alpha} K_\beta)_l$ generated by all o-subgroups $K_\beta, \quad \beta \in I \setminus \alpha$. Every element $x \in S$ is a product of finite set of elements $x_1, \ldots, x_n$ such that $x_i \in K_{\alpha_i}$ where $\alpha_{i_1}, \ldots, \alpha_{i_n}$ are mutually different elements of I. Moreover, $x > e$ in S iff $x_i > e$ for all $i = 1, \ldots, n$. In fact, if $x > e$, $x = x_1 \cdots x_k \cdot x_{k+1} \cdots x_n$ and $x_1, \ldots, x_k > e, \quad x_{k+1}, \ldots, x_n < e$ then $x_1 \cdot x_k > x_{k+1}^{-1} \cdots x_n^{-1} > e$ and $x_1 \cdots x_k \wedge x_{k+1}^{-1} \cdots x_n^{-1} > e$. The last inequality contradicts our assumption about orthogonality of the elements $x_1, \ldots, x_n$ and Proposition 2.2.11. So $S = (\bigcup_{\alpha\in I} K_\alpha)_l = \prod_{\alpha\in I} K_\alpha$ is a direct product of totally ordered groups $K_\alpha, \quad \alpha \in I$. The inner automorphisms of l-group G permute minimal polars of G, so the convex l-subgroup S is an ideal in G.□

Now we assume that each l-group G in this section has finite Boolean algebra $\mathcal{P}(G)$ of polars. Let $K_1, \ldots, K_n$ denote all minimal polars in $\mathcal{P}(G)$ and S denote the ideal of G generated by $K_1, \ldots, K_n$.

LEMMA 3. *For every $b \in G \setminus S, \ b > e$ there are two different positive integers $i, j, \quad 1 \leq i, j \leq n$ such that $b > x$ for all $x \in K_i \cup K_j$.*

Proof. We prove this statement by induction on n. For $n = 2$ the assertion follows from Lemma 1. Assume the result holds for all $k \leq n-1$ and let $k = n$.

Let $K = (b)^{\perp\perp}$ be a polar of G generated by the element b. If $K \neq G$, then $\mathcal{P}(K)$ contains less or equal then $n-1$ atoms and the results is true. Let $K = G = (b)^{\perp\perp}$ and G^0 be a convex l-subgroup of G generated by all non-trivial polars of G. If $b \notin G^0$, then the result holds by Lemma 1. Let now $b \in G^0$. Then $b = b_1 \cdots b^s$ where b_i is contained in a proper polar of G. Since $b > e$, then $b = |b_1 \cdots b_s| \leq |b_1| \cdots |b_{s-1}| \cdot |b_s| \cdot |b_{s-1} \cdots |b_1|$ and by Corollary 1 of Theorem2.2.1 $b = c_1 \cdots |c_r|$ where $c_i \geq e$ for all $i = 1, \ldots, r$ and each c_i is contained in a proper polar of G. As $b \notin S$ there is some $c_i \notin S, \quad 1 \leq i \leq r$. Then $b > c_i > e$ and $c \in H$ for some proper polar H of G. It is clear that each atom of $\mathcal{P}(H)$ is also an atom of $\mathcal{P}(G)$ and $\mathcal{P}(H)$ has less or equal than $n-1$ atoms. Thus by induction assumption the result holds.□

LEMMA 4. *The Boolean algebra $\mathcal{P}(G/S)$ is finite and the number of atoms $P(G/S)$ is less than the number of atoms of $\mathcal{P}(G)$.*

Proof. Let the Boolean algebra $\mathcal{P}(G/S)$ be infinite or contains more than $n-1$ atoms. Then there are n multially orthogonal polars $\bar{B}_1, \dots, \bar{B}_n$. Without loss of generality, we may assume that $\bar{B}_i$, $1 \leq i \leq n$ is generated by $\bar{b}_i = b_i S > S$ in G/S, $b_i > e$ in G. Hence $s_{ij} = b_i \wedge b_j \in S$ for all $i, j, i \neq j$. By Lemma 3 for every b_i the subset M_i of indeces j, $1 \leq j \leq n$ such that $b_i > x$ for all $x \in K_i$ contains at least two elements. Since $M_1 \cup \dots \cup M_n \subseteq \{1, \dots, n\}$ and each M_i contains more than one element, then there are M_p and M_q such that $M_p \cap M_q \neq \emptyset$. Let $t \in M_p \cap M_q$. Then for all $x \in K_t$, $b_q > x$, $b_p > x$ holds. Hence $s_{pq} = b_p \wedge b_q > x$ for all $x \in K_t$. But $S = \prod_{i=1}^{n} K_i$ is a direct product of o-groups K_i and there is no element which is greater than all elements of some direct factor K_i. This is a contradiction. □

THEOREM 1. (Conrad [2].) *A lattice ordered group G has a finite Boolean algebra of polars of G if and only if G is constructed from the finite number of o-groups with the use of direct products and lexicographical extensions.*

Proof. We prove this statement by induction on the number of polars of l-group G. If the Boolean algebra of polars of G contains only two elements, then G is an totally ordered group and the result holds in this case. Let us assert that the result holds for all l-groups G for which $\mathcal{P}(G)$ contain less than 2^n polars. Let now G be a l-group and the Boolean algebra polar of G contains 2^n elements and $K_1, \dots, K_n$ be the minimal polars of G. Let G^0 be the product of all maximal polars of G. By Lemma 1, G is a lexicographic extension of G^0 by o-group G/G^0. We claim that G^0 is the direct product of two polars A and B and $\mathcal{P}(A)$, $\mathcal{P}(B)$ contain less than 2^n elements.

If $n = 2$, then $G^0 = S = K_1 \times K_2$ and result holds for this case. Now let $n > 2$ and let $S = K_1 \times \dots \times K_n$. Then, by Lemma 4, the Boolean algebra $\mathcal{P}(\bar{G})$ the quotient group $\bar{G} = G/S$ contain less than n atoms and thus we can apply our inductive hypothesis. So $\bar{G}^0 = \bar{A} \times \bar{B}$. There are three different cases: a) $G^0 = E$, $\bar{A} = E$, $\bar{B} = E$; b) $\bar{G}^0 \neq E$, $\bar{A} = E$, $\bar{B} \neq E$ and $\bar{B}$ is totally ordered group; c) $\bar{A} \neq E$, $\bar{B} \neq E$.

In case a) $\bar{G}^0 = E$ hence $G^0 = S$ and the result holds.

In case b) $\bar{B}$ is an 0-group. Let $H_1, \dots, H_t$ be the maximal polars of G. Then $G^0 = H_1 \cdots H_t$ and $G/S = \bar{H}_1 \cdots \bar{H}_t$ where $\bar{H}_i$ is the image of H_i under natural l-homomorphism of G onto G/S for all $i = 1, \dots, t$.

Since G^0 is an o-group and $\bar{H}_1, \ldots, \bar{H}_t$ are convex subgroups of $\bar{G}^0$, then $\bar{G}^0 = \bar{H}_i$ for some i, $1 \leq i \leq t$. Let us assume that $\bar{G}^0 = \bar{H}_1$, then $G^0 = H_1 \cdot S$. Since $H_1 \cap S$ is a polar of S, then $H_1 \cap S = K_{j_1} \times \cdots \times K_{j_r}$ for some $r < n$. Let $T = K_{j_1} \times \cdots \times K_{j_{n-r}}$ be the direct product of all other minimal polars of G. Then T is a convex l-subgroup of G such that $T \cap H_1 = E$ and $H_1 \cdot T = G^0$. Hence G^0 is the direct product of convex l-subgroups H_1 and T, each of whose contains less than 2^n polars. Thus by inductive hypothesis the result holds in this case.

In case c) $\bar{G}^0 = \bar{A} \times \bar{B}$ and $\bar{A} \neq E \neq \bar{B}$. Let A, B be inverse images of $\bar{A}, \bar{B}$ under the natural homorphism of G onto G/S correspondingly. Then $G = A \cdot B$ and $A \cap B = S$. Let J_A be the subset of indeces i, $1 \leq i \leq n$ such that there is an element $a \in A \backslash S$ for which $a > x$ for all $x \in K_i$. Let J'_A be the subset of all indeces j, $1 \leq j \leq n$, $j \notin J_A$ and let K_A be a polar of G generated by all polars K_i, $i \in J_A$. Now we claim that $A = K_A \cdot \prod_{j \in J'_A} K_j$. In fact, if $a \in A$, $a > e$ then for every $j \notin J_A$ there is an element $c_j \in K_j$, $c_j > e$ such that $a \wedge c_j = b_j$, $b_j \neq c_j$ and the element $a \cdot \prod_{j \in J'_A} b_j^{-1}$ is orthogonal to all c_j, $j \in J'_A$. Therefore

$$a \prod_{j \in J'_A} b_j^{-1} \in (\bigvee_{j \in J'_A}^{\mathcal{P}} (c_j)^{\perp})^{\perp} = \prod_{i \in J_A} K_i = A.$$

This means that $a \in K_A \cdot \prod_{j \in J'_A} K_j$ and therefore $A = K_A \cdot \prod_{j \in J'_A} K_j$. By similar arguments $B = K_B \cdot \prod_{j \in J'_B} K_j$ where J'_B is a subset of indeces $j \in \{1, \ldots, n\}$ such that there is an element $b \in B \setminus S$ for which $b > x$ for all $x \in K_j$. Then $J'_B = \{1, \ldots, n\} \setminus J_B$, $K_B = K_{j_1} \times \cdots \times K_{j_s}$ where $j_1, \ldots, j_s \in J_B$.

It is clear that $J_A \cap J_B = \emptyset$. In fact, if there is $i \in J_A \cap J_B$, then there are elements $a, b > e$, $a \in A$, $b \in B$ and $a > x$ $b > x$ for all $x \in K_i$. Then $a \wedge b > x$ for all $x \in K_i$, too. It contradicts to $a \wedge b \in S$. Therefore $K_A \cap K_B = E$. Let $J_0 = \{1, \ldots, n\} \setminus (J_A \cup J_B)$ then $G^0 = K_A \times K_B \times \prod_{i \in J_0} K_i$ and by the inductive hypothesis the resul is valid in this case.□

Chapter 4

Ordered permutation groups

In this chapter and some others (if necessary) capital Latin letters are used to denote groups, lattices, l-groups, o-groups; small Latin letters $a, b, c, x, y, z, k, f, g, h, \ldots$ their elements. Capital Greek letters are used for totally ordered sets and small Greek letters $\alpha, \beta, \gamma, \xi, \eta, \zeta$ and others are used for elements of totally ordered sets. We use small Greek letters also in other contexts, i.e., as members of index sets, ordinals. Small Greek letters $\varphi, \psi, \vartheta, \tau, \ldots$ are usually reserved for mappings. Letters I, J are reserved for general index sets and i, j for their elements. We employ brackets only in extreme cases, we omit them where it is possible. For example, let Ω_1, Ω_2 be totally ordered sets, $\alpha \in \Omega_1$, $\varphi : \Omega_1 \to \Omega_2$, $g \in A(\Omega_1)$ and $\psi : A(\Omega_1) \to A(\Omega_2)$, then $\alpha\varphi \cdot g\psi$ denotes the element $((\alpha)\varphi)((g)\psi)$ of the totally ordered set Ω_2.

4.1 Holland's Embedding Theorem

Let us recall Example 4.1.2.: Let Ω be a totally ordered set and $A(\Omega)$ be a set of all order permutations of Ω. Then $A(\Omega)$ is an l-group under composition, and lattice operations $\vee, \wedge$ are defined by the rules:

(1) $$\alpha(f \vee g) = \max\{\alpha f, \alpha g\},$$

(2) $$\alpha(f \wedge g) = \min\{\alpha f, \alpha g\},$$

for $f, g \in A(\Omega)$, $\alpha \in \Omega$.

The following result provided a new intuitive standpoint from which one can view l-groups, making many of their basic properties obvious.

THEOREM 1. (Holland's embedding theorem. Holland [2].) *Any l-group is isomorphic to an l-subgroup of the l-group of order permutations of a totally ordered set.*

Proof. Let G be an l-group and for each $g \in G$, $g \neq e$ let $V = V(g)$ be the value of g. Now $V(g)$ is prime (Theorem 3.3.3) and therefore the set $R(G : V)$ of right cosets of V is totally ordered under the order relation defined by $Vx \leq Vy$ iff $x \leq zy$ for some $z \in V$.

It is easy to see that each element $a \in G$ induces an order permutation $R_V(a)$ of $R(G : V)$ defined by $(Vx)R_V(a) = Vxa$ (Proposition 3.3.2). The mapping $R_V : G \to A(R(G : V))$ is an l-homomorphism from G into transitive l-subgroup $A(R(G : V)$. We need only to show that $(a \vee b)R_V(a \vee b) = R_V(a) \vee R_V(b)$, $R_V(a \wedge b) = R_V(a) \wedge R_V(b)$ for every $a, b \in G$. It follows from Proposition 3.3.1 that for any right coset Vx,

$$(Vx)(R_V(a \vee b)) = Vx(a \vee b) = Vxa \vee Vxb,$$

$$(Vx)(R_V(a \wedge b)) = Vx(a \wedge b) = Vxa \wedge Vxb.$$

Hence

$$R_V(a \vee b) = R_V(a) \vee R_V(b), \quad R_V(a \wedge b) = R_V(a) \wedge R_V(b).$$

It is easy to see that the kernel $K = \ker R_V$ of homomorphism R_V is continued in V. It is clear that the element $g \notin K_V \subseteq V(g) = V$. Since $\bigcap_{g \in G, g \neq e} K_{V(g)} = E$, the l-group G is isomorphic to an l-subgroup of the Cartesian product $\bar{\prod}_{g \in G, g \neq e} A(R(G : V))$ of l-groups $A(R(G : V))$. Choose the total order of the index set for this product (i.e., $G \setminus \{e\}$). Now we totally order $\Omega = \bigcup_{g \in G, g \neq e} R(G : V(g))$ lexicographically: setting $x < y$ if $x, y \in R(G : V)$ for some $g \in G$, $x < y$ in $R(G : V(g))$ or $x \in R(G : V(g))$, $y \in R(G : V(h))$ where $g < h$ under the total ordering given above to $G \setminus \{e\}$. Now we can define the isomorphism φ from the l-group G into the l-group $A(\Omega)$ as follows:

$(V(g)x)((a)\varphi) = V(g)xa$ for every $V(g)x \in \Omega$, $a \in G$.

The detail can be easily checked.□

The *action* of l-group G on the totally ordered set Ω is the l-homomorphism $\psi : G \to A(\Omega)$. The kernel $\ker\psi$ of ψ is called the *lazy subgroup* (if $g \in \ker\psi$ then $\alpha \cdot (g)\psi = \alpha$ for all $\alpha \in \Omega$). If the kernel ψ is trivial, we call the action *faithful* and say that ψ is a *representation* of the l-group G.

Holland's embedding theorem provides the utilization of the theory of groups, of order permutations of totally ordered sets.

4.2 Groups of Order Permutations of Totally Ordered Sets

1^0. Let Ω be a totally ordered set under the order $\leq$ and $A(\Omega)$ be the l-group of all order permutations of Ω. If group G is a subgroup of $A(\Omega)$ we will endow G with the ordering of $A(\Omega)$ and say that G is an *ordered permutation group*. We denote it (G, Ω). Thus, an ordered permutation group (G, Ω) is a subgroup of $A(\Omega)$ with its inherited action on Ω. If G and H are ordered permutation groups of totally ordered set Ω and T respectively, then (G, Ω) and (H, T) are isomorphic as ordered permutation groups if there is an isomorphism π of the group G into the group H and an order isomorphism τ of the totally ordered set Ω into T such that $((\xi)g)\tau = (\xi g)\tau = (\xi\tau)(g\pi)$ for all $g \in G$, $\xi \in \Omega$. This isomorphism of ordered permutation groups (G, Ω) and (H, T) will be denoted as (π, τ).

If G and H are po-groups with partial orders induced by "usual" lattice order of $A(\Omega)$ and $A(T)$ accordingly and (π, τ) is an isomorphism of the ordered permutation group (G, Ω) onto the ordered permutation group (H, T), then π is also an isomorphism of po-groups G and H. Then π is also an isomorphism of po-groups G and H. If G, H are l-subgroups of $A(\Omega)$ and $A(T)$ accordingly and (π, τ) is an isomorphism of ordered permutation groups (G, Ω) and (H, T), then π is an l-isomorphism of l-groups G and H.

Let $G \subseteq A(\Omega)$ be a group of order permutations of totally ordered set Ω. A *convex G-congruence* on Ω is an equivalence relation ϑ on Ω such that:

(1) $\alpha \equiv \beta \pmod{\vartheta}$ implies $\alpha \equiv \beta \pmod{\vartheta}$ for all $\alpha, \beta, \gamma \in \Omega$ such that $\alpha \leq \beta \leq \gamma$,

(2) $\alpha \equiv \beta \pmod{\vartheta}$ implies $\alpha g \equiv \beta g \pmod{\vartheta}$ for all $\alpha, \beta, \gamma \in \Omega$ and $g \in G$.

If ϑ is a congruence on Ω, $\alpha \in \Omega$ then

$$\alpha\vartheta = \{\xi \in \Omega | \ \xi \equiv \alpha \pmod{\vartheta}\}$$

is a congruence class of ϑ.

The set of all convex G-congruences of an ordered permutation group (G, Ω) will be denoted by $\mathcal{K}$. We define a partial order on $\mathcal{K}$ by letting $\zeta \leq \vartheta$ in $\mathcal{K}$ iff $\alpha\zeta \subset \alpha\vartheta$ for all $\alpha \in \Omega$. It is evident that $\mathcal{K}$ contains the greatest and the least convex congruences υ and ε. These convex G-congruences υ, ε are called *trivial* G-congruences. The convex G-congruence is non-trivial if one of the congruence classes contains more than one element and is not Ω. It follows from the definition that the totally ordered set Ω is the union of disjoint convex congruent classes (partition Ω). If $\alpha_1\vartheta \neq \alpha_2\vartheta$ and $\alpha_1 < \alpha_2$ in Ω, then $\beta_1 < \beta_2$ for all $\beta_1 \in \alpha_1\vartheta$, $\beta_2 \in \alpha_2\vartheta$. It follows that the factor-set Ω/ϑ is a totally ordered set under the induced from Ω order. It is easy to see that the subset $(\alpha\vartheta)g = \{\beta g | \beta \in \alpha\vartheta\}$ of Ω is a class of congruent elements and if $\alpha_1\vartheta < \alpha_2\vartheta$ in Ω/ϑ then $(\alpha_1\vartheta)g < (\alpha\vartheta)g$. By this reason every element $g \in G$ defines an order permutation on the factor-set Ω/ϑ.

A convex subset $\Delta \subseteq \Omega$ is a *G-block* if for each $g \in G$ either $\Delta g = \Delta$ or $\Delta g \cap \Delta = \emptyset$. Any non-empty convex G-invariant subset of Ω is G-block, of course.

PROPOSITION 1. *Let (G, Ω) be an ordered permutation group. The classes of G-congruences of (G, Ω) are G-blocks and conversely, every G-block is a class of G-congruence of (G, Ω).*

Proof. Let Δ be a G-block. Let $\alpha\zeta\beta$ if $\alpha = \beta$ or $\alpha, \beta \in \Delta g$ for some $g \in G$. Since Δ is a G-block, then ζ-classes are convex and partition Ω, so ζ is an equivalence relation on Ω. If $\zeta\xi\beta$ and $f \in G$, either $\alpha = \beta$ or $\alpha, \beta \in \Delta g$. In the first case $\alpha f = \beta f$, in the second case $\alpha f, \beta f \in \Delta g$. Hence ζ is a G-congruence of (G, Ω) having Δ as a congruent class. The converse statement is trivial.□

Proposition 1 determines the correspondence between G-congruences of (G, Ω) and G-blocks of Ω.

Let ϑ be a G-congruence and $L_\vartheta = \{g \in G | \ (\alpha\vartheta)g = \alpha\vartheta$ for all $\alpha \in \Omega\}$. Let $e \leq g_1 \leq g_2$, $g_1, g_2 \in G, g_2 \in L_\vartheta$. Then $\alpha \leq \alpha g_1 \leq \alpha g_2$ and $\alpha\vartheta \leq (\alpha\vartheta)g_1 \leq (\alpha\vartheta)g_2$ for every $\alpha \in \Omega$. But $(\alpha\vartheta)g_2 = \alpha\vartheta$ and so $(\alpha\vartheta)g_1 = \alpha\vartheta$ for every $\alpha \in \Omega$ and $g_1 \in L_\vartheta$. Thus L_ϑ is a convex subgroup of G. L_ϑ is a sublattice of G because $(\alpha\vartheta)(g \vee e) = \alpha(g \vee e)\vartheta = \max\{\alpha, \alpha g\}\vartheta = \alpha\vartheta$ for every $g \in L_\vartheta$. For arbitrary $f \in G, g \in L_\vartheta, \alpha \in \Omega$ is valid $\alpha(f^{-1}gf)\vartheta = ((\alpha f^{-1}\vartheta)g)f = (\alpha f^{-1})\vartheta f = \alpha f^{-1}f)\vartheta = \alpha\vartheta$ and L_ϑ is a normal subset of G. So, L_ϑ is an ideal of l-group G and for arbitrary element $g \in G$ every element of coset gL_ϑ defines some order permutation of the totally ordered factor-set Ω/ϑ. By these reasons we can identify l-group G/L_ϑ with l-subgroup of $A(\Omega/\vartheta)$. Let $\varepsilon_\vartheta : G \to G/L_\vartheta$ be the natural l-homomorphism from G onto G/L_ϑ and $\varphi_\vartheta : \Omega \to \Omega/\vartheta$ be the natural map from Ω onto Ω/ϑ. Then the pair $(\varepsilon_\vartheta, \varphi_\vartheta)$ defines l-homomorphism from ordered permutation group (G, Ω) onto $(G/L_\vartheta, \Omega/\vartheta)$. The l-group of order permutation $(G/L_\vartheta, \Omega/\vartheta)$ is called a *quotient group of ordered permutation l-group* (G, Ω) *by G-congruence* ϑ, and ideal L_ϑ is called a *lazy subgroup* of G. The pair $(\varepsilon_\vartheta, \varphi_\vartheta)$ is called a *natural homomorphism* from (G, Ω) onto $(G/L_\vartheta, \Omega/\vartheta)$.

THEOREM 1. (The Homomorphism theorem.) *For every homomorphism* (π, τ) *of ordered permutation l-group* (G, Ω) *on the ordered permutation l-group* (H, T)*, the map* τ *defines the G-congruence* ϑ *on* Ω *(defined by the rule:* $\alpha \equiv \beta \pmod{\vartheta}$ *if and only if* $(\alpha)\tau = (\beta)\tau$*). Then there is an isomorphism* $(\pi_\vartheta, \tau_\vartheta)$ *of quotient group* $(G/L_\vartheta, \Omega/\vartheta)$ *onto the* (H, T) *such that* $\pi = \varepsilon_\vartheta\pi_\vartheta, \ \tau = \varphi_\vartheta\tau_\vartheta$.

Proof is straightforward.□

Of course, the notions of quotient group, G-congruence, homomorphism of ordered permutation group can be extended to arbitrary subgroups of $A(\Omega)$ and the Homomorphism theorem will be true (see, for example, the book of A.M.W.Glass [6]).

EXAMPLE 1. Let $\Omega = \bigcup_{n \in \mathbf{Z}} Q_n$, where totally ordered sets Q_n are isomorphic to the totally ordered set $\mathbf{Q}$ of rationals. Fixing the isomorphism from $\mathbf{Q}$ onto Q_n we denote elements of Ω as q_n, $q \in \mathbf{Q}$, $n \in \mathbf{Z}$. We insert the set Ω into totally ordered set if we define $q < r$ if $n < m$ for $n = m$, $q < r$ in $\mathbf{Q}$. For arbitrary $\varphi \in A(\mathbf{Q})$ let us consider the set G of order permutations $\varphi_n \in A(\Omega)$, $n \in \mathbf{Z}$ defined by the relations $\varphi_n(q_m) = ((q)\varphi)_{n+m}$. It is easy to see that the set G is an l-subgroup of $A(\Omega)$ and on Ω there is only one nontrivial congruence ϑ defined by the rule:$q \equiv r \pmod{\vartheta}$ iff $n = m$. The mapping (π, τ) where $(\vartheta_n)\pi = \psi^n$, $(g_m) = m$ is a homomorphism of ordereded permutation l-group (G, Ω) onto the order permutation group $(H, \mathbf{Z})$ of totally ordered set $\mathbf{Z}$ where $H = \mathrm{gp}\{\psi\}$, $(k)\psi = k + 1$ for any $k \in \mathbf{Z}$.

2^0 . Let (G, Ω) be an ordered permutation l-group of Ω and Δ be a subset of Ω such that $(\Delta)g = \Delta$ for every $g \in G$. Then Δ is called *G-invariant.* Any element $g \in G$ induces the order permutation on the G-invariant subset of Δ. This action of G on Δ is not faithful, in general.

If $\Delta \subseteq \Omega$, then the set $G_{(\Delta)} = \{g \in G | \Delta g = \Delta\}$ is called a *stabilizer* of Δ. In other words, G is the set of all elements of G which map Δ onto Δ (not necessarily fixing Δ pointwise). A *pointwise stabilizer* of Δ is the set of all elements $g \in G$ such that $(\alpha)g = \alpha$ for all $\alpha \in \Delta$. We denote it as G_Δ. If $\Delta = \{\alpha\}$ we write $G_{(\alpha)} = G_\alpha$. It is evident that $G_{(\Delta)}$, G_Δ, G_α are l-subgroups of G and G_Δ is a convex l-subgroup. If the subset Δ is G-invariant, then G_Δ is an ideal of G and every two elements $g, h \in G$ induce on Δ the same permutation iff $G_\Delta g = G_\Delta$. In this case, we consider $(G/G_\Delta, \Delta)$ as an ordered permutation l-group of the set Δ. This group is called a *restriction* of (G, Ω) on Δ and the mapping $\varepsilon_\Delta : G \to G/G_\Delta$ such that $(g)\varepsilon_\Delta = gG_\Delta$ is an action of G on Δ.

If Δ is a convex G-invariant subset of Ω, then we can define the convex G-congruence ϑ on Ω by the rule: $\alpha \equiv \beta \pmod{\vartheta}$ if α, β are less than every $\delta \in \Delta$, or $\alpha, \beta \in \Delta$, or α, β are bigger than every $\delta \in \Delta$. The factor-set Ω/ϑ is order isomorphic to one of four sets: $\{\xi\} \cup \Delta \cup \{\zeta\}$, $\{\xi\} \cup \Delta$, $\Delta \bigcup \{\zeta\}$, Δ where $\xi < \delta < \zeta$ for any $\delta \in \Delta$ (if the elements ξ, ζ there exist, of course). The mapping $(\varepsilon_\vartheta, \pi_\vartheta)$: $(G/G_\Delta, \Delta) \to (G/L_\vartheta, \Omega/\vartheta)$ where $(gG_\Delta)\varepsilon = gL_\vartheta$, $(\alpha)\pi_\vartheta = \alpha$ is an l-isomorphism of ordered permutation l-groups. It is also obvious that the mapping $\pi : (g)\pi = g$ is an l-homomorphism of l-group G in l-group $A(\Delta)$ for any G-invariant subset Δ of Ω.

3^0. Let us recall that an ideal of totally ordered set Ω is a subset A of Ω bounded from above such that $\xi \leq \alpha$ implies $\xi \in A$ for all $\xi \in \Omega$, $\alpha \in A$. The set of all ideals of the totally ordered set Ω is denoted by $\overline{\Omega}$. The set $\overline{\Omega}$ is a totally ordered set under inclusion. The mapping

$\varepsilon : \Omega \to \overline{\Omega}$, $(\alpha)\varepsilon = A_\alpha = \{\xi \in \Omega | \xi \leq \alpha\}$ is an order isomorphism. We identify the set Ω with $(\Omega)\varepsilon$ and call $\overline{\Omega}$ a *Dedekind completion* of Ω. The elements of $\overline{\Omega}$ will be denoted by $\bar{\alpha}$. For every $\bar{\alpha} \in \overline{\Omega}$, $\bar{\alpha} = \sup\{\xi \in \Omega | \xi \leq \alpha\}$ and any bounded above subset $\Delta \subseteq \Omega$ has the greatest lower bound in Ω.

PROPOSITION 2. (Holland [3].) *Every order permutation g of totally ordered set Ω can be extended uniquely to the order permutation $\bar{g}$ of the Dedekind completion $\bar{\Omega}$ of Ω.*

Proof. Let us define $(\bar{\alpha})\bar{g} = \sup\{(\alpha)g | \xi \leq \bar{\alpha},\ \xi \in \Omega\}$ for each $\bar{\alpha} \in \bar{\Omega}$, $g \in A(\Omega)$. It is easy to verify that $\bar{g} \in A(\bar{\Omega})$ and $(\xi)\bar{g} = (\xi)g$ for arbitrary $\xi \in \Omega$.□

THEOREM 2. (Holland [3].) *There is an isomorphism (π, ε) from the ordered permutation l-group $(A(\Omega), \Omega)$ into the ordered permutation l-group $(A(\bar{\Omega}), \bar{\Omega})$.*

Proof. As above, let us $(g)\pi = \bar{g}$ for $g \in A(\Omega)$ and $\varepsilon : \Omega \to \bar{\Omega}$. We have proved that $\bar{g} \in A(\bar{\Omega})$. If $\alpha \in \Omega$, $f, g \in A(\Omega)$ then $(\alpha)((g\pi)(f\pi)) = ((\alpha)\bar{g})\bar{f} = (\alpha g)f = (\alpha)gf = (\alpha)\overline{gf}$. Thus, since $(g\pi)(f\pi) = \bar{g} \cdot \bar{f}$ and $\overline{gf}$ agree on Ω, they are equal. Thus it is proved that π is an isomorphism of the group $A(\Omega)$ into the group $A(\bar{\Omega})$. It is evident that $((\alpha)g)\varepsilon = (\alpha)(g\pi)$ for every $\alpha \in \Omega$, $g \in A(\Omega)$. So, (π, ε) defines the l-isomorphism from $(A(\Omega), \Omega)$ into $(A(\bar{\Omega}), \bar{\Omega})$.□

Due to Theorem 2 we can identify $A(\Omega)$ and $(A(\Omega))\pi$.

EXAMPLE 2. The ordered permutation group $A(\mathbf{Q})$ of rationals is an l-subgroup of ordered permutation group $A(\mathbf{R})$ of reals.

4.3 Transitive Ordered Permutation Groups

1^0. The ordered permutation group $G \subseteq A(\Omega)$ is *transitive* on Ω if for every $\alpha, \beta \in \Omega$ there is an element $g \in G$ such that $\alpha g = \beta$. The totally ordered set Ω is *homogeneous* if $A(\Omega)$ is transitive on Ω . The subset Δ of Ω is a *G-orbit* if:

1) Δ is G-invariant,
2) for every two elements $\alpha, \beta \in \Delta$ there exists $g \in G$ such that $\alpha g = \beta$.

PROPOSITION 1. *For every element $\alpha \in \Omega$ the subset $\alpha G = \{\alpha g \mid g \in G\}$ is G-orbit.*

Proof. If $\beta \in \alpha G$, then $\beta = \alpha g$ for some $g \in G$. Hence $\beta h = \alpha g h \in \alpha G$ for arbitrary $h \in G$. If $\beta, \gamma \in \alpha G$, then $\beta = \alpha g,\ g = \alpha h$ for some $g, h \in G$ and $\beta(g^{-1}h) = \alpha g(g^{-1}h) = \alpha h = \gamma$.□

The subset αG is called a *G-orbit* of element α.

PROPOSITION 2. *Let (G, Ω) be an ordered permutation l-group. Then the totally ordered set Ω is the union of mutually disjoint G-orbits and l-group G is a subdirect product of l-groups $\bar{G}_\alpha = G/G_\alpha, \alpha \in \Omega$.*

Proof. Let Δ be an arbitrary G-orbit and $\alpha \in \Delta$. It is easy to see that $\Delta = \alpha G$ and for any $\alpha, \beta \in \Omega$ either $\alpha G = \beta G$ or $\alpha G \cap \beta G = \emptyset$. The pointwise stabilizer $G_{\alpha G}$ is an ideal of G for every G-orbit αG and $\bigcap_{\alpha \in \Omega} G_{\alpha G} = E$ because $g \in \bigcap_{\alpha \in \Omega} G_{\alpha G}$ implies $\alpha g = \alpha$ for all $\alpha \in \Omega$. Proposition 4.2.2 completes the proof.□

Of course, a G-orbit may be not convex (see Example 4.2.2). Let G_α be a stabilizer of $\alpha \in \Omega$ and $\bar{\Omega}$ be the Dedekind completion of Ω. By Proposition 4.2.2 G is an l-subgroup of $A(\bar{\Omega})$. For every $\alpha \in \bar{\Omega}$ let $G_\alpha = \{g \in G \mid \alpha g = \alpha\} = G \cap A(\bar{\Omega})_\alpha$ It is obvious that G_α is a convex l-subgroup.

PROPOSITION 3. (Holland [3].) *Let (G, Ω) be an ordered permutation l-group. Then the l-subgroup G_α is prime for all $\alpha \in \Omega$. If (G, Ω) is transitive on Ω, then the subgroups G_α, G_β are conjugate in G for every $\alpha, \beta \in \Omega$.*

Proof. By Theorem 3.3.1, it is sufficient to prove that for all subset $g, h \in G,\ g, h \geq e$ from $g \wedge h \in G_\alpha$ it follows $g \in G_\alpha$ or $h \in G_\alpha$. Suppose $g \wedge h \in G_\alpha$ for $g, h \geq e,\ g, h \in G$ and $g \notin G_\alpha$. Hence $\alpha g > \alpha$. On the other hand, $\alpha(g \wedge h) = \min\{\alpha g, \alpha h\} = \alpha$, then $\alpha h = \alpha$ and $h \in G_\alpha$.

Now, let l-group G be transitive on Ω, $\alpha, \beta \in \Omega$ and $\alpha g = \beta$. Then for arbitrary $h \in G_\alpha$, $\beta(g^{-1}hg) = \alpha h g = \alpha g = \beta$, therefore $g^{-1}G_\alpha g \subseteq G_\beta$. By similar arguments, $g^{-1}G_\alpha g \subseteq G_\beta$.□

Now, we investigate transitive representations of l-groups by ordered permutation l-groups of totally ordered sets. The action φ of an l-group G by order permutations of totally ordered set Ω is a *transitive action* if $G\varphi$ is a transitive subgroup of $A(\Omega)$.

We assert that $R(G:H)$ is the set of all right cosets of l-group G by convex l-subgroup H with natural order and R_H is the map from $R(G:H)$ onto $R(G:H)$ which is defined by the equality: $(Hx)R_H(g) = Hxg$. If H is a prime subgroup of l-group G, then $R(G:H)$ is totally ordered under the relation: $Hx \leq Hy$ in $R(G:H)$ iff there is $h \in H$ such that $x \leq hy$.

PROPOSITION 4. (Holland [3].) *For any prime subgroup H of an l-group G the action R_H of G by order permutations of totally ordered set $R(G:H)$ is transitive. Conversely, for any transitive action ψ of l-group G by order permutations of totally ordered set Ω there is a prime subgroup H in G and an order isomorphism χ from the totally ordered set $R(G:H)$ to Ω such that $(\beta)(g\psi) = ((\beta\chi^{-1})g)\chi$ for every $\beta \in \Omega$, $g \in G$.*

Proof. It follows from the proof of Theorem 4.1.1 that R_H is a homomorphism from G into $A(R(G:H))$. Let $Hx, Hy \in R(G:H)$. Then $(Hx)R_H(x^{-1}y) = Hxx^{-1}y = Hy$ and R_H is a transitive action.

Conversely, let ψ be a transitive action of l-group G into $A(\Omega)$ for some totally ordered set Ω and α be an element of Ω. By Proposition 3 the inverse image H of $(G\psi)_\alpha$ is a prime subgroup of G and the set $R(G:H)$ is totally ordered. Let χ be the map which is defined by the rule: $(Hg)\chi = (\alpha)g\psi$. If $Hg = Hf$, then $g = hf$ for some $h \in H$. Hence $(Hg)\chi = (\alpha)g\psi = (\alpha)(hf)\psi = (\alpha)((h\psi)(g\psi)) = (\alpha)(f\psi) = (Hf)\chi$. If $Hg \leq Hf$, then $g \leq hf$ for some $h \in H$ and $(Hg)\chi = (\alpha)g \leq (\alpha)(hf\psi) = (\alpha)(h\psi)(f\psi) = (\alpha)(f\psi) = (Hf)\chi$. In this case $(H\gamma)\chi = (Hf)\chi$ and the following equalities are valid in $G:\quad (\alpha)((g\psi)(f^{-1}\psi))$, $(\alpha)(gf^{-1})\psi = \alpha$, $(gf^{-1})\psi \in (G\psi)$, $Hgf^{-1} = H$, $Hg = Hf$. Therefore the map χ is an isomorphism from totally ordered set $R(G:H)$ into Ω. As ψ is a transitive action G on Ω, then for any $\beta \in \Omega$ there exists an element $g \in G$ such that $(\alpha)g\psi = \beta$ and $\beta = (Hg)\chi$ i.e., χ is an isomorphism from $R(G:H)$ onto Ω. Now, for any $g \in G$, $\beta \in \Omega$ there is the element $f \in G$ with the property $Hf = (\beta)\chi^{-1}$. Hence $(\alpha)f\psi = \beta$, $(\beta)g\psi = (\alpha)(f\psi)(g\psi) = (\alpha)(fg)\psi = (Hfg)\chi = (Hf \cdot g)\chi = (((\beta)\chi^{-1})g)\chi$ and this concludes the proof of the properties.□

Now, it is clear that for any l-group G there exist "wide" transitive actions G on a totally ordered set. These actions correspond to minimal prime subgroups of G (by Proposition 2.3.4 the minimal prime subgroups exist in every l-group).

Let us consider the question on existence of transitive representation of l-groups. The convex l-subgroup H of the l-group G is *representing* if H is prime and it does not contain any nontrivial ideal of G.

THEOREM 1 (Holland [2].) *If G is an l-group, then there exists a totally ordered set Ω and isomorphism φ of G onto a transitive subgroup of $A(\Omega)$ if and only if G contains a representing subgroup.*

Proof. Let φ be a transitive representation of G into $A(\Omega)$ for some totally ordered set Ω. By Proposition 4, we can assume that $\Omega = R(G:H)$ where H is a prime subgroup of G and $(Hx)(g\varphi) = Hxg$ for $x, g \in G$. Let us suppose that there exists a nontrivial ideal N such that $N \subseteq H$. Then $(Hx)(g)\varphi = Hxg = Hxgx^{-1}x = Hx$ for every $x \in G$, $g \in N$. As φ is a representation then $(g)\varphi = e$ and $g = e$ for each $g \in N$. Thus $N = E$ and H is representing.

Conversely, let H be a representing subgroup of G. Then the action R_H of l-group G by order permutations of totally ordered set $R(G:H)$ is faithful. In fact, let g be an arbitrary element of the kernel N of the action R_H. Then $H \cdot R_H(g) = Hg = H$ and $g \in H$. Therefore, $N \subseteq H$. But H is representing , hence $N = E$ and R_H is a representation.□

COROLLARY 1. (Holland [2].) *An Abelian transitive order permutation l-group (G, Ω) is totally ordered.*□

An interesting related fact due to P.M.Cohn [1] is that $A(\Omega)$ is an o-group iff $A(\Omega)$ is Abelian.

COROLLARY 2. (Holland [2].) *Let G be a simple l-group. Then it has a transitive representation by order permutations of some totally ordered set Ω.*□

An l-group G is *subdirectly irreducible* if $\bigcap_{\alpha \in I} N_\alpha = E$ for any set $\{N_\alpha | \alpha \in I\}$ of ideals of G implies $N_\alpha = E$ for some $\alpha \in I$. G is *finitely subdirectly irreducible* if $H \cap N = E$ for some ideals H, N implies $H = E$ or $N = E$.

COROLLARY 3. (Holland [2].) *Any subdirectly irreducible l-group G has a transitive representation by order permutation of some totally ordered set Ω.*

Proof. Let K be a minimal proper ideal of a subdirectly irreducible l-group G. Choose $g \in K$, $g \neq e$ and let $V(g)$ be a value of g. By Theorem 3.3.3, the l-subgroup $V(g)$ is a prime subgroup of G. Hence the action $R_{V(G)}$ is transitive on $R(G : V(g))$. But the kernel L of this action is contained in $V(g)$ and $g \notin L$. It is a contradiction, then $L = E$ i.e., the action $R_{V(g)}$ is a transitive representation.□

COROLLARY 4. *A transitive ordered permutation l-group (G, Ω) is finitely subdirectly irreducible.*

Proof. Let A, B be ideals of l-group G and $A \cap B = E$. By previous statements there exists a representing l-subgroup $V \subseteq G$ such that the totally ordered sets Ω and $R(G : V)$ are order isomorphic and $A \cap V = B \cap V = E$. But for any $a \in A$, $b \in B$, $a \neq e \neq b$ is valid $a \wedge b = e$. As V is prime it follows $a \in V$ or $b \in V$. Therefore, $a = e$ or $b = e$ which is impossible. This contradiction completes the proof.□

Note that E is a representing subgroups of l-group G iff G is an o-group.

3⁰. Now we investigate the properties of convex G-congruences of transitive ordered permutation l-groups $G, \Omega)$.

PROPOSITION 5. *Let (G, Ω) be a transitive ordered permutation l-group and $\alpha \in \Omega$. Then the mapping $\varphi : \zeta \to \alpha\zeta$ from the set $\mathcal{K}$ of all convex G-congruences of G into the set of G-blocks containing α has the following properties:*

1) φ *is one-to-one mapping,*
2) $\zeta_1 \leq \zeta_2$ *if and only if* $\alpha\zeta_1 \subseteq \alpha\zeta_2$.

Proof. 1) By Proposition 1, φ maps $\mathcal{K}$ onto the set of G-blocks containing α. Let us suppose that there are convex G-congruences ζ_1, ζ_2 such that $\zeta_1 \neq \zeta_2$ and $\alpha\zeta_1 = \Delta = \alpha\zeta_2$. Then there exist $\alpha, \beta \in \Omega$ such that $\alpha\zeta_1\beta$ but $\alpha\zeta_2\beta$ is not true. By transitivity (G, Ω) there is an element $g \in G$ such that $\alpha g \in \Delta$ and, consequently, $\alpha g\zeta_1\beta g$ and by our assumption $\alpha g, \beta g \in \Delta = \alpha\zeta_2$. Therefore, $\alpha g\zeta_2\beta g$ and since ζ_2 is a convex G-congruence, then $\alpha\zeta_2\beta$. This correlation contradicts to assumption.

2) is obviously true.□

Let us $A < B$ for $A, B \subseteq \Omega$ if $\alpha < \beta$ for all $\alpha \in A,\ \beta \in B$.

PROPOSITION 6. *Let (G, Ω) be a transitive ordered permutation l-group and $\alpha \in \Omega$. Then the set of all G-blocks containing α is totally ordered under inclusion.*

Proof. Let $\Delta_1,\ \Delta_2$ be G-blocks and $\alpha \in \Delta_1 \cap \Delta_2$. If $\Delta_1 \neq \Delta_2 \cap \Delta_2 \neq \Delta_2$ then there exist $\delta_1, \delta_2 \in \Omega$ such that $\delta_1 \in \Delta_1 \setminus \Delta_2$ and $\delta_2 \in \Delta_2 \setminus \Delta_1$. We can assume that $\delta_1 < \beta$ for all $\beta \in \Delta_2$ and $\delta_2 > \beta$ for all $\beta \in \Delta_1$. By transitivity of l-group (G, Ω) there is an element $g \in G$ such that $\alpha g = \delta_2$. Hence $\Delta_2 g \cap \Delta_2 \neq \emptyset$ and $\Delta_2 g = \Delta_2,\ \Delta_2 g^{-1} = \Delta_2$. If $\Delta_1 g \cap \Delta_1 = \emptyset$, then $\Delta_1 g < \Delta_1$ or $\Delta_1 g > \Delta_1$. But the first inequality is impossible because Δ_1 contains $\alpha g = \delta_2$ and $\delta_2 > \beta$ for all $\beta \in \Delta_1$. Thus, $\Delta_1 g > \Delta_1$ and all elements of $\Delta_1 g$ are bigger than α. But $\delta_1 < \Delta_2 g = \Delta_2 = \Delta_2 g^{-1}$, and consequently $\delta_1 g < \Delta_2,\quad \delta_1 g < \alpha$. Therefore, $\Delta_1 g = \Delta_1, \alpha \in \Delta_1$ and $\alpha g = \delta_2 \notin \Delta_1$ by the choice of the element $g \in G$. This contradicts our assumption.□

COROLLARY 1. *Let (G, Ω) be a transitive ordered permutation l-group. Then the set $\mathcal{K}$ of all convex G-congruences of Ω is totally ordered.*

Proof. Immediate from Propositions 5,6.□

COROLLARY 2. *Let (G, Ω) be a transitive ordered permutation l-group and $\{\zeta_\iota | \iota \in I\}$ be an arbitrary family of convex G-congruences of Ω. Then $\bigcap_{\iota \in I} \zeta_\iota$ and $\bigcup_{\iota \in I} \zeta_\iota$ are convex G-congruences.*

Proof. It is easy to see from Proposition 6 that the union and intersection of the family of G-blocks containing element $\alpha \in \Omega$ is again a G-block. Now, the proof of proposition is immediate from this fact.□

Let H be a convex l-subgroup of (G, Ω). Then the subset $\alpha H = \{\alpha h | h \in H\}$ is an H-orbit containing α.

PROPOSITION 7. *If (G, Ω) is a transitive ordered permutation l-group and $\alpha \in \Omega$ then arbitrary G-block containing α is an H-orbit for the unique convex l-subgroup H of G such that $H \geq G_\alpha$. Conversely, for any convex l-subgroup $H \geq G_\alpha$, the H-orbit αH is a G-block containing α. There is one-to-one correspondence between G-blocks and H-orbits* *which preserves the set-theoretic inclusions.*

Proof. Let Δ be a G-block containing α. By transitivity of (G, Ω), for arbitrary $\delta \in \Delta$ there is an element $g \in G$ such that $\alpha g = \delta$. Hence $\Delta \cap \Delta g \neq \emptyset$ and $\Delta = \Delta g, \quad g \in G_{(\Delta)}$. So, $\delta = \alpha g \in \alpha G_{(\Delta)}$ for every $\delta \in \Delta$. Therefore, $\Delta = \alpha G_{(\Delta)}$. It is evident that $G_{(\Delta)} \supseteq G_\alpha$ and $G_{(\Delta)}$ is a convex l-subgroup of G. Now, let us suppose that there are convex l-subgroups H, K such that $\alpha H = \alpha K, \quad H, K \supseteq G_\alpha$. Then for any element $h \in H$ there is an element $k \in K$ that $\alpha h = \alpha k$. So, $\alpha h k^{-1} = \alpha, \quad hk^{-1} \in G_\alpha$ and $k \in H$. Then $H \subseteq K$ and by similar arguments $K \subseteq H$. Therefore $H = K$ and the uniqueness is proved.

Conversely, let H be a convex l-subgroup of G and $H \supseteq G_\alpha$. If $\alpha f_1 \leq \beta \leq \alpha f_2$ for some $f_1, f_2 \in H$, let $g \in G$ such that $\alpha g = \beta$. Hence $\beta = \alpha((f_1 \vee g) \wedge f_2)$. But $f_1 \wedge f_2 \leq (f_1 \vee g) \wedge f_2 \leq f_2$ and therefore αH is a convex subset of Ω. Now, let us assume that $\alpha H \neq \alpha H f$ for some element $f \in G$ and $\beta \in \alpha H \cap \alpha H f \neq \emptyset$. Then $\alpha h_1 = \alpha h_2 f = \beta$ for some $h_1, h_2 \in H$ and $h_2 f h_1 \in G_\alpha, h_2 f h_1^{-1} \in H$. Therefore, $f \in H$ and $\alpha H f = \alpha H$. It means that αH is a G-block.□

4.4 Doubly Transitive Ordered Permutation Groups

1^0. An ordered permutation group (G, Ω) is *o-2-transitive* if for every $\alpha_1 < \alpha_2, \beta_1 < \beta_2$ in Ω there exists $g \in G$ such that $\alpha_1 g = \beta_1, \alpha_2 g = \beta_2$. More generally, (G, Ω) is *o-n-transitive* if for every $\alpha_1 < \alpha_2 < \ldots < \alpha_n, \beta_1 < \beta_2 < \ldots < \beta_n$ in Ω there exists $g \in G$ such that $\alpha_i g = \beta_i$ for each $i = 1, 2, \ldots n$.

PROPOSITION 1. *If (G, Ω) is an o-2-transitive ordered permutation l-group, then it is o-n-transitive for all $n \geq 2$.*

Proof. Let (G, Ω) be o-n-transitive ($n \geq 2$) and let

$$\alpha_1 < \alpha_2 < \ldots < \alpha_n < \alpha_{n+1}, \beta_1 < \beta_2 < \ldots < \beta_n < \beta_{n+1}$$

be given. By induction assumption, there exists $f \in G$ such that $\alpha_i f = \beta_i$ for $i = 1, 2, \ldots, n$. We can assume that $\alpha_{n+1} f \geq \beta_{n+1}$ (otherwise we consider $f^{-1}, \beta_i f^{-1} = \alpha_i$, for $i = 1, 2, \ldots, n, \beta_{n+1} f^{-1} > \alpha_{n+1}$). As (G, Ω)

is o-2-transitive there exists $g \in G$ such that $\alpha_1 g = \beta_n$ and $\alpha_n g = \beta_{n+1}$. Then for each $i = 1, 2, ..., n$ is valid $\alpha_i g >> \alpha_1 g \geq \beta_n$ and $\alpha_i(f \wedge g) = \beta_i$ for $i = 1, 2, ..., n, n+1$.□

An ordered permutation group (G, Ω) is called *o-semitransitive* if there exists $\bar{\gamma} \in \bar{\Omega}$ such that $\gamma g = \gamma$ for all $g \in G$ and for all $\alpha_1, \alpha_2, \beta_1, \beta_2 \in \Omega$ such that $\alpha_1 < \beta_1 < \gamma, \gamma < \alpha_2 < \beta_2$ there exist $g, h \in G$ such that $\alpha_1 g = \beta_1, \alpha_2 h = \beta_2$. In other words, l-group (G, Ω) is o-semitransitive if $G \subseteq A(\Omega)_\gamma$ for some $\gamma \in \Omega$ and G is transitive on the sets $\Omega_\gamma^+ = \{\alpha \in \Omega | \alpha > \gamma\}$ and $\Omega_\gamma^- = \{\alpha \in \Omega | \alpha < \gamma\}$.

THEOREM 1. (Holland [3].) *An ordered permutation l-group (G, Ω) is o-2-transitive if and only if it is transitive and the stabilizer subgroup G_α is o-semitransitive for every $\alpha \in \Omega$.*

Proof. Let (G, Ω) be o-2-transitive. Then (G, Ω) is transitive and for every two pair $(\alpha, \beta_1), (\alpha, \beta_2)$ where $\alpha < \beta_1 < \beta_2$ there exists $g \in G$ such that $\alpha g = \alpha, \beta_1 g = \beta_2$ i.e., G_α is o-semitransitive.

Conversely, let (G, Ω) be transitive and G_α be o-semitransitive for every $\alpha \in \Omega$. Let $\alpha_1 < \alpha_2, \beta_1 < \beta_2$ in Ω and $\alpha \in \Omega$. Then there exist elements $g, h \in G$ such that $\alpha_1 g = \beta_1 h = \alpha$ and $\alpha_2 g, \beta_2 h \in \Omega_\alpha^+$. Let us suppose that $\alpha_2 g \leq \beta_2 h$. By semitransitivity of G_α, there is an element $f \in G_\alpha$ such that $\alpha_2 g f = \beta_2 h$. Hence $\alpha_1 g f h^{-1} = \alpha f h^{-1} = \alpha h^{-1} = \beta_1, \alpha_2 g f h^{-1} = \beta_2 h h^{-1} = \beta_2$ and (G, Ω) is o-2-transitive.□

Let g be an element of an ordered permutation group (G, Ω). A *support* of g is the set $\mathrm{supp}(g) = \{\alpha \in \Omega | \alpha g \neq \alpha\}$. If there exists $\alpha \in \Omega$ such that $\alpha > \beta (\alpha < \beta)$ for all $\beta \in \mathrm{supp}(g)$ then $\mathrm{supp}(g)$ is bounded from above (below, accordingly). If $\mathrm{supp}(g)$ is bounded from above and from below, then it is bounded. The order permutation g of totally ordered set Ω is *bounded* if $\mathrm{supp}(g)$ is bounded.

An ordered permutation group (G, Ω) is called *o-primitive* if there are only trivial convex G-congruences on Ω : the least G-congruence ε and the greatest G-congruence v.

Let Δ, Ω be totally ordered sets and $\Delta \subseteq \Omega$. If for all $\alpha, \beta \in \Omega$, such that $\alpha < \beta$, there exists $\delta \in \Delta$ such that $\alpha < \delta < \beta$, then the subset Δ is *dense* in Ω.

THEOREM 2. (Holland [3].) *If a transitive o-primitive ordered permutation l-group (G, Ω) contains a bounded from above or below el-*

ement g, then (G,Ω) is o-2-transitive.

Proof. Let us suppose that $g_0 \in G, g_0 > e, \mathrm{supp}(g_0)$ is bounded from above and $\beta = \inf_{\bar{\Omega}}(\mathrm{supp}(g_0))$. Now, we will prove that the subset $G(\beta) = \{\beta g | g \in G\}$ is a dense subset of $\bar{\Omega}$. If $\beta \in \Omega$, then by the transitivity of $(G,\Omega), G(\beta) = \Omega$ and, of course, dense in $\bar{\Omega}$. Now, let $\beta \in \bar{\Omega} \setminus \Omega$ and there exist $\alpha, \delta \in \bar{\Omega}, \alpha < \delta$ such that the inequality $\alpha < \beta g < \delta$ is not valid for all $g \in G$. Then there exist $\alpha', \beta' \in \Omega$ such that $\alpha \leq \alpha' < \delta' \leq \delta$. Now we define the relation ϑ on the set Ω by the rule: $\gamma \equiv \omega \pmod{\vartheta}$ iff $\beta g < \min\{\gamma, \omega\}$ or $\beta g > \max\{\gamma, \omega\}$ for all $g \in G$. It follows from the definition that $\gamma \equiv \gamma \pmod{\vartheta}$ and $\gamma \equiv \omega \pmod{\vartheta}$ implies $\omega \equiv \gamma \pmod{\vartheta}$. Let $\gamma_1 \equiv \gamma_2 \pmod{\vartheta}$ and $\gamma_2 \equiv \gamma_3 \pmod{\vartheta}$. We can assume that one of two inequalities is valid: a) $\gamma_1 \leq \gamma_2 \leq \gamma_3$ or b) $\gamma_1 \leq \gamma_3 \leq \gamma_2$.

In case b), $\gamma_1 \equiv \gamma_3 \pmod{\vartheta}$ is evident. As there does not exist any element of the kind $\beta g (g \in G)$ between γ_1, γ_2 and between γ_2, γ_3, then there does not exist element βg between γ_1, γ_3 . Therefore $\gamma_1 \equiv \gamma_3 \pmod{\vartheta}$ and ϑ is an equivalence relation on Ω. It is evident that ϑ is convex and ϑ is not trivial. Let $\gamma_1 \equiv \gamma_2 \pmod{\vartheta}, g \in G$ and $\gamma_1 g \not\equiv \gamma_2 g \pmod{\vartheta}$. If $\gamma_1 < \gamma_2$, then $\gamma_1 g < \gamma_2 g$. If there exists an element $h \in G$ such that $\gamma_1 g < \beta h < \gamma_2 g$ then $\gamma_1 < \beta h g^{-1} < \gamma_2$. But these inequalities contradict our assumption $\gamma_1 \equiv \gamma_2 \pmod{\vartheta}$. Therefore ϑ is a nontrivial convex G-congruence on Ω which is impossible since (G,Ω) is o-primitive. Then $G(\beta)$ is dense in $\bar{\Omega}$.

Now, fix an element $\gamma \in \Omega$ and let $\alpha < \gamma,\ \alpha \in \Omega$. Then there exists an element $g \in G$ such that $\alpha < \beta g < \gamma$. The element $\beta g = \inf_{\bar{\Omega}} \mathrm{supp} g^{-1} g_0 g$. From the inequality $\alpha < \beta g$ it follows $\alpha g^{-1} g_0 g = \alpha$. From the inequality $\beta g < \gamma$ it follows that there exists an element $\delta \in \Omega$ such that $\alpha < \delta < \gamma$ and $\delta g^{-1} g_0 g > \delta$. The subset $\delta G = \{\delta h | h \in G\}$ is dense in $\bar{\Omega}$ hence there exists $h \in G, h \geq e$ such that $\delta \leq \gamma h < \delta g^{-1} g_0 g$. Let $k = h g^{-1} g_0 g h^{-1}$ and from the inequalities $\alpha h \leq \alpha < \beta g$ it follows that

$$\alpha k = \alpha h g^{-1} g_0 g h^{-1} = ((((\alpha)h)g^{-1})g_0)g)h^{-1} = \alpha h h^{-1} = \alpha,\ k \in G_\alpha$$

and

$$\gamma k = \gamma h g^{-1} g_0 g h^{-1} = (\gamma h) g^{-1} g_0 g h^{-1} \geq (\delta) g^{-1} g_0 g h^{-1} > \gamma h h^{-1} = \gamma.$$

Hence it is proved that for all $\alpha \in \Omega$ there exists an element $k \in G_\alpha$ such that $\gamma k > \gamma$.

Now we will prove that for $\alpha < \gamma < \delta, \alpha, \gamma, \delta \in \Omega$ there is an element $h \in G_\alpha$ such that $\gamma h > \delta$. Let us suppose that the subset γG_α is bounded from above in Ω and $\bar{\gamma} = \sup_{\bar{\Omega}} \gamma G_\alpha$. By previous arguments there is $k \in G_\alpha$ and $\gamma < \gamma k, \alpha < \gamma k^{-1} < \bar{\gamma}$. From the definition of $\bar{\gamma}$ it follows that there is an element $r \in G_\alpha$ such that $\gamma r > \bar{\gamma} k^{-1}$ and therefore $(\gamma) rk > \bar{\gamma}$. By the transitivity of (G, Ω) there exist an element $h \in G, h \geq e$ and $\gamma h = \delta$. For the element $x = h \wedge k$ the following equalities are true:

$$\alpha x = \min\{\alpha h, \alpha k\} = \min\{\alpha, \alpha k\} = \alpha, \gamma x = \gamma(h \wedge k) = \min \gamma h, \gamma k = \delta.$$

Therefore the stabilizer l-subgroup G_α is semitransitive for any $\alpha \in \Omega$. By Theorem 1 (G, Ω) is o-2-transitive.□

2^0. A group G is called a *divisible* group if the equation $x^n = g$ has solution in G for all $n \in \mathbf{N}, g \in G$. Let Ω be a totally ordered set. An element $f \in A(\Omega)$ is called a *positive convex cycle* if for some $\alpha \in \Omega, \alpha < \alpha f$ and $\beta \neq \beta f$ imply $\alpha f^n \leq \beta \leq \alpha f^{n+1}$ for some $n \in \mathbf{Z}$. In other words, supp(f) consists of a single interval and the images of every point of this interval under the powers of f are coterminal in this interval. An element $f \in A(\Omega)$ is a *negative convex cycle* if $\alpha f < \alpha$ for some $\alpha \in \Omega$ and $\beta \neq \beta f$ implies that $\alpha f^n \leq \beta \leq \alpha f^{n+1}$ for some $n \in \mathbf{Z}$. We say that convex cycle is a *bump*. Each positive element $f \in A(\Omega)$ is the unique supremum in l-group $A(\Omega)$ of a disjoint set of bumps and the support of each of these bumps is called a *supporting interval* of f or a *orbital* for short. More formally, a supporting interval of f containing $\alpha \in \Omega$ is $o(f, \alpha) = \{\beta \in \Omega | \alpha f^k \leq \beta \leq \alpha f^t$ for some $k, t \in \mathbf{Z}$. It is evident that for all $\alpha, \beta \in \Omega, f \in A(\Omega)$ the set $o(f, \alpha)$ is convex and either $o(f, \alpha) = o(f, \beta)$ or $o(f, \alpha) \cap o(f, \beta) = \emptyset$ and $\Omega = \bigcup_{\alpha \in \Omega} o(f, \alpha)$.

THEOREM 3. (Holland [3].) *If $A(\Omega)$ is o-2-transitive, then it is divisible.*

Proof. Let $g \in A(\Omega)$ and n be an integer number. It is suffice to assume that g is one positive bump. We choose any point $\alpha \in \text{supp}(g)$, then the points $\ldots, \alpha g^i, \ldots$ are coterminal in $o(g, \alpha)$. Now we choose points $\beta_1, \beta_2, \ldots, \beta_n$ with $\alpha g < \beta_1 < \beta_2 < \ldots < \beta_n = \alpha g$. Since $A(\Omega)$

is o-2-transitive it is o-n-transitive by Proposition 1. Hence we can find $x \in A(\Omega)$ such that $\alpha x = \beta_1, \beta_i x = \beta_{i+1}$ for $i = 1, 2, ..., n-1$. Taking x as defined only on the interval $[\alpha, \beta_{n-1}]$ we can extend its action to $[\beta_{n-1}, \beta_n]$ by letting $x = x^{-n+1}g$. Now we extend its action to the right and to the left by the rule: if $\beta \notin \text{supp}(g)$ then $\beta x = \beta$, if $\beta \in \text{supp}(g)$ then there exists an integer k such that $\alpha g^k \leq \beta < \alpha g^{k+1}$, then $\beta x = \beta g^{-k} x g^k$. It is not difficult to verify that $x \in A(\Omega)$ and $\beta x^n = \beta g$ for all $\beta \in \Omega$.□

4.5 o-Primitive Ordered Permutation l-Groups

Let us recall that the ordered permutation group (G, Ω) is o-primitive if there are only trivial convex G-congruence on Ω. Our aim in this part is to give a complete classification of o-primitive transitive ordered permutation l-groups. This classification will prove extremely useful in later chapters and it is the cornerstone in the study of ordered permutation l-groups because all ordered permutation l-groups can be "constructed" from primitive ordered permutation l-groups.

1^0. LEMMA 1. *Let (G, Ω) be a transitive ordered permutation group , Δ be a convex subset in Ω such that $\Delta g = \Delta$ for all $g \in G, \alpha \in \Delta$ such that $\alpha g \in \Delta$. Then there exists a convex G-congruence ϑ on Ω defined by the rule: $\alpha \equiv \beta \pmod{\vartheta}$ if and only if $\alpha, \beta \in \Delta g$ for some $g \in G$.*

Proof. For all $g \in G$ the subset Δg is convex. If $\alpha \in \Delta g \cap \Delta f$ for some $f, g \in G$, then there exist $\beta \in \Delta$ such that $\alpha = \beta g$. For arbitrary $\gamma \in \Delta$ we can find an element $h \in G$ such that $\gamma h = \beta$. Thus $\gamma h g^{-1} f = \beta g f^{-1} = \alpha f^{-1} \in \Delta, \Delta h g f^{-1} = \Delta$ and $\Delta f = \Delta h f = \Delta g$. Therefore ϑ is a convex equivalence on Ω. If $\alpha \equiv \beta \pmod{\vartheta}$, then $\alpha, \beta \in \Delta g$ for some $g \in G$ and $\alpha f, \beta f \in \Delta g f$. From the definition of ϑ it follows that ϑ is a convex congruence on Ω.□.

A transitive ordered permutation l-group (G, Ω) is said to satisfy the *separation property* if for every non-empty bounded subset Δ of Ω and for every $\alpha, \beta \in \Omega, \alpha \neq \beta$ there is $g \in G$ such that exactly one of $\alpha g, \beta g$ belongs to Δ $(\alpha, \beta$ are separated by $\Delta)$.

Let $\Delta \subseteq \Omega$, and $\alpha, \beta \in \Omega$. We say that α and β are Δ-*connected* if for every $g \in G$ either $\alpha g, \beta g \in \Delta$ or $\alpha g, \beta g \notin \Delta$. Let us recall that if H is a subgroup of (G, Ω), then the subset $\alpha H = \{\alpha h | h \in H\}$ of Ω is a H-orbit of the element α .

LEMMA 2. *Let (G, Ω) be an ordered permutation l-group and ϑ be a convex G-congruence. Then for all $\alpha \in \Omega$ the subset $G_{\alpha\vartheta} = \{g \in G | \alpha g \equiv \alpha \pmod{\vartheta}\}$ is a prime l-subgroup of G and $G_{\alpha\vartheta} \supseteq G_\alpha$. Conversely, if (G, Ω) is a transitive ordered permutation l-group containing the stabilizer l-subgroup $G_{\alpha\vartheta}$, then the relation ϑ defined by the rule: $\alpha \equiv \beta \pmod{\vartheta}$ if and only if $\alpha g, \beta g \in \alpha H$ for some $g \in G$, is a convex G-congruence on Ω .*

Proof. Let ϑ be a convex G-congruence on Ω . If $g_1, g_2 \in G_{\alpha\vartheta}$, then $\alpha g_1 g_2 \equiv \alpha g_1 \equiv \alpha \pmod{\vartheta}$ and $g \vee e \in G_{\alpha\vartheta}$. Let $e \leq h \leq g, g, h \in G, g \in G_{\alpha\vartheta}$. Hence $\alpha \leq \alpha h \leq \alpha g$ and it follows from $\alpha g \equiv \alpha \pmod{\vartheta}$ that $\alpha h \equiv \alpha \pmod{\vartheta}$ and $h \in G_{\alpha\vartheta}$. Therefore, $G_{\alpha\vartheta}$ is a convex l-subgroup of G. It is clear that $G_{\alpha\vartheta} \supseteq G_\alpha$. Proposition 4.3.3 shows that G_α is a prime in G. Hence $G_{\alpha\vartheta}$ is a prime l-subgroup of G, too.

Conversely, let (G, Ω) be a transitive order permutation l-group and H be a convex l-subgroup of G containing $G_{\alpha\vartheta}$. Now, let us suppose that $\beta \leq \gamma \leq \delta, \beta, \delta \in \alpha H$. Then there exist $h_1, h_2 \in H$ such that $\alpha h_1 \leq \gamma \leq \alpha h_2$. Then $\alpha \leq \gamma h_1^{-1} \leq \alpha h_2 h_1^{-1}$. By the transitivity of (G, Ω) we may choose an element $g \in G$ such that $\alpha g = \gamma h_1^{-1}, g \geq e$. Then $g_1 = g \wedge (h_2 h_1^{-1} \vee e) \in H, g_1 h_1 \in H$ and $\alpha g_1 = \gamma h_1^{-1}$. Hence $\alpha g_1 h_1 = \gamma, \gamma \in \alpha H$ and we conclude that αH is a convex subset of Ω . If $\alpha f \in \alpha H$ for some $f \in G$, then $\alpha f = \alpha h$ for suitable element $h \in H$ and $\alpha f h^{-1} = \alpha,\ f h^{-1} \in G_\alpha$. Therefore $f \in H, (\alpha H) f = \alpha H$. So, Δ satisfies the hypotheses of Lemma 1 and the application of this lemma completes the proof.□

THEOREM 1. (Holland [3].) *If (G, Ω) is a transitive ordered permutation l-group, then following statement are equivalent.*

(1) *(G, Ω) is o-primitive on Ω.*
(2) *For each $\alpha \in \bar{\Omega}$ G_α is a maximal convex l-subgroup of G.*
(3) *For each $\alpha \in \Omega$ G_α is a maximal convex l-subgroup of G.*
(4) *If $\alpha \in \bar{\Omega}$, then αG is dense in $\bar{\Omega}$.*
(5) *(G, Ω) satisfies the separation property.*

Proof. (1) implies (2). Let $\alpha \in \Omega$. By Proposition 4.3.3 G_α is a prime l-subgroup of G. Let C be a convex l-subgroup of G properly containing G_α. Let $e \le g \in C \setminus G_\alpha$. Then $\alpha < \alpha g$. Hence there exists $\beta \in \Omega$ such that $\alpha \le \beta \le \alpha g$. Let $e \le f \in G$. Then $\alpha f \le \beta f = \beta \le \alpha g$, so $\alpha f g^{-1} \le \alpha$ and hence $fg^{-1} \vee e \in G_\alpha$. Therefore $e \le f \le fg \le (f \vee g)g^{-1} = fg^{-1} \vee e \in G_\alpha \subseteq C$. As C is convex $f \in C$ and hence $G_\beta \subseteq C$.

By Lemma 2, C determines a convex G-congruence ϑ on Ω and $\beta < \beta g, \beta g \equiv \beta \pmod{\vartheta}$. By (1) G is o-primitive, then for every $\gamma \in \Omega$ is true $\gamma \equiv \beta \pmod{\vartheta}$. So, for every $\gamma \in \Omega$ there exists $h \in C$ such that $\beta h = \gamma$. We claim that each element $g_1 \in G, g_1 \ge e$ is contained in G. Let $h \in C$ such that $\beta h = \beta g_1$. Let $g_2 = (g_1 \vee h)h^{-1}$. Then $\beta g_2 = \beta(g_1 \vee h)h^{-1} = \beta$ and therefore $g_2 \in C$. As $h \in C$, then $g_1 \vee h = g_2 h \in C$. But $e \le g_1 \le g_1 \vee h$ and $g_1 \in C$ since C is convex. Then $C = G$. Thus G_α is a maximal convex l-subgroup of G for all $\alpha \in \Omega$.

(2) implies (3). It is trivial.

(3) implies (4). Define the relation ϑ on Ω by $\beta \equiv \gamma \pmod{\vartheta}$ if there is no $g \in G$ such that αg lies between β and γ . It is easy to verify that ϑ is a convex G-congruence on Ω . Let $\delta \in \Omega$. It follows from Lemma 2 that $C = \{g \in G | \delta g \equiv \delta \pmod{\vartheta}\}$ is a convex l-subgroup of G containing G_δ. It is clear that $C \ne G$. Therefore, since G_δ is maximal, then $C = G_\delta$. Now, if $\delta \equiv \beta \pmod{\vartheta}, \alpha, \beta \in \Omega$ then, there exists $g \in G$ such that $\delta g = \beta$ and so $g \in C = G_\delta$. Hence $\delta = \beta$. Therefore αG is dense in Ω .

(4) implies (5). Let Δ be a convex bounded subset of Ω and $\alpha = \inf_{\bar{\Omega}} \Delta, \alpha \in \Omega$. If $\gamma, \beta \in \Omega$ and the open interval (γ, β) is non-empty, then since δG is dense in Ω there exists $g \in G$ such that $\gamma < \alpha g < \beta$. Then $\gamma g^{-1} < \alpha < \beta g^{-1}$. Hence there is $\delta \in \Delta$ such that $\delta < \beta g^{-1}$. As G is transitive, then there exists $f \in G$ such that $f \le e$ and $\delta = \beta g^{-1} f$. Thus $\gamma g^{-1} f \le \gamma g^{-1} < \alpha \le \delta = \beta g^{-1} f$. Therefore $\gamma g_1^{-1} f \notin \Delta$ but $\beta g^{-1} f \in \Delta$, and so γ and β are separated by Δ .

(5) implies (1). Let there exists a nontrivial convex G-congruence. Then there are at least two congruent classes, say Δ_1 and Δ_2, where $\delta < \gamma$ for all $\delta \in \Delta_1$ and $\gamma \in \Delta_2$. Thus any $\delta \in \Delta_1$ is a lower bound for Δ_2. Because G is transitive, then for some $\delta \in \Delta_1$ and $g \in G$ is valid $\delta g \in \Delta_2$. Hence δg^2 is an upper bound for Δ_2. Thus Δ_2 is a bounded convex subset of Ω . Since some congruence class contain

more than one element, by transitivity of G the set Δ_2 must contains some $\gamma_1, \gamma_2, \gamma_1 \neq \gamma_2$. But now for all $f \in G$ either $\gamma_1 f, \gamma_2 f \in \Delta_2$ or $\gamma_1 f, \gamma_2 f \notin \Delta_2$. Hence γ_1, γ_2 are not separated by Δ_2.□

COROLLARY 1. (Holland [3].) *If* $\bar{\Omega} = \Omega$ *and* (G, Ω) *is transitive on* Ω, *then* (G, Ω) *is o-primitive.*

Proof. Immediate from (4) since $\alpha G = \Omega$ for all $\alpha \in \Omega$ in this case.□

PROPOSITION 1. (Holland [3].) *If* (G, Ω) *is o-2-transitive on* Ω *then* (G, Ω) *is o-primitive on* Ω .

Proof. Clearly, any o-2-transitive ordered permutation l-group satisfies the separation property. By Theorem 1, (G, Ω) is o-primitive.□

2⁰. Let (G, Ω) be a transitive ordered permutation l-group, $\zeta \leq \vartheta$ be the convex G-congruences on Ω and Δ be a congruence class of ϑ. Then $(\bar{G}_{(\Delta)}, \Delta)$ is a transitive ordered permutation l-group of totally ordered set Δ where $\bar{G}_{(\Delta)} = G_{(\Delta)}/G_{\Delta}$. The restriction $\zeta' = \zeta_{|\Delta}$ of ζ on Δ is a convex $G_{(\Delta)}$-congruence on Δ and $(\hat{G}_{(\Delta)}, \Delta/\zeta')$ is a transitive ordered permutation l-group of the totally ordered set Δ/ζ' where $\hat{G}_{(\Delta)} = \bar{G}_{(\Delta)}/L_{\zeta'}, L_{\zeta'}$ is a lazy subgroup of $\bar{G}_{(\Delta)}$ under ζ' (see Chapter 4 Section 2). The ordered permutation l-group $(\hat{G}_{(\Delta)}, \Delta/\zeta')$ is called a *component* of (G, Ω) at (ζ, ϑ) on Δ or, for short, (ζ, ϑ)-component of (G, Ω) on Δ . We claim that the component is independent under the choice of the congruence class Δ .

LEMMA 3. *Let* (G, Ω) *be a transitive ordered permutation l-group and* $\zeta \leq \vartheta$ *be the convex* G*-congruences on* Ω . *Let* Δ, T *be congruence classes of* ϑ . *Then* (ζ, ϑ)*-components of* (G, Ω) *at* Δ *and* T *are order isomorphic.*

Proof. By the transitivity of (G, Ω) there is an element $g \in G$ such that $\Delta g = T$. Hence $G_T = g^{-1} G_\Delta g$. Let $\zeta'_\Delta = \zeta_{|\Delta}, \zeta'_T = \zeta_{|T}$ and L is a lazy subgroup of $(\bar{G}_{(\Delta)}, \Delta/\zeta'_\Delta), L_T = g^{-1} L_\Delta g$ is the lazy subgroup of $(\bar{G}_{(T)}, T/\zeta'_T)$. For $\delta \in \Delta$ let $(\delta\zeta'_\Delta)\varphi = \delta g \zeta'_T$ and $(L_\Delta f)\psi = L_T g^{-1} f g$. Then $\psi : G_{(\Delta)} \to \hat{G}_{(T)}$ is an l-isomorphism and $\varphi : \Delta/\zeta'_\Delta \to T/\zeta'_T$ is an order isomorphism. It is clear that $((\delta\zeta'_\Delta)\varphi(L_\Delta f)\psi = (\delta\zeta T')(L_T g^{-1} f g) = (\delta\zeta'_\Delta)(L_\Delta f)\varphi$ and hence the pair (ψ, φ) is the isomorphism from the order permutation l-group $(G_{(\Delta)}, \Delta/\zeta'_\Delta)$ onto

$(\hat{G}_{(T)}, T/\zeta'_T)$.□

Now we will omit the references to congruence class in the component of the transitive ordered permutation group relative to the pair of congruences. We will also write Δ/ζ for $\Delta/\zeta_{|\Delta}$ if Δ is an union of ζ-classes. Let (G,Ω) be a transitive ordered permutation group and $\zeta < \vartheta$ be G-congruences. We say that ϑ covers ζ if there is no G-congruence ξ such that $\zeta < \xi < \vartheta$.

PROPOSITION 2. *Let (G,Ω) be a transitive order permutation l-group and $\zeta < \vartheta$ be a G-congruence on Ω . Then each (ζ,ϑ)-component of (G,Ω) is primitive if and only if ϑ covers ζ .*

Proof. Let Δ be an arbitrary congruence class of G-congruence ϑ . It is evident that any $G_{(\Delta)}$-block has the form T/ζ where $T \subseteq \Delta$ and it is a $G_{(\Delta)}$-block in Ω . For all elements $g \in G \quad \Delta g \cap \Delta \neq \emptyset$ implies $Tg \cap T \neq \emptyset$. But Δ is a G-block. Then $\Delta g = \Delta$. Hence $g \in G_{(\Delta)}$ and $Tg = T$. Therefore T is a G-block. Now the proof of proposition arises from Proposition 4.2.1.□

EXAMPLE 1. Let $\Omega = \mathbf{R} \overleftarrow{\times} \mathbf{Q}$ where $\mathbf{R}, \mathbf{Q}$ are totally ordered sets of real and rational numbers, respectively. Let ϑ be an equivalence relation defined by the rule: $(\alpha,\beta) \equiv (\alpha',\beta') \pmod{\vartheta}$ for $\alpha \in \mathbf{R}, \quad \beta \in \mathbf{Q}$ iff $\beta = \beta'$ and $G = \{g \in A(\Omega) | \sigma \equiv \tau \pmod{\vartheta}\}$ for $\sigma, \tau \in \Omega$ iff $\sigma g \equiv \tau g$ (mod ϑ). It is easy to see that ϑ is a G-congruence and (G,Ω) is a transitive ordered permutation l-group. The G-congruence ϑ is a unique nontrivial G-congruence on Ω . Now we describe (ε,ϑ)-components of (G,Ω) where ε is the least G-congruence. Let us consider an arbitrary congruence class Δ , for example, $\Delta = \{(\alpha,\beta_0) | \alpha \in \mathbf{R}, \quad \beta_0 \in \mathbf{Q}\}$. It is evident that Δ is an order isomorphic to $\mathbf{R}$ and $A(\Delta)$ is l-isomorphic to $A(\mathbf{R})$. For arbitrary $s \in A(\mathbf{R})$ let us define $g_s \in G_\Delta$ such that $(\alpha,\beta)g_s = (\alpha s, \beta_0)$ if $\beta = \beta_0$ and $(\alpha,\beta)g_s = (\alpha s, \beta)$ if $\beta \neq \beta_0$.

The mapping $\varphi : s \to L_\varepsilon g_s$ is an l-isomorphism from $A(\mathbf{R})$ onto $\hat{G}_{(\Delta)}$. But $A(\mathbf{R})$ is o-2-transitive and (ε,ϑ)-component of (G,Ω) is o-primitive.

Now, let us describe (ζ,v)-component of (G,Ω) where v is the greatest G-congruence on Ω . It is evident that Ω is the unique class of$v$ and $L_\zeta = \{g \in G | \Delta g = \Delta$ for all classes of G-congruence $\zeta\}$. For arbitrary $s \in A(\mathbf{Q})$ let us define $\bar{g}_s \in G$ such that $(\alpha,\beta)\bar{g}_s = (\alpha, \beta s)$. It is clear that φ maps $A(\mathbf{Q})$ into G/L_ζ. By previous reasons (ζ, v)-component of (G, Ω) is primitive.

Let (G, Ω) be a transitive ordered permutation l-group, ζ, ϑ be G-congruences on Ω, $\zeta < \vartheta$ and ϑ covers ζ . Then the pair (ζ, ϑ) is called a *covering pair of G-congruences.* Let $k(G, \Omega)$ denote the set of all covering pairs of G-congruences on Ω . We will write $(\zeta_k, \zeta^k) = k \in k(G, \Omega)$ where ζ^k covers ζ_k. By Corollary 1 of Proposition 4.2.6 the set $\mathcal{K}$ of all G-congruences is totally ordered. By letting $k < q$ iff $\zeta^k < \zeta_q$ we obtain the total order on the set $k(G, \Omega)$. Let $\alpha, \beta \in \Omega$ and $\alpha \neq \beta$. Let ζ_k be the union of all G-congruences ζ on Ω for which $\beta \notin \alpha\zeta$ and ζ^k be the intersection of all G-congruences ζ for which $\alpha \equiv \beta \pmod{\zeta}$. By Corollary 2 of Proposition 4.2.6, ζ_k and ζ^k are G-congruences on Ω and ζ^k covers ζ_k. Hence $k = (\zeta_k, \zeta^k) \in k(G, \Omega)$ and it is determined by α, β uniquely. In this case k is called a *value of pair* (α, β) and is denoted by $k = \mathrm{val}(\alpha, \beta)$.

PROPOSITION 3. *Let (G, Ω) be a transitive ordered permutation l-group. Then for any pair of elements $\alpha, \beta \in \Omega$, $\alpha \neq \beta$, is valid* $\mathrm{val}(\alpha, \beta) \in k(G, \Omega)$. *Conversely, for arbitrary $k \in k(G, \Omega)$ there exist $\alpha, \beta \in \Omega$ such that* $k = \mathrm{val}(\alpha, \beta)$.

Proof. The first part of the proposition is clear. Let $k \in k(G, \Omega)$. As far as $\zeta_k \neq \zeta^k$ is concerned, there exist $\alpha, \beta \in \Omega$, such that $\alpha \in \beta\zeta^k$, $\alpha \notin \beta\zeta_k$. It follows from $\alpha \in \beta\zeta^k$ that $k \leq \mathrm{val}(\alpha, \beta)$. From $\alpha \notin \beta\zeta_k$. we obtain $k \geq \mathrm{val}(\alpha, \beta)$, therefore $k = \mathrm{val}(\alpha, \beta)$.□

PROPOSITION 4. *Let (G, Ω) be a transitive ordered permutation l-group and ζ be an arbitrary G-congruence on Ω . Then*

$$\bigcap_{k\in k(G,\Omega),\zeta\leq\zeta_k} \zeta_k = \zeta = \bigcup_{k\in k(G,\Omega),\zeta^k\leq\zeta} \zeta^k.$$

It is clear that $\delta = \bigcup_{k\in k(G,\Omega),\zeta^k\leq\zeta} \zeta^k \leq \zeta$. Let us suppose that there exist $\alpha, \beta \in \Omega$ such that $\alpha \in \beta\zeta, \alpha \notin \beta\delta$. Let $q = \mathrm{val}(\alpha, \beta)$. Then $\delta \leq \zeta_k < \zeta \leq \zeta^k$. As $\zeta^k \leq \zeta$ then $\zeta^k \leq \delta$ and $\zeta^q \leq \delta \leq \zeta^q, \delta \neq \zeta^q$. It is a contradiction. By the similar arguments $\zeta = \bigcap_{k\in k(G,\Omega),\zeta\leq\zeta_k} \zeta_k = \zeta$.□

3⁰. The following results are obtained by S.McCleary [4],[5] and are useful in the further investigations of o-primitive transitive ordered permutation l-groups.

LEMMA 4. *Let (G,Ω) be a transitive ordered permutation l-group and $\alpha,\beta \in \Omega$. Then the set $\Omega(\alpha,\beta) = \beta G_\alpha = \{\beta g | g \in G_\alpha\}$ is convex. For all $\alpha,\beta,\beta' \in \Omega$ is valid $\Omega(\alpha,\beta) = \Omega(\alpha,\beta')$ or $\Omega(\alpha,\beta) \cap \Omega(\alpha,\beta') = \emptyset$ and $(\Omega(\alpha,\beta))g = \Omega(\alpha g,\beta g)$ for all $g \in G$.*

Proof. Let $\delta,\gamma \in \Omega(\alpha,\beta), \xi \in \Omega$ and $\delta < \xi \leq \gamma$. There exist $g_1,g_2 \in G_\alpha$ such that $\delta = \beta g_1, \gamma = \beta g_2$. We may assume that $\alpha < \beta$. Then $\alpha < \beta g_1 < \xi < \beta g_2$ and $\alpha < \beta < \xi g_1^{-1} < \beta g_1^{-1} g_2$. By the transitivity of G there is an element $h \in G, h > e$ such that $\beta h = \xi g_1^{-1}$. Hence $\alpha(h \wedge g_2 g_1^{-1}) = \min\{\alpha h, \alpha\} = \alpha$ and $h \wedge g_2 g_1^{-1} \in G_\alpha$. But $\beta(h \wedge g_2 g_1^{-1}) = \min\{\beta h, \beta g_2 g_1^{-1}\} = \min\{\xi g_1^{-1}, \beta g_2 g_1^{-1}\} = (\min\{\xi, \beta g_2\}) g_1^{-1} = \xi g_1^{-1}$ and $\xi g_1^{-1}, \xi \in \Omega(\alpha,\beta)$. Therefore $\Omega(\alpha,\beta)$ is a convex subset of Ω . It is evident that $\Omega(\alpha,\beta)$ is the G_α-orbit of the element β . By Proposition 4.3.2 for all $\beta,\beta' \in \Omega$ is valid either $\Omega(\alpha,\beta) = \Omega(\alpha,\beta')$ or $\Omega(\alpha,\beta) \cap \Omega(\alpha,\beta') = \emptyset$.

Let $g \in G$. Then

$$\begin{aligned}\Omega(\alpha,\beta) = (\beta G_\alpha)g = \{\beta h g | h \in G_\alpha\} = \{\beta g g^{-1} h g | h \in G_\alpha\} = \\ = \{(\beta g)(g^{-1} h g) | h \in G_\alpha\}.\end{aligned}$$

By Proposition 4.3.3, the subset $\{g^{-1} h g | h \in G_\alpha\}$ of G coincides with stabilizer l-subgroup $G_{\alpha g}$. Hence $(\Omega(\alpha,\beta))g = (\alpha g) G_{\alpha g} = \Omega(\alpha g, \alpha g)$.□

LEMMA 5. *Let (G,Ω), be a transitive ordered permutation l-group and stabilizer l-subgroups $G_\alpha \neq E$ for all $\alpha \in \Omega$. Then for every $\alpha \in \Omega$ there exists $\beta \in \Omega$ such that $\alpha = \inf \Omega(\alpha,\beta)$.*

Proof. Let $\alpha,\xi \in \Omega, \xi < \alpha, \Omega(\alpha,\xi) \neq \{\xi\}$, and $\mu = \sup_{\bar{\Omega}} \Omega(\alpha,\xi)$. There is an element $g \in G$ such that $g > e$ and $\xi g = \alpha$. Then $\alpha = \Omega(\alpha,\xi)g$ and $\nu = \sup_{\bar{\Omega}} \Omega(\alpha,\xi) = \mu g > \mu$. Now, consider the subset $\Omega(\alpha,\nu) = \nu G_\alpha$. We claim that there are no elements of Ω between α and $\Omega(\alpha,\nu)$. Conversely, assume that $\alpha < \gamma < \Omega(\alpha,\nu)$ for some $\gamma \in \Omega$. By Theorem 1, the subset μG is dense in Ω , so there is an element $h \in G, h > e$ such that $\alpha < \mu h < \gamma$. We can suppose that $e < h < g$. As $\xi g = \alpha, h \leq g$ we must have $\xi h \leq \alpha \leq \mu h$ and $\alpha \in \Omega(\alpha,\xi)h, \alpha h^{-1} \in \Omega(\alpha,\xi)$. Because G_α is transitive on $\Omega(\alpha,\xi)$ there is $k \in G_\alpha$ such that $\xi k = \alpha g^{-1} k > \alpha h^{-1}$ or equivalent $\alpha < \alpha g^{-1} k h$. Then $\mu g g^{-1} k h = \mu k h = \mu h \leq \gamma$.

Now, we can find $f \in G, f \leq e$ such that $\alpha g^{-1}kh = \alpha$. Then for $q = g^{-1}kgh$ is valid $\alpha q = \alpha, \nu q = \mu g q = \mu g g^{-1} k h f = \mu h f \leq \gamma f \leq \gamma$. It follows that for arbitrary $\gamma \in \Omega, \alpha < \gamma < \nu$ there is an element $q \in G_\alpha$ such that $\nu q \leq \gamma \leq \nu$. It means that there is not any element of Ω between $\Omega(\alpha, \nu)$ and α.

Finally, we claim that $\Omega(\alpha, \nu) \neq \{\nu\}$. Otherwise α is an immediate predecessor of ν and hence μ has an immediate predecessor. But it is impossible as $\mu = \sup_{\bar{\Omega}} \Omega(\alpha, \xi)$ and $\Omega(\alpha, \xi)$ has not the greatest element. If $\beta \in \Omega, \beta \in \Omega(\alpha, \nu)$ then $\Omega(\alpha, \nu) = \Omega(\alpha, \beta)$ and $\alpha = \inf_{\bar{\Omega}} \Omega(\alpha, \beta)$.

It is easy to see that the set $\{\Omega(\alpha, \gamma) | \gamma \in \Omega\}$ is a totally ordered set under the rule: $\Omega(\alpha, \gamma) \leq \Omega(\alpha, \delta)$ iff $\gamma \leq \delta$. The subset $\Omega(\alpha, \beta) = \Omega_{+1}(\alpha)$ is the smallest element of the set $\{\Omega(\alpha, \gamma) | \gamma \in \Omega, \gamma > \alpha\}$.

By analogous arguments, an o-primitive transitive ordered permutation l-group (G, Ω) has the greatest element $\Omega_{-1}(\alpha)$ in the set $\{\Omega(\alpha, \gamma) | \gamma \in \Omega, \gamma < \alpha\}$.

An ordered permutation l-group (G, Ω) is said to be *periodic* (or a *periodic ordered permutation l-group*) if there exists an order permutation $t \in A(\bar{\Omega})$ such that following conditions hold:

(6) $t = tg$ for all $\in G$

(7) t has no fixed points in $\bar{\Omega}$,

(8) for some $\beta \in \Omega$ the open interval $(\alpha, \alpha t)$ coincides with $\Omega(\alpha, \beta) = \beta G_\alpha$.

In this case the ordered permutation t is called *a period* of l-group (G, Ω).

This definition generalizes the notation of periodic permutation of reals: function $f : \mathbf{R} \to \mathbf{R}$ has period $t \in \mathbf{R}$ if $t > 0$ in $\mathbf{R}$ and t is the smallest number such that $f(x + t) = f(x) + t$ for all $x \in \mathbf{R}$.

An o-2-transitive ordered permutation l-group (G, Ω) is called *pathological* if supp(f) is unbound in Ω for every $f \in \Omega, f \neq e$.

THEOREM 2. (McCleary [2], [3]) *Let (G, Ω) be a transitive o-primitive ordered permutation l-group. Then one from the following conditions is valid:*

(9)	G	*is an arhimedean totally ordered group and* (G,Ω) *is the right regular representation of* G,
(10)	(G,Ω)	*is o-2-transitive on* Ω ,
(11)	(G,Ω)	*is periodic and the stabilizer subgroup* G_α *is an o-2-transitive nonpathological ordered permutation l-group of the totally ordered set* $\Omega(\alpha,\beta)$ *for all* $\alpha,\beta\in\Omega,\alpha\neq\beta$.

Proof. Let (G,Ω) be a transitive o-primitive l-group. If $\alpha g\neq\alpha$ for all $\alpha\in\Omega, g\in G$, then $G=E$ for every α . By Theorem 1, the trivial l-subgroup E is a maximal convex l-subgroup of G. Hence the lattice $\mathcal{C}(G)$ of all convex l-subgroup of G has only two elements: E and G. By Proposition 3.2.1, G is an Archimedean o-group.

Now, let us suppose that $G\neq E$ for some $\alpha\in\Omega$ and (G,Ω) is not o-2-transitive. By Theorem 4.4.2, supp(f) is unbound in Ω for every $f\in\Omega, f\neq e$. It follows From Lemma 4 that there exists the smallest convex subset $\Omega_{+1}(\alpha)=\Omega(\alpha,\beta)=\beta G_\alpha$ in the totally ordered set $\{\Omega(\alpha,\xi)|\xi\in\Omega,\xi>\alpha\}$ where $\Omega(\alpha,\gamma)\leq\Omega(\alpha,\delta)$ iff $\gamma\leq\delta$. The subset $\Omega_{+1}(\alpha)$ contains two or more elements. We claim that there are no smallest and greatest elements in $\Omega_{+1}(\alpha)$ and $\Omega_{+1}(\alpha)$ is bounded from above in Ω . If $\Omega_{+1}(\alpha)$ is not bounded from above then $\Omega_{+1}(\alpha)=\{\beta\in\Omega|\beta>\alpha\}$ and by transitivity of G_α on $\Omega_{+1}(\alpha)$ l-group (G,Ω) is o-2-transitive. So, $\Omega_{+1}(\alpha)$ is bounded from above. Let $\alpha=\sup_{\bar\Omega}\Omega_{+1}(\alpha)$.

Now, we set $\alpha t=\bar\alpha$ for $\alpha\in\Omega$ and $\zeta t=\sup_{\bar\Omega}\{\delta t|\delta\in\Omega,\delta<\zeta\}$ for $\zeta\in\bar\Omega\setminus\Omega$. We claim that $t\in A(\bar\Omega)$ and t is a period of (G,Ω). By definition, t is a mapping from Ω into Ω . Let $\alpha,\beta\in\Omega,\alpha\leq\beta$ and $g\in G, g\geq e, \alpha g=\beta, \Omega_{+1}(\alpha)g=\Omega_{+1}(\beta)$. Then $\alpha tg=\beta t$ and $g\geq e$ implies $\alpha t\leq\beta t$. If $\alpha t=\beta t$, then $\alpha th=\alpha t$ for all $h\in G_\alpha$ and $\beta tf=\beta t$ for all $f\in G_\alpha$. Hence $G_\alpha\subseteq G_{\alpha t}, G_\beta\subseteq G_{\beta t}$. By Theorem 1, $G_\alpha=G_\beta=G_{\alpha t}$ and $\Omega_{+1}(\alpha)=\Omega_{+1}(\beta),\alpha=\beta$. Therefore t is one-to-one mapping from Ω into Ω .

For every $\xi\in\Omega$ the set ξG is dense in Ω . Hence for $\alpha<\beta,\alpha,\beta\in\bar\Omega$ there are elements $\eta,\zeta\in\Omega$ such that $\alpha<\eta<\zeta<\beta$, and $\alpha t=\sup_{\bar\Omega}\{\xi t|\xi<\alpha,\xi\in\Omega\}<\eta t<\zeta t<\sup_{\bar\Omega}\{\xi t|\xi<\beta,\xi\in\Omega\}=\beta t$, i.e., t is an ordered mapping.

If $\alpha\in\Omega$ let $\alpha_1=\inf\Omega_{-1}(\alpha)$ where $\Omega_{-1}(\alpha)$ is the greatest element in the set $\{\Omega(\alpha,\xi)|\xi\in\Omega,\xi<\alpha\}$. It is easy to see that $\alpha_1 t=\alpha$ and t

maps Ω onto Ω . So, $t \in A(\bar{\Omega})$.

Now, let $g \in G$. Then for $\xi \in \Omega$ is valid $\xi tg = (\sup_{\bar{\Omega}} \Omega_{+1}(\xi))g = \sup_{\bar{\Omega}} \Omega_{+1}(\xi)g = \sup_{\bar{\Omega}} \Omega_{+1}(\xi g) = \xi gt$. We claim that t has no fixed points in Ω . Suppose that there is $\gamma \in \Omega$ such that $\gamma t = \gamma$. Then let us consider the subset $T = \{\xi \in \Omega$ iff there is integer $n \in \mathbf{Z}$ such that $vt^n < \xi < vt^{n+1}$ where $v \in \Omega, v < \gamma, vt \neq v\}$. It is evident that T is convex and $T \neq \Omega$. Let $g \in G, \xi \in T$ and $\xi g \in T$. Then $vt^n \leq \xi \leq \xi t^{n+1} \leq \gamma t, \xi t^{-(n+1)} \leq v \leq vt^{-n}$ and $\xi t^{-(n+1)}g \leq vg \leq vt^{-n}g$. Then $\xi gt^{-(n+1)} \leq vg \leq vgt^{-n}$ and $vg \in T$. Then for the set T all conditions of Lemma 1 are valid . This statement contradicts the o-primitivity of (G, Ω).

It is easy to see that for arbitrary $\alpha \in \Omega$ there exists $\beta \in \Omega$ such that $\alpha = \inf \Omega(\alpha, \beta)$ and $\alpha t = \sup \Omega(\alpha, \beta)$. Therefore, it is proved that non o-2-transitive and nonarchimedean o-primitive l-group (G, Ω) is periodic.

Now, let $\alpha \in \Omega$. We claim that G_α acts faithfully on $\Omega(\alpha, \beta)$ for all $\beta \in \Omega, \beta \neq \alpha$, and $\alpha t = \sup \Omega_{+1}(\alpha)$. The order permutation t has no fixed point in Ω, hence for arbitrary $\xi \in \Omega$ there exists a integer n such that $\alpha t^n \leq \xi \leq \alpha t^{n+1}$. Therefore the action of l-group G_α on Ω is uniquely defined by the action on $\Omega_{+1}(\alpha)$: if $\alpha t^n \leq \xi \leq \alpha t^{n+1}$, then $\xi t^{-n} \in \Omega_{+1}(\alpha)$ and $\xi g = \xi t^{-n} gt$. It follows that if $g_\alpha \in G_\alpha$ acts trivially on $\Omega_{+1}(\alpha)$, then is a trivial permutation on Ω . Hence G_α acts faithfully on $\Omega_{+1}(\alpha)$ and on arbitrary $\Omega(\alpha, \beta)$. Then for any $\beta \in \Omega, \beta \neq \beta G_\alpha$ there exists a integer n such that $\Omega(\alpha, \beta) = \Omega_{+1}(\alpha)t^n$.

Now, we claim that G_α is o-2-transitive order permutation l-group of the totally ordered set $\Omega_{+1}(\alpha)$. By Theorem 4.5.1, it is enough to prove that the l-subgroup $G_\alpha \cap G_\beta$ acts transitively on the set $\{\gamma \in \Omega_{+1}(\alpha) | \gamma > \beta\}$. Let $\delta > \gamma > \beta, \gamma, \delta \in \Omega_{+1}(\alpha)$. The l-subgroup G_α is transitive on $\Omega_{+1}(\alpha)$, then there exists $h \in G_\alpha, h > e$ such that $\gamma h = \delta$. But $\Omega_{+1}(\beta) = \Omega_{+1}(\alpha g) = \Omega_{+1}(\alpha)g$. The l-subgroup G_β acts transitively on $\Omega_{+1}(\beta)$, then there is an element $f \in G_\beta, f > e, \gamma f = \delta$. Hence $h \wedge f \in G_\alpha \cap G_\beta, \gamma(h \wedge f) = \delta$ and G_β acts o-2-transitively on $\Omega_{+1}(\alpha)$ and on all $\Omega(\alpha, \beta)$.

Finally, we claim that G_α is a nonpathological ordered permutation l-group. For arbitrary $\alpha \in \Omega$ the l-group G_α acts o-2-transitively and faithfully on $\Omega_{+1}(\alpha)$. Hence for any $\xi, \eta \in \Omega_{+1}(\alpha), \xi < \eta$ there exists $g \in G_\alpha$ such that $\xi g < \xi, \eta < \eta g$. Then there is some interval $[\beta, \gamma] =$

$\{\delta \in \Omega | \beta \leq \delta \leq \gamma\}$ such that $\delta g < \delta$ is valid for all $\delta \in [\beta, \gamma]$. Therefore $\delta(g \vee e) = \delta$ for all $\delta \in [\beta, \gamma]$ and $\eta(g \vee e) \neq \eta, g \vee e \neq e$. Let $\delta \in \Omega, \beta \leq \delta \leq \gamma$. It is evident that $g \vee e \in G_\alpha$ and $\mu(g \vee e) = \mu$ for all $\nu \in \Omega, \alpha \leq \mu \leq \gamma$ or $\beta t \leq \mu < \alpha t$. Hence $\mathrm{supp}(g \vee e)$ is bounded from above and below in $\Omega_{+1}(\alpha)$. Therefore the l-subgroup G_α is non pathological ordered permutation l-group of $\Omega_{+1}(\alpha)$ (and automatically of $\Omega(\alpha, \beta)$ for all $\alpha, \beta \in \Omega$ such that $\beta G_\alpha \neq \beta$).□

4.6 Simple Lattice-Ordered Groups

A number of properties of o-primitive ordered permutation l-groups (G, Ω) is very close to the properties of simple l-groups. In general these notions are different but a lot of interesting examples of o-primitive l-groups (G, Ω) are simple and the theory of ordered permutations of totally ordered sets is a nice tool for the construction of new infinite simple groups.

1^0. The following result on permutation groups of infinite sets was due to G.Higman [1]. Let G be a permutation group of infinite (non ordered) set Ω . For $g \in G$, let $\mathrm{supp}(g) = \{\alpha \in \Omega | \alpha g \neq \alpha\}$. The subset $\Delta \subseteq \Omega$ is *unstable* with respect to g if $\Delta g \cap \Delta = \emptyset$. For group G the derived subgroup or commutator subgroup G' of G is the subgroup which is generated by all commutators $[x, y] = x^{-1}y^{-1}xy$, $x, y \in G$.

THEOREM 1. (Higman[1].) *Let G be a permutation group of infinite set Ω. If for any $g, h, f \in \Omega, f \neq e$, there exists $r \in G$ such that the subset $(\mathrm{supp}(g) \cup \mathrm{supp}(h))r$ is unstable with respect to f then derived subgroup G' of G is a simple group.*

Proof. Let N be a nontrivial normal subgroup of G and let $e \neq f \in N$. Let g, h be arbitrary elements of G, r be an element of G such that $(\mathrm{supp}(g) \cup \mathrm{supp}(h))r$ is unstable with respect to f and $d = rfr^{-1} \in N$. Then $(\mathrm{supp}(g) \cup \mathrm{supp}(h))rf \cap (\mathrm{supp}(g) \cup \mathrm{supp}(h))r = \emptyset$, hence $(\mathrm{supp}(g) \cup \mathrm{supp}(h))d \cap (\mathrm{supp}(g) \cup \mathrm{supp}(h)) = \emptyset$. Then $(\mathrm{supp}(g))d \cap \mathrm{supp}(h) = \emptyset$. But $(\mathrm{supp}(g))d = \mathrm{supp}(d^{-1}gd)$, then $\mathrm{supp}(d^{-1}gd) \cap \mathrm{supp}(h) = \emptyset$. But $(\mathrm{supp}(g))d = \mathrm{supp}(d^{-1}gd)$ then $\mathrm{supp}(d^{-1}gd) \cap \mathrm{supp}(h) = \emptyset$.

It means that the elements $d^{-1}gd$ and h are commute and $[g, h] =$

$g^{-1}\cdot d^{-1}gd\cdot h^{-1}\cdot d^{-1}g^{-1}d\cdot g\cdot h = g^{-1}d^{-1}g\cdot d\cdot h^{-1}d^{-1}h\cdot h^{-1}g^{-1}\cdot d\cdot gh \in N$. So, $G' \subseteq N$.

Now, we claim that $G'' = G'$. Let us suppose that $G' \neq E$. By the previous argument, it is enough to prove that $G'' \neq E$. Let $g = h = f \neq e, f \in G'$. According to our assumption there is an element $d \in G$ such that $\mathrm{supp}(d^{-1}fd) \cap \mathrm{supp}(f) = \emptyset$. Therefore $f \neq d^{-1}fd, f^{-1}d^{-1}fd \neq e$ and $[f,d] \in G'' \neq E$.

Let $g,h,f \in G', f \neq e$ and $r \in G, (\mathrm{supp}(g)\cup\mathrm{supp}(h))rf\cap(\mathrm{supp}(g)\cup \mathrm{supp}(h))r = \emptyset$. Then there exists $s \in G$ such that $(\mathrm{supp}(f)\cup\mathrm{supp}(r))s$ is unstable with respect to f.

Then $(\mathrm{supp}(f)\cup\mathrm{supp}(r))s \cap (\mathrm{supp}(f)\cup\mathrm{supp}(r))sf = \emptyset, (\mathrm{supp}(f)\cup \mathrm{supp}(r))\cap(\mathrm{supp}(f)\cup\mathrm{supp}(r))sfs^{-1} = \emptyset$. From $(\mathrm{supp}(g)\cup\mathrm{supp}(h))r \subseteq \mathrm{supp}(f)$ it follows that $\mathrm{supp}(f)\cap\mathrm{supp}((sfs^{-1})^{-1}r(sfs^{-1})) = \emptyset$. Then $(\mathrm{supp}(g)\cup\mathrm{supp}(h))r(sfs^{-1})^{-1}r^{-1}(sfs^{-1}) = (\mathrm{supp}(g)\cup\mathrm{supp}(h))r$ and the set $(\mathrm{supp}(g)\cup\mathrm{supp}(h))r(sfs^{-1})^{-1}r^{-1}(sfs^{-1})$ is unstable with respect to f. Hence for any $g,h,f \in G', f \neq e$ there is an element $q = [r^{-1}, sfs^{-1}] \in G'$ such that $(\mathrm{supp}(g)\cup\mathrm{supp}(h))q$ is unstable respect with to f and G' satisfies all conditions of the theorem.

Now, let K be an normal subgroup of G'. Then by previous argument $G'' \subseteq K$, $G' = G''$ implies $K = G'$ and G' is a simple group.□

2^0. Let $R(\Omega)$ be the set of all $g \in A(\Omega)$ such that support of g is bounded below (g "lives on the right"), $L(\Omega)$ is defined dually and $B(\Omega) = R(\Omega)\cap L(\Omega)$ is the set of all $g \in A(\Omega)$ such that $\mathrm{supp}(g)$ is bounded. It is evident that $B(\Omega), R(\Omega), L(\Omega)$ are ideals of the l-group $A(\Omega)$. Hence ordered permutation l-group (G,Ω) is simple iff $G \subseteq B(\Omega)$ or G contains no elements with bounded above or below support.

Let G be an l-group. The element $g \in G$ is *regular* if $g \neq e$ and $|g| \wedge h^{-1}gh| \neq e$ for all $g \in G$. If for $g \in G, g \neq e$ there exists an element h such that $|g| \wedge |f^{-1}h^{-1}ghf| = e$ for all $f \in G,\ f > e$ then g is *insular.*

PROPOSITION 1. (Holland [5].) *An element g of a transitive ordered permutation l-group (G,Ω) is insular if and only if $g \in (\Omega)$.*

Proof. Let $g \in B(\Omega), g \neq e$. Then $\mathrm{supp}(g) \subseteq [\alpha,\beta]$ for some $\alpha,\beta \in \Omega$. By transitivity (G,Ω), there is $k \in G$ such that $\alpha k > \beta$. Then $\mathrm{supp}(k^{-1}gk) = (\mathrm{supp}(g))k$ and by definition of k is valid $\mathrm{supp}(k^{-1}gk) >$

supp(g). Therefore supp$(f^{-1}k^{-1}gkf) = (\text{supp}(k^{-1}gk))f >$ supp(g) and $|g| \wedge |f^{-1}k^{-1}gkf| = e$ for all $f > e, f \in (G, \Omega)$, i.e., the element g is insular.

Conversely, assume that g is an insular element of the transitive ordered permutation l-group $(G, \Omega), g > e$. Then there is an element $h \in (G, \Omega)$ such that $|g| \wedge |f^{-1}h^{-1}ghf| = e$ for all $f > e, f \in (G, \Omega)$. Let $\alpha_1 \in \Omega$ and $\alpha_1 h^{-1}gh > \alpha_1$. If $\beta \in \Omega$, $\alpha_1 < \beta$, $\beta g > \beta$, then there is an element $f_1 \in (G, \Omega), f_1 > e$ and $\alpha f_1 = \beta$. Therefore $\beta f_1^{-1}h^{-1}ghf_1 = (\alpha_1 h^{-1}gh)f > \alpha_1 f = \beta$. So, $\beta(g \wedge f^{-1}h^{-1}ghf) = \min\{\beta g, \beta f_1^{-1}h^{-1}ghf_1\} > \beta$. This contradicts our assumptions. Thus for all $\beta > \alpha_1, \beta \in \Omega$ is valid $\beta g = \beta$ and supp(g) is bounded above.

Now, let $\alpha_2, \beta \in \Omega$ such that $\alpha_2 g > \alpha_2$ and $\beta \leq \alpha h^{-1}$. Then $\alpha_2 \geq \beta h$ and there exists an element $f_2 \in G$ such that $f_2 > e, \beta h f_2 = \alpha_2$. From our assumption it follows that $e = g \wedge f_2^{-1}h^{-1}ghf_2$. Hence $\alpha_2 = \alpha_2(g \wedge f_2^{-1}h^{-1}ghf_2) = \min\{\alpha_2 g, \alpha_2(f_2^{-1}h^{-1}ghf_2)\}$. But since $\alpha_2 g > \alpha_2$ then $\alpha_2(f_2^{-1}h^{-1}ghf_2) = \alpha_2$ and $(\beta g)hf_2 = \alpha_2, \beta g = \alpha_2(hf_2)^{-1} = \beta$ which is a contradiction . So, for all $\beta < \alpha_2, \beta \in \Omega$ is valid $\beta qg = \beta$ and supp(g) is bounded below.□

THEOREM 2. (Holland [5].) *Let G be a lattice ordered group. Then G is a simple l-group which contains an insular element if and only if G is isomorphic to a transitive o-primitive ordered permutation l-group (G, Ω) of bounded permutation of some totally ordered set Ω.*

Proof. Let G be a simple l-group and $g \in G, g \neq e$ is an insular element. By Corollary 2 of Theorem 4.3.1, the l-group G is a transitive ordered permutation l-group (G, Ω) for some totally ordered set Ω. By Proposition 1, the element g is bounded. The set $B(\Omega) \cap G$ of all bounded elements of G is a nontrivial ideal of G. Since G is a simple l-group we conclude that $B(\Omega) \cap G = G$. Let $e \neq g \in G$ and supp$(g) \subseteq [\alpha, \beta]$. If ϑ is a G-congruence on Ω and $\alpha \equiv \beta \pmod{\vartheta}$, then there is an l-homomorphism (G, Ω) onto l-group $(G/L_\vartheta, \Omega/\vartheta)$ (see Theorem 4.2.1) and $g \in \ker\vartheta = L_\vartheta$. Then $G = L_\vartheta$ and all elements of G are congruent. It follows that for any nontrivial G-congruence ϑ is valid $\alpha \equiv \beta \pmod{\vartheta}$. Since the union of the family of G-congruences is a G-congruence, then G has a maximal proper G-congruence ζ. Let $\Lambda = \Omega/\zeta$. Then $L_\zeta = E$ and (G, Λ) is a transitive o-primitive representation of l-group G. It is evident that all elements of (G, Λ) are bounded.

Conversely, suppose that (G,Ω) is a transitive o-primitive ordered permutation l-group of totally ordered set Ω and all elements of G are bounded. Let N be a nontrivial ideal of G. Let us define the relation ϑ on Ω by the rule: $\alpha \equiv \beta \pmod{\vartheta}$ iff there exists $f \in N$ such that $f \geq e, \quad \alpha f \geq \beta, \quad \beta f \geq \alpha$. It is evident that ϑ is reflexive and symmetric. Let $\alpha \equiv \beta \pmod{\vartheta}, \quad \beta \equiv \gamma \pmod{\vartheta}$. Then there are $f, g \in N$ such that $f \geq e, \quad g \geq e, \quad \alpha f \geq \beta, \quad \beta f \geq \alpha$ and $\beta g \geq \gamma, \quad \gamma g \geq \beta$. If $\alpha \leq \beta \leq \gamma$, then $\alpha f g \geq \beta g \geq \gamma$ and $\gamma f \geq \gamma \geq \beta$. If $\alpha \leq \gamma \leq \beta$ then $\alpha f \geq \beta \geq \gamma$ and $\gamma f \geq \gamma \geq \beta$. Hence $\alpha \equiv \gamma \pmod{\vartheta}$ and ϑ is an equivalence relation on Ω.

Let us suppose that $\alpha \equiv \beta \pmod{\vartheta}$ and $g \in G$. Then there exists an element $f \in N$ such that $f \geq e, \alpha f \geq \beta$ and $\beta f \geq \alpha$. Hence $\alpha g g^{-1} f g = (\alpha g)(g^{-1} f g) \geq \alpha g, (\beta f)(g^{-1} f g) = (\beta f) g \geq \alpha g$. Therefore ϑ is G-congruence on Ω.

Now, assume that $\alpha \leq \beta \leq \gamma$ and $\alpha \equiv \beta \pmod{\vartheta}$. Then there is an element $f \in N$ such that $f \geq e, \alpha f \geq \gamma$ and $\beta f \geq \gamma$. So, $\beta f \geq \beta \geq \alpha$ and $\alpha \equiv \beta \pmod{\vartheta}$ and ϑ is convex G-congruence. It follows from $N \neq E$ that $\vartheta \neq \varepsilon$. But (G,Ω) is o-primitive, then $\vartheta = \upsilon$ and for all elements $\alpha, \beta \in \Omega$ is valid $\alpha \equiv \beta \pmod{\vartheta}$.

Now, let us consider $g \in G, g \geq e$. By our assumption, supp(g) $\subseteq [\alpha, \beta]$ for some $\alpha, \beta \in \Omega, \alpha < \beta$ and $\alpha \equiv \beta \pmod{\vartheta}$. Then there is $f \in N, f \geq e$ and $\beta \leq \alpha f$. For all $\gamma \in [\alpha, \beta]$ is valid $\gamma g \leq \beta \leq \alpha f$. If $\gamma \notin [\alpha, \beta]$, then $\gamma g = \gamma = \gamma f$ and $e \leq g \leq f$ in G. But N is a convex l-subgroup hence $g \in N$. Therefore, $N = G$.□

Now, we will prove that derived subgroup G' of a transitive o-primitive ordered permutation l-group (G,Ω) of bounded permutations of a totally ordered set Ω is a simple group. In fact, for all $g, h, f, f > e$ there exists $\alpha \in \Omega, \alpha f \neq \alpha$. Let $\beta = \alpha f > \alpha$ and $\alpha_1 < \text{supp}(g) \cup \text{supp}(h) < \beta_1$. By Theorem 4.4.2, l-group (G,Ω) is o-2-transitive. So, there exists $r \in G$ such that $\alpha_1 r = \alpha, \beta_1 r = \beta$. Then $(\text{supp}(g) \cup \text{supp}(h)) r f \cap (\text{supp}(g) \cup \text{supp}(h)) r = \emptyset$ and it follows by Theorem 1, that G' is a simple group.□

THEOREM 3. (Holland [5], McCleary [2].) *If an ordered permutation l-group (G,Ω) is periodic and o-primitive, then G is a simple l-group.*

Proof. Let (G, Ω) is a periodic o-primitive ordered permutation l-group and the element t is a period of G. If $g \in G, \alpha \in \Omega$ and the element γg is known for all $\gamma, \gamma \in \Omega_{+1}(\alpha) = \beta G_\alpha = \{\xi \in \Omega | \alpha < \xi < \alpha t\}$, β is a suitable element of Ω (see Lemma 4.5.4 and Theorem 4.5.2), then the element δg is defined for all $\delta \in \Omega$ by the rule: if $\alpha t^n \leq \delta \leq \alpha t^{n+1}$ then $\delta g = \delta t^{-n} g t^n$. For all $\delta \in \Omega$ the subset δG is dense in Ω . Then there is an element $h \in G$ such that $\mathrm{supp}(h^{-1}gh) \cap \Omega_{+1}(\alpha) \neq \emptyset$. Let N be an ideal of G which is generated by g. Now we claim that $\Delta = \Omega$ where $\Delta = \{\delta \in \Omega | \alpha g_1 \leq \delta \leq \alpha g_2$ for some $g_1, g_2 \in N\}$.

It is easy to see that Δ is a convex subset of Ω and Δ contains more than one element as $\alpha \in \mathrm{supp}(h^{-1}gh)$. Let $k \in G$ and $\alpha k \in \Delta$. For the element $\delta \in \Delta$ we can find elements $g_1, g_2 \in N$ such that $\alpha g_1 \leq \delta \leq \alpha g_2$. Thus, $\alpha g_1 k \leq \delta k \leq \alpha g_2 k, \alpha k(k^{-1} g_1 k) \leq \delta k \leq \alpha k(k^{-1} g_2 k)$. But $\alpha k \in \Delta$. Then for some $g_3, g_4 \in N$ is valid $\alpha g_3 \leq \alpha k \leq \alpha g_4, \alpha g_3 k^{-1} g_1 k \leq \alpha k(k^{-1} g_1 k) \leq \alpha k \leq \alpha k(k^{-1} g_2 k) \leq \alpha g_4 k^1 g_2 k$ and $\delta k \in \Delta$.

Now, by Lemma 4.5.1, the subset Δ defines an G-congruence ϑ on Ω and $\vartheta \neq \varepsilon$. From the o-primitivity of (G, Ω) it follows that $\Delta = \Omega$. Hence for any $\alpha \in \Omega$ there is an element $q \in N, \alpha q > \delta$.

Let now $f \in G, f > e$. Then $\delta f \leq \alpha t f$ holds for all $\alpha \in \Omega_{+1}(\Omega), \alpha < \delta \leq \alpha t$. For the element $\alpha t f$ we can find $q \in N$ such that $\alpha q > \alpha t f$. Hence $\delta \leq \delta f \leq \alpha t f \leq \alpha q \leq \delta q$. By the periodicality of G with period t, these inequalities are true for all $\delta \in \Omega$ and $f \in N$. Therefore, $N = G$ and G is a simple l-group.□

3^0. Let $A(\Omega)$ be an l-group of all order permutations of a totally ordered set Ω and $A(\Omega)$ is o-2-transitive. It is evident that there are three nontrivial ideals of $A(\Omega)$: $B(\Omega)$, $R(\Omega)$, $L(\Omega)$.

THEOREM 4. (Higman [1].) *If a group $B(\Omega)$ of bounded permutations of a totally ordered set Ω is o-2-transitive, then $B(\Omega)$ is a simple l-group.*

Proof. It was mentioned after the proof of Theorem 2 that $B(\Omega)$ satisfies all conditions of Theorem 1. Therefore, it is enough to prove that $B(\Omega) = B(\Omega)'$. We claim that every element $g \in B(\Omega)$ is a commutator of two elements of $B(\Omega)$.

Let $g \in B(\Omega), g \neq e$ and $\alpha_0 < \mathrm{supp}(g) < \alpha_1$ for some $\alpha_0, \alpha_1 \in \Omega$. Then there are elements $\beta_0, \beta_1 \in \Omega$ such that $\beta_0 < \alpha_0, \beta_1 < \alpha_1$. From o-2-transitivity of l-group $B(\Omega)$ it follows that there exist the infinite sequences $\alpha_1 < \alpha_2 < \ldots < \alpha_n < \ldots < \beta_1$ and $\alpha_0 > \alpha_{-1} > \ldots > \alpha_{-n} > \ldots > \beta_0$. Let $\gamma_0 = \inf\{\alpha_{-n}|n \in \mathbf{N}\} \in \Omega$ and $\gamma_1 = \sup\{\alpha_n|n \in \mathbf{N}\} \in \Omega$. For any integer i let φ_i be an order isomorphism from the interval $[\alpha_i, \alpha_{i+1}]$ onto the interval $[\alpha_{i+1}, \alpha_{i+2}]$. The existence of this order isomorphism is guaranteed by o-2-transitivity of l-group $B(\Omega)$. Now, we define elements $h, \hat{g} \in B(\Omega)$ by the rules:

$$\alpha h = \begin{cases} \alpha & \text{if} \alpha \leq \gamma_0 \text{ or } \alpha \geq \gamma_1, \\ \alpha\varphi_i & \text{if } \alpha \in [\alpha_i, \alpha_{i+1}], \end{cases}$$

and

$$\alpha\hat{g} = \begin{cases} \alpha & \text{if } \alpha \leq \alpha_0 \text{ or } \alpha \geq \gamma_1, \\ \alpha(\varphi_1\varphi_2\cdots\varphi_i)^{-1}g(\varphi_1\varphi_2\cdots\varphi_i) & \text{if } \alpha \in [\alpha_{i+1}, \alpha_{i+2}],\ i \geq 0, \\ \alpha g & \text{if } \alpha \in [\alpha_0, \alpha_1]. \end{cases}$$

Now, let us consider the element $k = h^{-1}\hat{g}^{-1}h\hat{g}$. We claim that $k = g$. For $\alpha \leq \alpha_0$ is valid $\alpha k = \alpha h^{-1}\hat{g}^{-1}h\hat{g} = \xi\hat{g}^{-1}h\hat{g}$ where $\xi = \alpha h^{-1} < \alpha_1$. Then $\alpha k = \xi h\hat{g} = \alpha\hat{g} = \alpha$. If $\alpha_0 \leq \alpha \leq \alpha_1$, then $\alpha k = \xi\hat{g}^{-1}h\hat{g}, \xi = \alpha h^{-1} < \alpha_0$. Hence $\alpha k = \alpha\hat{g} = \alpha g$. If $\alpha_{i+1} \leq \alpha \leq \alpha_{i+2} (i \geq 0)$, then $\alpha k = \alpha h^{-1}\hat{g}^{-1}h\hat{g} = (\alpha\varphi_i^{-1})^{-1}\bar{g}h\bar{g} = \alpha\varphi_1^{-1}\varphi_{i-1}^{-1}\cdots\varphi_1^{-1}g^{-1}\varphi_1\varphi_2\cdots\varphi_{i-1}h\hat{g}$. From $\zeta = \alpha\varphi_i^{-1}\varphi_{i-1}^{-1}\cdots\varphi_1^{-1}g^{-1}\varphi_1\cdots\varphi_{i-1}\varphi_i \in [\alpha_i, \alpha_{i+1}]$ it follows

$$\alpha k = \zeta\hat{g} = \zeta\varphi_i^{-1}\varphi_{i-1}^{-1}\cdots\varphi_1^{-1}\varphi_1\cdots\varphi_{i-1}\varphi_i =$$
$$= \alpha(\varphi_i^{-1}\varphi_{i-1}^{-1}\cdots\varphi_1^{-1}g\varphi_1\cdots\varphi_{i-1}\varphi_i)\varphi_i^{-1}\varphi_{i-1}^{-1}\cdots\varphi_1^{-1}g\varphi_1\cdots\varphi_i = \alpha.$$

So, $\alpha k = \alpha$ for all $\alpha \notin [\alpha_0, \alpha_1], \alpha \in \Omega$ and $\alpha k = \alpha g$ for all $\alpha \in [\alpha_0, \alpha_1]$. Hence $k = g$ and $g \in B(\Omega)'$. From Theorem 1 it follows that $B(\Omega)'$ is simple.□

COROLLARY 1.(Higman [1].) *Every nontrivial normal subgroup of o-2-transitive l-group $A(\Omega)$ contains group $B(\Omega)$ of bounded order permutations of Ω .*

Proof. Let N be a normal subgroup of $A(\Omega\ )$ and $f \in N, f \neq e$. Then there exists a bounded element $g \in B(\Omega)$ such that $fg \neq gf$. Hence $[f, g] = f^{-1}gf^{-1}g \in N \cap B(\Omega)$. As $B(\Omega)$ is simple, then $N \cap B(\Omega) = B(\Omega)$.

Let Ω be a totally ordered set, $\alpha \in \Omega, g \in A(\Omega)$ and (g) be an infinite cyclic subgroup which is generated by the element g. Let $o(g, \alpha)$ be an orbital of g (see Chapter 4, Secton 4, $\mathbf{2^0}$) that is the convexification of the (g)-orbit $\alpha(g)$. It is evident that $o(g,\alpha)$ does not depend on the choice of representative $\xi \in o(g,\alpha)$. The set of all orbitals of g is totally ordered: $o(g,\alpha) \le o(g,\beta)$ for orbitals $o(g,\alpha), o(g,\alpha)$ of g iff $\xi \le \eta$ for all $\xi \in o(g,\alpha), \eta \in o(g,\beta)$. It is easy to see that $o(g,\alpha) = \alpha$ iff $\alpha g = \alpha$. In this case we say that g acts on $o(g,\alpha)$ *identically.* In other cases g acts on $o(g,\alpha)$ without fixed points. If $\alpha g > \alpha$, then $\xi g > \xi$ for all $\xi \in o(g,\alpha)$. In this case we say that g acts on $o(g,\alpha)$ *positively.* If $\alpha g < \alpha$, then we say that g acts on $o(g,\alpha)$ *negatively.*

LEMMA 1. (Holland [2].) *The elements g, f of o-2-transitive l-group $A(\Omega)$ are conjugate in $A(\Omega)$ if and only if there exists an one-to-one correspondence φ between the set of all orbitals of g and the set of all orbitals of f such that*

(1) *if $o(g,\alpha) < o(g,\beta)$, then $o(g,\alpha)\varphi < o(g,\beta)\varphi$,*
(2) *if g acts on $o(g,\alpha)$ positively (negatively, identically), then f acts on $o(g,\alpha)\varphi$ positively (negatively, identically accordingly).*

Proof. Let $f = h^{-1}gh$. For any orbital $o(g,\alpha)$ let $o(g,\alpha)\varphi = o(g,\alpha)h = \{\beta h | \beta \in o(g,\alpha)\}$. We claim that $o(g,\alpha)\varphi = o(f,\alpha h)$. If $\xi \in o(g,\alpha)$, then there exists $n \in \mathbf{Z}$ such that $\alpha g^n \le \xi \le \alpha g^{n+1}$, $(\alpha h)f^n = \alpha h(h^{-1}gh)^n = \alpha h h^{-1} g^n h = \alpha g^n h \le \xi h \le \alpha g^{n+1} h = (\alpha h) f^{n+1}$. Then $o(g,\alpha)\varphi \subseteq o(f,\alpha h)$. Analogously $o(f,\alpha h) \subseteq o(g,\alpha)\varphi$. Thus, φ maps the set of orbitals of g on the set of orbitals of f. All others properties of φ are easy to verify.

Conversely, let φ be a map between sets of orbitals of g and f which satisfies conditions 1), 2). We define an order permutation $h \in A(\Omega)$ such that $f = h^{-1}gh$. If g acts on $o(g,\alpha)$ identically, then $o(g,\alpha) = \{\alpha\}$ and f acts on $o(g,\alpha)\varphi$ identically, too, and $o(g,\alpha)\varphi = \alpha\varphi = \beta$. Then, let us suppose that $\alpha h = \beta$. If g acts on $o(g,\alpha)$ positively, then we fix $\alpha \in o(g,\alpha)$ and $\beta \in o(g,\alpha)\varphi$. Since $A(\Omega)$ is o-2-transitive on Ω, then there is $k \in A(\Omega)$ such that $\alpha k = \beta$, $\alpha g k = \beta f$. For every $\xi \in o(g,\alpha)$ there is an unique integer n such that $\alpha g^n \le \xi < \alpha g^{n+1}$. Let $\xi h = \xi g^{-n} k f^n$. It is evident that h is an order isomorphism of $o(g,\alpha)$ onto $o(g,\beta)$. Next we show that $\xi f = \xi h^{-1} g h$ for all $\xi \in o(f,\beta)$. Let $\xi \in o(f,\beta), \beta f^n \le \xi < \beta f^{n+1}$. Then $\xi h^{-1} = \xi f^{-n} k^{-1} g^n, \alpha g^n \le \xi h^{-1} < \alpha g^{n+1}$ and the element $\xi h^{-1} g = \xi f^{-n} k^{-1} g^{n+1}$ lies between the elements αg^{n+1} and αg^{n+2}. Hence $\xi h^{-1}gh = (((\xi f^{-n})k^{-1})g^{n+1})h = \xi f^{-n}k^{-1}g^{n+1}g^{-(n+1)}kf^{n+1} = \xi f$.

The case of negative action of g on $o(g,\alpha)$ is similar.

Now, the set Ω is a disjoint union of orbitals of element g and every $\xi \in \Omega$ is contained in some orbital $o(g,\alpha)$. Therefore h is defined everywhere on the set Ω and $f = h^{-1}gh$.□

Let us recall that $R(\Omega)$ is the set of all $f \in A(\Omega)$ such that support of f is bounded below, $L(\Omega)$ is defined dually and $B(\Omega) = R(\Omega) \cap L(\Omega)$.

THEOREM 5.(Holland [2].) *If a totally ordered set Ω contains a countable unbounded from above and below subset and the l-group $A(\Omega)$ is o-2-transitive on Ω, then the l-groups $A(\Omega)/B(\Omega)$ and $A(\Omega)/R(\Omega)$ are simple l-groups.*

Proof. First, we prove that $A(\Omega) = L(\Omega) \cdot R(\Omega)$. If $f \in A(\Omega)$ and $\alpha, \beta \in \Omega, \alpha < \beta$ and g is the map defined by rule:

$$\xi g = \begin{cases} \xi f, & \text{if} \quad \xi \le \alpha, \\ \xi h, & \text{if} \quad \alpha \le \xi \beta, \\ \xi & \text{if} \quad \xi \le \beta, \end{cases}$$

where h is an element of $A(\Omega)$ such that $\alpha h = \alpha f, \beta h = \beta$. Then $g \in A(\Omega), g \in L(\Omega)$ and $fg^{-1} \in R(\Omega)$.

Let $f \in R(\Omega), g \in L(\Omega), f > e, g > e$ and N be an ideal of $A(\Omega)$ such that $g \in N$. We claim that $f \in N$, too. Let $\beta \in \Omega$ and $\xi f = \xi$ for all $\xi \in \Omega, \xi \le \beta$.

First, we assume that the set of fixed points of the permutation g is bounded above by some element α . Now, we fix $h \in A(\Omega)$ such that $\alpha h = \beta$. It is clear that the set of fixed points of the element $h^{-1}gh$ is bounded by the element β and the elements $h^{-1}gh, h^{-1}gh \vee f$ have the same sets of orbitals. By Lemma 1 elements $h^{-1}gh$ and $h^{-1}gh \vee f$ are conjugate. Hence if $g \in N$ then $h^{-1}gh \vee f \in N$. From the inequality $e < f < h^{-1}gh \vee f$ it follows that $f \in N$.

Now, let us suppose that the set of fixed points of the order permutation g is unbounded above. Then there is an unbounded above sequence $\alpha_1 < \beta_1 < \alpha_2 < \beta_2 < \ldots < \alpha_n < \beta_n < \ldots$ in Ω such that every open interval $(\alpha_i, \beta_i) = \{\xi \in \Omega | \alpha_i < \xi < \beta_i\}$ is contained in some orbital $o(g,\gamma_i), \gamma_i \in \Omega$. Then $\xi g \ne \xi$ for all $\xi \in \Omega, [\alpha_i, \beta_i]$. Let $h \in A(\Omega)$ such that $\alpha_i h = \beta_i, \beta_i h = \alpha_{i+1} (i = 1, 2, ...)$. Then the orbital

$o(h^{-1}gh, \gamma_i h)$ contains the open interval (β_i, α_{i+1}) and it is clear that for all $\xi \in [\beta_i, \alpha_{i+1}], \xi \in \Omega$ is valid $\xi h^{-1}gh \neq \xi$. The subset $\{\xi \in \Omega | \xi > \alpha_1\}$ is the union of closed intervals $[\alpha_i, \beta_i]$ and $[\beta_i, \alpha_{i+1}]$. Therefore the element $g \vee h^{-1}gh$ has not fixed points greater than α_1. By previous arguments, $f \in N$. Hence l-group $A(\Omega)/L(\Omega)$ is l-simple.

Now, we claim that every normal subgroup of $A(\Omega)$ is ideal in $A(\Omega)$. Let N be a normal subgroup of $A(\Omega)$ and $e < g < f, g, f \in A(\Omega),)f \in N$. Then $\mathrm{supp}(g) \subseteq \mathrm{supp}(f)g$ and the set of orbitals for f and fg are the same. The elements f and fg act positively on all orbitals. By Lemma 1 the elements f and fg are conjugate in $A(\Omega)$. Hence there is $h \in A(\Omega)$ such that $fg = h^{-1}fh$. Therefore $g = f^{-1}h^{-1}fh$ and $g \in N$. Thus, the normal subgroup N is convex. Let now $f \in N$. Since $A(\Omega)$ is o-2-transitive, then $A(\Omega)$ is divisible by Theorem 4.4.3. Choose an element $f_2 \in A(\Omega)$ such that $f_1^2 = f \vee e$. Let $f_2 \in A(\Omega)$ be defined by the rule:

$$\xi f_2 = \begin{cases} \xi f & \text{if } \xi f \leq \xi, \\ \xi f_1 & \text{if } \xi f \geq \xi. \end{cases}$$

The permutations f_2, f have the same sets of orbitals and the same actions on all orbitals. Hence the elements f_2 and f are conjugate and $ff_2^{-1} \in N$. The elements $ff_2^{-1}, f \vee e$ have the same sets of orbitals. Their action on the sets of orbitals is positive. By Lemma 1, $ff_2^{-1}, f \vee e$ are conjugate. Therefore, $f \vee e \in N$ and N is an l-subgroup of $A(\Omega)$. Now it is clear that the group $A(\Omega)/L(\Omega)$ is a simple group because $A(\Omega)/L(\Omega)$ is a simple l-group.□

In the book of A.M.W.Glass [6] it is proved that if $A(\Omega)$ is o-2-transitive and Ω has a countable unbounded above and below subset, then $R(\Omega), L(\Omega)$ and $B(\Omega)$ are the only proper normal subgroups of $A(\Omega)$. In contrast, this book gives an example of totally ordered set Ω such that $A(\Omega)$ is o-2-transitive and $A(\Omega)$ has a proper normal subgroup which is different from $R(\Omega), L(\Omega), B(\Omega)$. In the case when every countable subset of Ω has an upper bound, the situation is more complex. This problem was investigated in the works of R.N.Ball [1], R.N.Ball and M.Droste [1], M.Droste and S.Shellah [1].

The simple l-groups $B(\mathbf{R})$ and $L(\mathbf{R})/B(\mathbf{R})$ are not isomorphic since every positive element of $B(\mathbf{R})$ is disjoint from one of its conjugates, while this is not true for $L(\mathbf{R})/B(\mathbf{R})$. S.McCleary in [5] proved that simple l-groups $A(\mathbf{R})/L(\mathbf{R})$ and $A(\mathbf{R})/R(\mathbf{R})$ are not isomorphic or even elementarily equivalent.

4.7 Examples of Simple l-Groups

1^0. Now we apply the previous results to some concrete groups. Let us recall that a totally ordered field is a field K with a total order $\leq$ which is compatible with $+$ and for $a, b \in K$ such that $a \leq b, \quad c \geq 0$ it valid $ac \leq bc$.

PROPOSITION 1. *The group of all bounded order permutations of totally ordered field K is o-2-transitive.*

Proof. Let K be a totally ordered field and $a, b, c, d \in K, \quad a < b, \quad c < d$. Then the function $f : f(x) = (d-b)(c-a)^{-1}(x-a)+b$ is an order permutation of totally ordered set K. Now it is easy to construct a finitely piecewise linear function $g \in A(K)$ with bounded support such that g coincides with f on the closed interval $[a, c]$.□

EXAMPLE 1. (Dlab [1].) Let $\mathbf{R}$ be a totally ordered field of real numbers and Γ be the set of all well-ordered sequences $\{a_\xi | \xi \leq \tau, \quad a_\xi \in \mathbf{R}\}$ of all order types having the greatest elements and such that $a_\sigma = \sup\{a_\xi | \xi < \sigma\}$ for every limit ordinal $\sigma, \quad \sigma \leq \tau$. For every sequence $\{a_\xi | \xi \leq \tau, \quad a_\xi \in \mathbf{R}\}$ in Γ we denote a_0 the first member and a_τ the last member of this sequence. Let $D = D(\mathbf{R})$ be the set of all order permutations of $\mathbf{R}$ such that for $f \in D$ there exists the sequence $\{a_\xi | \xi \leq \tau\}$ in Γ such that f is a linear function on every closed interval $[a_\xi, a_{\xi+1}]$ and $f(x) = x$ for all $x \notin [a_0, a_\tau]$.

It is easy to see that D is an l-subgroup of $A(\mathbf{R})$, it is o-2-transitive and every element of D is bounded. We claim that every element of D is a commutator in D.

Let the closed interval $[a_0, a_\tau]$ be convexification of supp(f) where $f \in D$ and $a_0 = \inf \operatorname{supp}(f), \quad a_\tau = \sup\{\operatorname{supp}(f)\}$. As in the proof of Theorem 4.6.4, there are points $b, b_i \in \mathbf{R} \quad (i = 1, 2, ...)$ such that $b_i < b_{i+1}, \quad b_1 = a_0, \quad b_2 = a_\tau, \quad b = \sup\{b_i | i = 0, 1, 2, ...\}$. Now, we define element $g \in D$ by the rule:

$$g(x) = \begin{cases} x & \text{if } x \notin [b_0, b], \\ \frac{(b_2-b_0)(x-b_0)}{(b_1-b_0)} + b & \text{if } x \in [b_0, b_1], \\ \frac{(b_{i+2}-b_{i+1})(x-b_i)}{(b_{i+1}-b_i)} + b_{i+1} & \text{if } x \in [b_i, b_{i+1}], i \geq 1. \end{cases}$$

Let $h \in A(\mathbf{R})$ such that

$$h(x) = \begin{cases} x & \text{if } x \notin [b_1, b], \\ g^k(f^\varepsilon(g^{-k}(x))) & \text{if } x \in [b_{k+1}, b_{k+2}], \quad \varepsilon = (-1)^k, \quad k \in \mathbf{Z}. \end{cases}$$

It is clear that $h \in D$ and $h(x) = f(x)$ for $x \in [b_1, b_2]$, $f = g^{-1}h^{-1}gh$. Hence $D = D'$ and D is a simple group by Theorem 4.6.1. We denote this group by D_1. It is clear that D_1 is not totally ordered. But, it is possible to define a total order on the group D.

EXAMPLE 2. (Dlab [1].) Let D be a group from Example 1. We define an order on D via: $f \geq e$ for $f \in D$ if $f(x) = x$ for all $x \leq a$ and $(f(b) - f(a))(b-a)^{-1} > 1$, where $[a, b]$ is the first interval of supp(f) and f is linear on this interval, $f(x) \neq x$ for $x \in [a, b]$. In other words, we order D by calling an element f $(f \in D)$ positive if its rightmost nonidentity linear piece has slope greater than 1.The direct calculations show that $P = \{f \in D | f \geq e\}$ is a total order on D. We will denote the group D by this total order by D_0.

The group D_1^0 of all $f \in D$ such that f has only finite number linear pieces is an l-subgroup of $A(\mathbf{R})$ and it is also an orderable group. It was proved by C.Chehata [1] that D_1^0 is a simple group. This fact follows from Theorem 4.6.1, too. We need only to prove by direct calculation that the group D_1^0 is generated by the functions with two linear pieces and every function with two linear pieces is a commutator of some elements of D_1^0. Another proof of this fact (with direct calculations, too) can be found in the book of A.Kokorin and V.Kopytov [2].

EXAMPLE 3. (Holland [3], McCleary [5].) Let $ZP(\mathbf{R}) = \{f \in A(\mathbf{R}) |$ there is a positive integer n such that $f(x+n) = f(x)+n\}$. It is evident that $ZP(\mathbf{R})$ is an l-subgroup of $A(\mathbf{R})$. We claim that $ZP(\mathbf{R})$ is a simple group.

For each $a \in \mathbf{R}$, let $\mathcal{F}(a)$ be the filter of neighborhoods of a (with respect to standard order topology on $\mathbf{R}$). Let $G(a)$ be the set of all $g \in$

$ZP(\mathbf{R})$ such that $g(x) = x$ for all $x \in X$, where X is some neighborhood of a : $G(a) = \bigcup_{X \in \mathcal{F}(a)} ZP(\mathbf{R})_X$ where $ZP(\mathbf{R})_X$ is a pointwise stabilizer of X in $ZP(\mathbf{R})$ (see Chapter 2, Section 2). It is clear that $G(a)$ is an l-subgroup of $ZP(\mathbf{R})$ for every $a \in \mathbf{R}$.

We claim that the group $ZP(\mathbf{R})$ is generated by the subgroups $G(a), a \in \mathbf{R}$. All functions f :

$$f(x) = x + \frac{1}{n}$$

are in $ZP(\mathbf{R})$, then it follows that $ZP(\mathbf{R})$ is generated by functions $g > e$ such that $0 \leq g(0) \leq \frac{1}{2}$. If g has period n, then $g(n) = g(0) + n$. Let us suppose that permutation g does not stabilize any interval on $[0, n]$. Then we fix points c_1, c_2, c_3, c_4 from $[0, n]$ such that $0 < c_1 < g(c_2) < \frac{1}{2} < c_2 < c_3 < c_4 < n$ and define the function $g_1 \in ZP(\mathbf{R})$ by the rule:

$$g(x) = \begin{cases} g(x) & \text{if } 0 \leq x \leq c_1, \\ (g(c_1) - c_2)(c_1 - c_2)^{-1}(x - c_2) + c_2 & \text{if } c_1 \leq x \leq c_2, \\ x & \text{if } c_2 \leq x \leq c_3, \\ (g(c_4) - c_3)(c_4 - c_3)^{-1}(x - c_3) + c_3 & \text{if } c_3 \leq x \leq c_4, \\ g(x) & \text{if } c_4 \leq x \leq n, \\ g(x - kn) + kn & \text{if } kn \leq x \leq (k+1)n. \end{cases}$$

It is clear that the functions g_1 and gg_1^{-1} stabilize some intervals on $\mathbf{R}$. Hence $g = g_1^{-1}g_1 \in G(a)$.

Now we will prove that for every $g \in ZP(\mathbf{R})$ the subgroup $H = (g)^{ZP(\mathbf{R})} = \text{gp}h^{-1}gh|h \in ZP(\mathbf{R})$ contains $G(a)$ for any $a \in \mathbf{R}$. Let $a \in \mathbf{R}$. If $g(a) = a$ then $H \cap ZP(\mathbf{R}) \neq E$. If $g(a) > a$ then there are $s \in \mathbf{R}$ and integers p, q such that $s \neq g^{-1}(a)$, $s < a < s + n$ where n is a period of g, $s < a < g^{-1}(a + np)$, $g(a + np) < s + np$. There are $b, c \in \mathbf{R}$ such that $s < b < c < s + n$, $b < g(a + np) < c$ and $a, g^{-1}(a + np)$ are not contained in open interval (b, c). Let $b_1, c_1 \in \mathbf{R}$ and $b < b_1 < g(a+np) < c_1 < c$. There is a positive element $h \in ZP(\mathbf{R})$ such that $h(b_1) = c_1$, $h(x) = x$ for all $x \in (s, b) \cup (c, s + n)$. Then for $h_1 = g^{-1}h^{-1}gh$ is valid $h(a) + pn = h(a + pn) = h(g(h^{-1}(g^{-1}(a + pn)))) = h(g(g^{-1}(a + pn))) = h(a + pn) = h(a) + pn = a + pn$. So, $h(a) = a$ and $H \cap ZP(\mathbf{R})_a \neq E$ for all $g \in ZP(\mathbf{R})$.

Now, let $f_1 \in G_a$ and m be a period of f_1. Then f_1 stabilizes some segment $[u, v]$ containing a. Let t be the least common multiple of n and m (n is a period of g), then the representation $\bar{f}_1$ of the permutation f_1 on $[a, a+t]$ is an order permutation from $B(\mathbf{R})$. By Theorem 4.6.5, it follows that normal subgroup of $A(\mathbf{R})$ which is generated by the element $\bar{f}_1$ contains the subgroup $B(\mathbf{R})$ and hence the element $\bar{f}_1$ is contained in normal subgroup $H = (g)^{ZP(\mathbf{R})}$. Therefore $ZP(\mathbf{R})$ is a simple group.□

EXAMPLE 4. (McCleary [4].) Let $GP(\mathbf{R})$ be the group of all order permutations f of $\mathbf{R}$ such that for every $a \in \mathbf{R}$ there is a positive integer $n = n(f, a)$ such that $f(a + nk) = f(a) + nk$ for all integers k. It is clear that $ZP(\mathbf{R}) \subseteq GP(\mathbf{R})$ and $GP(\mathbf{R})$ is a pathological l-group. We claim that $GP(\mathbf{R})$ is a simple l-group.

It is easy to see that every element $f \in GP(\mathbf{R})$ is bounded from above by some linear function $x+c$. Let us suppose that N is a nontrivial ideal of $GP(\mathbf{R})$, $f \in GP(\mathbf{R})$, $f > e$ and $a \in \operatorname{supp}(f)$. Let k be a linear function such that $k(x) = x + f(a) - a$ and p a natural number such that $p \cdot (f(a) - a) > n(f, a)$. For every integer q is valid $k^{-1}(f(a + q \cdot n(f, a))) = k^{-1}(a + q \cdot n(f, a)) = f(a) + q \cdot n(f, a) - f(a) \neq a = a + q \cdot n(f, a)$. Let $g = (fk^{-1})^p k^p = f \cdot k^{-1} f k \cdot k^{-2} f k^2 \cdots k^{-p+1} f k^{p-1} \in N$, then $g(a) = a + p \cdot (f(a) - a)$ and $g(a + q \cdot n(f, a)) = k^p(a + q \cdot n(f, a)) = a + q \cdot n(f, a) + p \cdot (f(a) - a) = g(a) + q \cdot n(f, a)$. It follows that $n(g, a) \leq n(f, a)$ and $a + n(g, a) < a + n(f, a) < a + p \cdot (f(a) - a) = g(a)$. Let $c = g(a) - (a + n(g, a))$, $u(x) = x + c$. It is evident that $u \in GP(\mathbf{R})$.

For arbitrary $y \in GP(\mathbf{R})$ there is an integer q such that $a + q \cdot n(g, a) \leq y < a + (q + 1) \cdot n(g, a)$. Then $u(y) = y + c = y + g(a) - a - n(g, a) < a + (q + 1) \cdot n(g, a) + g(a) - a - n(g, a) = a + q \cdot n(g, a) = g(a + q \cdot n(g, a)) \leq g(y)$ and $e \leq u \leq g$ in $GP(\mathbf{R})$. Thus, the linear function $u(x)$ is contained in N. It is easy to see that every linear function is contained in N, too. As every positive element of $GP(\mathbf{R})$ is bounded above by suitable linear function then $N = GP(\mathbf{R})$. Therefore $GP(\mathbf{R})$ is a simple l-group.□

3⁰. All Examples 1-4 of simple l-groups are pathological. Of course, not every pathological l-group is ac simple group. For example, l-group $P(1, \mathbf{R})$ of all ordered permutations $f \in A(\mathbf{R})$ such that $f(x + 1) = f(x) + 1$ is pathological. The function $g : g(x) = x + 1$ is a central element in $P(1,\mathbf{R})$ hence $P(1,\mathbf{R})$ is not a simple group. But this group is periodic, o-2-transitive and by Theorem 4.6.3, it is a simple l-group.

Now, we prove that there are no simple pathological l-groups of order permutation of totally ordered sets.

EXAMPLE 5. (Glass, McCleary [1].) Let $ZP(\mathbf{R})_0$ be the stabilizer subgroup of the point 0 in the group $ZP(\mathbf{R})$ from Example 3 and $ZP(\mathbf{R}^+)_0$ be the restriction of the l-group $ZP(\mathbf{R})_0$ on the subset of positive reals. It is clear that the l-group $ZP(\mathbf{R}^+)_0$ is o-2-transitive on $\mathbf{R}^{+1}$ and pathological. But $ZP(\mathbf{R}^+)_0$ is not a simple l-group because of the restriction of the subgroup G_0 (see Example 3) on $\mathbf{R}^+$ is a nontrivial ideal of $ZP(\mathbf{R}^+)_0$.

Now we know a lot of pathological non-simple ordered permutation l-groups. The first example of these lattice-ordered groups was given by A.M.W.Glass [2]. Other examples can be found in the work of Glass and McCleary [1].

Chapter 5

Right-ordered groups

5.1 Definition of Right-Ordered Group

A *right-ordered* (*ro-group*) group G is a group that is at the some time a totally ordered set under relation $\leq$ such that $xz \leq yz$ whenever $x \leq y$. If group G can be right-ordered, then G is called *right-orderable* group (RO-*group*). Of course, every o-group a is ro-group and every O-group is a RO-group.

By analogy we can define a left-ordered group G as a group that is at the some time a totally ordered set and the order $\leq$ is preserved by multiplication on the left ($zx \leq zy$ whenever $x \leq y$). The theory of left-ordered groups is analogous to that of right-ordered groups because the antiisomorphism $\varphi : x \to x^{-1}$ transforms a right-ordered group into a left ordered group and vice versa. The theory of ro-groups has a lot applications in the l-group theory because the class of right-orderable groups coincides with the class of groups which are embedded into lattice ordered groups and every lattice order of group G is the intersection of some family of right orders of G.

1^0. The following characterization of right orders on group G is analogous to Theorem 2.1.1. If G is a ro-group then $P = \{x \in G | x \geq e\}$ is a *positive cone* of G.

THEOREM 1. *The positive cone P of ro-group G has the following properties:*

(1) $P \cdot P \subseteq P$ (*P* is a semigroup),
(2) $P \cap P = \{e\}$ (*P* is a pure subset),
(3) $P \cup P^{-1} = G$ (*P* is a linear subset).

Conversely, if there is a pure linear subsemigroup P in group G, then it is possible to transform G into a ro-group such that P will be the positive cone of G.□

Similarly to po-groups, $x \leq y$ in ro-group G iff $yx^{-1} \in P$ where $P = P(G)$ is a positive cone of G. We identify the right order on the group G with its positive cone $P(G)$. Theorem 1 ensures that this identification is correct.

A subgroup H of ro-group G is convex if H is a convex subset in totally ordered set G. A homomorphism φ from ro-group G into ro-group H is a homomorphism from group G into group H such that $x\varphi \leq y\varphi$ for all $x, y \in G$, $x \leq y$.

THEOREM 2. *A quotient group G/N of ro-group G by a convex normal subgroup N can be right ordered such that the natural homomorphism of group G onto group G/N is a homomorphism of ro-groups. If φ is a homomorphism from ro-group G into ro-group H, then the kernel N of its homomorphism is a convex normal subgroup of ro-group G and there is an isomorphism ψ from the naturally right-ordered quotient group G/N into H such that $x\varphi = (xN)\psi$ for all $x \in G$.*

The proof of this theorem is similar to those of Theorems 2.3.2 and 2.3.3 and can be easily restored.□

2^0. In some cases the usage of ro-groups is easier than of totally ordered groups. The following statement illustrates this fact.

PROPOSITION 1. *If a normal subgroup N and a quotient group $\bar{G} = G/N$ of group G are ro-groups with positive cones P_N and $\bar{P}$ respectively, then there is a right order on G such that $P(G) \cap N = P_N$, the subgroup N is convex , the right order P of G induces on the quotient group G/N a right order $\bar{P}$.*

We leave the details for the reader.□

COROLLARY 1. *Let $S(G)$ be complete with respect to joins and intersections and totally ordered by set-theoretical inclusion system of subgroup of group G. If for every jump $A \prec B$ in $S(G)$, the subgroup A is normal in B and quotient group B/A is a ro-group, then there is a* *right order P on G such that all subgroups $H \in S(G)$ are convex with respect P and P induces on all quotient groups B/A of jumps $A \prec B$ in $S(G)$ the given right orders.*

Proof. Let $\{A_\alpha \prec B_\alpha | \alpha \in I\}$ be the set of all jumps of the system $S(G)$. Then A_α is a normal subgroup of B_α and the quotient group B_α/A_α is a ro-group with the positive cone P_α. Now, we define an order P on group G by the rule: if $x \in B_\alpha \setminus A_\alpha$ where $A_\alpha \prec A_\alpha$ the is suitable jump in $S(G)$, then we put $x \in P$ iff either $xA_\alpha \in P_\alpha$ or $x = e$.

The properties (1),(2),(3) of Theorem 1 are verified easily.□

COROLLARY 2. (Zaitseva [1].) *If group G has a system $S(G)$ of subgroups which is complete subnormal totally ordered by inclusion and quotient groups of all jumps of which are Abelian torsion-free groups, then the group G is right-orderable group.*

Proof. By Proposition 3.3.8, every Abelian torsion-free group is a O-group. By Corollary 1, G is a right-orderable group. □

3^0. Now we will indicate necessary and sufficient conditions for the right orderability of a group. For any subset M of group G, let sgr(M) be a subsemigroup of G generated by M.

We say that the partial order $\leq$ on group G is a *partial right order* on G if it is preserved by multiplication on the right ($xz \leq yz$ whenever $x \leq y$) . Every partial right order is defined by positive cone $P = \{x \in G | x \leq e\}$ uniquely.

LEMMA 1. *A partial right order P of group G can be extended to the total right order if and only if it satisfies the following property.*

(4) *For every finite set of nonidentity elements $x_1, \ldots, x_n$ in G , sgr$(P \setminus \{e\}, x_1^{\varepsilon_1}, \ldots, x_n^{\varepsilon_n})$ does not contain e for some choice of signs $\varepsilon_1, \ldots, \varepsilon_n$: $\varepsilon_i = +1$ or -1 $(i = 1, 2, \ldots, n)$.*

Proof. If partial right order P can be extended to the total right order Q, then choose signs ε_i such that $x_i^{\varepsilon_i} \in Q$. It is clear that sgr$(P \setminus E, x_1^{\varepsilon_1}, \ldots, x_n^{\varepsilon_n}) \subseteq Q \setminus E$.

Next we prove the second part of the lemma and obtain the sufficientl condition from it. Suppose, if possible, that P satisfies (4) but neither sgr$(P \setminus E, x)$ nor sgr$(P \setminus E, x^{-1})$ satisfies (4) for some $x \in G$. Then G contains nonidentity elements $y_1, \ldots, y_n, z_1, \ldots, z_m$ such that

$e \in \operatorname{sgr}(P \setminus E, x, y_1^{\varepsilon_1}, \dots y_n^{\varepsilon_n})$ and $e \in \operatorname{sgr}(P \setminus E, x^{-1}, z_1^{\delta_1}, \dots, z_m^{\delta_m})$ for every choice of signs ε_i, δ_j. Thus,

$$e \in \operatorname{sgr}(P \setminus E, x_1^{\varepsilon_1}, \dots, x_n^{\varepsilon_n}, z_1^{\delta_1}, \dots, z_m^{\delta_m})$$

for all choices of signs $\varepsilon_1, \dots \varepsilon_n$. This contradicts condition (4). Since (4) is satisfied by the union of the ascending chain of partial right orders provided it is satisfied by the members of the chain, there exists a maximal right partial order Q that is an extension of P and satisfies (4). If partial right order Q is not total order on G, then there is $x \notin Q \cup Q^{-1}$. Then for some sign $\varepsilon = \pm 1$ the subsemigroup $\operatorname{sgr}(Q \setminus E, x^{\varepsilon})$ satisfies (4). But $Q \subset \operatorname{sgr}(Q \setminus E, x^{\varepsilon}) \cup E$. This contradicts maximality of Q.□

THEOREM 3. (Conrad [1]) *A group G is right-orderable if and only if for any finite set of nonidentity elements $x_1, \dots, x_n$ in G the subsemigroup* $\operatorname{sgr}(P \setminus E, x_1, \dots x_n)$ *does not contain e for some choice of signs $\varepsilon_1, \dots \varepsilon_n : \varepsilon_i = \pm 1$ $(i = 1, 2, \dots, n)$.*

Proof.Theorem 3 follows from Lemma 1 if we set $P = E$.□

COROLLARY 3. *Group G is right-orderable if and only if every finite generated subgroup of G is right-orderable.*□

3⁰. Let RO be a class of RO-groups. Group G is called *locally* RO-*indicable* if every finitelly generated non-trivial subgroup of G can be mapped homomorphically onto a nontrivial RO-group.

THEOREM 4. (Burns and Hale [1].) *If group G is locally* RO-*indicable, then it is a RO-group.*

Suppose that it is not so and G is a counter-example. By Theorem 3, we can assume that $G = \operatorname{gp}(g_1, \dots, g_n)$ and the number n of generators is minimal. Then $e \in \operatorname{sgr}(g_1, \dots, g_n)$ for all choices of signs $\varepsilon_1, \dots \varepsilon_n : \varepsilon_i = \pm 1$ $(i = 1, 2, \dots, n)$ and there is a normal subgroup K in G such that G/K is a nontrivial RO-group. Since G/K is right orderable, then at least one of g_i belongs to K and since G/K is non-trivial not all of them do. Let $g_1, \dots, g_r \notin K$, $g_{r+1}, \dots, g_n \in K$ $(0 < r \leq n)$ and $\delta_1, \dots, \delta_r$ are signs, $\delta_i = \pm 1$ such that $\operatorname{sgr}(g_1, \dots, g_r) \cap K = \emptyset$. By the minimality of n there are δ_i $(i = r+1, \dots, n)$ such that $e \notin \operatorname{sgr}(g_{r+1}, \dots, g_n)$. But $e \in \operatorname{sgr}(g_{r+1}, \dots, g_n)$, hence there are positive integers $k_1, \dots, k_s$ such that $e = g_{i_1}^{k_1 \delta_{i_1}} \cdots g_{i_s}^{k_s \delta_{i_s}}$, $(1 \leq i_j \leq n)$ and at least one of the $i_j \leq r$. Therefore $K = g_1^{k_1\delta_{i_1}} \cdots g_{i_s}^{k_s\delta_{i_s}} K$, which contradicts that $\mathrm{sgr}(g_1^{\delta_1} \ldots g_r^{\delta_r}) \cap K = \emptyset$.□

Group G is *locally indicable* if every finitely generated non-trivial subgroup of G can be mapped homomorphically onto infinite cyclic group.

COROLLARY 4. (Burns and Halle [1].) *If group G is locally indicable, then it is a RO-group.* □

5.2 Representation of Right-Ordered Groups by Order Permutations

1^0. The following result is very important for the applications of RO-group theory to lattice ordered groups.

THEOREM 1. (Cohn [1], Zaitseva [1], Conrad [1].) *A group G is right-orderable if and only if it is isomorphic to a subgroup of the group of order permutations of some totally ordered set.*

Proof. Let G be a right-ordered group. Then the right regular representation $R : G \to A(G)$, where $xR(g) = xg$ for all $x \in G$ is an isomorphic embedding of the group G into the group of order permutations $A(G)$ of the totally ordered set G.

For the converse it is enough to right order arbitrary ordered permutation group $A(\Omega)$ of any totally ordered set Ω. At first we well order the set Ω in a certain way $\prec$:

$$\xi_1 \prec \xi_2 \prec \ldots \prec \xi_i \prec \xi_{i+1} \prec \ldots$$

In the group $A(\Omega)$ we consider the subset $P = \{g \in A(\Omega) | (\xi_i)g > \xi_i$ for the first ξ_i under $\prec$ in $\mathrm{supp}(g) \cup \{e\}\}$. It follows from standard arguments that P is a pure linear subsemigroup in $A(\Omega)$ and $A(\Omega)$ is right ordered.□

COROLLARY 1. *A class of all right orderable groups coincides with the class of groups which are embeddable into l-groups.*□

This statement follows from Theorem 1 and Holland's embedding theorem (Theorem 4.1.1). Theorem 1 was refined by Smirnov [7].

We recall some notations of the module theory. Let K be associative ring with the unit e. Unitary right K-module M is a *free right K-module* if there is a subset $X \subset M$ such that every element $a \in M$ can be uniquely expressed in the form $a = x_1k_1 + \cdots + x_nk_n$ where $x_1, \ldots, x_n \in X$, $k_1, \ldots k_n \in K$. The subset X is a basis of M over K. Let $\mathbf{Z}[G]$ be a group ring over the ring $\mathbf{Z}$ of integers. Any right module over $\mathbf{Z}[G]$ is called *G-module.* If M is a right module over ring K, the additive group of M is totally ordered and this order is preserved under multiplication by elements of some subset $\Gamma \subset K$, then M is called *Γ-ordered K-module.* If $K = \mathbf{Z}[G]$, $\Gamma = G$, then M is called for short *G-ordered module.*

THEOREM 2. (Smirnov [7].) *Let G be a right-ordered group. Then G is isomorphically embedded into the group of order permutations of some totally ordered free Abelian group.*

Proof. Let G be a right-ordered group, A be the additive group of group ring $\mathbf{Z}[G]$. It is evident that A is a free Abelian group with the basis G. We claim that A is G-ordered $\mathbf{Z}[G]$-module. An arbitrary element $a \in A, a \neq 0$ has unique representation $a = n_1g_1 + \cdots + n_sg_s$ where $g_1, \ldots, g_s \in G$, $g_1 < \ldots < g_s$ in right order of group G and $n_1, \ldots, n_s$ are nonzero integers. Set $a \in P$ if $n_s > 0$. Direct calculations show that $P \cup E$ defines on A a total order and this order is preserved under multiplication to the right by elements of G. Let R is the mapping which corresponds to every element $g \in G$ the right multiplication $R(g) : xR(g) = xg$ for all $x \in A$. Then R is a isomorphism of the group G into the group of order automorphisms of the totally ordered Abelian group A.$\square$

2^0. Now we investigate some properties of right-ordered groups which are very useful for its applications to the theory of totally ordered groups and lattice ordered groups. Let G be a group and Γ be a group of endomorphisms of G. The group G is a *Γ-orderable* group if there is a total order $\leq$ of group G and this total order is preserved under the actions of the endomorphisms of Γ.

LEMMA 1. (Smirnov [7].) *Let F be a non-abelian free group, A be a normal subgroup of F. Then quotient group F/A is right-orderable if and only if $F/[A, A]$ is $F/[A, A]$-orderable.*

Proof. Let us suppose that group $G = F/[A, A]$ is right-orderable. We define the right multiplication of elements of free Abelian group $M = A/[A, A]$ by elements of G by the rule: $ag = f^{-1}af$, where $a \in M$ and $g = Af$, $Af \in G$. Standard arguments shows that M is right G-module. By Magnus's theorem (see the book of M. Hall [1]), the module M is a submodule of the free G-module S. Every free G-module is a direct sum of free cyclic G-modules and every cyclic G-module is isomorphic to the additive group of the integral group ring $\mathbf{Z}[G]$. From right orderability of group G it follows that G-module $\mathbf{Z}[G]$ is G-orderable. Hence G-modules S and M are G-orderable, too. Therefore, free Abelian group $A/[A, A]$ has a total order which is invariant under inner automorphisms of the group $A/[A, A]$.

Conversely, let the group $A/[A, A]$ have some total order P and P is invariant under inner automorphisms of the group $F/[A, A]$. we claim that the group F/A is right-orderable. Let C be a centralizer of $A/A[A, A]$ in $F/[A, A]$. From $F/[A, A]$-orderability of the group $A/[A, A]$ it follows that the group F/C is right-orderable. We show that $C = A$. Let c be any element of C and $H/[A, A]$ is a subgroup of $F/[A, A]$ which is generated by A and c. It is evident that $H/[A, A]$ is Abelian. Therefore $[H, H] = [A, A]$ in free nonabelian group F and by result of Auslender and Lindon (see the book of H.Neumann [1]) it follows that $H = A$. Hence $c \in A$ and $C = A$.□

THEOREM 3. (Smirnov [7].) *Let F be a noncommutative free group, A be a normal subgroup of F and F/A have an infrainvariant system $S(F/A)$ of subgroups with torsion free Abelian quotient groups of jumps of $S(F/A)$. Then the group $F/[A, A]$ is orderable.*

Proof. Let us suppose that F is a free group with two or more free generators and $A \neq E$. By hypothesis there exists an infrainvariant system $S(F)$ of subgroup F which connects A to F:

$$F = F_1 \supset \ldots \supset F_\alpha \supset F_{\alpha+1} \supset \ldots \supset F_\xi = A,$$

which torsion free Abelian factors $F_\alpha/F_{\alpha+1}$. Let $F'_0 = F$ and $F'_\alpha = [F_\alpha, F_\alpha]$ for all $\alpha \neq 0$. By the result of Dunwoody (see Dunwoody [1]) the system $S'(F)$:

$$F = F_1 \supset F'_1 \supset \ldots \supset F'_\alpha \supset F'_{\alpha+1} \supset \ldots \supset F'_\xi = [A, A]$$

is closed under arbitrary joins and intersections of its members. As F is a free group, then $F_\alpha \supset F_{\alpha+1}$ implies $F'_\alpha \supset F'_{\alpha+1}$, $F'_\alpha \neq F'_{\alpha+1}$. Thus $S(F)'$ does not contain repetitions. From $F_\alpha \supset F_{\alpha+1} \supseteq F'_\alpha \supset F'_{\alpha+1}$ it follows that $F'_\alpha/F'_{\alpha+1}$ if a free Abelian group. Let N_α be a normalizer of F'_α in F. Then N_α is a normalizer of F'_α, too. By Theorem 3.2.4, we need only to prove that for all α a quotient group $F'_\alpha/F'_{\alpha+1}$ has an invariant under inner automorphisms of N_α total order. For $\alpha = 0$, $F'_0 = F \supset F'_1 = F'$ and this statement is evident. If $\alpha \neq 0$, then $F \supseteq N_\alpha \supset F_{\alpha+1} \supseteq F'_\alpha \supset F_{\alpha+1}$. Let $N(F_\alpha)$ be a normalizer of F_α in F. We claim that $N(F_\alpha) = N_\alpha$. Since F'_α is fully invariant in F_α, then $N(F_\alpha) \supseteq N_\alpha$. To prove the converse, let $g \in N_\alpha$. Then $F_\beta = g^{-1}F_\alpha g \in S(F)$. Now, by the result of Auslender and Lyndon, from equality $F'_\beta = g^{-1}F_\alpha g = F_\alpha$ it follows that $F_\beta = F_\alpha$ and $g \in N(F_\alpha)$.

So, N_α is a normalizer in F of subgroups F_α, $F_{\alpha+1}$, F'_α, $F'_{\alpha+1}$. By hyporhesis, there exists an infrainvariant system of subgroup connecting $F_{\alpha+1}$ to F with torsion-free Abelian factors. Then the quotient group $N_\alpha/F_{\alpha+1}$ has a soluble subnormal system of subgroups with torsion-free quotient groups. By Corollary 4.1.2, quotient group $N_\alpha/F_{\alpha+1}$ is right-orderable and by Lemma 1, quotient group $F_{\alpha+1}/F'_{\alpha+1}$ has an invariant under inner automorphisms of N_α total order. Therefore, quotient group $F'_\alpha/F_{\alpha+1}$ has an invariant under inner automorphisms of N_α total order, too.□

COROLLARY 2 (Smirnov [7].) *If F is a free group, A is a normal subgroup of F and F/A has an infrainvariant soluble system of subgroups, then F/A' is right-orderable and F/A'' is orderable.*
Here as usual $A' = [A, A]$, $A'' = [A', A']$.

Proof. Let $S(F)$:

$$F = F_1 \supset \ldots \supset F_\alpha \supset F_{\alpha+1} \supset \ldots F_\xi = A$$

be a soluble infrainvariant system of subgroups containing A to F. Then $S'(F)$:

$$F'_0 = F \supset F'_1 \supset \ldots \supset F'_\alpha \supset F'_{\alpha+1} \supset \ldots \supset F_\xi = A'$$

is soluble infrainvariant system of subgroups with torsion-free Abelian factors connecting A' to F. By Corollary 1, the quotient group F/A' is right-orderable and by Theorem 3, quotient group F/A'' is orderable.□

5.3 Convex Subgroups in Right-Ordered Group

It is clear that the system of convex subgroups of right-ordered group is totally ordered by inclusion. Properties of the system of convex subgroups of ro-group and o-groups are some times quite different.

1⁰. The following Smirnov's example establishes the existence of non-abelian ro-groups without non-trivial convex subgroups.

EXAMPLE 1 (Smirnov [7].) Let G be a group of all matrices of the form

$$\begin{pmatrix} k & a \\ 0 & 1 \end{pmatrix}$$

where k, a are rational numbers and $k > 0$. We denote this matrice compactly by (k, a). For any positive irrational number ε, let P denote the subset of all matices $(k, a) \in G$ such that $k + a\varepsilon \geq 1$. If (k_1, a_1), $(k_2, a_2) \in P$, then $(k_1, a_1)(k_2, a_2) = (k_1k_2, k_1a_2 + 1)$, $k_1k_2 + \varepsilon(k_1a_1 + a_1) = k_1(k_2 + \varepsilon a_2) + \varepsilon a_1 \geq 1$ and so P is a semigroup. It is clear that $P \cap P^{-1} = \{(1, 0)\}$ and P defines a right total order on G by the rule:

$$(k_1, a_1) \leq (k_2, a_2) \text{ if and only if } (k_2, a_2)(k_1, a_1)^{-1} \in P.$$

We claim that G does not have any non-trivial convex subgroup. Let A be the subgroup of all matrices of the form $(1, a)$, $a \in \mathbf{Q}$. The right order P induces a total order on A and $(1, a) > (1, 0)$ if and only if $a > 0$. It is evident that this total order on A is Archimedean and A has not non-trivial convex subgroups. For every $g = (k, a) > (1, 0)$ there are rational numbers x, y $\;x > (1 - (k^{-1} - k^{-1}a\varepsilon))\varepsilon^{-1}$, $0 < y < (k + a\varepsilon^{-1})k^{-1}\varepsilon^{-1}$. Therefore, for matrices $(1, x)$, $(1, y)$ the following inequalities are valid, $(1, x) \geq (k, a) \geq (1, y) > (1, 0)$. If C be a convex subgroup of G then $C \cap A$ is a convex subgroup of Archimedean totally ordered group A. Hence $C \cap A = A$ or $C \cap A = E$. If $A \subseteq C$, then there is an element $g > (1, 0)$, $g \notin A$. By above arguments, there is an element $(1, x) \in A$ such that $(1, x) > g > (1, 0)$ which contradicts to our assumption and $C = G$. If $A \cap C = E$, then for any element $c \in C$, $c > (1, 0)$ there is an element $(1, y) \in A$ such that

$c > (1, y) > (1, 0)$. The subgroup C is convex, hence $(1, y) \in C \cap A$ which is impossible.

A right-ordered group G is an *Archimedean right-ordered group* if for every positive elements $a, b \in G$, $a > e$, $b > e$ there are positive integers m, n such that $a^n > b$, $b^m > a$. The following result is very unexpected.

THEOREM 1. (Conrad [1].) *An Archimedean right-ordered group is Abelian and therefore it is isomorphic to a subgroup of the additive group of reals.*

Proof. By Theorem 3.2.1, it is enough to prove that right order P is normal in Archimedean group G. Let $x, y \in G$, $x > e$, $y > e$. By assumption, there is a minimal positive integer n such that $x < y^n$. Then $xyx^{-1} > e$ for otherwise $xy < x < y^n$ would imply $x < y^{n-1}$. Thus $xyx^{-1} > e$ for all $x > e$, $y > e$. Then $x^{-1}y^{-1}x > e$, hence $x(x^{-1}y^{-1}x)x^{-1} = y^{-1} > e$, a contradiction. Then $P^x \subseteq P$ for all $x \in G$ and P is a normal subsemigroup in G.□

2^0. Throughly investigated ro-groups are the ro-groups G whose system of convex subgroups is a subnormal system with torsion free Abelian factors or, equivalently, if for every element $g \in G$, $d \neq e$ the jump of convex subgroups $A \prec B$ determined by element $g \in B \setminus A$ is subnormal and the quotient group B/A is order isomorphic to a subgroup of reals. This class of right-ordered groups was investigated by Conrad [1]. This ro-groups are called *Conradian* ro-groups.

LEMMA 1. (Conrad [1].) *Let G be a right-ordered group. Then the following statements are equivalent:*

(1) *for all $a, b \in G$, $a > e$, $b > e$ there exists a positive integer n such that $(ab)^n > ba$;*

(2) *for all $a, b \in G$, $e < a < b$ there exists a positive integer n such that $ab^n a^{-1} > b$;*

(3) *for all $a, b \in G$, $a > e$, $b > e$ there exists a positive integer n such that $a^n b > a$.*

Proof.(1) ⇒ (2). From $e < a$ it follows $b < ab$ and by (1) $ab^n a^{-1} = (aba^{-1})^n > baa^{-1} = b$ for some positive integer n.

(2) ⇒ (1). From $e < a$ it follows $b < ab$ and by (2) $(ba)^n = b(ab)^n b^{-1} > ab$ for some positive integer n.

(3) $\Rightarrow$ (1). Let us suppose that $(ab)^n < ba$ for all n. Then $(ba)^n b < a(ba)^n b = (ab)^{n+1} < ba$ for all n, contradicting (3).

(1) $\Rightarrow$ (3). If $e < a \leq b$, then $e < a \leq b \leq ab$. If $a > b$, then $a = cb$ where $c = ab^{-1} > e$. Thus by (1) $(cb)^n > bc$ for some n. Hence $a^n b = (cb)^n > bcb = ba > a$.□

THEOREM 2. (Conrad [1].) *A right-ordered group G is Conradian if and only if it has one of the properties (1)-(3).*

Proof. Let us suppose that a ro-group G has one of the properties (1)-(3) and $A \prec B$ be any jump in the system of convex subgroups of G. We claim that for every elements $x, y \in B \setminus A$, $x, y > e$ there exists a positive integer n such that $x^n > y$. For the contradiction let $a, b \in B \setminus A$, $a, b > e$ and $e < a^n < b$ for all positive integere n. By S we denote the subset of all elements x of G, $x > e$ such that $x < a^n$ for some positive integer n and $N = S \cup S^{-1} \cup E$. If $x, y \in N$ and $x = e$ or $y = e$ then $xy \in N$. Suppose that $x, y \in N$ and $x \neq e$, $y \neq e$, $xy \neq e$. If $x, y \in S$ then $e < x < a^n$, $e < y < a^n$ for some positive integer n. Suppose that $a^r y^{-1} < a^n$ for all positive integers r. Then $a^{n+q} y^{-1} < a^n$ for all positive integers q. But $y < a^n$, $ya^q < a^{n+q}$ and $ya^q y^{-1} < a^{n+q} < a^n$ for all positive integers q. This contradicts property (2). Therefore, there exists a positive integer r such that $a^n \leq a^r y^{-1}$, $a^n y \leq a^r$. Hence $e < xy < a^n y \leq a^r$ and $xy \in S$, $xy \in N$. For $x, y \in S^{-1}$ the same arguments show that $xy \in S^{-1}$, $xy \in N$.

Now consider the case $x \in S$, $y \in S^{-1}$. By definition, $e < x < a^n$ and $e < y < a^n$ for some integer n. Suppose that $e < xy$ and $xy \notin S$. Then $xy > a^r$, $x > a^r y^{-1}$ for all integers r. By property (1), there is a positive integer q such that $a^{nq} y^{-1} > a^n$. Then $x > a^{nq} y^{-1} > a^n$ and this contradicts the choice of integer n. If $xy < e$ then $y^{-1} \in S$, $x^{-1} \in S^{-1}$, $e < y^{-1} x^{-1}$ and by above argument $y^{-1} x^{-1} \in S$, $xy \in N$.

Suppose that $x \in S^{-1}$, $y \in S$ and $e < x^{-1} < a^n$, $e < y < a^n$. If $xy > e$ and $xy \notin S$ then $xy > a^r$, $x > a^r y^{-1}$ for all integers r. From $y < a^n$ it follows $x > a^n y^{-1} > e$. These inequalities contradict our assumption. If $xy < e$ then $y^{-1} \in S^{-1}$, $x \in S$, $y^{-1} x^{-1} > e$ and by above arguments, $y^{-1} x^{-1} \in S$, $xy \in S$ and $xy \in N$. Hence N is a subgroup of G. It is evident that the semigroup S is convex, so are the semigroup S^{-1} and the subgroup N. But now $a \in N$, $b \notin N$ and therefore $A \subset N$, $N \subset B$ properly, a contradiction. Hence for every elements $x, y \in B \setminus A$, $x, y > e$ there exists a positive integer n such that $x^n > y > e$.

Next we show that the subgroup A is normal in B. Let $b > e$, $b \in B \setminus A$. Then $a < b$ for all $a \in A$ and $ba^{-1} > e$, $ba^{-1} \in B \setminus A$. By what we just proved, $(ba^{-1})^n > b$ for some n. Let n be the least integer for which this holds, then $(ba^{-1})^{b^{-1}} > e$ for $(ba^{-1})^{b^{-1}} \leq e$ would imply $b(ba^{-1}) \leq (ba^{-1})^n$ and $b < (ba^{-1})^{n-1}$. Therefore, $(ab^{-1})^{b^{-1}} < e$ and $a^{b^{-1}} < b$ for all $a \in A$. If $a^{b^{-1}} \in B \setminus A$ for some $a \in A$, then there is an integer n such that $(b^{-1}ab)^n > b$. But again it is impossible because $a^n \in A$. Hence for all $a \in A$ is valid $bab^{-1} \in A$ or $bab^{-1} < e$. In the last case $ba^{-1}b^{-1} > e$ and $a^{-1} \in A$. Therefore, $bAb^{-1} \subseteq A$ for all $b \in B$.

Now let us suppose that for some $b \in B \setminus A$, $b > e$ and $a \in A$ is valid $b^{-1}ab \in B \setminus A$. For simplicity we can assume that $b^{-1}ab > e$. By property (2) there exists a positive integer m such that $b(b^{-1}a^n)^m b^{-1} > b^{-1}a^n > e$. But $b(b^{-1}a^n)^m b = (a^n b^{-1})^m$, then $(a^n b^{-1})^m > e$, $a^n b^{-1} > e$. So $a^n > b$ and this inequality contradicts to the choice of the element b. Therefore $b^{-1}ab \in A$ for all $b \in B$, $b > e$, $a \in A$ and the subgroup A is invariant in B.

The group B/A is an Archimedean ro-group. By Theorem 1, B/A is order isomorphic to a subgroup of the additive group of reals.

Conversely, let G be a Conradian ro-group and $e < a < b$, $a, b \in G$. let $A \prec B$ be a convex jump determinated by b. Then the quotient group B/A is Archimedean and $b \notin A$. Therefore, $ab = cba$ and $ab^r a^{-1} = c'b^r$ for some $c, c' \in A$. If $c'b^r < b$ then $c' < b^{-1} < e$, $b \in A$. This contradicts the choice of the convex jump $A \prec B$. Therefore, $ab^r a^{-1} > b$ and G has property (2).□

5.4 Locally Nilpotent Right-Ordered Groups

1^0. A finite normal series $G = G_0 \supseteq G_1 \supseteq \ldots \supseteq G_n = E$ is called a *central series* of G if G_{i-1}/G_i is contained in the centre $\zeta(G/G_i)$ of the quotient group G/G_i for every i. A group G is *nilpotent* if it has at least one central series, a *class of nilpotence* of G in this case is the least of the lenghts n of its central series. The *lower central series* $G = \gamma_1 G \supseteq \gamma_2 G \supseteq \ldots$ of any group G is defined by the rule: $\gamma_{i+1}(G) = [\gamma_i G, G]$. An *upper central series* $E = \zeta_0(G) \subseteq \zeta_1(G) \subseteq \ldots$ is defined inductively by the rule:$\zeta_{i+1}(G)/\zeta_i(G)$ is the center of G/ζ_i. Thus $\zeta_1(G)$ is the center of G. The members of upper central series of G are *hypercentres* of G. It is well known that group G is nilpotent of class n iff there exists a positive integer m such that $\zeta_m(G) = G$ and n is the smallest integer with this property. Equivalently, group G is nilpotent of class n iff $\gamma_m G = E$ for some positive integer m and n is the smallest integer with this property.

PROPOSITION 1. *A torsion-free nilpotent group is orderable.*

Proof. It is well known (see, for example, the book of M.Hall [1]) that all factors $\zeta_{i+1}(G)/\zeta_i(G)$ of the upper central series of a nilpotent torsion-free group G are Abelian torsion-free groups. By Theorem 4.2.3, G is orderable.□

From Proposition 1 and Corollary 3 of Theorem 5.1.3 it follows that any locally nilpotent torsion-free group is right-orderable. It is not difficult to see that there are right orders of nilpotent torsion-free groups which are not total orders.

EXAMPLE 1. Let $G = \mathrm{gp}(a, b, c | [a, b] = c,\ [a, c] = [b, c] = e).$. The group G is a torsion-free nilpotent group of class 2 and infinite cyclic subgroup (c) is a centre of G. Then the series $E \subset (b) \subset \mathrm{gp}(b, c) \subset G$ is subnormal and all factors of this series are infinite cyclic groups. Every element $x \in G$ can be represented uniquely in the form $x = a^p b^q c^r$ where p, q, r are integers. Let $x > e$ iff $p > 0$ or $p = 0,\ r > 0$ or $p = r = 0,\ q > 0$ It is easy to verify that this order $<$ is right order of G. This order is not invariant in G because $b > e$ and $a^{-1}ba = bc^{-1} < e$. Hence G with this order is not an o-group.

2^0. Now we describe the system of convex subgroups of locally nilpotent right-ordered groups.

LEMMA 1. (Neumann B.H.) *Let G be a locally nilpotent group and S be a subsemigroup of G. Then for any $u, v \in S$ there are $z, t \in S$ such that $zu = tv$.*

Proof. Use induction on the nilpotency class of the subgroup $L = \mathrm{gp}(u, v)$. If L is Abelian take $z = v,\ t = u$. If L is nilpotent of class $k + 1$, then consider the subgroup $M = \mathrm{gp}(uv, vu)$. Since $vu =$

$uv[v,u]$, $[v,u] \in \zeta_k(G)$ the the quotient group $M/\zeta_k(L) \cap M$ is cyclic and $\zeta_k(M) \supseteq \zeta_k(M) \cap M$. Therefore, M is nilpotent of class k. By inductive assumption, there are elements $x, y \in M$ such that $xvu = yuv$. Take $z = xv$, $t = yu$.□

THEOREM 1. (Ault [1], Rhemtulla [1].) *Right-ordered locally nilpotent groups are Conradian.*

Proof. Let G be a right-ordered locally nilpotent group. We claim that condition (3) of Lemma 5.3.1 holds. Let $a, b \in G$, $a, b > e$ and suppose that $a^n b \leq a$ for all positive integers n. Let S be a subsemigroup of G generated by the elements ab, a. By Lemma 1, there are $z, t \in S$ such that $zab = ta$ and $z > e$, $t > e$ and $ta > a$. Then $zab = a^{k_1} b a^{k_2} b \cdots a^{k_r} b$ and $k_1, k_2, \ldots, k_r \geq 0$. But $a^n b < a$ for all positive integers n and hence $zab \leq a^{k_2+1} b \cdots a^{k_r} b \leq a^{k_r+1} b \leq a$. This contradicts the inequality $ta > a$.□

COROLLARY 1. *A right-ordered locally nilpotent group without nontrivial convex subgroup is Archimedean and order isomorphic to a subgroup of ordered additive group of reals.*

3^0. Now we consider the problem of extensions of partial right orders of locally nilpotent groups.

THEOREM 2. (Rhemtulla [1].) *Every partial right order of a torsion-free nilpotent group can be extended to a total right order.*

Proof. Assume, by contradiction, that P is the positive cone of a maximal partial right order of G and P is not a total right order. Then for some $x \in G$, $x \notin P \cup P^{-1} \cup \{e\}$. Since P is maximal, neither $\mathrm{sgr}(P, x)$ nor $\mathrm{sgr}(P, x^{-1})$ can be the positive cone of partial right order and we conclude that $x^{s_1} p_1 \cdots x^{s_m} p_m = e$ and $x^{-r_1} q_1 \cdots x^{-r_n} q_n = e$ where $p_i, q_j \in P$, $s_i, s_j > 0$, $1 \leq i \leq m$, $1 \leq j \leq n$.

If $x \in \zeta_1(G)$ then these equalities reduce to $x^s p = x^{-r} q = e$ for some $s, r > 0$, $p, q \in P$. Hence $x^{sr} \in P$, $x^{-sr} \in P$, $x^{sr} \in P \cap P^{-1} = \{e\}$ and $x^{sr} = e$ $x = e$. So $\zeta_1(G) \subseteq P \cap P^{-1} \cap \{e\}$. Assume that $\zeta_i(G) \nsubseteq P \cap P^{-1} \cap \{e\}$ but $\zeta_{i-1}(G) \subseteq P \cap P^{-1} \cap \{e\}$.

Let W be the set of all words $w = x^{s_1} p_1 \cdots x^{s_m} p_m$, $p_1, \ldots, p_m \in P$, $s_1, \ldots, s_m > 0$ such that $m \geq 1$ and $w = e$ in G. Define function μ

from P to the set of non-negative integers by the rule:

$$\mu(p) = \begin{cases} 0 & \text{if } [p,x] = e \\ j & \text{if } [p,x] \in \zeta_j(G) \setminus \zeta_{j-1}(G). \end{cases}$$

For any $w = x^{s_1}p_1 \cdots x^{s_m}p_m$ in W let $l(w) = \max\{\mu(p_i) | i = 1, \ldots, m\}$. Note that $l(w) = 0$ implies $[x, p_i] = e$ for $1 \le i \le m$, $x^{s_1}p_1 \cdots x^{s_m}p_m = x^s p$ with $s = \sum_{i=1}^{m} s_i > 0$ and $p = p_1 \cdots p_m \in P$. Hence $x^{-s} \in P$. Now we will show that there exists $w \in W$ with $l(w) = 0$. Suppose that $l(w) > 0$ for every $w \in W$.

Let W_1 be a subset of W consisting of those elements of W with $l(w)$ minimal. We call p_i a dominant component of $w = x^{s_1}p_1 \cdots x^{s_m}p_m$ if $\mu(p_i = l(w)$. Since $w = e$ then there are at least two dominant components of w.

Let W_2 be the set of those elements of W_1 with the least number of dominant components.

Let W_3 be the set of those elements $w = x^{s_1}p_1 \cdots x^{s_m}p_m$ in W_2 with $\mu(p_1) = |w|$. Let $j > 1$ be the smallest integer such that for all $w = x^{s_1}p_1 \cdots x^{s_m}p_m$ in W_3, $\mu(p_i) < \mu(p_1)$ for all i, $1 < i < j$ and $\mu(p_j) = \mu(p_1)$ for some $w \in W_3$.

Let W_4 be the set of those $w = x^{s_1}p_1 \cdots x^{s_m}p_m$ in W_3 with $\mu(p_j) = \mu(p_1)$.

Finally, let W_5 be the set of all $w = x^{s_1}p_1 \cdots x^{s_m}p_m$ in W_4 with m minimal. By our assumption, it follows that $m > 1$ for all $w \in W_5$, $w = x^{s_1}p_1 \cdots x^{s_m}p_m$. Pick any $w = x^{s_1}p_1 \cdots x^{s_m}p_m$ in W_5. Since $x^{s_j}p_j = p_j(x[x,p_j])^{s_j}$, we must have $[x,p_j] < e$, for otherwise $w' =$

$$= x^{s_1}p_1 \cdots x^{s_m}p_m = x^{s_1}p_1 \cdots x^{s_{j-1}}p_{j-1}p_j(x[x,p_j])^{s_j}x^{s_{j+1}} \cdots x^{s_m}p_m$$

and $w' \in W$, $l(w') = l(w) = \mu(p_1) \ge \mu(p'_{j-1})$ where $p'_{j-1} = p_{j-1}p_j$. If $j = 2$ or $\mu(p_1) > \mu(p'_{j-1})$, then we contradict the choice of W_2 and if $j \ne 2$ and $\mu(p_1) = \mu(p'_{j-1})$ then we contradict the choice of W_4. Thus $[p_j, x] > e$ and for $j < m$ we consider

$$w_1 = x^{s_1}p_1 \cdots x^{s_{j-1}}p_{j-1}x^{s_j+1}p_{j1}x^{s_{j+1}-1} \cdots x^{s_m}p_m \in W_5,$$

where $p_{j1} = p_j^x = p_j[p_j,x] \in P$ and $\mu(p_{j1}) = \mu(p_1$. Repeated application of the above arguments yields w_i, $i = 1, 2, \ldots, s_{j+1}$ where

$$w_i = x^{s_1}p_1 \cdots x^{s_{j-1}}p_{j-1}x^{s_j+i}p_{ji}x^{s_{j+1}-i} \cdots x^{s_m}p_m$$

for $1 \le i \le s_{j+1}$, $p_{j\,i} = p_{j\,i-1}[p_{j\,i-1}, x] \in P$ and $\mu(p_1) = \mu(p_{j\,i})$. Now $w_i \in W_5$ for $i < s_{j+1}$, and

$$w_{s_{j+1}} = x^{s_1} p_1 \cdots x^{s_{j-1}} p_{j-1} x^{s_j + s_{j+1}} p' x^{s_{j+2}} \cdots x^{s_m} p_m$$

where $p' = p_{j\,s_{j+1}} p_{j+1} \in P$ and $\mu(p') \le \mu(p_1)$ contradicts the choice of W_2 and $\mu(p') = \mu(p_1)$ contradicts the choice of W_5.

If $j = m$ we conjugate $x^{s_1} p_1 \cdots x^{s_m} p_m \in W_5$ by $x^{s_m} p_m$ and in this case we have $w' = x^{s_1} p_1 \cdots x^{s_{m-1}} p_{m-1} \in W_5$. By similar arguments (as in the case $j < m$), there is an element

$$w'' = x^{s_m + s_1} (p_m [p_m, x])^{s_1} p_1 \cdots x^{s_{m-1}} p_{m-1}$$

with one dominant component $(p_m[p_m, x])^{s_1} p_1$ (if $\mu((p_m[p_m, x])^{s_1} p_1) = \mu(p_1\ )$ or $l(w'') < l(w)$ which contradicts to the choice of W_2.

We have thus established that $w = x^s p = e$ for some positive integer s and some $p \in P$. Similarly, we obtain $x^{-r} q = e$ for some $r > 0$ and $q \in P$. These two equations imply that $x^{rs} \in P \cap P^{-1} = \{e\}$ with $rs \ne 0$, a contradiction.□

A group G is *fully right-orderable* if every partial right order of G can be extended to a right order. Let us denote by RO^* the class of fully right-orderable groups. The class $\mathcal{X}$ of groups is *locally closed* if $G \in \mathcal{X}$ whenever every finitelly generated subgroup of G belongs to $\mathcal{X}$.

PROPOSITION 2. (Rhemtulla [1].) *The class of RO^*-groups is locally closed.*

Proof. Let every finitely generated subgroup of G be a RO*-group and P be a maximal partial right order of G. Suppose that there is $x \in G$ such that $x \notin P \cup P^{-1} \cup \{e\}$. Then neither $\mathrm{sgr}(x, P)$ nor $\mathrm{sgr}(x^{-1}, P)$ is a partial right order and $e = x^{r_1} p_1 \cdots x^{r_n} p_n = x^{-s_1} q_1 \cdots x^{-s_m} q_m$ for some positive integers $r_1, \dots, r_n, s_1, \dots, s_m$ and $p_1, \dots, p_n, q_1, \dots, q_m \in P$. Let $H = \mathrm{gp}(x, p_1, \dots, p_n, q_1, \dots, q_m)$. By our hypothesis, $H \in RO^*$. But it is evident that the partial order $H \cap P$ cannot be extended to a right order of H.□

COROLLARY 2. (Rhemtulla [1].) *Any torsion-free locally nilpotent group is right fully orderable.*□

Of course, there exist right-orderable groups which are not RO*-groups.

EXAMPLE 2. (Mura and Rhemtulla [2].) Let $G = \text{gp}(a, b|b^{-1}ab = a^{-1})$. It is clear that G is right-ordered by the rule: $x = b^k a^s > e$ iff $k > 0$ or $k = 0,\ s > 0$. But the group G is not fully right-orderable. For example, the semigroup $P = \text{sgr}(b^2, ab^{-2})$ defines a partial right order which cannot be extended to a right order of G. For any extensions of P we must have $b > e$ since $b^2 > e$. Also $ab^{-2} > e$ implies $ab^{-1} > b > e$ and hence $bab^{-1} = a^{-1} > e$, a contradiction.□

5.5 Right-Orderable Groups

1⁰. Let us recall that formulae φ of the first-order language of the signature σ is an *implication* (or *quasi-identity*) of σ if φ is of the form

$$(\forall x_1, \ldots, x_n)((A_1 = B_1 \& A_2 = B_2 \& \ldots \& A_k = B_k) \Rightarrow (A = B))$$

where $A_1, \ldots, A_k, B_1, \ldots, B_k, A, B$ are terms of the signature σ of variables $x_1, \ldots, x_n$. The class $\mathcal{K}$ of algebraic systems of the signature σ is a *quasivariety* if there is a set Φ of implications such that an algebraic system G belongs to $\mathcal{K}$ iff all implications of Φ are valid on G.

Let I be a set. The family $\mathcal{F}$ of subsets I is called a *filter* over I if and only if the following conditions are satisfied,

a) $\emptyset \notin \mathcal{F}$,

b) if $A, B \in \mathcal{F}$, then $A \cap B \in \mathcal{F}$,

c) if $A \in \mathcal{F}$ and $A \subseteq B$, then $B \in \mathcal{F}$.

Filter $\mathcal{F}$ is called an *ultrafilter* if for any subset A of I either $A \in \mathcal{F}$ or $I \setminus A \in \mathcal{F}$.

Let $\{A_\gamma | \gamma \in I\}$ be the set of algebraic systems of signature σ and $\mathcal{F}$ be a filter over I. Then we can define the equivalence relation $\equiv_{\mathcal{F}}$ on the Cartesian product $\bar{A} = \bar{\prod\limits_{\gamma \in I}} A_\gamma$:

$$f \equiv_{\mathcal{F}} g \text{ for } f, g \in \bar{A} \text{ iff } \{\gamma \in I | f(\gamma) = g(\gamma)\} \in \mathcal{F}.$$

It is easy to see that the set $\bar{A}/\mathcal{F}$ of equivalence classes of $\bar{A}$ is an algebraic system of the signature σ again. It is called a *reduced product* and denoted by $\bar{\prod\limits_{\gamma \in I}} A_\gamma/\mathcal{F}$. If $\mathcal{U}$ is a ultrafilter over I then the reduced product $\bar{\prod_{\gamma \in I}} A_\gamma / \mathcal{U}$ is called an *ultraproduct*. Let H, S, P, P_U are operators of homomorphic images, subalgebras, direct products and ultraproducts respectively. It is known (see, for example, the book of P.Cohn [2]) that the class $\mathcal{K}$ of algebraic systems is a quasivariety if and only if it is closed under operators S, P, P_U and contains a trivial algebra.

PROPOSITION 1. *The class of right-orderable groups is closed under ultraproducts.*

Proof. Let G_γ, $\gamma \in I$ ba a family of right-orderable groups and $\mathcal{U}$ is an ultrafilter over I. We consider the right-ordered group as algebraic system of the signature $r = \{\cdot\ ,\ ^{-1}\ , e, \leq\}$. It is clear that the ultraproduct $G = \bar{\prod_{\gamma \in I}} G_\gamma / \mathcal{U}$ is a group. Now we define on G the order $\leq$ by the rule: $f \leq g$ for $f, g \in G$ iff the subset $K = \{\alpha \in I | f(\alpha) \leq g(\alpha)\} \in \mathcal{U}$. It is evident that $\leq$ is a right order on group G. □

PROPOSITION 2. *The class RO of all right-orderable groups is a quasivariety.*

Proof. It is clear that class *RO* is closed under operators S, P and by Proposition 1 it is closed under P_U.□

2^0. By Theorem 5.2.1, every right-ordered group is embeddable into a lattice ordered group. Therefore every ro-group is torsion-free. Here we will prove that the class of right-orderable group and the class of all torsion-free group are different.

Subgroup H is called an *isolated* subgroup of group G if $g^n \in H$, $n \in \mathbf{Z}$ imply $g \in H$ for any $g \in G$. An *isolator* of subgroup H in group G is the smallest isolated subgroup of G containing H.

THEOREM 1. (Rhemtulla [2].) *Let $G \neq E$ be a finitely generated group such that G/G' is finite and G has the subgroup K of finite index such that every right order of K is Conradian. Then group G cannot be right-ordered.*

Proof. Suppose that G is a ro-group under right order $\leq$ and K_1 is an isolator of K' in K. Then $K_1 = \{x \in K | x^n \in K' \text{ for some } n \in \mathbf{Z}\}$. The subgroup K is a finitely generated Conradian ro-group. Hence $K/K_1 \neq E$ is torsion-free finitely generated Abelian group. Choose the coset represertatives $e < x_1 < x_2 < \ldots < x_n$ of K in G and consider the transfer map $\tau: G \to K/K_1$ given by:

$$\tau(g) = x_n g(\overline{x_n g})^{-1} \cdot x_{n-1} g(\overline{x_{n-1} g})^{-1} \cdots x_1 g(\overline{x_1 g})^{-1} \cdot K_1,$$

where $\overline{x_i g}$ is the coset representative of $x_i g K_1$. It is clear that τ is a homomorphism. Since K/K_1 is a torsion-free Abelian group and G/G' is finite then τ is the trivial homomorphism. Since right-ordered group K is Conradian and K if finitely generated, then there is the greatest convex subgroup C of K such that K/C is an Archimedean totally ordered group. Since subgroup K has a finite index in G, then there exists a positive integer p such that $x_n^p \in K$. Let $g \in K \setminus C$, $g > e$ and $g > x_n^p$. Since $g > x_n > x_{n-1} > \ldots > x_2 > x_1 = e$ we have $gx_i^{-1} > e$ and $x_i g x_i^{-1} > e$ for all $i = 1, 2, \ldots, n$. So $x_n g x_n^{-1} \cdot x_{n-1} g x_{n-1}^{-1} \cdots x_1 g x_1^{-1} > e$. On the other hand, $\overline{x_i g} = x_i$ and $x_n g x_n^{-1} \cdots x_1 g x_1^{-1} = \tau(g) \in K_1$. But $x_n g x_n^{-1} \cdots x_1 g x_1^{-1} \in C$ and $x_n g x_n^{-1} \cdots x_2 g x_2^{-1} > e$ Then $\tau(g) = x_n g x_n^{-1} \cdots x_2 g x_2^{-1} > g$ yield $g \in C$, a contradiction. □

EXAMPLE 1. (Wigold [1], Smirnov [8].) Let

$$G = \mathrm{gp}(a, b, c \mid [a, b, a] = [a, b, b] = e,\ c^{-1}ac = a^{-1},\ c^{-1}bc = b^{-1},\ [a, b] = c^{2k},\ k \in \mathbf{Z},\ k > 1),$$

$K = \mathrm{gp}(a, b)$, $K \subset G$. Then K has a finite index in G and it is a nilpotent torsion-free group of class 2. By Theorem 5.4.1, every right order of K is Conradian. A derived subgroup G' of G is equal to group $\mathrm{gp}(a^2, b^2, [a, b])$ and has a finite index in G, too. Therefore, by Theorem 1, group G cannot be right-ordered. Simple verification shows that group G is torsion-free.

3⁰. Orderability properties of a group is useful tool in investigations of group rings.

LEMMA 1. *Let G be a right-ordered group and $A = \{a_1, \ldots, a_n\}$, $B = \{b_1, \ldots, b_m\}$ are non-empty finite subsets of G such that $n+m > 2$. Then there are at least two different elements in $A \cdot B$ which can be uniquely represented in the form $a_i b_j$, $1 \le i \le n$, $1 \le j \le m$.*

Proof. Let $a_1 < a_2 < \ldots < a_n$, $b_1 < b_2 < \ldots < b_m$ and $a_1 b_p$ is the least element of $\{a_1 b_j | 1 \le j \le m\}$ and $a_n b_q$ be the greatest element of the set $\{a_n b_j | 1 \le j \le m\}$. Then $a_1 b_p < a_i b_j$ for all i, j, $2 \le i \le$

n, $1 \leq j \leq m$ and $a_1 b_p < a_i b_j$ for all i, j $(i,j) \neq (1,p)$. Similar arguments show that $a_n b_q$ is the greatest element of the set $A \cdot B$.

PROPOSITION 3. (Bowdi [1], Smirnov [8], Higman [1].) *Let K be a ring without zero divisors and G be a RO-group. Then the group ring $K[G]$ has no zero divisors and if x is a unit of $K[G]$, then $x = \lambda a$ for some $a \in G$ and λ is a unit of K.*□

PROPOSITION 4. (LaGrange and Rhemtulla [1].) *If $\mathbf{Z}[G]$ is isomorphic to $\mathbf{Z}[H]$ and G is a RO-group, then H is isomorphic to G.*

Proof. Since $\mathbf{Z}[G]$ is isomorphic to $\mathbf{Z}[H]$, then the group U_G of units of $\mathbf{Z}[G]$ is isomorphic to the group U_H of units $\mathbf{Z}[H]$. By Proposition 3, it follows that U_G is the direct product $G \times (c)$ where $c^2 = e$ and H is a subgroup of $G \times (c)$. The group ring $\mathbf{Z}[G]$ has no zero divisors. Therefore $\mathbf{Z}[H]$ has no zero divisors, too, and H is torsion-free. Hence H is isomorphic to the subgroup of G and also is a RO-group. Again, by Proposition 3, $U_H = H \times (c')$, $c^2 = e$. Thus H is isomorphic to G.□

Chapter 6

Totally ordered groups

6.1 Tests for Orderability of Groups

Some properties of totally ordered groups and their convex subgroups were discussed in Chapters 3 and 5. In this section we discuss group-theoretic conditions for orderability of groups. For these purposses the terminology of semigroups is very useful.

1⁰. LEMMA 1. *Let P, Q be pure normal subsemigroups (partial orders) of group G and $P \cap Q = \{e\}$. Then the subset PQ^{-1} is a pure normal subsemigroup (partial order) of G.*

The proof of Lemma is staightforward. □

By $S(a_1, \ldots, a_n)$ we denote normal subsemigroups of group G generated by $a_1, \ldots, a_n \in G$.

THEOREM 1. (Fuchs [1].) *A partial order P of a group G can be extended to total order of G if and only if it satisfies the following property.*

(1) *For every finite set of non-identity elements $a_1, \ldots, a_n$ in G $\quad P \cap S(a_1^{\varepsilon_1}, \ldots, a_n^{\varepsilon_n}) = \emptyset$ for some choice of signs $\varepsilon_i = \pm 1$, $i = 1, 2, \ldots, n$.*

Moreover, if P satisfies condition (1), then for any $a \in G$, $a \neq e$ either $P \cdot S(a, e)$ or $S(a^{-1}, e)$ is a partial order of G satisfying condition (1).

Proof. Suppose P can be extended to the total order Q on G.

Choose ε_i such that $a_i^{\varepsilon_i} \notin Q^{-1}$. Then

$$P \cap S(a_1^{\varepsilon_1}, \dots, a_n^{\varepsilon_n} \subseteq Q \cap S(a_1^{\varepsilon_1} \dots a_n^{\varepsilon_n}) = \emptyset.$$

Conversely, let the partial order P satisfy condition (1) and $a \in G$, $a \neq e$. Then the subsemigroup $P \cdot S(a, e)$ or $P \cdot S(a^{-1}, e)$ is a partial order on G satisfying condition (1). Indeed, for any $a \in G$, $a \neq e$ there is a sign $\varepsilon = \pm 1$ such that $P \cap S(a^{\varepsilon}, e) = \emptyset$. Suppose that $\varepsilon = +1$. By Lemma 1, the subsemigroup $P \cdot S^{-1}(a, e) = P \cdot S(a^{-1}, e)$ is a partial order on G. Now, we have two possibilities: a) $P \cap S(a^{-1}, e) \neq \emptyset$; b) $P \cap S(a^{-1}, e) = \emptyset$. In case a) the subsemigroup $P \cdot S(a^{-1}, e)$ satisfies condition (1). Suppose that $P \cdot S(a^{-1}, e)$ does not satisfy (1). Then there exist non-identity elements $a_1, \dots, a_n$ such that for every choice of signs $\varepsilon_i = \pm 1$, $i = 1, 2, \dots, n$ the intersections $P \cdot S(a^{-1}, e)$ and $S(a_1^{\varepsilon_1}, \dots, a_n^{\varepsilon_n})$ are non-empty. Thus $P \cap S(a, a_1^{\varepsilon_1}, \dots, a_n^{\varepsilon_n}) \neq \emptyset$ and $P \cap S(a^{-1}, a_1^{\varepsilon_1}, \dots, a_n^{\varepsilon_n}) \neq \emptyset$. This contradicts condition (1).

In case b) the subsemigroups $P \cdot S(a, e)$ and $P \cdot S(a^{-1}, e)$ are partial orders. We claim that $P \cdot S(a, e)$ or $P \cdot S(a^{-1}, e)$ satisfies (1). Suppose, if possible, that P satisfies (1) but neither $P \cdot S(a, e)$ nor $P \cdot S(a^{-1}, e)$ satisfies (1). Then there are non-identity elements $a_1, \dots, a_n, b_1, \dots, b_m$ such that for every choice of signs ε_i, δ_j the intersections of $S(a, a_1^{\varepsilon 1}, \dots, a_n^{\varepsilon_n})$ and $S(a^{-1}, b_1^{\delta_1}, \dots, b_m^{\delta_m})$ with P are non-empty. Thus,

$$P \cap S(a^{\varepsilon}, a_1^{\varepsilon 1}, \dots, a_n^{\varepsilon_n}, b_1^{\delta_1}, \dots, b_m^{\delta_m}) \neq \emptyset$$

for all choices of signs $\varepsilon, \varepsilon_i, \delta_j$. This contradicts (1). Since condition (1) is satified by the union of the ascending chain of partial orders, then there exists a maximal partial order Q that is an extension of P and satisfies (1). Let $a \in G$, $a \neq e$. By above arguments, the sugsemigroups $Q \cdot S(a, e)$ or $Q \cdot S(a^{-1}, e)$ define a partial order with condition (1). By the maximality of Q it follows that $Q = Q \cdot S(a, e)$ or $Q = Q \cdot S(a^{-1}, e)$. Hence $a \in Q$ or $a^{-1} \in Q$ and Q is a total order of G.□

The following two theorems are corollaries of Theorem 1 if we set $P = \{e\}$.

THEOREM 2. (Ohnishi [2], Łoš [1].) *A group G is orderable if and only if for any finite set of non-identity elements $a_1, \dots, a_n \in G$ is valid $e \notin S(a_1^{\varepsilon_1}, \dots a_n^{\varepsilon_n})$ for some choice of signs $\varepsilon_i = \pm 1$.*□

THEOREM 3. (Lorenzen [2].) *A group G is orderable if and only if for any finite set of non-identity elements $a_1, \dots, a_n$ the intersection of 2^n subsemigroups $S(a_1^{\varepsilon_1}, \dots, a_n^{\varepsilon_n})$ obtained by taking all choices of signs $\varepsilon_i = \pm 1$, $i = 1, 2, \dots, n$, is empty.*□

COROLLARY 1. (Mal'cev [4], Neumann [3].) *A group G is orderable if and only if every finitely generated subgroup of G is orderable.*□

COROLLARY 2. (Mal'cev [4].) *A class of orderable groups is a quasivariety of groups.*□

2⁰. In Section 2 of Chapter 3 we established some tests for orderability of groups in terms of subgroup systems and in § 3 of Chapter 5 some sufficient conditions for the orderability of groups.

Group G is a *direct product* of subgroups G_α, $\alpha \in I$, *with amalgameted subgroup* H iff the following conditions hold:

(a) G is generated by subgroups G_α, $\alpha \in I$;

(b) $G_\alpha \cap G^\alpha = H$ for all $\alpha \in I$, where G^α is a subgroup of G generated by all G_β, $\beta \neq \alpha$;

(c) elements of subgroups G_α and G_β commute for all $\alpha, \beta \in I$, $\alpha \neq \beta$.

By definition it follows that H is a central subgroup of G.

Let $\{G_\alpha | \alpha \in I\}$ be a set of groups and each group G_α contains a subgroup H_α which is isomorphic to some group H and φ_α is the isomorphism from H onto H_α. Then the quotient group of the direct product $\prod_{\alpha \in I} G_\alpha$ by the subgroup generated by elements $x = \varphi_\alpha(h)\varphi_\beta(h)^{-1}$ where $h \in H$ is a direct product of subgroups G_α, $\alpha \in I$ with amalgameted subgroups H and with isomorphisms φ_α. We denote it by $G = \prod_{\alpha \in I}(\varphi_\alpha(H) = H_\alpha)$. If the index set if finite, then we denote it by $G = G_1 \times G_2 \times \cdots \times G_n(\varphi_1^{-1}(H) = \varphi_2^{-1}(H) = \ldots = \varphi_n^{-1}(H))$ or for short $G = G_1 \times G_2 \times \cdots \times G_n(H_1 = H_2 = \ldots = H_n)$.

LEMMA 2. (Kopytov [1].) *If c is a central element of an o-group G and for some element $c \in G$ is valid $a^n > c$ then, $g_1^{-1}ag_1 \cdots g_n^{-1}ag_n > c$ for all $g_1, \dots, g_n \in G$.*

Proof. Suppose that $g_i^{-1}ag_i \leq g_j^{-1}ag_j$ for all $i = 1, 2, \dots, n$. Then $g_1^{-1}ag_1 \cdots g_n^{-1}ag_n \geq g_i^{-1}a^n g_i > g_i^{-1}cg_i = c$.□

THEOREM 4. (Kopytov [1].) *Every orderable group G with centre*

Z can be embedded into orderable group G^* with following properties:
(2) *the centre Z^* of G^* is divisible,*
(3) *$G^* = G \cdot Z^*$, $G \cap Z^* = Z$ and G^*/Z^* is isomorphic to G/Z.*
Every total order of G is extended to total order of G^*.

Proof. Let Z^* be a minimal divisible Abelian group containing Z and $G^* = G \times Z^*(Z = Z)$ be a direct product of the groups G and Z^* with amalgameted subgroup Z and natural embeddings of the group Z into G and Z^*. It is evident that properties (2),(3) of the theorem are true for group G^*. By Proposition 2.2.4, every orderable group is a group with unique extraction of roots. Then group G^*/Z^* is isomorphic to G/Z and it is a torsion-free group, G is a normal subgroup of G^* and G^*/G is a periodic Abelian group because G^*/G is isomorphic Z^*/Z. Thus, for every $x \in G^*$ is valid $x^n \in G$ for some positive integer n.

Let P be a positive cone of some total order of G and $P^* = \{x \in G^* | x^n \in P$ for some positive integer $n\}$. It is clear that $P \subseteq P^*$, $P^* \cup (P^*)^{-1} = G^*$, $x^{-1}P^*x \subseteq P^*$ for all $x \in G^*$ and $P^* \cap (P^*)^{-1} = \{e\}$. We claim that P^* is a semigroup. Let $x, y \in P^*$. Then $x^n, y^m \in P$ for some positive integers n, m. By definition of group $G^* = G \times Z^*(Z)$, it follows that $x = ua$, $y = vb$ where $u, v \in G$, $a, b \in Z^*$. Hence, $c = b^m \in Z$ and $y^m = v^m b^m \in P$. Therefore, $v^m > b^{-m} = c^{-1}$ and by Lemma 2, it follows that $u^{-mn+1}vu^{mn-1} \cdots u^{-1}vu \cdot v \cdot c^n \in P$ and $(xy)^{mn} = x^{mn} \cdot x^{-mn+1}yx^{mn-1} \cdots x^{-1}yx \cdot y = x^{mn} \cdot u^{-mn+1}vbu^{mn-1} \cdots u^{-1}vbu \cdot vb = x^{mn} \cdot u^{-mn+1}vu^{mn-1} \cdots u^{-1}vu \cdot v \cdot b^{mn} = x^{mn} \cdot u^{-mn+1}vu^{mn-1} \cdots u^{-1}vu \cdot v \cdot c^n \in P$. This shows that P^* is a total order of G^* extending a total order P of G.□

A subgroup H of the orderable group G is *relatively convex* in G if it is convex under some total order of G.

LEMMA 3. (Kokorin [3].) *Let Z be a divisible central subgroup and H be a relatively convex subgroup in G. Then the product HZ is relatively convex in G.*

Proof. Let H be a convex subgroup in G under total order on G and $S(G) = \{H_\gamma | \gamma \in \Gamma\}$ be a system of convex subgroups in G under this total order. From $S(G)$ we produce the system $S'(G) = \{H_\gamma Z | \gamma \in \Gamma\}$. Let $S''(G)$ be a completion of $S'(G)$ under intersections and joins. Then $S''(G)$ satisfies the hypothesis of Therem 3.2.3. We prove only that all subgroups $H_\gamma Z$ ($\gamma \in \Gamma$) are strongly isolated in G. Suppose that there is an element $a \in G$ such that $a \cdot g_1^{-1}ag_1 \cdots g_n^{-1}ag_n = hz$ where $h \in H_\gamma$, $z \in Z$. Let $z_1 \in Z$ and $z_1^{n+1} = z$. Then $(az_1^{-1}) \cdot g_1^{-1}(az_1^{-1})g_1 \cdots g_n^{-1}(az_1^{-1})g_n = h$. But H_γ is strongly isolatad in G, so $az_1^{-1} \in H_\gamma$, $a \in H_\gamma Z$. Thus $H_\gamma Z$ is strongly isolated in G.□

Group G is *fully orderable* if every its partial order can be extended to total order.

THEOREM 5. (Kokorin [3], Kopytov [1].) *Let G be an orderable (fully orderable) group and Z be a centre of G. Then the quotient group G/Z is orderable (fully orderable).*

The proof follows from Theorem 4 and Lemma 3. □

COROLLARY 3. *Let G be an orderable group and H be a hypercentre of G. Then H is relatively convex.*□

COROLLARY 4. *If the direct product $G = \prod_{\alpha \in I} G_\alpha(\varphi_\alpha(H) = H_\alpha)$ of orderable groups G_α is torsion-free, then G is orderable, too.*□

COROLLARY 5. *A group G is orderable if and only if both its hypercentre H and the quotient group G/H are orderable.*□

6.2 Weakly Abelian Total Orders

1^0. The system of convex subgroups $S(G)$ of totally ordered group G is an infrainvariant soluble system (see Theorem 3.2.3). Hence, in all totally ordered groups the identity

$$(1) \qquad |[x,y]| \le \max\{|x|,|y|\} \le |x| \cdot |y|$$

is valid. It is not difficult to see that a stronger identity

$$(2) \qquad |[x,y]| \le |x|$$

is not valid in all totally ordered groups.

EXAMPLE 1. Let G be a group of all matrices of the form

$$x = \begin{pmatrix} \xi & \alpha \\ 0 & 1 \end{pmatrix}$$

where ξ, α are rational numbers and $\xi > 0$. We denote $x = (\xi, \alpha)$. Let $(\xi, \alpha) > (1, 0)$ iff $\xi > 1$ or $\xi = 1, \quad \alpha > 0$. Direct calculations show that G is a totally ordered group. Let $x = (1, -1), \quad y = (\frac{1}{3}, 0)$. Then $|[x, y]| = |(1, -2)| = (1, 2) > |x| = (1, 1)$.

Total order P on group G is *weakly abelian* if identity (2) is valid in o-group G.

THEOREM 1. *A totally ordered group G is weakly abelian if and only if the system of convex subgroup of G is central i.e., $[B, G] \subset A$ for every jump $A \prec B$ in the system of all convex subroups of G.*

Proof. Let G be a weakly abelian totally ordered group and $A \prec B$ be a jump in the system of convex subgroups of G. Let $x \in B \setminus A, \quad x > e, \quad y \in G$. We claim that y lies in the normalizer $N_G(A)$ of subgroup A. Conversely, let $y \notin N_G(A)$. Then $y^{-1}By \supset B, \quad y^{-1}By \neq B$, or $y^{-1}By \subset B, \quad y^{-1}By \neq B$. In the first case $y^{-1}xy > z$ for all $z \in B$ and in the other case $yxy^{-1} > z$ for all $z \in G$. Hence $y^{-1}xy > x^2$ and $x^{-1}y^{-1}xy = [x, y] > x$ or $yxy^{-1} > x^2$ and $x^{-1}yxy^{-1} > x > e$. These inequalities lead to contradiction.

Now we need to verify that element y centralizes the jump $A \prec B$, i.e., $[x, y] \in A$. A quotiuent group B/A is Archimedean then by Theorem 3.2.1, B/A is a subgroup of additive group of reals. We will use for the group B/A multiplicative notations and identify coset xA with real number 1. Then every element of group B/A can be uniquely represented as $(xA)^r$, where r is a real number. By Proposition 3.2.2, the inner automorphisms of group G defined by the element y induces on the group B/A raising to a some real power. If the element y does not centralize a jump $A \prec B$, then without loss of gerelality we assume that $y^{-1}xAy = (xA)^s$ and $s > 2$. Therefore, $y^{-1}xAy > x^2, \quad x^{-1}y^{-1}xy > x > e$, a contradiction (2). Hence, $[x, y] \in A$ and $[y, B] \subseteq A$.

The converse statement is evident.□

COROLLARY 1. *A class of groups having weakly abelian total orders coincides with a class of group having a central series with torsion-free quotient groups.*□

Let $A_x \prec B_x$ and $A_y \prec B_y$ be jumps in the system of convex subgroup of totally ordered group G defined by elements $x, y \in G, \quad x, y > e$. If $B_x \subseteq A_y$ then $x^n < y$ for all integers n, we write $x \ll y$. It is easy to see that an o-group G is weakly abelian iff for all $x, y \in G$ is valid

$$|[x,y]| \ll |x|, |y|. \tag{3}$$

It is equivalent to identity for the signature $l = \{\cdot, e, ^{-1}, \vee, \wedge\}$:

$$y^{-1}|x|y \wedge |x^2| = y^{-1}xy, \tag{4}$$

or

$$e \vee x \vee x^2 \vee y^{-1}xy = x^2 \vee e. \tag{5}$$

PROPOSITION 1. (Smirnov [1].) *Any total order of locally nilpotent torsion-free group G is weakly abelian.*

Proof. It is enough to prove Proposition for torsion-free nilpotent groups with two generators. In this case group G has a *finite special rank*. Therefore all convex subgroups of G are normal. Let $A \prec B$ be any jump in the system of convex subgroups of G, $x \in B \setminus A$, $x > e$ and $y \in G$. If y does not centralize the quotient group B/A, then there is a real number $s \neq 1$, $s > 0$ such that $y^{-1}xAy = (xA)^s$. (Here we identify the quotient group B/A with a subgroup of reals and use multiplicative notations). So $[x,y]A = (xA)^{s-1}$ and $[x,\underbrace{y,\ldots,y}_{n}]A = (xA)^{(s-1)^n} \neq A$ for all integers n which contradicts nilpotence of G. Hence $[x,y] \in A$ and the system of convex subgroups of G is central.□

A group G is called an *Engel* group if for each pair x, y of elements in G there is an integer $n = n(x,y) > 0$ such that $[x,\underbrace{y,\ldots,y}_{n}] = e$.

COROLLARY 1. *Any total order of an Engel torsion-free group G is weakly abelian.*

The proof of Corollary 1 is similar to that of Proposition 1.□

PROPOSITION 2. *Let a group G be locally approximated by groups having weakly abelian total orders. Then G also has weakly abelian total order.*□

PROPOSITION 3. *Free groups, free soluble groups and free polinilpotent groups have weakly abelian total orders.*□

2^0. Here we characterize weakly abelian orderable groups in terms of normal subsemigroups.

LEMMA 1. (Kopytov and Medvedev [1]) *Let partial order P of group G satisfy the following property.*

(6) *For every finite set of non-identity elements $a_1, \dots, a_n$ and $x_1, \dots, x_n$ in G is valid for some choice of signs $\varepsilon_i = \pm 1$, $i = 1, 2, \dots, n$*

$$P \cap S(a_1^{\varepsilon_1}[x_1, a_1], \dots, a_n^{\varepsilon_n}[x_n, a_n]) = \emptyset.$$

Then for any $a \in G$, $a \neq e$ either $P \cdot S(a, e)$ or $P \cdot S(a^{-1}, e)$ is a partial order on G satisfying condition (6).

Proof. Suppose that neither $P \cdot S(a, e)$ nor $P \cdot S(a^{-1}, e)$ satisfies (6). Then there are non-identity elements u_i, x_i $(1 \leq i \leq n)$ and v_j, y_j $(1 \leq j \leq m)$ such that $P \cdot (a, e) \cap S(u_1^{\varepsilon_1}[x_1, u_1], \dots u_n^{\varepsilon_n}[x_n, u_n]) \neq \emptyset$ and $P \cdot (a^{-1}, e) \cap S(v_1^{\delta_1}[y_1, v_1], \dots v_m^{\delta_m}[y_m, v_m]) \neq \emptyset$ for every choice of signs ε_i, δ_j. Thus

$$P \cap S(a^{\varepsilon}[e, a], u_1^{\varepsilon_1}[x_1, u_1], \dots u_n^{\varepsilon_n}[x_n, u_n]) v_1^{\delta_1}[y_1, v_1], \dots v_m^{\delta_m}[y_m, v_m]) \neq \emptyset$$

for all choices of signs $\varepsilon, \varepsilon_i, \delta_j = \pm 1$, $1 \leq i \leq n$, $1 \leq j \leq m$. This contradicts condition (6) for P.□

THEOREM 2. (Kopytov and Medvedev [1].) *A partial order P on group G can be extended to weakly abelian order on G if and only if it satisfies property (6).*

Prof. We use notations from Lemma 1. Supose that P can be extended to a weakly abelian total order Q on G. Choose ε_i such that $a_i^{\varepsilon_i} < e$, $i = 1, 2, \dots, n$. But Q is weakly abelian; then $a_i^{\varepsilon_i} < e$ and

$$P \cap S(a_1^{\varepsilon_1}[x_1, a_1], \dots, a_n^{\varepsilon_n}[x_n, a_n]) \subseteq$$
$$Q \cap S(a_1^{\varepsilon_1}[x_1, a_1], \dots, a_n^{\varepsilon_n}[x_n, a_n]) = \emptyset.$$

Conversely, let the partial order $\leq$ satisfy condition (6). Since (6) is satisfied by the union of an ascending chain of partial orders there exists a maxmal partial order Q that is an extension of P and satisfies (6). If Q is not total order, then there is an element $a \in G$, $a \notin Q \cup Q^{-1}$. By Lemma 1, the semigroup $Q \cdot S(a, e)$ or $Q \cdot S(a^{-1}, e)$ defines a partial order which satisfies condition (6). By the maximality of partial order

Q, it follows that $Q = Q \cdot S(a, e)$ or $Q = Q \cdot S(a^{-1}, e)$. Hence $a \in Q$, or $a^{-1} \in Q$ and Q is a total order on G. It is easy to see that (G, Q) is a weakly abelian o-group.□

THEOREM 3. (Kopytov and Medvedev [1].) *Group G has weakly abelian total order if and only if for any finite set of non-identity elements $a_1, \dots, a_n$ and elements $x_1, \dots, x_n$ of G*

$$e \notin S(a_1^{\varepsilon_1}[x_1, a_1], \dots, a_n^{\varepsilon_n}[x_n, a_n])$$

for some choice of signs $\varepsilon_i = \pm 1, \quad i = 1, \dots, n$.

Theorem 3 follows from Theorem 2 when $P = \{e\}$.□

COROLLARY 2. *A class of groups having weakly abelian total orders is a quasivariety.*□

These results have interesting applications for the group theory. In [2] A.Rhemtula proved that if group G is locally residually finite p-group for infinite many primes p, then G is orderable. Appling Theorem 2 we can prove that this groups has weakly abelian orders i.e., it has central series of subgroups with torsion-free quotient groups.

LEMMA 2. (Rhemtulla [2].) *Let α_{ij} $(i = 1, 2, \dots, 2^n;\ j = 1, 2, \dots, n)$ be non-negative integers such that $\sum_{j=1}^{n} \alpha_{ij} > 0$ for all i. Let p be a prime number such that $p \geq (2\alpha)^{2^n}$ where $\alpha = \max\{\alpha_{ij} | 1 \leq i \leq 2^n,\ 1 \leq j \leq n\}$. Then in any solution set $\{x_1, \dots, x_n\}$ to the system of 2^n equations $\sum_{j=1}^{n} \varepsilon_{ij}\alpha_{ij}x_j \equiv 0 \pmod p$ associated with signs $\varepsilon_{ij} = \pm 1$ such that $x_j \equiv 0 \pmod p$ for some j.*

Proof. We use induction on n. The result is true when $n = 1$. Assume that the result holds for $n < m$ and let $n = m$. Let $d = 2^{m-1}$ and for any integer r let $r' = d + r$. By renumbering 2^m equations, if necessary, assume that $\varepsilon_{i1} = 1$, $\varepsilon_{i'j} = -1$ and $\varepsilon_j = \varepsilon_{i'j}$ for all $i = 1, 2, \dots, d$ and $j > 1$. We produce a system of d equations in $m - 1$ unknowns satisfying the hypotheses of the Lemma as follows. For any $r = 1, \dots, d$ we have

$$\alpha_{r1}x_1 + \sum_{j=2}^{m} \varepsilon_{rj}\alpha_{rj}x_j \equiv 0 \pmod p, \tag{7}$$

$$(8) \qquad \alpha_{r'j}x_1 + \sum_{j=2}^{m} \varepsilon_{rj}\alpha_{r'j}x_j \equiv 0 \pmod p.$$

As the r-th equation of the new system pick: (7) if $a_{r1} = 0$; (8) f $\alpha_{r1} \neq 0$ but $\alpha_{r'1} \neq 0$; $\sum_{j=1}^{m} \alpha_{rj}(\alpha_{r1}\alpha_{r'j} + \alpha_{r1}\alpha_{rj})x_j \equiv 0 \pmod p$ if $\alpha_{r1} \neq 0$ and $\alpha_{r'1} \neq 0$. Notice that for any r if $\sum_{j=2}^{m} = 0$ then $x_1 \equiv 0 \pmod p$ and the Lemma is proved. Thus we may assume that at least one coefficient of the r-th equation of new system is positive. Let β be the maximum of the coefficients of the new system of equations. Then $\beta \leq 2\alpha^2$ so that $p \geq (\alpha)^{2^m} = (4\alpha^2)^{2^{m-1}} \geq (2\beta)^{2^{m-1}}$. Thus the hypotheses of the Lemma are satisfied and, by induction, $x_j \equiv 0 \pmod p$ for some j. □

THEOREM 4. (Kopytov and Medvedev [1].) *If group G is a locally residually finite p-group for infinitely many primes p, then G has a weakly abelian total order.*

Proof. Let G' satisfy the hypotheses of the Theorem, but have no weakly abelian total order. By Corollary 2, we can assume that G is finitely generated and a residually finite p-group for infinitely many primes p. By Theorem 3, there are non-identity elements $a_i, x_i \in G$ $(1 \leq i \leq n$ such that $e \in S(a_1^{\varepsilon_1}[x_1, a_1], \ldots, a_n^{\varepsilon_n}[x_n, a_n])$ for all choices of signs $\varepsilon_1, \ldots, \varepsilon_n$. Let n be a positive integer with this property. Now, for every choice of signs $\varepsilon_{i1}, \ldots, \varepsilon_{in}$ $i = 1, 2, \ldots, 2^n$ there are elements $\sigma_{ij} = g_{ij}^{(1)} + \cdots + g_{ij}^{(m_{ij})}$, $m_{ij} \geq 1$ of integral group ring $\mathbf{Z}[G]$ of group G such that

$$(9) \qquad (a_1^{\varepsilon_{i1}}[x_1, a_1])^{\sigma_{i1}} \cdot (a_2^{\varepsilon_{i2}}[x_2, a_2])^{\sigma_{i2}} \cdots (a_n^{\varepsilon_{in}}[x_n, a_n])^{\sigma_{in}} = e$$

for $i = 1, 2, \ldots, 2^n$. Here $b^{g_1 + \cdots + g_n} = (g_1^{-1}bg_1)(g_2^{-1}bg_2)\cdots(g_n^{-1}bg_n)$. Let X be a normal subgroup of G generated by elements $a_1, \ldots, a_n$. Choose a prime $p > (2m)^{2^n}$, where $m = \max\{m_{ij} | 1 \leq i \leq 2^n, 1 \leq j \leq n\}$. By our assumption, there is a normal subgroup K such that G/K is a finite p-group and $X \neq K$. We can assume that $A = XK/K$ is a minimal normal subgroup of G/K. Thus XK/K is cyclic of order p. Let a_i' be the image of a_i under natural homomorphism τ of group XK on the group A. By reordering the suffices, if necessary, assume that

$A = (a'_1)$, $a'_j = a_1'^{k_j}$ where $k_1 = 0$, $0 \le k_j < p$ $(j = 2, \dots, r; r \le n)$ and $k_j = 0$ for $r < j \le n$.

The system (9) of 2^n equations now is reduced to the following.

$$(a'_1)^{(\sum_{j=1}^{n} \varepsilon_{ij} m_{ij} k_j)} = e, \quad i = 1, 2, \dots, 2^n.$$

If $m < n$, then from 2^n equations above choose 2^m equations associated with the 2^m choices of signs $\varepsilon_{ij} = \pm 1$ $(j = 1, 2, \dots, m)$ in such a way that for each $i \in \{1, 2, \dots, 2^m\}$ at last one $\alpha_{ij} \ne 0$ in this set of 2^m equations. This is possible by our choice of n. The above equations now are reduced to

$$\sum_{j=1}^{m} \varepsilon_{ij} \alpha_{ij} k_j \equiv 0 \pmod p, \quad i = 1, 2, \dots, 2^m.$$

The above equations satisfy the hipotheses of Lemma 2, and hence $k_j \equiv 0 \pmod p$ for some $j \in \{1, 2, \dots, m\}$. This is contradiction and the Theorem is proved.□

COROLLARY 3. (Robinson [1].) *Let G be a soluble group of finite special rank and G be a residually finite p-group for infinitelly many primes p. Then G is a torsion-free nilpotent group.*

Proof. Note that G has only finite number convex subgroups for any total order on G. Hence, by Theorem 4, G has a finite central series.□

6.3 Total Orders on Abelian Groups of Finite Rank

Now we describe all total orders on Abelian groups of finite rank. This descripton was found by Zaitseva [1], Teh [1], Trevisan [1], Hion [1]. In this part we use additive notaions.

Let G be a torsion-free Abelian group of finite rank and $d_1, \dots, d_r$ be a maximal linear independent over $\mathbf{Q}$ subset of elements of G. Then every element $g \in G$ has the unique representation of the form $g = k_1 d_1 + \cdots k_r d_r$ where $k_i \in \mathbf{Q}$. Let G^* be a divisible hull of G. We can regard G^* as a vector space of dimension r over $\mathbf{Q}$ with the basis

$d_1, \ldots, d_r$. It is clear that for every $x \in G^*$ there is a positive integer $n = n(x)$ such that $n \cdot x \in G$. By this reason, any total order of G can be uniquely extended to a total order of G^*. Hence the description of all total orders of G is eqivalent to that of all total orders of G^*.

Let $\leq$ be a certain total order of G^*. By finiteness of rank of G, it follows that G^* has only a finite number of convex subgroups. The group G^* is divisible, hence $G^* = A_1 \overrightarrow{\oplus} A_2 \overrightarrow{\oplus} \cdots \overrightarrow{\oplus} A_n$ is a lexicographic sum of quotient groups $A_i\,(1 \leq i \leq n)$ of convex subgroups of G^*, all subgroups A_i are Archimedean and $x = a_1 + \cdots + a_n > 0$ iff $a_k > 0$ in totally ordered group A_k where k is the least index such that $a_k \neq 0$ in totally ordered group A_k. By Hölder's theorem (Theorem 3.2.1), any totally ordered group A_i is order isomorphic to a subgroup of the totally ordered additive group of reals. Now, fix an isomorphism of group A_i into $\mathbf{R}$ and let $a_{i1}, \ldots, a_{ik_i}$ be the basis of vector space A_i over $\mathbf{Q}$.

Therefore, every total order of group G^* defines the decomposition of G^* into direct sum of subgroups $A_1, \ldots, A_n$ and determines the family of real numbers $\{c_{ij} = (a_{ij})\varphi_i | 1 \leq i \leq n,\ 1 \leq j \leq k_i\}$ such that numbers $c_{i1}, \ldots, c_{ik_i}$ are linearly independent over $\mathbf{Q}$. It is clear that any decomposition of this kind of group G^* into the direct sum is unique to within isomorphism and subgroups $B_2 = A_a \overrightarrow{\oplus} A_3 \overrightarrow{\oplus} \cdots \overrightarrow{\oplus} A_n,\quad B_3 = A_3 \overrightarrow{\oplus} \cdots \overrightarrow{\oplus} A_n, \ldots, B_n = A_n$ are defined uniquely, too. If the basis $a_{i1}, \ldots, a_{ik_i}$ of A_i are fixed then the mappings $\varphi_i : A_i \to \mathbf{R}$ are uniquely determined to within positive real multiplier. More exactly, if $c_{i1}, \ldots, c_{ik_i}$ and $c'_{i1}, \ldots, c'_{ik_i}$ correspond to the same total order of A_i and the same bases of A_i, then by Proposition 3.2.2 there is a positive real number c such that $c'_{ij} = c_j \cdot c$. If we assume that $(a_{i1})\varphi_i = 1$, then the series of subgroups $B_1 = G \supset B_2 \supset \cdots \supset B_n \supset E$ and isomorphisms φ_i defines a total order of the group G^* unquely to within the choice of basis $a_{i1}, \ldots, a_{ik_i}$ of the group B_i over B_{i+i}.

Now we fix the basis $a_1, \ldots, a_r$ of group G^* and represent the integer r as a sum of positive integers $r = k_1 + \cdots + k_s$. For every this sum there is a decomposition of G^* into a direct sum of subgroups $C_1, \ldots, C_s$ where C_i is a subspace of vector space G^* over $\mathbf{Q}$ generated by the elements $u_{i1} = a_{k_1 + \cdots k_{i-1}+1}, \ldots, u_{ik_i} = a_{k_1 + \cdot k_i}$. Then we fix isomorphism $\varphi_i : C_i \to \mathbf{R}$ such that $(u_{i1})\varphi_i = 1,\quad (u_{ij})\varphi_i = c_{ij}$ and real numbers $c_{i1}, \ldots, c_{ik_i}$ are the linearly independent system over $\mathbf{Q}$ for every $i =$

$1, 2, \ldots, s$. It is clear that the set

$$P = \{x \in G^* |\ \ x{=}\sum_{i,j} t_{ij}u_{ij},\ t_{ij} \in \mathbf{Z} \text{ and } t_{i1}c_{i1} + \cdots t_{ik_i}c_{ik_i} > 0 \text{ in } \mathbf{R} \text{ where } i \text{ is the least index with the property } \sum_{j=1}^{k_i} t_{ij}c_{ij} \neq 0\}$$

is a pure linear subsemigroup in G^*. The positive cone P defines a total order on group G^*, all subgroups $C_k + \cdots C_s$, $k = 1, 2, \ldots, s$ are convex and isomorphisms φ_i define Archimedean orders on factors. It is evident that these orders describe all total orders of group G^* to within the automorphisms of group G^*.

COROLLARY 1. *Any finitely generated Abelian totally ordered group is a lexicographic sum of Archimedean totally ordered groups.*□

COROLLARY 2. (Matsusita [1].) *Every free group with two or more free generators has uncountably many total orders.*□

6.4 Orderable Groups with Finite Number of Total Orders

1⁰. It is evident that any torsion-free Abelian group of rank 1 has only two different total orders. It is not yet known if there are orderable groups with $2n$ total orders for all positive integers n. Here we can describe all numbers of total orders of soluble orderable groups. Let us recall that a partially ordered group is order simple if there are no nontrivial convex normal subgroups.

LEMMA 1. (Kopytov [4].) *Let G be a group of order permutations of totally ordered set Ω. Suppose that*

(1) *for every element $g \in G$ there is an element $\alpha \in \Omega$ such that $\beta g = \beta$ for all $\beta \in \Omega$, $\beta \geq \alpha$;*

(2) *for every elements $\alpha, \beta \in \Omega$ there is $g \in G$ such that $\alpha g > \beta$.*

Then group G is non-abelian.

Proof. Let $F(g)$ be a set of $\alpha \in \Omega$ such that $\beta g = \beta$ for all β, $\beta \geq \alpha$. Let $g \in G$, $g \neq e$, $\alpha \in \Omega$ and $\alpha g > \alpha$. Then there are $\beta \in F(g)$ and $h \in G$ such that $\alpha h^{-1} = \gamma > \beta$. Then $\gamma h = \alpha < \gamma$ and $\gamma g h = \gamma h = \alpha$. But $\gamma h g = \alpha g > \alpha$ and $hg \neq gh$.□

THEOREM 1. (Kopytov [4].) *If totally ordered group G is order-simple, then the derived subgroup G' is also order-simple under induced total order.*

Proof. Assume, by way of contradiction, that the derived subgroup G' of order-simple group G has a non-trivial convex and normal in G' subgroup H. Then, of course,the group G is non abelian. Let Ω be the set of all convex in G' and normal in G' subgroups of G'. The set Ω is totally ordered by inclusion. Now we define the mapping φ of group G into the order permutation group of Ω by the rule: $H \cdot (g)\varphi = g^{-1}Hg$ for all $H \in \Omega$, $g \in G$. If $H \in \Omega$, then $H = \bar{H} \cap G'$ for some convex subgroup $\bar{H}$ of the group G. Therefore, $H(g)\varphi = g^{-1}(\bar{H} \cap G')g = g^{-1}(\bar{H} \cap G')g = H_1$ and H_1 is a convex subgroup of G'. It is clear that H_1 is normal in G'. If $g, h \in G$, $H \in \Omega$, then $H \cdot (gh)\varphi = h^{-1}g^{-1}Hgh = h^{-1}(H \cdot (g)\varphi)h = (H \cdot (g)\varphi)(h)\varphi = H \cdot (g\varphi)(h\varphi)$ and φ is an isomorphism of group G. It is clear also that if $A, B \in \Omega$, $A \subseteq B$, then $A \cdot (g)\varphi \subseteq B \cdot (g)\varphi$ for all $g \in G$. Hence φ is a representation of G by order permutations of totally ordered set Ω.

We claim that G satisfies the hypotheses of Lemma 1. Let $A = \bar{A} \cap G'$ and $B = \bar{B} \cap G'$, where $\bar{A}, \bar{B}$ are convex in G and $\bar{A} \subseteq \bar{B}$. We can assume that $\bar{A}$ is a maximal convex subgroup of G such that $\bar{A} \cap G' = A$. If $\bar{A}^g \subseteq \bar{B}$ for all $g \in G$ then $\bigcup\limits_{g \in G} \bar{A}^g$ is a proper convex subgroup of G which is contrary to the assumption. Hence $\bar{A}^h \supset \bar{B}$ for some $h \in G$ and $A(h)\varphi \supseteq B$, $A(h)\varphi \neq B$. So, (2) is true for $G\varphi$.

Now, let $g \in G$, $(g)\varphi \neq e$ and $\bar{A} \prec \bar{B}$ be a convex jump determined by g in G. If $C \in \Omega$, $C = \bar{C} \cap G' \supseteq B$ for some convex in G subgroup $\bar{C}$ then $C \cdot g\varphi = C$. By order-simplicity of G, there is a subgroup $C, C \in \Omega$ such that $C \supset \bar{B} \cap G'$. If $C_1 \in \Omega$, $C_1 = \bar{C}_1 \cap G'$ and $\bar{C}_1 \subset \bar{B}$, then there is an element $h \in G$ such that $h^{-1}\bar{C}_1 h \supseteq \bar{B}$. Therefore, $C_1 \cdot (h)\varphi \supseteq B$. By order-simplicity of G, $\bar{B} \cap G' \neq E$ and $\bar{B} \cap G' \neq G'$. Hence the hypothesis (1) of Lemma 1 is valid for $G\varphi$, too. So, $G\varphi$ is non-abelian. On the other hand, the kernel of the homomorphism φ contains G', which is a contradiction.□

COROLLARY 1. (Kopytov [4].) *Let G be a soluble totally ordered group. Then either G is Archimedean or it has a proper non-trivial normal subgroup.*

Proof. Let G be a conterexample of the minimal solvable lenght. Then $G' \neq E$. By Theorem 1, G' is order-simple, and by our choise of G, G' is Archimedean. Therefore, every convex subgroup C of G either contains G' or $C \cap G' = E$ and C is central in G. In both cases C is normal in G which is a contradiction. □

Corollary 1 is no longer true if the hypothesis of solvability is replaced by local solvability or the existence of ascending normal series in group G (Kopytov [4].).

THEOREM 2. (Kopytov [4].) *Let G be a solvable totally ordered group. If G is not an Abelan group or rank 1 and has a finite number of total orders, then this number of total orders is a multiple of 4. For every positive integer n there is a soluble group with $4n$ orders.*

Proof. It is clear that if P is a total order of group G, then P^{-1} is also a total order of G. Therefore, this finite number of total orders is even and there is only a finite number of convex normal subgroups under each total order. By Lemma 1, all numbers of non-trivial convex subgroups are greater than or equal to one. Let A be a minimal convex subgroup of G under some total order. Then the group G/A has only $2k$ total orders. Subgroup A has only a finite number of G-invariant total orders. Let this number be $2s$. We claim that the number of all total orders of the group G under which A is a minimal convex subgroup is equal to $4s$. Let $P_1, \ldots, P_{2s}$ be all different G-invariant total orders of A and $\bar{Q}_1, \ldots, \bar{Q}_{2k}$ be all different total orders of the quotient group G/A. Let Q_i be a pre-image in G of positive cone $\bar{Q}_i$ of G/A. Then $Q_i \cup P_j$, $1 \leq i \leq 2$, $1 \leq j \leq 2s$ are all different total orders of group G under which A is convex. Therefore the number of all total orders of G is a multiple of 4.

Now we show that for every positive integer n there exists a soluble group of finite special rank with $4n$ different total orders. Let $f(x)$ be a polynomial with integer coefficients irreducible over $\mathbf{Z}$ which has n different positive real roots. It is easy to verify that, for example, the polynomial

$$f(x) = \sum_{i=0}^{n} \gamma_i x^{n-i} = \sum_{i=0}^{n} (x - x_0)^{n-i} p(p-1) \cdots (p-i+1) C_n^{n-i}$$

where p is a prime number, x_0 is an integer greater than all roots of polynomial

$$x^n + pC_n^{n-1} + \cdots + p(p-1)\cdots(p-n+1).$$

Now, let A be a vector space over the field $\mathbf{Q}$ of rationals and φ be the linear transformation of A such that the characteristic polynomial of φ is $f(x)$. Let G be a semidirect product of A and infinite cyclic group (b) such that the multiplication in G is defined by the rule :$(a, b^k) \cdot (a', b^s) = (a + a'\varphi^k, b^{k+s})$ where $(a, b^k), (a', b^s) \in G, \quad a, a' \in A, \quad b^k, b^s \in (b)$. We claim that group G has $4n$ different total orders. Let $\alpha_i \quad (1 \leq i \leq n)$ be a positive real root of $f(x)$ and $\mathbf{Q}(\alpha_i)$ be an extension of rationals with the number α_i. The dimension of the field $\mathbf{Q}(\alpha_i)$ is n and there is an isomorphism τ_i of the vector space $\mathbf{Q}(\alpha_i)$ over $\mathbf{Q}$ onto A such that : $(1)\tau_i = a_0, \quad (\alpha_i\xi)\tau_i = (\xi)\tau_i\varphi$ where a_0 is a fixed element of A and $\xi \in \mathbf{Q}(\alpha_i)$. It is clear that $\mathbf{Q}(a_i)$ is a subfield of the field $\mathbf{R}$ of reals and total order of $\mathbf{Q}(a_i)$ induced from $\mathbf{R}$ is stable under multiplication by α_i. Let R_i be the image of the positive cone of the totally order additive group of $\mathbf{Q}(\alpha_i)$ under isomorphism τ_i. Then R_i is an invariant set under linear transformation φ.

Now from the total order R_i we construct two total orders P_{+i} and P_{-i} of group G by the rule :

$$P_{+i} = \{(a, b^k) | k > 0 \text{ or } k = 0, a \in R_i\}$$

$$P_{-i} = \{(a, b^k) | k < 0 \text{ or } k = 0, a \in R_i\}.$$

All orders $P_{\pm 1}, i = 1, 2, \ldots, n$ are different. From $b \in P_{+i}, \quad b \notin P_j \quad 1 \leq i, j \leq n$ it follows that $P_{+i} \neq P_{-j}$ for all i, j. Each total order $R_i, \quad 1 \leq i \leq n$, is Archimedean. Hence conjugation by b of group G with total order R_i induces on A the multiplication on positive real number

$$\alpha_i = \frac{((a_0)\varphi)\tau_i^{-1}}{(a_0)\tau_i^{-1}}.$$

As $\alpha_i \neq \alpha_j$ for $i \neq j$, then all total orders $R_i, \quad i = 1, 2, \ldots, n$ are different. So, G has $4n$ different total orders $P_{\pm i}^{\pm 1}$.

Now we claim that every total order of group G is equal to some $P_{\pm i}^{\pm 1}$. Let P be any total order of G and $a_0 > e$ under P. Group G is soluble and has a finite rank. So every convex subgroup of G is normal in G.

Let H be a proper convex subgroup of G under P. From irreducibility of A under φ it follows that $H = A$ or $H \cap A = E$. The latter is impossible because centralizer $C(A)$ of A in G is equal to A. Therefore, A is convex under all total orders of G. Let τ be an order embedding of totally ordered group $(A, P \cap A)$ into totally ordered additive group of reals. Then $(b^{-1}a_0b)\tau \cdot ((a_0)\tau)^{-1} = (b^{-1}ab)\tau \cdot ((a)\tau)^{-1} = \alpha > 0$ for all $a \in A$. Without less of generality, we may assume that $(a-0)\tau = 1$. Then $(1)\tau^{-1}, (\alpha)\tau^{-1}, \ldots, (\alpha^{n-1})\tau^{-1}$ is a basis of vector space A over $\mathbf{Q}$. As the characteristic polynomial of φ is $f(x) = \sum_{j=0}^{n} \gamma_j x^{n-j}, \quad \gamma_j \in \mathbf{Z}$, then in totally ordered group A the equality $b^{-n}a_0b^n + \gamma_1 b^{-n+1}a_0b^{n-1} + \cdots + \gamma_n a_0 = 0$ is valid. Hence $(b^{-n}a_0b^n + \gamma_1 b^{-n+1}a_0b^{n-1} + \cdots + \gamma_n a_0)\tau = \alpha^n + \gamma_1\alpha^{n-1} + \cdots + \gamma_n = 0$ in the field $\mathbf{R}$. Therefore, α is a real root of the polynomial $f(x)$ and $\alpha = \alpha_i$ for some $i, \quad 1 \leq i \geq n$. Then the total order $P \cap A$ is equal to some R_i and P is equal to some $P_{\pm i}$.□

6.5 Fully Orderable Groups

Recall that group G is fully orderable if every partial order of G can be extended to a total order of G. The class of fully orderable groups is a subclass of class O-groups and is considerably smaller. The examples illustrating the situation are presented in Chapter 6. On the other hand, this class is large enough to contain some interesting classes such as torsion-free nilpotent groups, metabelian O-groups. This class is very important for the theory of l-groups because many different questions have solutions for a class of fully orderable groups.

Now we single out the cornestone condition of the extension of partial orders.

(O^*) If $b, c \in S(a), \quad a \in G$ then $S(b) \cap S(c) \neq \emptyset$.

Here $S(x)$ is a minimal normal subsemigroup of G generated by x.

The group G is a group with *strongly isolated identity* or a Γ*-torsion-free group* if $x^{g_1} \cdot x^{g_2} \cdots x^{g_n} = e$ implies $x = e$. From the results of Chapter 6 §1, any O-group is a Γ-torsion-free group.

LEMMA 1. *Let G be a Γ-torsion-free group and $a \in G$. Then the* *element a is comparable with e under every maximal partial order on G if and only if a has property (O^*).*

Proof. Let a be comparable with e under every maximal partial order and there are $b, c \in S(a)$ such that $S(b) \cap S(c) = \emptyset$. By Lemma 4.1.1, the normal subsemigroup $S(b) \cdot S(c^{-1}, e)$ defines some partial order of G. Suppose Q is a maximal partial order extending $S(b) \cdot S(c^{-1}, e)$. Then $a \in Q \cup Q^{-1}$ and either $a \in Q$ or $a \in Q^{-1}$. If $a \in Q$ then $S(a) \subseteq Q$, $c \in Q$ which contradicts to our assumption. Similar arguments show that $a \notin Q^{-1}$.

Conversely, let G be a Γ-torsion-free group, a have property (O^*), P be a maximal partial order, and $a \notin P \cup P^{-1}$. If $S(a) \cap P^{-1} = \emptyset$, then $P \cdot S(a, e)$ is a partial order extending order P on G. This contradicts maximality of P. Similarly, if $S(a) \cap P = \emptyset$, then $P \cdot S(a^{-1}, e)$ is an extinsion of P. Thus, $S(a) \cap P \neq \emptyset$ and $S(a) \cap P^{-1} \neq \emptyset$. Take $b \in S(a) \cap P$, $c \in S(a) \cap P^{-1}$, $d \in S(b) \cap S(c)$. Since G is Γ-torsion-free, then $d \neq e$. Thus, $e \neq d \in P \cap P^{-1}$, a contradiction.□

THEOREM 1. (Ohnishi [1].) *A group G is a fully orderable group iff G is a Γ-torsion-free group and every element of G has property (O^*).*□

COROLLARY 1. (Mal'cev [4].) *If every finitely generated subgroup of group G is fully orderable, then G is fully orderable.*□

2^0.A subgroup H of group G is *G-fully orderable* if for every maximal partial order P of G, the restriction $P \cap H$ of P to H is a total order of H or, equivalently, $H \subseteq P \cup P^{-1}$.

COROLLARY 2. *A subgroup H of group G is G-fully orderable if every element of H has property* (O*).□

LEMMA 2. (Kokorin [6].) *Let G be an o-group and*

$$a = g^{x_1 + \cdots + x_n} = x_1^{-1} g x_1 \cdots x_n^{-1} g x_n,$$

$g, x_i \in G$ $(i = 1, \ldots, n)$. Then there are elements $y, z \in G$ such that $a^z \leq g^n \leq a^y$.

Proof. Let x_k, x_r $(1 \leq k, r \leq n)$ be the elements of G such that $x_k^{-1} g x_k \leq x_i^{-1} g x^i \leq x_r^{-1} g x_r$ for all $i = 1, \ldots, n$. Then $(g^n)^{x_k} \leq a \leq (g^n)^{x_r}$ and $z = x_k^{-1}$, $y = x_r^{-1}$.□

LEMMA 3. (Kokorin [6].) *Let G be an orderable group, H be a normal G-fully orderable relatively convex subgroup of G. If*

$$a^{g_1+\cdots+g^n} \cdot a^{-n} \in H$$

for all $g_1, \ldots, g_n$ *in* G, *then the element a is comparable with e under every maximal partial order on G.*

Proof. Let P be a maximal partial order on G and element a is noncompareble with e under P. Then $S(a) \cap P \neq \emptyset$ and $S(a) \cap P^{-1} \neq \emptyset$. Hence there are elements $s_1, s_2 \in P$ such that $s_1 = a^{g_1+\cdots+g_n}$, $s_2 = (a^{-1})^{h_1+\cdots+h_n}$. By orderability of the group G/H and Theorem 2.3.4, there is a total order Q of G such that $H \cap Q = H \cap P = P_1$ and subgroup H is convex under Q. By Lemma 2, there are $y, z \in G$ such that $s_1^y \leq (s_2^{-1})^z$. Then $s_1^y \leq (s_2^{-1})^z$ and $h = s_1^y s_2^z \leq e$. This contradicts $h \in P_1 \subseteq Q$.□

THEOREM 2. (Mal'cev [4].) *A torsion-free localy nilpotent group is a fully orderable group.*

Proof. Let G be a torsion-free locally nilpotent group. By Corollary 1, we may assume that G is a finitely generated nilpotent group. Let

$$E \subset Z_1 \subset \cdots \subset Z_n = G$$

be the upper central series of G. It is clear that Z_1 is G-fully orderable subgroup of G. Suppose that Z_k is G-fully orderable for some k $(1 \leq k \leq n-1)$. By definition of central series, $a^{x_1+\cdots x_m} \cdot a^{-m} \in Z_k$ for all $a \in Z_{k+1} \setminus Z_k$, $x_1, \ldots, x_m \in G$. Then element a is comparable with e under every maximal order on G by Lemma 3. Therefore, Z_{k+1} ia a G-fully orderable subgroup of G and, by inductioin, G is fully orderable.□

COROLLARY 3. (Kokorin [6].) *Every member of the upper central series of orderable group G is G-fully orderable.*□

COROLLARY 4. (Kokorin [6].) *Let G be an orderable group and H be an normal G-fully orderable subgroup such that G/H is a nilpotent torsion-free group. Then G is fully orderable.*

THEOREM 3. (Kokorin [6].) *Every metabelian Γ-torsion-free group is a fully orderable.*

Proof. By Lemma 3.2.1, the isolator A of derived subgroup G' of group G is a G-fully orderable subgroup of G. The subgroup A is Abelian and the proof of Theorem 3 follows from Corollary 4. □

THEOREM 4. (Kokorin [5], Kargapolov [1].) *Let A, B be fully orderable groups. Then the direct product $A\times B$ is also a fully orderable group.*

Proof. Let A, B be fully orderable groups, $G = A \times B$ and P be a maximal partial order on G. All elements of subgroups A, B have the property (O^*) in group G. Therefore, the subgroups A, B are totally ordered under P. We claim that for every elements $s = ab, \quad a \in A, \quad b \in B$ either $s \in P$ or $s \in P^{-1}$ is valid.

Suppose, for reductio ad absurdum, that $S(s) \cap P \neq \emptyset$ and $S(s) \cap P^{-1} \neq \emptyset$. Then there are elements $s_1, s_2^{-1} \in P$ such that

$$s_1 = (ab)^{x_1+\cdots+x_n} = a^{x_1+\cdots+x_n} \cdot b^{x_1+\cdots+x_n}$$

and

$$s_2^{-1} = ((ab)^{-1})^{y_1\cdots+y_n} = a^{y_1+\cdots+y_n} \cdot b^{y_1+\cdots+y_n}.$$

From total orderability of groups A, B under P by Lemma 2, there are elements $a_1, a_2 \in A \quad b_1, b_2 \in B$ such that $(a^{x_1+\cdots+x_n})^{a_1} \leq (a^{y_1+\cdots+y_n})^{a_2}$ and $(b^{x_1+\cdots+x_n})^{b_1} \leq (b^{y_1+\cdots+y_n})^{b_2}$. Therefore, $s_3 = s_1^{b_1a_1}(s_2^{-1})^{b_2a_2} \leq e$. This contradicts $s_1, s_2^{-1} \in P$.□

The class of fully orderable groups is closed unger direct products and it is not closed under Cartesian products (Kargapolov [1]) and therefore the class of fully orderable groups cannot be axiomatized by formulae of first-order logic. The class of fully orderable groups is not closed under subgroups (Kopytov [5]), subdirect products (Mura and Rhemtulla [2]). On the other hand, every orderable soluble group of finite special rank is fully orderable (Mura and Rhemtulla [2]).

3⁰. The classes of orderable groups and fully orderable groups are different. The first example of this kind was given by Holland [1]. In this section we prove Holland's test for it.

Let $G_H = (\mathbf{R} \oplus \mathbf{R}) \cdot A$ be a splitting extension of direct sum $K = \mathbf{R} \oplus \mathbf{R}$ of additive groups of reals by the subgroup $A = \mathrm{gp}(\alpha, \beta)$ of the automorphisms group of K generated by automorphisms α, β defined by rules:

$$(x,y)^{\alpha}=(y,x),\quad (x,y)^{\beta}=(x,\frac{y}{2}) where x,y\in\mathbf{R},\quad (x,y)\in K.$$

In group G_H consider elements $h_1=\alpha,\ h_2=\beta$ and $h_3=(-3,4)$. Let $P=\{(x,y)\in K|x>0,y>0 \text{ in } \mathbf{R}\}$. It is easily verified that P is a pure normal subsemigroup of G_H and $e=(0,0)\notin P$. We then note that $h=h_3\cdot h_1^{-1}h_3h_1=h_3h_3^{\alpha}=(1,1)\in P$ and $(h_2h_1)^{-1}h_3^{-1}(h_2h_1)\cdot(h_2^{-1}h_3^{-1}h_2)=(-2,3)(3,-2)=(1,1)=h$. Since $h\in P$ the normal subgroup of G_H generated by h does not contain $e=(0,0)$.

THEOREM 5. (Holland [1].) *Let X be a group and N be a normal subgroup of X such that the quotient group X/N is isomorphic to the group G_H. Then X is not fully orderable.*

Proof. Suppose, for reductio ad absurdum, that X is fully orderable. Then there is an homomorphism $\varphi: X\to G_H$ which maps X onto group G_H. Hence there are elements $a,b,c\in X$ such that $a\varphi=h_1,\ b\varphi=h_2\ c\varphi=h_3$. Let $S(f_1,f_2)$ be a normal subsemigroup of X generated by f_1,f_2 where $f_1=c\cdot a^{-1}ca,\quad f_2=(ba)^{-1}c(ba)\cdot(b^{-1}cb)$. It is easily verified that $f_1\varphi=h_3\cdot h_1^{-1}h_3h_1=h,\quad f_2\varphi=(h_2h_1)^{-1}h_3^{-1}(h_2h_1)\cdot h_2^{-1}h_3^{-1}h_2=h$ and $S(f_1,f_2)\varphi\subseteq P$. Therefore, $S(f_1,f_2)$ is a pure normal subsemigroup of X and it defines a partial order on X. We claim that this partial order cannot be extended to total order of X. Suppose that Q is a total order of X extending P. As $f_1\in Q$, then $c\in Q$ and $f_2^{-1}=b^{-1}cb\cdot(ba)^{-1}c(ba)\in Q$, a contradiction. □

COROLLARY 1. *A free group with three or more free generators is not fully orderable.*□

COROLLARY 2. *A free soluble group with derived lenght more than 3 and with three or more free generators is not fully orderable.*

Other examples of not fully orderable groups were constructed by Kargapolov [1], Fuchs and Sasiada [1], Kopytov [3]. The readers can find these examples in the book of Kokorin and Kopytov [2].

Chapter 7

Embeddings of lattice-ordered groups

7.1 Embeddings into Lattice-Ordered Groups

In this section we describe the partially ordered groups which are order isomorphic to subgroups of l-groups or Cartesian product of totally ordered groups. The results of Chapters 4 and 5 are very usefull for this purpose.

THEOREM 1. (Hollister [1], Conrad [5].) *Let G be a group and P be a lattice order on G. Then P is the intersection of right orders of G.*

Proof. Let G be an l-group and P be a positive cone of G. By Theorem 4.1.1, we can assume that G is an l-subgroup of l-group $A(\Omega)$ of order permutations of totally order set Ω. We have $g \in P$ iff $\alpha g \geq \alpha$ for all $\alpha \in \Omega$. If $g \notin P$, then there is $\alpha_g \in \Omega$ such that $\alpha_g g < \alpha_g$. With element α_g we construct a total right order P_g on group G such that $g \notin P_g$, $P \subseteq P_g$. For this purpose we well-order the set Ω such that the first element is α_g: $\alpha_g = \alpha_1 \prec \alpha_2 \prec \ldots \prec \alpha_i \prec \alpha_{i+1} \prec \ldots$ $(i \in I)$. Then we define the right order P_g as we did in Theorem 5.2.1 : $h \in P_g$ iff $\alpha_i g > \alpha_i$ where α_i is the least (under $\prec$) element of Ω such that $\alpha_i g \neq \alpha_i$. It is evident that P_g is a total right order on G and

$P \subseteq P_g$, $g \notin P_g$. Therefore, $P = \bigcap_{g \notin P} P_g$.□

The question of embeddings of partially ordered groups into l-groups is connected with the notion of an l-group free over partially ordered group.

Let $\mathcal{K}$ be a class of l-groups and G be a partially ordered group. The lattice-ordered group $F_{\mathcal{K}}(G, \pi) \in \mathcal{K}$ is *$\mathcal{K}$-free l-group over po-group* G if there is an order isomorphism $\pi : G \to F_{\mathcal{K}}(G, \pi)$ such that

(1) π^{-1} is also an order isomorphism of $\pi(G)$ onto G;

(2) $\pi(G)$ generates $F_{\mathcal{K}}(G, \pi)$ as an l-group;

(3) if H is an l-group, $H \in \mathcal{K}$ and $\sigma : G \to H$ is an order homomorphism G into H, then there exists an l-homomorphism $\tau : F_{\mathcal{K}}(G, \pi) \to H$ such that $\sigma = \pi\tau$.

The notion of $\mathcal{K}$-free l-group over po-group was introduced by Weinberg [2] for Abelian l-groups and by Conrad [5] for the class of all lattice-ordered groups $\mathcal{L}$. In this case $F_{\mathcal{L}}(G, \pi)$ is called a *free l-group over po-group* G and is denoted by $F(G, \pi)$. It is clear that if the free over po-group G l-group $F(G, \pi)$ exists, then partial order of G is isolated.

THEOREM 2. (Conrad [5].) *Let G be a po-group and a partial order P of G. Then an l-group $F(G, \pi)$ free over po-group G exsists if and only if P is the intersection of some family of total right orders of G. If $F(G, \pi)$ for G exists, then it is unique.*

Proof. Let $F(G, \pi)$ be an l-group free over G . We leave to the reader the standard proof that $F = F(G, \pi)$ is unique. We can assume that $G \subseteq F(G, \pi)$ and $P = Q \cap G$ where Q is a lattice order on F. By Theorem 1, $Q = \bigcap_{i \in I} Q_i$ where $\{Q_i | i \in I\}$ is a certain set of right total orders on F. Therefore $P = G \cap Q = G \cap (\bigcap_{i \in I} Q_i) = \bigcap_{i \in I} (G \cap Q_i)$.

Conversely, let G be a po-group, P be a positive cone of G and $P = \bigcap_{i \in I} P_i$ where $\{P_i | i \in I\}$ the set of all right orders on G extending P. For each $i \in I$ let G_i be the right ordered group obtained when group G is endowed with the right order P_i. Then the right regular represention $R_i : G \to A(G_i)$, $xR_i(g) = xg$ for all $x \in G_i, g \in G$ is an isomorphic embedding of group G into the l-group of order permutations $A(G_i)$ of totally ordered set G_i. Let B_i be an l-subgroup of $A(G_i)$ generated by

$R_i(G)$ and $B = \bar{\prod_{i\in I}} B_i$. Then we can embed G into l-group B via the map π whose i-th component is just the right regular represention R_i of G into $A(G_i)$, $(g\pi)_i = R_i(g)$ for $g \in G$. Let F be an l-subgroup of B generated by $G\pi$. We assert that (F,π) is a free l-group over G.

Let $g \in P$. Then $R_i(g) > e$ for all $i \in I$. (If there is $i \in I$ such that $xR_i(g) < x$ for some $x \in G_i$, then $xg < x$ in ro-group G_i. Hence $xgx^{-1} < e$ in G_i and $xgx^{-1} \notin P_i$. But $g \in P$ and therefore, $xgx^{-1} \in P$. This contradicts our assumption $P \subseteq P_i$). So, $(g\pi)_i = R_i(g) > e$ in B_i for all $i \in I$ and $g\pi > e$ in B.

Conversely, let $g\pi > e$ in B, then $(g\pi)_i = R_i(g) > e$ in $A(G_i)$. Then $eR_i(g) = eg = g$ in the totally ordered set G_i and $g \in P_i$ for all $i \in I$. Therefore, $g \in \bigcap_{i\in I} P_i$, $g \in G$ and π is an order isomorphism G into F.

To prove that the l-group (F,π) constructed above is an l-group free over po-group G , we suppose that σ is an order homomorphism of G into l-group H. Now every element x of F can be written in the form $x = \bigvee_{\delta\in\Delta}\bigwedge_{\gamma\in\Gamma}(g_{\delta\gamma})\pi$ where Γ,Δ are finite sets and $g_{\delta\gamma} \in G$. We define the mapping $\tau : F \to H$ by the rule: $x\tau = (\bigvee_{\delta\in\Delta}\bigwedge_{\gamma\in\Gamma}(g_{\delta\gamma})\pi)\tau = \bigvee_{\delta\in\Delta}\bigwedge_{\gamma\in\Gamma}(g_{\delta\gamma})\sigma$. We claim that $x\tau \neq e$ in H implies $x \neq e$ in F. If $x\tau \neq e$, then by Holland's Embedding Theorem (Theorem 4.1.1.), H is an l-subgroup of l-group $A(\Omega)$ of order permutations of some totally ordered set Ω. As $x\tau = \bigvee_{\delta\in\Delta}\bigwedge_{\gamma\in\Gamma}(g_{\delta\gamma})\sigma \neq e$, then there is $\zeta \in \Omega$ such that

$$\zeta \neq \zeta(\bigvee_{\delta\in\Delta}\bigwedge_{\gamma\in\Gamma}(g_{\delta\gamma})\sigma) = \max_{\delta\in\Delta}\{\min_{\gamma\in\Gamma}(g_{\delta\gamma})\sigma\}.$$

As in Theorem 1 we well-order the set Ω such that ζ is the least element of Ω under this well-ordering. Now we define a total right order Q on group H by the rule: $h \in Q$ iff $\xi h > \xi$ where ξ is the least element of Ω under well-ordering of Ω such that $\xi h \neq \xi$. It is clear that the right order Q extends lattice order of H, $P\sigma \subseteq G\sigma$ and every right order of the po-group $H\sigma$ is induced by some right order of the po-group G. Therefore, there is a right order P_i $(i \in I)$ on G such that $\bigvee'_{\delta\in\Delta}\bigwedge'_{\gamma\in\Gamma} g_{\delta\gamma} \neq e$ (here $\vee', \wedge'$ are a join and a meet in the right ordered group G_i). Then in $A(G_i)$ the following inequalities are valid:

$$(e)x = \max_{\delta\in\Delta}\{\min_{\gamma\in\Gamma} eg_{\delta\gamma}\} = \max_{\delta\in\Delta}\{\min_{\gamma\in\Gamma} g_{\delta\gamma}\} = \mathop{\dot{\bigvee}}_{\delta\in\Delta}\mathop{\dot{\bigwedge}}_{\gamma\in\Gamma} g_{\delta\gamma} \neq e$$

and $(e)(\bigvee_{\delta\in\Delta}\bigwedge_{\gamma\in\Gamma} g_{\delta\gamma}\pi_i) \neq e$. Hence

$$(\bigvee_{\delta\in\Delta}\bigwedge_{\gamma\in\Gamma} g_{\delta\gamma}\pi_i) \neq e, \quad x = (\bigvee_{\delta\in\Delta}\bigwedge_{\gamma\in\Gamma} g_{\delta\gamma}\pi) \neq e.$$

It follows that τ is well-defined and it is an l-homomorphism of F into H. The equality $\sigma = \pi\tau$ is evident. □

COROLLARY 1. (Conrad [5].) *Let G be a po-group and P be a positive cone of G. Then G is order isomorphic to a subgroup of an l-group iff P is intersection of right orders of G.*□

2^0. In this connection we give the following result.

THEOREM 3. (Kopytov [6].) *Po-group G with positive cone P is order isomorphic to a transitive subgroup of l-group $A(\Omega)$ of order permutations of the totally ordered ser Ω if and only if $P = \bigcap_{x\in G} x^{-1}Qx$ for some total right order Q of G.*

Proof. Let $P = \bigcap_{x\in G} x^{-1}Qx$, where Q is a certain total right order of G. Now consider the right regular represntion $R_Q : G \to A(G)$ of G by order permutation of the totally ordered under right order Q set G. As $xg = xgx^{-1}\cdot x > e$ for all $g \in P$, then $R_Q(g) > e$ in $A(G)$ for all $g \in P$. If element g is incomparable with e in the partial order P and $g \in Q$, then there is $y \in G$ such that $ygy^{-1} \notin Q$. Therefore, $eg > e$, $yg = ygy^{-1}\cdot y < y$ and $R_Q(g)$ is incomparable with e in l-group $A(G)$. This means that G is order isomorphic to a transitive subgroup of l-group $A(\gamma)$.

Conversely, let G be a transitive subgroup of l-group $A(\Omega)$ of order permutation of the totally ordered set Ω and P be a partial order on G induced by lattice order of $A(\Omega)$. As in Theorem 1 we well-order the set Ω : $\xi_0 \prec \cdots \prec \xi_i \prec \xi_{i+1} \prec \cdots$, $i \in I$, and define a total right order Q on G by the rule : $g \in Q$ iff $\zeta g > \zeta$ where ζ is the least element under $\prec$ in Ω such that $\zeta g \neq \zeta$. It is clear that $P \subseteq \bigcap_{f \in G} f^{-1}Qf$. Let $g \notin P \cup P^{-1}$. Then there are $\xi_i, \xi_j \in \Omega$ such that $\xi_i g > \xi_i$, $\xi_j g < \xi_j$. By transitivity G, there are $f, g \in G$ such that $\xi = \zeta f, \xi_j = \zeta h$. Then $\zeta hf < \zeta h$ and $\zeta fgf^{-1} > \zeta$, $\zeta hgh^{-1} < \zeta$. This means that $fgf^{-1} \in Q$ and $hgh^{-1} \notin Q$ or, equivalently, $g \in f^{-1}Qf$, $g \notin h^{-1}Qh$. Hence $g \in P$ iff $g \in \bigcap_{x \in G} x^{-1}Qx$.□

7.2 Vector Groups. Projectable l-Groups

Partially ordered groups which are embedable into Cartesian products of totally ordered groups have a lot applications in functional analysis, where spaces of various kinds of real-valued functions provide a momentum for investigating partially ordered vector spaces. In the past two decades the observation that Grothendick groups (such as K_0 of a ring or algebras) often possess a natural partially ordered Abelian group structure of this kinds, led to new directions of investigations. Another reason of our interest in Cartesian product of totally ordered groups is a possibility to apply developed techniques and methods of the theory of totally ordered groups for these partially ordered groups.

1⁰. A subgroup G of Cartesian product $\bar{G} = \bar{\prod}_{i \in I} G_i$ of family $\{G_i | i \in I\}$ of totally ordered groups under induced from $\bar{G}$ lattice order is called a *vector group*. Every element of G is a function f on I for which $f(i) \in G_i$ for each $i \in I$. Every G_i is called a component of vector group G. The element $f \in G$ is positive iff $f(i)$ is positive in G_i for all $i \in I$. By Example 2.3.1, there are vector groups which are not l-subgroups of $\bar{G}$ under the induced partial order. It is easy to see that the class of vector groups is closed under Cartesian products and subgroups.

THEOREM 1. *A partially ordered group is order isomorphic to a vector group exactly if its positive cone is an intersection of total orders.*

Proof. Let G be a vector group, $G \subseteq \bar{G} = \bar{\prod}_{i \in I} G_i$ and $P = \{f \in G | f(i) \geq e$ in G_i for all $i \in I\}$ be a positive cone of G. Let $i \in I$. Now we well-order the index set I by relation $\prec_i$ such that i is the least element of I under $\prec_i$. We define a lexicographical total order P_i on group $\bar{G}$ by the rule : $f \in P_i$ iff $f(j) > e$ in the totally ordered group G_j where j is the least element of I such that $f(j) \neq e$. It is evident that $P \subseteq P_i$ for arbitrary $i \in I$ and $P = \bigcap_{i \in I}(P_i \cap G)$.

Conversely, let a partial order P of G be an intersection of total orders P_i, $i \in I$, of group G. The same arguments as in the proof of Theorem 7.1.2 show that group G can be embedded into $H = \bar{\prod_{i \in I}} G_i$ where G_i is group G totally ordered by P_i for all $i \in I$. So, G is a vector group. □

2^0. Let us recall that a partial order P of a group G is strongly isolated if $g^{x_1 + \cdots x_n} \in P$ implies $g \in P$.

THEOREM 2. (Mal'cev [6].) *A partial order P of fully orderable group G is an intersection of total orders if and only if the order P is strongly isolated.*

Proof. It is clear that any total order is strongly isolated, then the intersection of any family of total orders is srtongly isolated, too.

Conversely, let P be a strongly isolated order of fully orderable group G and $a \notin P$. Then, by Lemma 6.1.1, the normal subsemigroup $P \cdot S(a^{-1}, e)$ is a partial order on G. Let Q_a be a total order on G extending P. So, $a \notin Q_a$ and $\bigcap_{a \in G \setminus P} = P$.□

COROLLARY 1. (Mal'cev [6].) *A partially ordered locally nilpotent group is isomorphic to a vector group if and only if its partial order is strongly isolated.*□

In this connection we mention that there are torsion-free groups G with isolated and not strongly isolated partial order (see Mal'cev [6] or Kokorin and Kopytov [1].).

3^0. A lattice ordered group G is *representable* if G is an l-subgroup of the Cartesian product of totally ordered groups. It is easy to see that l-group G is representable iff there is a family of prime ideals $\{N_\alpha | \alpha \in A\}$ of G such that $\bigcap_{\alpha \in A} N_\alpha = E$. The class of representable l-groups is very close to vector groups. The relation between these classes will be considered in Chapter 8.3.

An l-group G is called *strongly projectable* (*SP-group*) if every polar in G is a direct factor of G. Here we investigate the question on the embeddings of an l-group into a SP-group.

We say that l-group G^* is a SP-*hull* of l-group G if G^* is a strongly projectable l-group and there is an l-isomorphism ε of G into G^* such that:

(1) for every $g \in G^*$, $g > e$ there is an element $h \in G$ with $e < h\varepsilon \leq g$,

(2) no proper l-subgroup of G^* that contains $(G)\varepsilon$ is a strongly projectable l-group.

We need a few definitions from the theory of Booleal algebras. A *field of subsets* of the set A is a collection $\mathcal{B}$ of subsets of A such that:

$$\emptyset \in \mathcal{B}; \tag{3}$$

$$M \cap N \in \mathcal{B} \text{ if } M, N \in \mathcal{B}; \tag{4}$$

$$A \setminus M \in \mathcal{B} \text{ if } M \in \mathcal{B}. \tag{5}$$

Each field of subsets of A is a Boolean algebra under the partial-ordering of inclusion. On the other hand, suppose $\mathcal{B}$ is a collection of subsets A satisfing (3) and (4). Then, as it is well-known in the Boolean algebras theory, it is possible that $\mathcal{B}$ is a Boolean algebra under the partial-ordering of inclusion but not a field of subsets of A. (The collection of all regular open subsets of the real line is an example of this kind.)

PROPOSITION 1. *Suppose $\mathcal{B}$ is a collection of subsets of a set A with property* (3)-(4). *If $\mathcal{B}$ is the Boolean algebra under the partial-ordering of inclusion then there exists a Boolean isomorphism η of $\mathcal{B}$ onto field $\mathcal{F}$ of subsets of A with $M\eta \supseteq M$ for all $M \in \mathcal{B}$.*

Proof. Let E_ξ be the collection of all $M \in \mathcal{B}$ such hat $\xi \in M$. Then E_ξ is empty or E_ξ is a filter in $\mathcal{B}$. If E_ξ is empty, let U_ξ be any ultrafilter on $\mathcal{B}$; otherwse let U_ξ be an ultrafilter on $\mathcal{B}$ such that $U_\xi \supseteq E_\xi$. We define $M\eta = \{\xi \in A | M \subseteq U_\xi\}$ for all $M \in \mathcal{B}$. Now the direct verifcation shows that $\mathcal{B}\eta = \mathcal{F}$ is a field of subsets of A and η is an Boolean isomorphism.□

Let G be an l-subgroup of the Cartesian product $\bar{G} = \bar{\prod\limits_{\alpha\in\Lambda}} G_\alpha$ where G_α, $\alpha \in \Lambda$, are totally ordered groups. For every subset $K \subset \bar{G}$ let $\Lambda(K) = \{\alpha \in \Lambda | g(\alpha) \neq e \text{ for some } g \in K\}$. Now we define the map $\bar{\theta}$ from the Boolean algebra $\mathcal{P}(G)$ of all polars in l-group G into the family of subsets of Λ by the rule: $H\bar{\theta} = \Lambda(H)$ for all $H \in \mathcal{P}(G)$. It is clear that $\bar{\theta}$ is an isomorphism of $\mathcal{P}(G)$ onto the family $\mathcal{B}$ subsets of Λ and $\mathcal{B}$ is a Boolean algebra under inclusion: $\Lambda(M) \cap \Lambda(N) = \Lambda(M \cap N)$ where $M, N \in \mathcal{P}(G)$ and $\Lambda(E) = \emptyset$. By Proposition 1, there exists an isomorphism η of $\mathcal{B}$ on the field $\mathcal{F}$ of subsets of Λ. Then $\theta = \bar{\theta}\eta$ maps $\mathcal{P}(G)$ onto $\mathcal{F}$.

LEMMA 1. (Bleier [1].) *If $g \in G$, $M \in \mathcal{P}(G)$ and $\Lambda(g) \cap \Lambda(M) = \emptyset$, then $\Lambda(g) \cap (M)\theta = \emptyset$.*

Proof. $\Lambda(g) \cap \Lambda(M) = \emptyset$ implies $g^{\perp\perp} \cap M = E$ and hence $(g^{\perp\perp})\theta \cap (M)\theta = \emptyset$. Since $\Lambda(g) \subseteq \Lambda(g^{\perp\perp}) \subseteq (g^{\perp\perp})\theta$, then $\Lambda(g) \cap (M)\theta = \emptyset$.□

Let $\mathcal{F}$ be a field of subsets of Λ. Suppose $h \in \bar{G} = \bar{\prod\limits_{\alpha\in\Lambda}} G_\alpha$ is such that for some finite partition of Λ in $\mathcal{F}$, say $F_1, \ldots, F_n$ there are $g_1, \ldots, g_n \in G$ such that $h(x) = g_i(x)$ for all $x \in F_i$, $i = 1, \ldots, n$. Then we write $h = [g_i | F_i]$ and we denote the set of all such h by $G[\mathcal{F}]$. The direct verification shows that $G[\mathcal{F}]$ is an l-subgroup of $\bar{G}$, $G \subseteq G[\mathcal{F}]$ and the following statements are valid in $G[\mathcal{F}]$:

(6) if $B \in \mathcal{F}$, then the subset $G(B) = \{h \in G | \Lambda(h) \subseteq B\}$ is an ideal in $G[\mathcal{F}]$ and $G[\mathcal{F}] = G(B) \times G(\Lambda \setminus B)$;

(7) $\Lambda(h) \cap (M)\theta = \emptyset$ for all $h \in G[\mathcal{F}]$, $M \in \mathcal{P}(G)$ such that $\Lambda(h) \cap \Lambda(M) = \emptyset$;

(8) for every $h \in G[\mathcal{F}], h > e$ there is an element $g \in G$ with $e < g \leq h$;

(9) $G(M\theta)$ is a polar in $G[\mathcal{F}]$ for all $M \in \mathcal{P}(G)$.

THEOREM 3. (Bleier [1].) *Let G be an l-subgroup of $\bar{G} = \bar{\prod\limits_{\alpha\in\Lambda}} G_\alpha$, where each G_α is a totally ordered group. Then $G[\mathcal{F}]$ is a SP-hull of G.*

Proof. By (9) and (6), the l-group $G[\mathcal{F}]$ is strongly projectable, $G \subseteq G[\mathcal{F}]$ and G is dense in $G[\mathcal{F}]$. Suppose that H is an l-subgroup of $G[\mathcal{F}]$, $H \supseteq G$ and H is strongly projectable. If $K \subseteq H$, then by K^{∇} we denote the polar of K in H. Then $H = K^{\nabla} \times K^{\nabla\nabla}$ and $\Lambda(K^{\nabla}) \cap \Lambda(K^{\nabla\nabla}) = \Lambda(H) = \emptyset$ for every $K \subseteq H$. From desity of G in $G[\mathcal{F}]$ it follows that H is dense in $G[\mathcal{F}]$. Therefore, $M^{\nabla\nabla} \subseteq (M^{\nabla\nabla})^{\perp\perp} \subseteq M^{\perp\perp}$, $M^{\nabla} \subseteq (M^{\nabla})^{\perp\perp} = M^{\perp}$ for every subset M of H. Then $\Lambda(M^{\nabla\nabla}) \subseteq \Lambda(M^{\perp\perp})$ and $\Lambda(M^{\nabla}) \subseteq \Lambda(M^{\perp})$. But $\Lambda(M^{\perp\perp}) \cap \Lambda(M^{\perp}) = \emptyset$. So $\Lambda(M^{\nabla\nabla}) = \Lambda(M^{\perp\perp})$ and $\Lambda(M^{\nabla}) = \Lambda(M^{\perp})$.

Now if $g \in G$ and $M \in \mathcal{P}(G)$, then $g = g_1g_2$ where $g_1 \in M^{\nabla\nabla}$ and $g_2 \in M^{\nabla}$. We have then $g(\alpha) = g_1(\alpha)$ for all $\alpha \in \Lambda(M^{\perp\perp}) = \Lambda(M^{\nabla\nabla})$ and $g_1(\alpha) = e$ for all $\alpha \in \Lambda \setminus \Lambda(M^{\perp\perp})$. Also, $\Lambda(M^{\perp\perp}) = M\theta$, $\Lambda(M^{\perp\perp}) \in \mathcal{F}$. Thus $g_1 \in H$. But each element of $G[\mathcal{F}]$ is the product of finitely many elements like g_1. Thus $K = G[\mathcal{F}]$ and $G[\mathcal{F}]$ is a SP-hull of G.□

Now we prove the uniqueness of the SP-hull for every representable l-group.

THEOREM 4. (Bleier [1].) *Let G be an l-subgroup of the strongly projectable l-group M such that:*

(10) if N is an l-subgroup of M containng G and N is strongly projectable, then $N = M$;

(11) there exists an isomorphism τ of Boolean algebra $\mathcal{P}(G)$ onto Boolean algebra $\mathcal{P}(M)$ such that $H \subseteq H\tau$ for all $H \in \mathcal{P}(G)$.

Then there exists an isomorphism β of the l-group $G[\mathcal{F}]$ onto l-group M such that $g\beta = g$ for all $g \in G$.

Proof. Let M be an l-subgroup of Cartesian product $\bar{M} = \bar{\prod\limits_{\gamma\in\Gamma}} M_{\gamma}$ where each M_{γ} is a totally ordered group and $\Gamma(M) = \Gamma$ (here $\Gamma(X) = \{\gamma \in \Gamma | g(\gamma) \neq e$ for some $g \in X\}$). Then $\mathcal{F}^* = \{\Gamma(K) | K \in \mathcal{P}(M)\}$ is a field of subsets of Γ since M is strongly projectable. Let $\mathcal{B}^* = \{\Gamma(H) | H \in \mathcal{P}(G),\ G \subseteq H\}$. We define $\tau^* : \mathcal{B}^* \to \mathcal{F}^*$ by $(\Gamma(M))\tau^* = \Gamma(M\theta)$. Then τ^* is an isomorphism of Boolean algebra $\mathcal{B}^*$ onto $\mathcal{F}^*$ and $\Gamma(H) \subseteq \Gamma(H)\tau^*$ for every $H \in \mathcal{P}(G)$. By Theorem 3, $G[\mathcal{F}^*]$ is an SP-hull of G. We claim that $G[\mathcal{F}^*]$ is an l-subgroup of M. Let $g \in G$, $\Gamma(H) \in \mathcal{F}^*$ where $H \in \mathcal{P}(G)$ and $f \in G[\mathcal{F}^*$ such that $f(\gamma) = g(\gamma)$ for all $\gamma \in \Gamma(H)$ and $f(\gamma) = e$ for all $\gamma \in \Gamma \setminus \Gamma(H)$. Then $g = g_1g_2$ where $g_1, g_2 \in M$, $g_1 \in H, H \subseteq M$, $g_2 \in H_1 = \{x \in M | x \in H^{\perp}\}$. Therefore, $\Gamma(g_1) \subseteq \Gamma(H)$, $\Gamma(g_2) \subseteq \Gamma(H)$, $\Gamma(H) \cap \Gamma(H_1) = \emptyset$ and $g_1(\gamma) = g(\gamma)$ for all $\gamma \in \Gamma(H)$, $g_1(\gamma) = e$ for all $\gamma \in \Gamma \setminus \Gamma(H)$ and $f = g_1$, $f \in M$.

Each element of $G[\mathcal{F}^*]$ is the product of numerous elements like f. Thus $G[\mathcal{F}^*]$ an l-subgroup of M. Therefore, $G[\mathcal{F}^*] = M$. Now we are ready to determine $\beta; G[\mathcal{F}] \to G[\mathcal{F}^*]$ by $f\beta = g = [g_i|H_i\tau^*]$ for $f = [g_i|H_i\theta \in G[\mathcal{F}]$ We show that β is a well-defined map. Suppose $[g_i|H_i\theta] = [e|\Gamma] = e$. Then $\Gamma(g_i) \cap H_i\theta = \emptyset$ and $\Gamma(g_i^{\perp\perp}) \cap \Gamma(H_i) = \emptyset$, $g_i^{\perp\perp} \cap H_i = E$. Therefore, $\Gamma(g_i^{\perp\perp}) \cap \Gamma(H_i)\tau^* = \emptyset$, $\Gamma(g_i) \cap \Gamma(H_i)\tau^* = \emptyset$ and $[g_i|\Gamma(H_i)\tau^*] = e$.

Now suppose that $[a_i|A_i\theta] = [b_j|B_j\theta]$ where $A_i, B_j \in \mathcal{P}(G)$. Then $[a_ib_j^{-1}|(A_i \cap B_j)\theta] = [e|\Gamma$ and $[a_ib_j|(A_i \cap B_j)\tau^*] = [e|\Gamma]$, $[a_i|\Gamma(A_i)\tau^*] = [b_j|\Gamma(B_j)\tau^*]$ in $G[\mathcal{F}^*]$.

Thus β is well-defined. It is readily verified that β is an isomorphism of the l-group $G[\mathcal{F}]$ onto $G[\mathcal{F}^*]$ such that $g\beta = g$.□

COROLLARY 2. (Bleier [1].) *If G is a reprsentable l-group, then there is a unique SP-hull (up to the isomorphism over G).*□

COROLLARY 3. (Bleier [1].) *Let G be a representable l-group, $G[\mathcal{F}]$ is the SP-hull of G and $f \in G[\mathcal{F}]$, $f > e$. Then there exist $g, h \in G$ such that $e < g \leq f < h$.*

Proof. The existence of the element $g \in G$ with property $e < g \leq f$ follows from (8). If $f = [g_i|A_i] \in G[\mathcal{F}]$, then $h = \bigvee g_i$.□

4^0. Now we prove that every homomorphism of represnetable l-group G can be extended uniquely to the homomorphism of the SP-hull of G. The following Lemma was proved by Bleier [1] and can be easily verified.

LEMMA 2. *Let μ be an homomorphism of l-group G onto l-group H and $N = \ker\mu$, then*

(12) *if $S \subseteq G$ and $N \subseteq S^\perp$, then $S^\perp\mu = (S\mu)^\perp$,*

(13) *if $M \in \mathcal{P}(G)$ and $N \subseteq M$, then $M\mu = (M^\perp\mu)^\perp$ and $M\mu \in \mathcal{P}(H)$,*

(14) *if $N \in \mathcal{P}(G)$ and $K \in \mathcal{P}(G)$, then $K\mu^{-1} = \{g \in G | g\mu \in K\}$ is a polar in G,*

(15) *if $N \in \mathcal{P}(G)$, then the map $\bar{\mu} : \mathcal{P}(G) \to \mathcal{P}(H)$ where $K\bar{\mu} = (K \cup_{\mathcal{P}} N)\mu$ is a homomorphism of Boolean algebra $\mathcal{P}(G)$ onto $\mathcal{P}(H)$.* □

THEOREM 5. (Bleier [1].) *Suppose that G and H are represemtable l-groups, $\mu : G \to H$ is a homomorphism of G omto H. Let*

$N = \ker\mu$ be a polar in G. Then there is a homomorphism ν of the SP-hull $G[\mathcal{F}[G]$ of G onto SP-hull $H[\mathcal{F}^*]$ of homomorphism such that $g\nu = g\mu$ for all $g \in G$.

Proof. Let G be an l-subgroup of $\bar{G} = \bar{\prod\limits_{\alpha\in A}} G_\alpha$ and H be an l-subgroup of $\bar{H} = \bar{\prod\limits_{\beta\in B}} H_\beta$ where G_α $(\alpha \in A)$, H_β $(\beta \in B)$ are totally ordered groups. Let $\mathcal{B} = \{A(I)|I \in \mathcal{P}(G)\}$, η be a homomorphism of Boolean algebra $\mathcal{B}$ into the field $\mathcal{F}$ of subsets of A and $G[\mathcal{F}]$ be an SP-hull of G.

Similarly, let $\mathcal{B}^* = \{B(K)|K \in \mathcal{P}(H)\}$, η^* be a homomorphism of $\mathcal{B}^*$ into the field $\mathcal{F}^*$ of subsets of B and $H[\mathcal{F}^*]$ be an SP-hull of H. Let $\bar{\mu}$ be a homomorphism of $\mathcal{P}(G)$ onto $\mathcal{P}(H)$ as in (15). Now define the homomorphism ν of $G[\mathcal{F}]$ onto $H[\mathcal{F}^*]$ by the rule: $[g_i|A(I_i)\eta]\nu = [g_i\mu|B(I_i\bar{\mu})\eta^*]$ where $I_i \in \mathcal{P}(G)$. We show that ν is well-defined. Let $[g_i|A(I_i)\eta] = e$ Then $g_i^{\perp\perp} \cap I_i = E$, $(g_i^{\perp\perp} \cup \overset{\mathcal{P}}{\bigvee} N) \cap (I_i \overset{\mathcal{P}}{\bigvee} N) = N$, $(g_i^{\perp\perp} \cup \overset{\mathcal{P}}{\bigvee} N)\bar{\mu} \cap (I_i \overset{\mathcal{P}}{\bigvee} N)\bar{\mu} = E$. By (13), $(g_i^{\perp\perp} \cup \overset{\mathcal{P}}{\bigvee} N)\bar{\mu}$ is a polar in H and $g_i\mu \in (g_i^{\perp\perp} \cup \overset{\mathcal{P}}{\bigvee} N)\bar{\mu}$. Then $(g_i^{\perp\perp})\bar{\mu} \subseteq (g_i^{\perp\perp} \cup \overset{\mathcal{P}}{\bigvee} N)\bar{\mu}$. Therefore, $(g_i\mu)^{\perp\perp} \cap (I_i \overset{\mathcal{P}}{\bigvee} N) = E$, $B((g_i\mu)^{\perp\perp}) \cap B((I_i \overset{\mathcal{P}}{\bigvee} N)\bar{\mu}) = \emptyset$ and by (7), $B(g_i\mu) \cap (B(I_i \overset{\mathcal{P}}{\bigvee} N)\bar{\mu})\eta^* = \emptyset$. So ν is well-defined.

It is easy to verify that ν is a homomorphism. To see that ν maps $G[\mathcal{F}]$ onto $H[\mathcal{F}^*]$, it is enough to note that μ maps G onto H and that each finite partition of B in $\mathcal{F}^*$ is the image of the finite partition of α in $\mathcal{F}$. The latter is true because each finite partition of H in $\mathcal{P}(H)$ is the image under $\bar{\mu}$ of some partition of G in $\mathcal{P}(G)$.□

COROLLARY 4. (Bleier [1].) *Every automorphism of the representable l-group G is induced by some automorphism of SP-hull of G.*□

The most natural formulation of this concept was given by Conrad [7]. Vecksler in [1] produced the SP-hull for f-rings. The orthocompletion of Bernau [1] is a related concept. Ball [2] investigated the generalizations of SP-hull and different kinds of completeness.

7.3 Divisibility in l-Groups

Let us recall that group G is *divisible* if for every $g \in G$ and every positive integer n there is $x \in G$ such that $x^n = g$. In this section we will prove that every l-group can be embedded into a divisible l-group.

LEMMA 1. (Weinberg [4].)*Every totally ordered set X can be embedded into the totally ordered field F in such a way that the group $A(X)$ of order permutations of X is induced by any group of order automorphisms of the totally ordered field F.*

Proof. Let X be a totally ordered set and G be the free Abelian group with the basis X. We totally order G lexicographically with the use totally order of X. Then we consider the rational group ring $\mathbf{Q}[G]$; this can also be totally order lexicographically with the use the total order of G. This order has a unique extension to its quotient field F. It is straightforward to verify that every order permutation of X can be lifted to the order automorphism of the totally ordered field F.□

THEOREM 1. (Holland [2].) *Every l-group can be embedded into a divisible l-group.*

Proof. By Theorem 4.1.1, we can embed an l-group G into the l-group $A(X)$ of order permutations of some totally ordered set X. By Lemma 1, we can embed the totally ordered set X into some totaly ordered field F such that G is the l-subgroup of l-group of all order authomorphisms of the ordered field F. Now by Proposition 4.7.1, $A(F)$ is an o-2-transitive ordered permutation l-group and by Theorem 4.4.3, it is a divisible group. □

COROLLARY 1. *Every ro-group can be embedded into a divisible ro-group.*

2⁰. The question on embedding of totally ordered group into a divisible totally ordered group is still open. Here we discuss this problem for some classes of totally ordered groups.

THEOREM 2. (Mal'cev [2].) *Every locally nilpotent totally ordered group G can be embedded into a divisible locally nilpotent totally ordered group. The minimal divisible locally nilpotent totally ordered group G^* containing G is unique up to the isomorphism.*

Proof. It is a well-known Mal'cev Theorem (see, for example, the book of Kargapolov and Merslakov [1]) that minimal divisible torsion-free locally nilpotent group G^* containing G is unique up to the isomorphism. From definition of G^* it follows that for every $x \in G^*$ there is positive integer n such that $x^n \in G$. Let P be a total order of G and $P^* = \{x \in G^* | x^n \in P^* \text{ for some positive integer } n\}$. It is evident that P^* is a pure and linear subset in G^*. We claim that P^* is a normal subsemigroup in G^*. Let $x, y \in P^*$, $z \in G^*$. Then there are positive integers p, q, r such that $x^p, y^q \in P$, $z^r \in G$. Let $a = x^p$, $b = y^q$, $c = z^r$. Then the subgroup $A = \mathrm{gp}(a, b, c)$ of G is nilpotent and let $\mathcal{S}(A)$: $E = A_0 \subset A_1 \subset \cdots \subset A_n = A$ be the system of all convex subgroups of A under the total order $P \cap A$. It is clear that the system $\mathcal{S}(A)$ is finite and central in A. Let A_i^* be an isolator of A_i in G^*. Then the system $\mathcal{S}(A^*)$:

$$E = A_0^* \subset A_1^* \subset \cdots \subset A_n^* = A^*$$

is a central series of subgroups in A^*. Now we shall prove that $P^* \cap A^*$ is a total order of A^*.

Assume that the order $P \cap A$ can be uniquelly extended to the total order $P_k = P^* \cap B_k$ where $B_k = \mathrm{gp}(A, A^*)$. Then the system $\mathcal{S}(B_k)$:

$$E = A_0^* \subset A_1^* \subset \cdots \subset A_k^* \subset A_k^* \cdot A_{k+1} \subset \cdots \subset A_k^* A_n = B_k$$

is a system of convex subgroups of the group B_k under the total order P_k. It is easy to see that B_k is a normal subgroup in B_{k+1}. If $v \in P_k$, then $v \in A_k^* A_j \setminus A_k^* A_{j-1}$ for some $j > k$ or $v \in A_j^* \setminus A_{j-1}^*$ for some $j \leq k$. Now, by centrality of the system $\mathcal{S}(A^*)$ in A^* it follows that for every $u \in B_k$ the equality $u^{-1}vu = vw$ is valid for some $w \in A_{k-1}^*$. Therefore, the subsemigroup P_k is normal in B_{k+1}. By Theorem 4.4.2, the group B_{k+1} is fully orderable. Then the partial order P_k of B_{k+1} can be extended to the total order Q_{k+1} of the group B_{k+1}. It is clear that $Q_{k+1} = P_{k+1} = P^* \cap B_{k+1}$. Now by induction on k, it follows that $P^* \cap A^*$ is a total order of A^* uniquely extending P.□

COROLLARY 2. *If group G is approximated by torsion-free nilpotent groups, then G can be embedded into a divisible O-group.*□

COROLLARY 3. *Free, free solvable, and free polynilpotent groups can be embedded into a divisible O-group.* □

3^0. We need few definitions from the theory of ordered rings and ordered modules. Let M be a left K-module over ring K and $\Gamma \subseteq K$. Let us recall that module M is Γ-*ordered module* if it has an order relation which turns it into an ordered group and is preserved under multiplication by elements of Γ. Let Φ be a subset of K. We say that K-module is Φ-*complete* if the equation $x\varphi = a$ is solvable for all $a \in M$, $\varphi \in \Phi$. The subset N of M is called a Γ-*semimodule* if it is closed under addition and multiplication by the elements of Γ. We indicate by $S_\Gamma(a_1, \ldots, a_n)$ the Γ-semimodule of K-module M generated by elements $a_1, \ldots, a_n \in M$. The following statements can be easily verified.

LEMMA 2. *Let M be an K-module and Γ be a subset of K. Then M is Γ-orderable if and only if for any $a_1, \ldots, a_n \in M \setminus \{0\}$ there exist signs $\varepsilon_1, \ldots, \varepsilon_n$, ($\varepsilon_i = \pm 1$ for $i = 1, \ldots, n$) such that $0 \notin S_\Gamma(a_1, \ldots, a_n)$.*□

COROLLARY 4. *K-module is Γ-orderable if and only if it is locally Γ-orderable.*□

LEMMA 3. (Kopytov [2], Kokorin and Kopytov [1].) *Let M be an Γ -orderable K-module and Φ be a multiplicative subsemigroup of K acting on M as a commutative semigroup of monomorphisms and every $\varphi \in \Phi$ commute on M with every elements $k \in K$. Then M can be embedded in Φ-complete Γ-orderable K-module M^* satisfying the following condition: for each $a \in M^*$ there is $\varphi \in \Phi$ such that $a\varphi \in M$.*

Proof. Consider first the case when the semigroup Φ is generated by one element φ. In this case $M\varphi \subseteq M$ and if $M\varphi = M$ then M is Φ-complete. If $M\varphi \neq M$, then we can define the ascending chain of K-modules as follows: $M_0 = M$; assume that for each $i \leq n$ there has been defined a Γ-orderable K-submodule M_i such that every $k \in K$ induces a monomorphism on M_i and φ commutes on M with every $k \in K$. Now consider $M_n\varphi$: this is an K-submodule of M_n isomorphic to M_n by the isomorphism $\varepsilon_n : m\varepsilon_n = m\varphi$ for $m \in M_n$. Let M_{n+1} be an isomorphic copy of M_n and embed M_n in M_{n+1} using the isomorphism ε_n. It is clear that M_{n+1} is an Γ-orderable K-module and satisfies all the hypotheses necessary to the induction step. So, we have the infinite ascending chain of modules

$$M = M_0 \xrightarrow{\varepsilon_0} M_1 \xrightarrow{\varepsilon_1} \cdots \xrightarrow{\varepsilon_{n-1}} M_n \xrightarrow{\varepsilon_n} \cdots$$

Let $M^* = \bigcup_{n=0}^{\infty} M_n$. By Corollary 4, M^* is a Γ-orderable Φ-complete K-module. The proof can be extended to the case when Φ is generated by more than one element by using transfinite induction.□

Let us recall that a *wreath product* of groups A, B is the group $W = A\mathrm{Wr}B$ which is a semidirect product of Cartesian product $\mathrm{Fun}(B, A) = \overline{\prod_{b \in B}} A_b$ of the family of isomorphic copies $\{A_b | b \in B\}$ of the group A with the group B such that $b^{-1}fb(x) = f^b(x) = f(bx)$ is valid for all $f \in \mathrm{Fun}(B, A)$, $x, b \in B$. Any element $g \in W = A\mathrm{Wr}B$ can be uniquely represented in the form $g = bf$, $b \in B$, $f \in \mathrm{Fun}(B, A)$. The multiplication in the group W is defined by the rule: $bf \cdot b'f' = bb' \cdot f''$, where $f'' = b'^{-1}fb \cdot f' = f^{b'}f'$. Let $A\mathrm{wr}B$ be a subset of all elements $bf \in A\mathrm{Wr}B$ such that $f(x) \neq e$ only for finite number elements $x \in B$. It is readily verified that $A\mathrm{wr}B$ is a subgroup of $A\mathrm{Wr}B$. The subgroup $A\mathrm{wr}B$ is called a *restricted wreath product* of A and B. It is known that any extension G of group A by B is embedded into the wreath product $A\mathrm{Wr}B$. For example (H.Neumann [1]), the natural embeddig $\varphi_s : G \to A\mathrm{Wr}B$ can be defined by the rule $g\varphi_s = \bar{g}f_s$, where $\bar{g}$ is coset gA in the quotient group G/A and $f_s(b) = (\bar{g}b)_s^{-1}gb_s$, $\{h_s | n \in G\}$ is some fixed system of representatives of cosets hA ($h \in G$). It is easy to verify that $A\varphi_s \subseteq \mathrm{Fun}(B, A)$ and $G\varphi_s \cdot \mathrm{Fun}(B.A) = A\mathrm{Wr}B$.

THEOREM 3. (Bludov and Medvedev [1].) *A metabelian O-group G can be embedded into a divisible metabelian O-group.*

Proof. Let A be an isolator of the derived subgroup G'. By Proposition 3.2.4, A and $B = G/A$ are torsion-free Abelian groups. Let φ_s be a natural embedding G into $A\mathrm{Wr}B$ and H be the least strongly isolated subgroup of $A\mathrm{Wr}B$. It is clear that H is precisely the subset of all elements $h \in A\mathrm{Wr}B$ such that $h^{e+h_1+\cdots h_n} = h \cdot h^{h_1} \cdots h^{h_n} = e$. Suppose, for reductio ad absurdum, that $h \in G\varphi_s \cap H$, $h \neq e$. Then $h = bf$ where $b \in B$, $f \in \mathrm{Fun}(B, A)$, $f = a\varphi_s$ for some $a \in A$. But $H \subseteq \mathrm{Fun}(B, A)$, then $b = e$ and $h = f \in \mathrm{Fun}(B, A)$. So there are elements $b_1, \ldots, b_n \in B$ such $h^{e+b_1+\cdots b_n} = e$. In group G there are elements $g_1, \ldots, g_n$ such that

$g_i\varphi_s = b_i f_i$ $(i = 1, 2, \ldots, n)$ and $f_i \in \mathrm{Fun}(B, A)$. Then $e = h^{e+b_1+\cdots b_n} = (a\varphi_s)^{e+g_1\varphi_s,\ldots,g_n\varphi_s} = (a^{e+b_1+\cdots+g_n})$ and $a^{e+b_1+\cdots b_n} = e$ in orderable group G, which is a contradiction. Hence $a = e$ and $h = a\varphi_s = e$.

Let B^* be the least divisible Abelian group containing B. Now we define the map $\varphi^* : A\mathrm{Wr}B \to B^*$ by the rule: $(bf)\varphi^* = bf^*$ where $f^*(x) = f(x)$ for all $x \in B$ and $f^*(x) = e$ for all $x \in B^* \setminus B$. It is clear that φ^* is an isomorphic embedding of $A\mathrm{Wr}B$ into $A\mathrm{Wr}B^*$. We identify now the group $A\mathrm{Wr}B$ with the group $(A\mathrm{Wr}B)\varphi^*$.

Let H^* be the least strongly isolated subgroup of $A\mathrm{Wr}B^*$. Then by similar arguments $H^* \cap G\varphi_s = E$. By Theorem 3.2.3, the quotient group $G_1 = A\mathrm{Wr}B^*/H^*$ is orderable and G is embedded into G_1. In group G_1 there is an Abelian torsion-free subgroup $A_1 = \mathrm{Fun}(B^*, A)/H^*$ such that G_1 is a semidirect product of A_1 and B^*. Then A_1 can be considered as G_1-orderable $\mathbf{Z}[G_1]$-module. Let

$$\Phi = \{e + g_1 + \cdots + g_n | g_1, \ldots, g_n \in G\} \subseteq \mathbf{Z}[G_1].$$

Applying Lemma 3, A_1 can be embedded into Φ-complete G_1-orderable module A_1^*.

Now let G^* be a semidirect product of additive group of module A_1 with B^* such that inner automorphisms of G^* defined by $b \in B^*$ coincide with action $b \in \mathbf{Z}[G_1]$ on $\mathbf{Z}[G_1]$-module A_1^*. We claim that G^* is a divisible group. Consider some equation $x^n = g$, $g \in G^*$. Then $g = b_0 a_0$ for some $b_0 \in B^*$ and $a_0 \in A_1^*$. Assume that $x = bu$, $b \in B^*, u \in A_1^*$. Then the equation $x^n = g$ can be rewriten in equivalent form $x^n = b^n u^{b^{n-1}+\cdots+b+e} = b_0 a_0$. But in the group B^* there exists an element $b \in B^*$ such that $b^n = b_0$ and in the group A_1^* there exists an element $u \in A_1^*$ such that $u^{b^{n-1}+\cdots+b+e} = a_0$ (because A_1^* is Φ-complete). Therefore, $x = bu$ is the solution of the equation $x^n = b_0 a_0$.□

7.4 Embeddings into Simple l-Groups

1⁰. The *order union* $X \overleftarrow{\cup} Y$ of po-sets X, Y $(X \cap Y = \emptyset)$ is the union $X \cup Y$ partially ordered by the rule: $u \leq v$ iff either $u, v \in X$, $u \leq v$ in po-set X or $u \in X$, $v \in Y$ or $u, v \in Y$, $u \leq v$ in po-set Y. It is clear that order union of totally ordered sets also is a totally ordered set.

LEMMA 1. *For every totally ordered field K, an order-type of totally ordered set K is equal to order-types of totally ordered sets $K_+ = \{x \in K | x > 0\}$ and $K_- = \{x \in K | x < 0\}$.*

Proof. The map $\varphi : x \to x^{-1}$ is an order isomorphisn between K_+ and K_-. Hence the order-types of K_+ and K_- are the same. By similar arguments, the order-type of open interval $(1, \infty) = \{x \in K | x > 1\}$ is equal to the order-type of open interval $(0, 1) = \{x \in K | 0 < x < 1\}$ and is equal to order-type of the set K_+. But $K = K_- \overleftarrow{\cup} \{0\} \overleftarrow{\cup} K_+$ and $K_+ = (0, 1) \overleftarrow{\cup} \{1\} \overleftarrow{\cup} (1, \infty)$. Hence there exists an order isomorphism between totally ordered sets K and K_+.□

LEMMA 2. *Let every totally ordered set K_i $(i = 0, \pm 1, \pm 2, \ldots)$ be order isomorphic to the totally ordered set $\{K; \leq\}$ of the totally ordered field K and*

$$T = \overleftarrow{\bigcup_{-\infty < i < +\infty}} (K_i \overleftarrow{\cup} \{a_i\}) =$$

$$= \cdots \overleftarrow{\cup} K_{-1} \overleftarrow{\cup} \{a_{-1}\} \overleftarrow{\cup} K_0 \overleftarrow{\cup} \{a_0\} \overleftarrow{\cup} K_1 \overleftarrow{\cup} \cdots$$

Then totally ordered set T has a cofinal countable sequence and l-group $A(T)$ is o-2-transitive.

Proof of Lemma 2 follows directly from o-2-transitivity $A(K)$ (see Proposition 4.7.1) and the definition of the totally ordered set T.□

THEOREM 1. *Every lattice ordered group can be embedded into a simple lattice ordered group.*

Proof. By Theorem 4.3.1, l-group G is embedded into l-group $A(K)$ of all order permutations of the totally ordered field K. Let $T = \overleftarrow{\bigcup_{-\infty < i < +\infty}} (K_i \overleftarrow{\cup} \{a_i\})$ where K_i is order isomorphic to the totally ordered set K. Then by Lemma 2, the l-group group $A(T)$ is o-2-transitive. We can assert that G is an l-subgroup of $A(T)$ such that $\mathrm{supp}(g) \subseteq K_0$ for all $g \in G$. So, G is an l-subgroup of l-group $B(T) = \{g \in A(T) | g \text{ has bounded support}\}$. By Theorem 4.6.4, l-group $B(T)$ is a simple l-group.□

Other different embeddings of l-groups into simple l-groups can be constructed with the use of Lemma 2 and Theorem 4.6.5 or Examples

4.7.3, 4.7.4. For order embeddings of l-groups into simple l-groups Hall's method can be used (see, for example, the book of Kokorin and Kopytov [2]).

COROLLARY 1. *Every right-orderable group can be embedded into a simple right-orderable group.* □

The following result is similar to that of combinatorial group theory according to which every group can be embedded in one in which two elements of the same order are conjugate. It is due to K.R.Pierce and can be found in the book of A.M.W.Glass [6].

THEOREM 2. (Pierce [2].) *Every l-group can be embedded into the l-group in which any two strictly positive elements are conjugate.*

7.5 Free Products of l-Groups

The concept of the free product is fundamental to the study of any kind of algebraic systems. Intuitively, a free product is constructed by the family of algebras from a given class the new algebra in the "freest" way possible. By this, we mean that any other algebra generated by the given family must be a homomorphic image of the free product. Free products have been widely used in group theory and lattice theory to provide general methods of construction.

1⁰. First we consider free products of groups. Let $\{G_\alpha|\alpha \in A\}$ be a family of groups and suppose that every group G_α is a quotient group of free group $F_\alpha = \mathrm{gp}(x_{\alpha\beta}|\beta \in B_\alpha)$ with free generators $x_{\alpha\beta}$ by normal subgroup N_α. The *free product* $\prod\limits_{\alpha\in A}^{*} = G^*$ of groups G_α $(\alpha \in A)$ is the quotient group of the free group F with the generators $\{\bar{x}_{\alpha\beta}|\alpha \in A, \beta \in B_\alpha\}$ by normal subgroup N generated (as normal subgroup of F) by elements $w_\alpha = w_\alpha(\bar{x}_{\alpha\beta}), \beta \in B_\alpha$, where $w_\alpha(x_{\alpha\beta}) \in N_\alpha$ for every $\alpha \in A$. It is clear that the mapping φ_α defined by the rule:$(x_{\alpha\beta}N_\alpha)\varphi_\alpha = \bar{x}_{\alpha\beta}N$ is an isomorphic embedding of G_α onto the subgroup $\bar{G}_\alpha$ of G^* and $G^* = \mathrm{gp}(G_\alpha|\alpha \in A)$. If a group H is generated by subgroups H_α $(\alpha \in A)$ and for every $\alpha \in A$ there is a homomorphism ψ_α of G_α onto H_α, then there is a homomorphism ψ of G onto H such that $a\psi_\alpha = (a)\varphi_\alpha\psi$ for all $\alpha \in A$ and $a \in G_\alpha$. The free product of two groups G and H is denoted by $G * H$.

Now we identify every group G_α with the subgroup $G_\alpha\varphi_\alpha$ of G^* and $x_{\alpha\beta}N_\alpha$ with $\bar{x}_{\alpha\beta}N_\alpha$. The following important result was first proved by A.A.Vinogradov [1]. The proof presented below is due to A.Strojnowski and W.Wojas [1].

THEOREM 1. (Vinogradov [1].) *The free product* $G^* = \prod\limits_{\gamma\in\Gamma}^{*} G_\gamma$ *of orderable groups* G_γ, $(\gamma \in \Gamma)$ *is an orderable group and every total order of each group* G_γ *can be extended to a total order on* G^*.

Proof. It is evident that it is enough to prove the theorem for two orderable groups. So, let G, H be a totally ordered groups,$A = G \overleftarrow{\times} H$ be an antilexicographic product of G, H and $\mathbf{Q}[A]$ be a group ring of A over the field of rationals $\mathbf{Q}$. As in the proof of Lemma 7.3.1 we define the order on $\mathbf{Q}[A]$ by taking an element $\sum\limits_{i=1}^{n} r_i a_i$ to be positive (here $r_i \in \mathbf{Q}$, $\mathbf{a_i} \in \mathbf{A}$) if the coefficient r_i of the greatest a_i in the total order on A is positive in $\mathbf{Q}$. It is easy to prove that the group ring $\mathbf{Q}[A]$ with this order is a totally ordered ring and this total order of $\mathbf{Q}[A]$ extends total order of A.

Let T be the group of all infinite upper triangular matrices $t = [x_{ij}]$, $(i,j = 1,2,\ldots)$ over $\mathbf{Q}[A]$ where diagonal elements x_{ii} are elements of group A, and $x_{ij} = 0$ if $i > j$. Now we define the total order on the set of all pairs of indexes at the matrice elements which are on and over diagonal in the following way: $(i,j) \leq (k,l)$ iff $j - i < l - k$ or $j - i = l - k$ and $i < k$ or $i = k$, $j = l$. The set of all pairs of indexes is well-ordered under this order. Let $[x_{ij}], [y_{ij}] \in T$. Now we define the order on T by the rule: $[x_{ij}] < [y_{ij}]$ iff there exists a pair (k,l) such that $x_{kl} < y_{kl}$ in $\mathbf{Q}[A]$ and $x_{ij} = y_{ij}$ for all $(i,j) < (k,l)$. Because the set of all pairs of the indexes is well-ordered, then the pair (k,l) is uniquely determined for $[x_{ij}] \neq [y_{ij}]$. So, the group T is a totally ordered group under this order.

Let 0 and 1 be zero and identity of group ring $\mathbf{Q}[A]$. Now we consider the group homomorphism $\varphi : G * H \to T$ given by the rule:

$$g\varphi = \begin{pmatrix} g & 1-g & 0 & 0 & \dots \\ 0 & 1 & 0 & 0 & \dots \\ 0 & 0 & g & 1-g & \dots \\ 0 & 0 & 0 & 1 & \dots \\ \dots & \dots & \dots & \dots & \ddots \end{pmatrix},$$

$$h\varphi = \begin{pmatrix} 1 & 0 & 0 & 0 & \dots \\ 0 & h & 1-h & 0 & \dots \\ 0 & 0 & 1 & 0 & \dots \\ 0 & 0 & 0 & h & \dots \\ \dots & \dots & \dots & \dots & \ddots \end{pmatrix}$$

for $g \in G,\ h \in H$.

Thus, for the reduced word $x = g_1 h_2 g_3 \cdots h_n,\ x \in G * H$, the equality $x\varphi = g_1\varphi \cdot h_2\varphi \cdots h_n\varphi$ is valid. It is readily verified that for the reduced word x the matrix $x\varphi$ has the $(1, n+1)$-s entry equal to $y = (1-g_1)(1-h_2)(1-g_3)\cdots(1-h_n)$. Since the group ring $\mathbf{Q}[A]$ is totally ordered, then by Proposition 5.5.3, $\mathbf{Q}[A]$ contains no divisors of zero. Hence $y \neq 0$ and $x\varphi$ is not the unit matrix e. Therefore, the image of every non-empty word by φ is not e. Thus φ is isomorphism embedding $G * H$ into T. Then $G * H$ is an orderable group. It is easy to see that this total order of $G * H$ extends total orders of G and H.□

2^0. Vinogradov's theorem cannot be generalized directly for l-groups.

THEOREM 2. (Holland [10], Medvedev [4].) *Every lattice ordering of a free product of groups with more than one non-trivial factor is a total ordering.*

Proof. Let $G = A * B$ be a free product of nontrivial groups A, B and G be lattice ordered under some order P. By Kurosh's theorem on subgroups, in free products the centralizer of the element x which is not conjugate to any element of the subgroups A, B is infinite cyclic group (M.Hall [1]). Therefore by Proposition 2.2.5, the elements $[a,b] = a^{-1}b^{-1}ab$ and $a \cdot bab^{-1}$ are comparable with e in G for $a \in A,\ b \in B,\ a \neq e,\ b \neq e$. We have four different cases:

1) $[a,b] > e,\ a \cdot bab^{-1} > e$ 2) $[a,b] > e,\ a \cdot bab^{-1} < e$

3) $[a,b] < e,\ a\cdot bab^{-1} > e$ 4) $[a,b] < e,\ a\cdot bab^{-1} < e$.

Case 1. $a^{-1}b^{-1}ab > e,\ a\cdot bab^{-1} > e$. Then $b^{-1}ab > a,\ ba^{-1}b^{-1} > a^{-1}$ and $a > bab^{-1},\ ba^{-1}b^{-1}$. Hence $a^2 > (bab^{-1})(ba^{-1}b) = e$. Then $a > e$ by Proposition 2.2.2.

Case 2. Inequalities $abab^{-1} < e,\ a^{-1}b^{-1}ab > e$ are equivalent to $bab^{-1} < a^{-1},\ b^{-1}ab > a$. Then $a < b^{-1}a^{-1}b,\ a < b^{-1}ab$ and $a^2 < (b^{-1}a^{-1}b)(b^{-1}ab) = e$. Therefore, $a < e$.

By similar arguments, in case 3 $a > e$, and in case 4 $a < e$ hold.

It follows that in the l-group $G = A*B$ any element of the subgroup A is comparable with e. By similar arguments every element of the subgroup B is comparable with e, too. Hence, if $g \in \bigcup_{x\in G}(A\cup B)^x$, then g is comparable with e in the lattice ordered group G. Finally, if $g \in G\setminus \bigcup_{x\in G}(A\cup B)^x$, then g is comparable with e by Kurosh's theorem and Proposition 2.2.5. Then G isa totally ordered group. □

3^0. From Theorem 2 it follows that the free product of groups have bounded applications in the l-group theory. Now we consider the notion of free product in the category of l-groups. Let $\{G_\alpha|\alpha\in A\}$ be a family of l-groups. The *free product of l-groups* G_α $(\alpha\in A)$ is an l-group G^*, denoted by $\prod\limits_{\alpha\in A}^{l-*}$, together with the family of l-isomorphic embeddings $\{\varphi_\alpha : G_\alpha \to G^*|\alpha\in A\}$ such that

(1) $G^* = \text{l}-\text{gp}(G_\alpha\varphi_\alpha|\alpha\in A)$;

(2) if H is an l-group and $\{\psi_\alpha : G_\alpha \to H\}$ is a family of l-homomorphisms;

then there exists a unique l-homomorphism $\theta : G^* \to H$ satisying $\psi_a = \varphi_\alpha\theta$ for all $\alpha\in A$.

We often find it convenient to identify each free factor G_α with its image $G_\alpha\varphi_\alpha$ in $G^* = \prod\limits_{\alpha\in A}^{l-*} G_\alpha$ and thus view each φ_α as the inclusion map. It is easy to see that if the free product of l-groups G_α $(\alpha\in A)$ exists, then it is unique up to l-isomorphism.

We consider first the question of the existence of free products.

THEOREM 3. (Martinez [2].) *Let $\{G_\alpha|\alpha\in A\}$ be a family of l-groups. Then the free product $\prod\limits_{\alpha\in A}^{l-*} G_\alpha$ exists.*

Proof. Let for every $\alpha \in A$ $G_\alpha = l - \mathrm{gp}(x_{\alpha\beta}|\beta \in B_\alpha)$ be an l-group and $F_\alpha = l - \mathrm{gp}(\bar{x}_{\alpha\beta}|\beta \in B_\alpha)$ be an l-group free over trivially ordered free group with the free generators $\bar{x}_{\alpha\beta}$ ($\beta \in B\alpha$). Then G_α is the quotent group of l-group F_α by some ideal N_α. Let F be an l-group free over trivially ordered free group with the free generators $\{x_{\alpha\beta}|\alpha \in A, \beta \in B_\alpha\}$ and N be an ideal of F generated by all elements $w(\bar{x}_{\alpha\beta_1}, \ldots, \bar{x}_{\alpha\beta_n})$ where $w(x_{\alpha\beta_1}, \ldots, x_{\alpha\beta_n}) \in N_\alpha$. Let G^* be a quotient l-group F/N and φ_α be the mapping from G_α into G^* which is defined by the rule:

$$\text{if } x \in G_\alpha,\ x = \bigvee_{i\in I}\bigwedge_{j\in J} w_{ij}(x_{\alpha\beta}), \qquad \text{then } x \cdot \varphi_\alpha = \bigvee_{i\in I}\bigwedge_{j\in J} w_{ij}(\bar{x}_{\alpha\beta})N$$

where $w_{ij}(x_{\alpha\beta})$ are group words on the variables $x_{\alpha\beta}, \beta \in B_\alpha\ , \alpha \in A$. Every φ_α is a homomorphism of l-group G_α into l-group G^*.

Let G be an l-group and $\{\psi_\alpha|\alpha \in A\}$ be the system of homomorphisms of G_α into G. We define the map $\theta : \bigcup_{\alpha\in A, \beta\in B_\alpha} \{x_{\alpha\beta}N\} \rightarrow G$ by the rule : $\bar{x}_{\alpha\beta}N\theta = x_{\alpha\beta}N_\alpha\psi_\alpha$. It is easy to see that θ is uniquely extended to homomorphism of G^* into G and $\psi_\alpha = \varphi_{\alpha\theta}$ for all $\alpha \in A$.

Finally, for every $\alpha \in A$ we set $G = G_\alpha$ and define the system of homomorthisms $\{\psi_\gamma : G_\gamma \rightarrow G = G_\alpha|\gamma \in A\}$ by the rule: $G_\gamma\psi_\gamma = E$ if $\gamma \neq \alpha$, $\psi_\alpha = \varepsilon_\alpha$ is the identical mapping of G_α onto $G = G_\alpha$. The homomorphism θ for these G and the system $\{\psi_\gamma|\gamma \in A\}$ exists. As $\varepsilon_\alpha = \varphi_\alpha\theta$, then φ_α is an isomorphism of G_α into G^*. This means that l-group G^* is a free product of l-groups $G_\alpha, \alpha \in A$.□

Now we consider the class of l-groups as a category of partially ordered groups and the order homomorphisms as morphisms. Then the notion of the free product in the category of po-groups arises. Let $\{G_\alpha|\alpha \in A\}$ be a family of po-groups. A lattice-ordered group F and the family $\{\varphi_\alpha : G_\alpha \rightarrow F|\alpha \in A\}$ of order isomorphisms are called an *free product of po-groups* if $F = 1 - \mathrm{gr}(G_\alpha\varphi_\alpha|\alpha \in A)$ and there exists the (necessarily) unique l-homomorphism $\theta : F \rightarrow G$ such that $\psi_\alpha = \varphi_\alpha\theta$ for every l-group G and every family $\{\psi_\alpha : G_\alpha \rightarrow G|\alpha \in A\}$ of order homomorphisms of po-groups G_α into l-group G. Arguments similar to those in Section 1 show that the l-free product of po-groups exists if and only if each po-group G_α is a po-subgroup of some l-group. It is clear that the l-free product of po-groups is unique up to l-isomorphisms in this case.

Let $\{G_\alpha, P_\alpha | \alpha \in A\}$ be a family of po-groups, where P_α is a positive cone of G_α and P_α be the intersection of some set of right total orders of G_α for every $\alpha \in A$. Let group G^0 be a free product of groups G_α and φ'_α be a natural isomorphic embedding every G_α into G^0. Now let P^0 be a normal subsemigroup of G^0 generated by all elements $x\varphi_\alpha$ where $x > e$ in G_α, $\alpha \in A$. It is readily verified that the semigroup P^0 defines the partial order on G^0 and P^0 is the intersectoin of right total orders of convex l-subgroup G^0 extending P^0. Let $\tilde{G}$ be a free over G^0 l-group and π be an order embedding of partially ordered group G^0 into l-group $\tilde{G}$.

THEOREM 4. (Holland and Scrimger [1].) *A lattice-ordered group $\tilde{G}$ is a l-free product of partially ordered groups G_α $(\alpha \in A)$.*

Proof. Let ψ_α be an order homomorphism of G_α into l-group G for every $\alpha \in A$. Then there exists the unique order homomorphism ψ of G^0 into G such that $\psi_\alpha = \varphi'_\alpha\psi$ for all $\alpha \in A$. By definition of the l-group $\tilde{G}$, there is the unique l-homomorphism $\theta : \tilde{G} \to G$ such that $\psi = \pi\theta$. Then $\psi_\alpha = \varphi'_\alpha\pi\theta$. If we set $\varphi_\alpha = \alpha'_\alpha\pi$, then $\tilde{G}$ with the family $\{\varphi_\alpha | \alpha \in A\}$ is the l-free product of partially ordered groups G_α, $\alpha \in A$. The description of l-group F follows from the description of the free l-group over partially ordered group (see Theorem 7.1.2.).

Of course, l-free product of l-groups and free product of l-groups are different. The following statement establishes the connection between these products.

PROPOSITION 1. *Let $\{G_\alpha | \alpha \in A\}$ be a family of l-groups, $G^* = \prod^{l-*}_{\alpha \in A} G_\alpha$ and F be a free product and l-free product of l-groups G_α $(\alpha \in A)$ correspondingly. Let $\varphi_\alpha : G_\alpha \to G^*$, $\bar{\varphi}_\alpha : G_\alpha \to F$ for every $\alpha \in A$ be natural embeddings into G^* and F. Then G^* is a quotient group of l-group F by the ideal generated by elements $(x\bar{\varphi}_\alpha \vee y\bar{\varphi}_\alpha)^{-1}(x \vee y)\bar{\varphi}_\alpha$ where $x, y \in G_\alpha$, $\alpha \in A$.*□

3⁰. In Theorems 3-4, these free products of l-groups are intimately related to those of groups. The objective of this section is to show that there is also a significant link between free products of l-groups and free products of distributive lattices.

Let us recall that the lattice of any l-group is distributive (Proposition 2.2.2) and hence any l-group can be considered as a member of the class D_e of all distributive lattices (L; $\vee$, $\wedge$, e) with distinguished element (nulary operation) e (D_e-lattice).

Let $\{L_\alpha|\alpha \in A\}$ be a family of lattices in D_e. The *free product of distributive lattices* L_α with the distinguished element e (or a D_e-free product) is a lattice $L = \prod_{\alpha\in A}^{de-*} L_\alpha$ and a system $\{\varphi_\alpha : L_\alpha \to L|\alpha \in A\}$ of the embeddings of lattices L_α into L such that:

(3) L is generated by $\bigcup_{\alpha\in A} L_\alpha\varphi_\alpha$,

(4) if K is any lattice in D_e and $\{\psi_\alpha : L_\alpha \to K|\alpha \in A\}$ is the system of homomorphisms of lattices L_α into K,

then there exists a homomorphism θ of lattice L into K such that $\psi_\alpha = \varphi_\alpha\theta$ for every $\alpha \in A$. We identify each free factor L_α with its image $L_\alpha\varphi_\alpha$ in L.

The following result show the connection of the free product of l-groups with the free product of distributive lattices.

THEOREM 5. (Franchello [1].) *Let* $G^* = \prod_{i\in I}^{*} G_i$ *be the free product of l- group* $\{G_i|i \in I\}$. *Then the sublattice* L *of* G^* *generated by the free factors is the* D_e*-free product of the family of lattices* $\{G_i|i \in I\}$.

A reader can find the proof in the original paper.

7.6 Amalgamations of l-Groups

1⁰. The class of l-groups $\mathcal{U}$ has the *amalgamation property* if whenever H, G_1, $G_2 \in \mathcal{U}$ and $\sigma_i : H \to G_i$ are embeddings ($i = 1,2$), then there is an l-group G in $\mathcal{U}$ and embeddings $\varphi_i : G_i \to G$, $i = 1,2$, such that for all $g \in H$ the equality $g\sigma_1\varphi_1 = g\sigma_2\varphi_2$ holds. We call $(H, G_i, \sigma_i|i = 1,2) = (H, G_1, G_2, \sigma_1, \sigma_2)$ a *formation* in $\mathcal{U}$ and the triple $(\varphi_1, \varphi_2, G)$ an *amalgamation* in $\mathcal{U}$ of this formation or simply *amalgam* of G_1, G_2 over H. We can define the formation $(H, G_i, \sigma_i|i \in I)$ and amalgam (φ_i, G) similarly for every set of indexes I.

It is known that the class of groups the and class of distributive lattices satisfy the amalgamation property. However, the class of l-groups does not as the following example due to Pierce [2] shows.

EXAMPLE 1. Let $A = (a)$, $B = (b)$, $C = (c)$ be totally ordered infinite cyclic groups;$a, b, c > e$ and H be an l-group, $H = (A \times B) \overleftarrow{\times} C$,i.e., $x = a^m b^n c^k \geq e$ in H iff $k > 0$ or $k = 0, n, m \geq 0$. Let $G_1 = \mathrm{gp}(a_1, b_1, c_1 | a_1 b_1 = b_1 a_1,\ c_1^{-1} a_1 c_1 = b_1,\ c_1^{-1} b_1 c_1 = a_1)$. be an l-group under the order $a_1^m b_1^n c_1^k \geq e$ in G_1 iff $k > 0$ or $k = 0,\ m, n \geq 0$. It is clear that there is a unique l-isomorphism $\sigma_1 : H \to G_1$ such that $a\sigma_1 = a_1,\ b\sigma_1 = b_1,\ c\sigma_1 = c_1^2$.

Totally ordered subgroups A, B are disjoint prime l-subgroups of H. By Holland's embedding theorem (Theorem 4.1.1.), there is the embedding $\sigma_2 : H \to A(\Omega)$ where $\Omega = H/A \overleftarrow{\cup} H/B$. Let k be the element of $A(\Omega)$ defined by $(Ax)k = Ax$, $(Bx)k = Bxc$. Finally, let G_2 be the l-subgroup if $A(\Omega)$ generated by $H\sigma_2$ and k. Note that $b\sigma_2 \wedge k = e$ in $A(\Omega)$

We claim that the formation $(H, G_1, G_2, \sigma_1 \sigma_2)$ cannot be amalgamated in the class $\mathcal{L}$ of all l-groups. Suppose, for the proof by contradiction, $(\varphi_1, \varphi_2, F)$ amalgamates the above formation. Then $\varphi_1 : G_1 \to F$, $\varphi_2 :\to F$ are embeddings such that $g\sigma_1\varphi_1 = g\sigma_2\varphi_2$ for all $g \in H$. Let P be the smallest convex l-subgroup of F containing $b\sigma_2\varphi_2$ and $k' = (k(c^{-1}\sigma_2))\varphi_2$. Let V^0 be a subgroup of G_2 generated by the elements $b\sigma_2, k'$, V be a convex l-subgroup of G_2 generated by V^0. Evidently, $V = \{x \in G_2 |\ |x| \leq |y| \text{ where } y \in V^0\}$. Since elements $k', b_2\sigma_2$ containes in the stabilizer $A(\Omega)_B$ of the point $B \in \Omega$ in $A(\Omega)$, then $V \subseteq A(\Omega)_B \cap G_2$. But $a\sigma_2 \notin V$ as $B(a\sigma_2) = Ba \neq B$. Let N be a convex l-subgroup of F which includes $V\varphi_2$ and is maximal with respect to not containing $a\sigma_2\varphi_2$. By Theorem 3.3.3, N is a prime subgroup of F, so let π be the natural l-homomorphism of F into $A(F/N)$.

Now $N(c_1\varphi_1) < N(a_1c_1\varphi_1) \leq N(c^2\varphi_1)$ in F/N, for replacing the inequality by theequality would imply $N(c_1\varphi_1) = N(a_1c_1\varphi_1) = N(a_1\varphi_1)\cdot(c_1\varphi_1) = N(a\sigma_2)\cdot(c_1\varphi_1)$ and $a\sigma_2\varphi_2 \in N$ which contradicts to the definiton of N. Therefore, $N(c_1)((b\sigma_2\varphi_2)\pi) = N(c_1b_1)\varphi_1 = N(a_1c_1)\varphi_1 > N(c_1\varphi_1)$. Since $b\sigma_2 \wedge k = e$ and $b\sigma_2\varphi_2\pi \wedge k\varphi_2\pi = e$ in $A(\Omega)$, then

$$N(c_1\varphi_1)(k\varphi_2)\pi = N(c_1\varphi_1) < N(c_1^2\varphi_1) = N(c\sigma_2\varphi_2) =$$
$$N(k'\varphi_2)(c\sigma_2\varphi_2) = N(k\varphi_2) = (N)(k\varphi_2\pi) \leq (Nc\varphi_1)(k\varphi_2\pi)$$

which is impossible. Thus, the formation $(H, G_1, G_2, \sigma_1, \sigma_2)$ cannot be amalgameted in the class of all l-groups $\mathcal{L}$.

2^0. Now we investigate formations in $\mathcal{L}$ which can be amalgamated in $\mathcal{L}$.

LEMMA 1. (Kokorin and Kopytov [2].) *Let $(H, G_i, \sigma_i | i \in I)$ be a formation of the family of totally ordered groups $\{G_i | i \in I\}$ such that $H\sigma_i$ is a convex normal subgroup of G_i for all $i \in I$. Then $(H, G_i, \sigma_i | i \in I)$ can be amalgamated in the class of all totally ordered groups $\mathcal{O}$.*

Proof. Let G be a free product of G_i $(i \in I)$ with amalgamated subgroup H corresponding to embeddings σ_i $(i \in I)$. Then G is a quotient group of the free product $F = \prod\limits_{i\in I}^{*}$ of the groups G_i by the normal subgroup N generated by all elements $h\sigma_i(h\sigma_j)^{-1}$, $h \in H$. It is clear that $G/\bar{H} = H\sigma_i N$ is isomorphic to the free product $\bar{F} = \prod\limits_{i\in I}^{*} G_i/H\sigma_i$ of totally ordered groups G_i/H_i $(i \in I)$. By Theorem 7.5.1 group $\bar{G}$ has a total order $\bar{P}$ extending the total orders of all free factors $G_i/H\sigma_i$ $(i \in I)$. Let P_i be a total order of G_i. Then the semigroup $P_i \cap H\sigma_i$ defines the G-invariant total order on the normal subgroup $\bar{H}$. We define the total order P on G by the rule: $g > e$ in G iff $\bar{g} = g\bar{H} > e$ in the totally ordered quotient group $\bar{G} = G/\bar{H}$ or $g \in \bar{H}$, $g > e$ in $\bar{H}$. It is evident that P is a total order of G extending all totall orders P_i and G is an amalgam of formation $(H, G_i, \sigma_i | i \in I)$.□

LEMMA 2. (Conrad [4].) *Let G be an l-group and H be a convex l-subgroupof G. Then the mapping $\theta : X \to X \cap H$ gives a one-to-one correspondence between the prime subgroups of G not containing H and proper prime subgroup of H.*

Proof. It is obvious that if X is a prime subgroup of G not containing H, then $X \cap H$ is a proper prime subgroup of H.

Conversely, let A, B be prime subgroups of G not containing H such that $A \cap H = B \cap H$. Let $R(G : A)$ be a totally ordered set of right cosets. The convex l-subgroup H is not contained in A. Therefore, there is an element $h \in H \setminus A$, $h > e$. Let b be any positive element of the prime subgroup B. Then by Proposition 3.3.1, $Ah \wedge Ab = A(h \wedge b) = A$. But in $R(G : A)$ $Ah \wedge Ab = \min(Ah, Ab)$. From $Ah > A$ it follows

$Ab = A$ and $b \in A$. Hence all positive elements of B are contained in A and $B = A$.□

Let X be a prime ideal of G not containing H. Then it is clear that $X\theta$ is a prime ideal of H, too but the converse statement is not true.

Let us recall that l-group G is representable if G is an l-subgroup of the Cartesian product of totally ordered groups. G is representable iff there is a family of prime ideals $\{N_\alpha | \alpha \in A\}$ such that $\bigcap_{\alpha \in A} N_\alpha = E$. We use the following statement which will be proved in Chapter 9, Theorem 9.4.1: l-group G is representable iff every minimal prime subgroups of G is an ideal. The proof of this theorem is independent of this section. Now if X is a minimal prime subgroup of G not containing a convex l-subgroup H, then $H \cap X$ is a minimal prime subgroup of H.

THEOREM 1. (Reily [2], Medvedev [2].) *Let $(H, G_i, \sigma_i | i \in I)$ be a formation of representable l-groups G_i and $H\sigma_i$ be an ideal in G_i for every $i \in I$. Then formation $(H, G_i, \sigma_i | i \in I)$ can be amalgamated in the class of all representable l-groups $\mathcal{R}$.*

Proof. Let $\{H^\gamma | \gamma \in \Gamma\}$ be the set of all minimal prime subgroup of H. Then $\{H^\gamma \sigma_i | \gamma \in \Gamma\}$ is the set of all minimal prime subgroups of the convex l-subgroup $H_i = H\sigma_i$ in l-group G_i. By Lemma 2, there is the unique minimal prime subgroup H_i^γ of l-group G_i such that $H_i^\gamma \cap H_i = H^\gamma \sigma_i$. Let $\{H_i^\delta | \delta \in \Delta\}$ be the set of all minimal prime subgroups of G_i contained H_i. By our assumption all G_i $(i \in I)$ are representable. Therefore, $\bigcap_{\delta \in \Delta} H_i^\delta \cap \bigcap_{\gamma \in \Gamma} H_i^\gamma = E$ and l-group G_i is embedded into the Cartesian product $\bar{G}_i = \overline{\prod_{\gamma \in \Gamma}} G_i/H_i^\gamma \times \overline{\prod_{\delta \in \Delta}} G_i/H_i^\delta$ of totally ordered groups G_i/H_i^γ $(\gamma \in \Gamma)$ and G_i/H_i^δ $(\delta \in \Delta)$ (see Proposition 2.3.5.) by isomorphic embedding ρ_i. Let π_i^γ, π_i^δ $(\gamma \in \Gamma, \delta \in \Delta)$ be natural projections of $\bar{G}_i$ onto G_i/H_i^γ and G_i/H_i^δ respectively. By Lemma 1, every formation $(H/H^\gamma, G_i/H_i^\gamma, \sigma_i\rho_i\pi_i^\gamma | i \in I)$ of totally ordered groups G_i/H_i^γ has amalgam K_γ in the class of all totally ordered groups $\mathcal{O}$. Therefore, K_γ is a totally ordered group and there are order embeddings ψ_i^γ of totally ordered groups G_i/H_i^γ into K_γ. Let K be a Cartesian product of totally ordered groups K_γ $(\gamma \in \Gamma)$ and totally ordered groups G_i/H_i^δ $(i \in I, \delta \in \Delta)$. Now we define the map

$\varphi_i : G_i \to K$ $(i \in I)$ by the rule $x\varphi_i(\gamma) = (xH_i^\gamma)\psi_i^\gamma \in K$ for all $\gamma \in \Gamma$ and $x\varphi_i(\delta) = xH_i^\delta \in G_i/H_i^\delta$ for all $\delta \in \Delta$. It is readily verified that $(\varphi_i, K|i \in I)$ is an amalgam of G_i $(i \in I)$ over H.□

The following remarkable result is due to K.Pierce [1]. The short proof can be found in the work of Powell and Tsinakis's [1].

THEOREM 2. (K.Pierce [1].) *The class $\mathcal{A}$ of all Abelian lattice-ordered groups satisfies the amalgamation property.*

Chapter 8

Lattice properties in lattice-ordered groups

8.1 Order Completeness and Archimedean Property in l-Groups

1^0. The standard definition of the completeness of partially ordered sets is not usefull for po-groups. For example, let (a) be an infinite cyclic subgroup of po-group G generated by any positive element a. Then (a) cannot have the supremum (the least upper bound) in G. Therefore we modify this definition as follows: a partially ordered group G is *complete* (or *order complete*) if every non-empty bounded above subset M has a *supremum* $\sup M = \vee M = \bigvee\limits_{x \in M} \{x\}$.

The map $x \to x^{-1}$ is an anti-isomorphism of every po-group G. Therefore every bounded below subset M of po-group G has an *infinum* $\inf M = \wedge M = \bigwedge\limits_{x \in M} \{x\}$. In partially ordered group G infimum and supremum do not always exist, but there are useful computional relationship among those *infima* and *suprema* that do exist.

PROPOSITION 1. (Kantorovich [1], Birkhoff [1].) *Let G be a partially order group and $\{x_\alpha | \alpha \in A\}$ be a non-empty subset of G. The following identities hold in the sense that if either side exists, then the other exists too, in which case the two sides agree:*

(1) $x(\bigvee_{\alpha\in A}\{x_\alpha\}) = \bigvee_{\alpha\in A}\{xx_\alpha\}$; $\quad x(\bigwedge_{\alpha\in A}\{x_\alpha\}) = \bigwedge_{\alpha\in A}\{xx_\alpha\}$;

(2) $(\bigvee_{\alpha\in A}\{x_\alpha\})x = \bigvee_{\alpha\in A}\{x_\alpha x\}$; $\quad (\bigwedge_{\alpha\in A}\{x_\alpha\})x = \bigwedge_{\alpha\in A}\{x_\alpha x\}$;

(3) $(\bigvee_{\alpha\in A}\{x_\alpha)^{-1} = \bigwedge_{\alpha\in A}\{x_\alpha^{-1}\}$; $\quad (\bigwedge_{\alpha\in A}\{x_\alpha)^{-1} = \bigvee_{\alpha\in A}\{x_\alpha^{-1}\}$.

The following identities hold provided the relevant infima and suprema exist in G.

(4) $x\wedge(\bigvee_{\alpha\in A} x_\alpha) = \bigvee_{\alpha\in A}(x\wedge x_\alpha)$; $\quad x\wedge(\bigwedge_{\alpha\in A} x_\alpha) = \bigwedge_{\alpha\in A}(x\wedge x_\alpha)$;

(5) $x\vee(\bigvee_{\alpha\in A} x_\alpha) = \bigvee_{\alpha\in A}(x\vee x_\alpha)$; $\quad x\vee(\bigwedge_{\alpha\in A} x_\alpha) = \bigwedge_{\alpha\in A}(x\vee x_\alpha)$.

Proof. (1)-(3). Since the maps $a\to ax$, $a\to xa$ are order automorphisms of po-set G, then they preserve whatever infima or suprema exist in G. Since the map $a\to a^1$ is an order anti-isomorphism of G, it reverses whatever infima or suprema exsist in G.

(4). Set $a=\bigvee_{\alpha\in A} x_\alpha$. Then $a^{-1}\le x_\alpha^{-1}$, $x\wedge x_\alpha\le x\wedge a$ for all $\alpha\in A$. for each $\alpha\in A$ we complete

$$e\le(x\wedge a)(x\wedge x_\alpha)^{-1} = (x\wedge a)(x^{-1}\vee x_\alpha^{-1}) =$$
$$(x\wedge a)x^{-1}\vee(x\wedge a)x_\alpha^{-1} = (e\wedge ax^{-1})\vee(xx_\alpha^{-1}\wedge ax_\alpha^{-1})\le$$
$$\le e\vee ax_\alpha^{-1}\le a(a^{-1}\vee x_\alpha^{-1}) = ax_\alpha^{-1}.$$

By (3),

$$\bigwedge_{\alpha\in A} ax_\alpha = a(\bigwedge_{\alpha\in A} x_\alpha^{-1}) = a(\bigvee_{\alpha\in A} x_\alpha)^{-1} = e,$$
$$e\le\bigwedge_{\alpha in A}(x\wedge a)(x\wedge x_\alpha)^{-1}\le\bigwedge_{\alpha\in A} ax_\alpha^{-1} = e.$$

Therefore,

$$\bigwedge_{\alpha\in A}(x\wedge a)(x\wedge x_\alpha)^{-1} = e,\quad (x\wedge a)(\bigwedge_{\alpha\in A}(x\wedge x_\alpha)^{-1}) = e,$$
$$x\wedge a = (\bigwedge_{\alpha\in A}(x\wedge x_\alpha)^{-1})^{-1},\qquad x\wedge a = \bigvee_{\alpha\in A}(x\wedge x_\alpha).$$

(5) is proved by similar arguments. □

2^0. The subset T of partially ordered set X is *closed* if whenever $\{x_\alpha|\alpha\in A\}\subseteq T$ and $x=\bigvee_{\alpha\in A} x_\alpha$ exsist in X, then $x\in T$.

THEOREM 1. (Riesz [1], Birkhoff [1].) *Each closed convex l-subgroup A of complete l-group G is an ideal in G and $G = A \times A^{\perp}$. Moreover, each polar of complete l-group G is a closed subgroup in G.*

Proof. Let A be a closed convex l-subgroup of complete l-group G and $g \geq e$, $g \in G$. Let $M = \{m \in G | m = g \wedge |a|,\ a \in A\}$. It is clear that the subset M is bounded above in G. Therefore there is $t_1 = \sup M$. Since A is closed, then $t_1 \in A$ and $e \leq t_1 \leq g$. Let $t_2 = t_1^{-1} g$, $t_2 \geq e$, then $t_1 \geq g \wedge t_1|a| = t_1 t_2 \wedge t_1 |a| = t_1(t_2 \wedge |a|) \geq t_1$ where $a \in A$. Hence $t_2 \wedge |a| = e$ for all $a \in A$ and $g = t_1 t_2$ where $t_1 \in A$, $t_2 \in A^{\perp}$. Thus $G = A \times A^{\perp}$ and A is an ideal in G.

Let T be a polar of complete l-group G, $T = T^{\perp\perp}$ and a subset M be bounded above in G. Let $a = \sup M$. Then for all $x \in T^{\perp}$, $x > e$ by Proposition 1,

$$
\begin{gathered}
e \leq x \wedge |a| = x \wedge (a \vee a^{-1}) = \\
x \wedge ((\bigvee_{m \in M} m) \vee (\bigvee_{m \in M} m)^{-1}) = (x \wedge \bigvee_{m \in M} m) \vee (x \wedge \bigwedge_{m \in M} m^{-1}) = \\
(\bigvee_{m \in M} m) \vee (x \wedge \bigwedge_{m \in M} m^{-1}) = (\bigvee_{m \in M} (x \wedge m)) \vee (\bigwedge_{m \in M} (x \wedge m^{-1})) = \\
(x \wedge \bigvee_{m \in M} m) \vee (x \wedge \bigwedge_{m \in M} m^{-1}) = (\bigvee_{m \in M} (x \wedge m)) \vee (\bigwedge_{m \in M} (x \wedge m^{-1})) = \\
\bigvee_{m \in M} ((x \wedge m) \vee \bigvee_{n \in M} (x \wedge n^{-1})) = \bigvee_{m \in M} \bigwedge_{n \in M} ((x \wedge m) \vee (x \wedge n^{-1})) \leq \\
\bigvee_{m \in M} \bigwedge_{m \in M} ((x \wedge m) \vee (x \wedge m^{-1})) = \bigvee_{m \in M} \bigwedge_{m \in M} (x \wedge |m|) = e
\end{gathered}
$$

Therefore, $a \in T^{\perp\perp} = T$. Since M is bounded above in G, all suprema and infima in above calculations exist in G.□

COROLLARY 1. (Birkhoff [3].) *The set of all closed convex l-subgroups of lattice-ordered group G is a sublattice of the lattice $\mathcal{C}(G)$ of all convex l-subgroups of G. Moreover, this sublattice is the complete Boolean algebra.*□

It easy to see that for a complete l-group G the Boolean algebra $\mathcal{P}(G)$ is not a closed complete sublattice of the lattice $\mathcal{C}(G)$.

THEOREM 2. (Iwasawa [1].) *Each complete l-group G is Abelian.*

Proof (Birkhoff [3].) Let G be a complete l-group and $a, b \in G$, $a, b > e$. By Theorem 1, there are polars A_i ($i = 1, 2, \ldots, 8$) such that $G =$

$A_1 \times \cdots \times A_8$ and projections of the elements $a^{-1}b, [a,b], [[a,b],b]$ into each polars A_i are comparable with e. Here the projection of x into A_i is the element $x_i \in A_i$ for which $x = x_i y_i$ where $y_i \in A_i^{\perp}$. It is clear that all proections of the element x are uniquelly determined and $x = x_1 \cdots x_8$ (it is possible that $x_i = e$ for some indexes i). Thus it is sufficient to prove the commutativity of each factor A_i, $1 \le i \le 8$. Without loss of generality we can assume $(a^{-1}b)_i > e$. Let in the polar A_1 holds $[a,b]_1 \ge e$ and $[[a,b],b] \ge e$. Then $[a,b]_1^{-1} b^{-1}[a,b]_1 b \ge e$, $b^{-1}[a,b]_1 b \ge [a,b]_1$. Therefore, $b^{-k}[a,b]_1 b^k \ge [a,b]_1$ for all positive integers k. Further,

$$[a,b^n]_1 = [a,b]_1 \cdot b^{-1}[a,b]_1 b \cdots b^{-n+1}[a,b]_1 b^{n-1} \ge [a,b]_1^n$$

for all $n \ge 0$. As $(a^{-1}b)_1 > e$, then $(b^{-n}ab^n)_1 < b_1$, $[a,b^n]_1 = (a^{-1}b^{-n}ab^n)_1 \le (a^{-1}b)_1$ and $[a,b]_1^n \le (a^{-1}b)_1$ for all integers n. By completeness of the l-group G, there exists the element $c_1 = \bigvee_{-\infty}^{+\infty} [a,b]^n$.

By Proposition 1, $c_1[a,b]^{-1} = \bigvee_{-\infty}^{+\infty} [a,b]_1^{n-1} = c_1$ and $[a,b]_1 = e$.

Now, let A_2 be a polar of G for which $[a,b]_2 \ge e$ and $[[a,b],b]_2 \le e$. Then

$$b^{-1}[a,b]_2 b \le [a,b]_2, \quad b^{-k}[a,b]b^k \le [a,b]_2, \quad [a,b]_2 \le b^k[a,b]_2 b^{-k}$$

for all positive integers k. Since

$$[b^{-n},a]_2 = b^n[a,b]_2 b^{-n} \cdots b[a,b]_2 b^{-1} \ge [a,b]_2^n$$

for all positive integer n, then

$$[a,b]_2^n \le [b^{-n},a]_2 = (b^n a^{-1} b^{-n} a)_2 \le a_2.$$

By the completeness of the l-group G, there exists the element

$$c_2 = \bigvee_{n=-\infty}^{+\infty} [a,b]_2^n = (\bigvee_{n=-\infty}^{+\infty} [a,b]_2^{n+1})[a,b]_2^{-1} = c_2[a,b]_2^{-1}.$$

Hence $[a,b]_2 = e$ in A_2. Similar arguments shows that $[a,b]_i = e$ for all $i = 1,2,\ldots,8$ and consequently, $[a,b] = e$ for all $a,b \in G$.□

3°. It is clear that each complete o-group G is Archimedean o-group, i.e., the Archimedean axiom (see chapter 3.2) is valid in G:

(A) for any two elements $a, b \in G$, $a, b > e$ there are positive integers m, n such that $a^n > b$, $b^m > a$.

It is easy to see that for o-groups, the axiom (A) is equivalent to the axiom:

(A′) if for any two elements $a, b \in G$ the inequality $a^n < b$ holds in G for all integers n, then $a = e$.

The property (A′) is used for the generalization of Arhimedean property to po-groups. Po-group G is an *Archimedean po-group* if the property (A′) is valid in G. The following property is very closed to Archimedean property.

(A″) If for any two elements $a, b \in G$ the inequality $a^n \leq b$ holds in G for all positive integers n, then $a \leq e$.

Po-group G is called *integrally closed* when the property (A″) holds in G.

PROPOSITION 2. (Birkhoff [3].) *Any integrally closed partially ordered group G is Archimedean. Each Archimedean l-group is integrally closed.*

Proof. If $a^n \leq b$ for all integers n, then $a^n \leq b$ for all positive integers n, whence $a \leq e$. Also, $(a^{-1})^n \leq b$ for all positive integers n, whence $a^{-1} \leq e$ and $a = e$.

Let G be an Archimedean l-group and $a^n < b$ for some $a, b \in G$ and for all positive integers n. Then $(a \vee b)^n = a^n \vee a^{n-1} \vee \cdots \vee e \leq b \vee e$ and $(a \vee e)^{-n} \leq e < b \vee e$ for all positive integers n. Hence $(a \vee e)^n < b \vee e$ for all integers n and by (A′) it follows that $a \vee e = e$. □

The following example illustrates that in po-groups properties (A′) and (A″) are (in general) not equivalent.

EXAMPLE 1. (Fuchs [1].) Let C be an additive group of complex numbers partially ordered by the rule: $x + \mathbf{i}y \geq x' + \mathbf{i}y$ iff $x > x'$ and $y > y'$ or $x = x'$, $y = y'$. It is easy to see that po-group C is Archimedean. On the other hand, C is not integrally closed. Indeed, $n(-1+\mathbf{i}0) < 1+\mathbf{i}$ for all positive integers n and $-1+\mathbf{i}0$ is incomparable with 0 in this parial order.

The following remarkable result was obtained (in different variants) by Krull [1], Lorezen [1], Everett and Ulam [1], Birkhoff [3] and

Fuchs [1].

THEOREM 3. *Each integrally closed partially ordered group G is order embedded into complete partially ordered group G^*. If G is an l-group, then G^* is an l-group, too.*

Proof. Let G be an integrally closed po-group. For any nonempty subset X of G which is bounded above, let $U(X)$ be the nonempty set of its upper bounds. Let G^* be the set of all these subsets. On the set G^* we define a partial order $\leq$ by the rule: $U(X) \leq U(Y)$ iff $U(X) \supseteq U(Y)$. It is clear that the map $\varphi : G \to G^*$ where $x\varphi = U(\{x\})$ is the order map of po-set G into po-set G^*. Also, it is clear that G^* is a lattice if G is a lattice and φ is an lattice isomorphism of G in G^* in this case.

Now we define the group operations $\cdot$ on G^* by the rule:

$$U(X) \cdot U(Y) = U(XY) = \{z \in G | z \geq xy \text{ for all } x \in X,\ y \in Y\}.$$

Direct verification shows that operation $\cdot$ is well-defined, is associative and $U(e)$ is the identity element in G^*. It is readily verified that $U(Z)U(X) \leq U(Z)U(Y)$, $U(X)U(Z) \leq U(Y)U(Z)$ is valid in G^* for all $U(X), U(Y), U(Z) \in G^*$ such that $U(X) \leq U(Y)$.

We claim that G^* is a group. This is equivalent to the fact that for any $U(X) \in G^*$ there is $U(Y) \in G^*$ such that $U(Y)U(X) = U(e)$. Let $Y = U(X)^{-1} = \{z \in G | z^{-1} \in U(X)\}$. Then Y is bounded above for any element of $U(X)$. As $x \leq u$ for all element u in $U(X)$, then $U(Y)U(X) = U(U(X^{-1})X) \leq U(e)$.

On other hand, for all $x \in X$, $u \in U(X)$ and $v \in U(Y)U(X)$ is valid $u^{-1}x \leq v$, $x \leq uv$. Let $u_1 \in U(X)$, $v \in U(Y)U(X)$. Then $x \leq u_1 v$ for all $x \in X$. Therefore, $u_2 = u_1 v \in U(X)$. Assume that the elements $u_1, \ldots, u_k$ are defined. Then $u_{k+1} = u_k v \in U(X)$, $u_k \in U(X)$ and $u_{k+1} = u_1 v^k$, i.e., $v^{-k} = (v^{-1})^k \leq x^{-1}u_1$ for all positive integers k and for evry $x \in X$. As G is integrally closed, then $v^{-1} \leq e$, $v \geq e$ and $U(Y)U(X) \subseteq U(e)$, $U(Y)U(X) \geq U(e)$. Thus $U(Y)U(X) = U(e.)$ So, G^* is a po-group; if G is an l-group, then G^* is an l-group, too; and $\varphi : x\varphi = U(X)$ is an l-isomorphism from G into G^*. It is readily verified that G^* is integrally closed and we leave the details for the reader.□

COROLLARY 2. *Each Archimedean l-group is Abelian.*□

It is clear that the classs *Arch* of all Archimedean l-groups is closed under Cartesian products and l-subgroups and it is not closed under l-homomorphisms.

8.2 Completely Distributive l-Groups

1⁰. In Chapter 2 Section 2 it was proved that the lattice of each l-group is distributive. Let us recall that lattice L is called *completely distributive* iff for each family $\{x_{\alpha\beta} | \alpha \in A, \beta \in B\}$

(1) $$\bigwedge_{\alpha \in A} \bigvee_{\beta \in B} x_{\alpha\beta} = \bigvee_{\rho \in B^A} \bigwedge_{\alpha \in A} x_{\alpha\rho(\alpha)}$$

whenever all the indicated suprema and infima exist.

First we prove the following test of complete distributivity of an l-group.

THEOREM 1. (Weinberg [1].) *A lattice ordered group G is completely distributive if and only if*

(2) *for every element $g \in G$, $g > e$ there is an element $g^* \in G$ such that for each representation $g = \bigvee_{\alpha \in A} g_\alpha$ where $g_\alpha > e$ for all $\alpha \in A$ holds $g^* \leq g_\zeta$ for some $\zeta \in A$.*

Proof. Let G be a completely distributive l-group and $g > e$, $g \in G$. Let $\{g = \bigvee M_\alpha | \alpha \in A\}$ be the set of all representations of the element g as suprema of subsets M_α of G. Without loss of generality we can assume that this set of all representations for g has the form $\{g = \bigvee_{\beta \in B} x_{\alpha\beta} | \alpha \in A\}$ where the second index runs through the same index set B for every $\alpha \in A$ and the cardinality of B is less or equal to the cardinality of G. It is possibble since in the representation $g = \bigvee_{\beta \in B} x_{\alpha\beta}$ some elements $x_{\alpha\beta}$ may occur several times. Then $g = \bigwedge_{\alpha \in A} \bigvee_{\beta \in B} x_{\alpha\beta} = \bigvee_{\rho \in B^A} \bigwedge_{\alpha \in A} x_{\alpha\rho(\alpha)}$. It is clear that among elements $\bigwedge_{\alpha \in A} x_{\alpha\rho(\alpha)}$, $\rho \in B^A$, non-trivial elements of G or some infima do not exist in G. Otherwise we have $\bigwedge_{\alpha \in A} x_{\alpha\rho(\alpha)} = e$ for all $\rho \in B^A$ and therefore $\bigvee_{\rho \in B^A} \bigwedge_{\alpha \in A} x_{\alpha\rho(\alpha)} = e$, the contradiction with $g > e$. Let $\bigwedge_{\alpha\in A} x_{\alpha\bar{\rho}(\alpha)}$ exist in G for some $\bar{\rho} \in B^A$ and $\bigwedge_{\alpha\in A} x_{\alpha\bar{\rho}(\alpha)} = g^* \neq e$, then $g^* > e$. Since $\bar{\rho} \in B^A$, then for every $\alpha \in A$ holds $x_{\alpha\bar{\rho}(\alpha)} \geq g^*$.

If $\bigwedge_{\alpha\in A} x_{\alpha\bar{\rho}(\alpha)}$ do not exists in G, then the subset $\{x_{\alpha\rho(\alpha)} | \alpha \in A\}$ has no infimum in G. Since $e < x_{\alpha\bar{\rho}(\alpha)}$, then there is an element $g^* \in G$ such that $g^* > e$ and $g^* \leq x_{\alpha\bar{\rho}(\zeta)}$ for some $\zeta \in A$.

Conversely, let l-group G be not completely distributive. Then there are elements $x_{\alpha\beta}$, $\alpha \in A$, $\beta \in B$ such that

$$(3) \qquad x = \bigwedge_{\alpha\in A}\bigvee_{\beta\in B} x_{\alpha\beta} \neq \bigvee_{\rho\in B^A}\bigwedge_{\alpha\in A} x_{\alpha\rho(\alpha)} = y$$

where all the indicated suprema and infima exist. It is easy to see that $y < x$. Let $g = xy^{-1} > e$. Now let $g_{\alpha\beta} = (x \wedge x_{\alpha\beta})y^{-1} \vee e$, $\alpha \in A, \beta \in B$. Then by Proposition 8.1.1,

$$\bigvee_{\beta\in B} g_{\alpha\beta} = \bigvee_{\beta\in B} (x \wedge x_{\alpha\beta})y^{-1} \vee e = e \vee (\bigvee_{\beta\in B} (x \wedge x_{\alpha\beta})y^{-1}) =$$
$$= e \vee (\bigvee_{\beta\in B} (x \wedge x_{\alpha\beta}))y^{-1} = e \vee (x \wedge (\bigvee_{\beta\in B} x_{\alpha\beta}))y^{-1} =$$
$$= e \vee (x \wedge (\bigvee_{\beta\in B} x_{\alpha\beta}))y^{-1} = e \vee xy^{-1} = xy^{-1} = g$$

for all $\alpha \in A$.

Suppose, for reductio ad absurdum, that there exists an element $g^* \in G$, $g^* > e$ such that for every $\alpha \in A$ holds $g^* \leq g_{\alpha\rho(\alpha)}$ for some $\rho \in B^A$. On the other hand,

$$\bigwedge_{\alpha\in A} g_{\alpha\rho(\alpha)} = \bigwedge_{\alpha\in A} ((x \wedge x_{\alpha\rho(\alpha)})y^{-1} \vee e) =$$
$$e \vee (\bigwedge_{\alpha\in A} (x \wedge x_{\alpha\rho(\alpha)})y^{-1}) = e \vee (\bigwedge_{\alpha\in A} (x \wedge x_{\alpha\rho(\alpha)})y^{-1}) =$$
$$= e \vee (x \wedge (\bigwedge_{\alpha\in A} x_{\alpha\rho(\alpha)}))y^{-1} = e \vee (\bigwedge_{\alpha\in A} x_{\alpha\rho(\alpha)}) = e$$

Then $g^* \leq \bigwedge_{\alpha\in A} g_{\alpha\rho(\alpha)} = e$ and $g^* = e$, a contradiction.□

It is evident that in completely distributive l-group G the element g^* is not uniquely determined by the element g. Each pair (g, g^*) of elements of G such that $g, g^* > e$ and for every representation $g =$

$\bigvee_{\alpha \in A} g_\alpha$ where $g_\alpha > e$ for all $\alpha \in A$ holds $g_\zeta \geq g^*$ for some $\zeta \in A$ is called a *distributive pair* of elements of G.

2^0. Theorem 1 can be applied to some class of l-groups.

PROPOSITION 1. (Weinberg [1].) *Each totally ordered group G is completely distributive.*

Proof. Let g be a totally ordered group and $g \in G,\ g > e$. If the open interval $(e, g) = \{x \in G | e < x < g\}$ is empty or contains greatest element, then we set $g^* = g$. If (e, g) is not empty and has no greatest element, then we set g^* equal to any element of the interval (e, g).□

LEMMA 1. (Weinberg [1].) *A convex l-subgroup H of l-group G is closed if and only if the natural homomorphism φ of the lattice G on the lattice $R(G : H)$ of right cosets of H preserves infima and suprema.*

Proof. Let H be a closed convex l-subgroup of G and $x = \bigvee_{\alpha \in A} x_\alpha$ where $x, x_\alpha\ (\alpha \in A) \in G$. We claim that $x\varphi = \bigvee_{\alpha \in A} x_\alpha \varphi$, i.e., $Hx = \bigvee_{\alpha \in A} Hx_\alpha$. In fact, let $y \in G$ and for all $\alpha \in A$ holds $Hx_\alpha \leq Hy \leq Hx$ in the lattice $R(G : H)$. We can assume that $y \leq$ and $Hx_\alpha x^{-1} \leq Hyx^{-1} \leq H$. Then for every $\alpha \in A$ there exists an element $y_\alpha \in H$ such that $y_\alpha \leq e,\ y_\alpha x_\alpha x^{-1} \leq yx^{-1} \leq e$ and $e \leq xy^{-1} \leq xx_\alpha^{-1} y_\alpha^{-1}$. Since $xx_\alpha^{-1} \geq e$ for all $\alpha \in A$, then

$$e \vee xy^{-1} y_\alpha \leq xx_\alpha^{-1},$$

$$e \leq \bigwedge_{\alpha \in A} (e \vee xy^{-1} y_\alpha) \leq \bigwedge_{\alpha \in A} xx_\alpha^{-1} = x(\bigwedge_{\alpha \in A} x_\alpha^{-1}) = x(\bigvee_{\alpha \in A} x_\alpha)^{-1},$$

$$\bigwedge_{\alpha \in A} (e \vee xy^{-1}) y_\alpha) = e, \quad \bigwedge_{\alpha \in A} (yx^{-1} \vee y_\alpha) = yx^{-1},$$

$$\bigvee_{\alpha \in A} (xy^{-1} \wedge y_\alpha^{-1}) = xy^{-1}.$$

As $xy^{-1} \geq e,\ y_\alpha^{-1} \geq e$ and l-group H is convex and, closed then from the above equalities it follows that $x^{-1} \in H$, $Hx = Hy$ and $x\varphi = \bigvee_{\alpha \in A} x_\alpha \varphi$ in the lattice $R(G : H)$. The converse statement is obvious. For infima the proof is similar.□

PROPOSITION 2. (Fuchs [1].) *Let G be an l-group and $\{H_\alpha | \alpha \in A\}$ be the system of closed prime ideals of G such that $\bigcap_{\alpha \in A} H_\alpha = E$. Then G is completely distributive.*

Proof follows directly from Proposition 1 and Lemma 1.□

3⁰. Now, following the work of Byrd and Lloyd [1] we will characterize completely distributive l-groups. The important role in this characterization is played by Weingerg's characterization of complete distributivity in terms of distributive pairs (Theorem 1).

Let G be an l-group, M be a convex l-subgroup and $R(G:M)$ be the lattice of right cosets of M.

LEMMA 2. *A convex l-subgroup M of l-group G is closed if and only if the infimum of the set of all strictly positive elements of each right coset Mx of M for which $Mx > M$ in $R(G:M)$ is a strictly positive element of G.*

Proof. Let M be a closed convex l-subgroup of G and $Mx > M$. Let $\{x_\alpha | \alpha \in A\}$ is the set of all strictly positive elements of the right coset Mx. Assume, on the contrary, that $\bigwedge_{\alpha\in A} x_\alpha = e$. Since the natural mapping $\varphi : G \to R(G:M)$ of l-group G onto the lattice $R(G:M)$ preserves infima, then

$$M = (\bigwedge_{\alpha\in A} x_\alpha)\varphi = \bigwedge_{\alpha\in A} x_\alpha\varphi = \bigwedge_{\alpha\in A} Mx = Mx.$$

This contradicts $Mx > M$.

Conversely, let the infimum of the set of all strictly positive elements of each right coset Mx for which $Mx > M$ in $R(G:M)$ is a strictly positive element of G. Let $\{x_\alpha\}|\alpha \in A$ be a subset of strictly positive elements of M and $\bigvee_{\alpha\in A} x_\alpha = x$ exist in G. If $x \notin M$, then $x_\alpha^{-1}x > e$ and $M < Mx = Mx_\alpha^{-1}x$ for all $\alpha \in A$. Then there is an element $y \in G$ such that $y > e$, $x_\alpha^{-1}x \geq y$, $xy^{-1} \geq x_\alpha$ for all $\alpha \in A$ and $x > xy^{-1} \geq \bigvee_{\alpha\in A} x_\alpha = x$. This is the contradiction.□

Let M be a convex l-subgroup of l-group G. The *closure* $\bar{M}$ of M is the smallest closed l-subgroup of G containing M. Clearly, G is closed in G and $\bar{M}$ is the intersection of all closed l-subgroups of G containing M.

LEMMA 3. *Let M be a convex l-subgroup of an l-group G and M^* be the set of all suprema of subsets of positive elements of M. Then the closure $\bar{M}$ of M coinsides with the subgroup of G generated by M^*.*

Proof. Direct verification shows that M^* is a convex l-subsemigroup of G and $e \in M^*$. Let M_1 be a subgroup of G generated by M^* and $e \le x \le y$ where $y \in M_1$. Then $y = y_1^{\varepsilon_1} \cdots y_n^{\varepsilon_n}$ for some $y_i \in M^*$, $\varepsilon_i = \pm 1$ $(i = 1, \dots, n)$ and $e \le x \le y_1 \cdots y_n = y' \in M^*$. Then $x \in M^* \subseteq M_1$ and so M_1 is a convex subset of G. By similar arguments, M_1 is a directed subgroup of G and by Proposition 2.3.2, M_1 is an convex l-subgroup of G. Let now $\{a_\alpha | \alpha \in A\}$ be a subset of M_1 for which $\bigvee_{\alpha \in A} a_\alpha$ exists in G and let $a = \bigvee_{\alpha \in A} a_\alpha$ in G. Then $a \vee e = \bigvee_{\alpha \in A} (a_\alpha \vee e)$ and $a_\alpha \vee e \in M^*$ for all $\alpha \in A$. Since M^* is a closed subset then $a \vee e \in M^*$, too. The same arguments show that $a^{-1} \vee e \in M^*$ and $a = (a \vee e)(a^{-1} \vee e)^{-1} \in M_1$. Hence M_1 is closed in G and $M_1 = \bar{M}$.

LEMMA 4. *A prime subgroup containing a closed prime subgroup is closed.*

Proof. Suppose that H is a closed prime subgroup and M is a prime subgroup which properly contains it; choose $y \in M \setminus H$, $y > e$. Suppose further that $\{x_\alpha | \alpha \in A\} \subseteq M$, $x_\alpha \ge e$ for all $\alpha \in A$ and $x = \bigvee_{\alpha \in A} x_\alpha$. We will show that $x \in M$. Since H is closed we have by Lemma 1, that $Hx = \bigvee_{\alpha \in A} x_\alpha$. Now, if $x \wedge y \in H$ we have $x \in H \subseteq M$, since H is prime. So, we can assume that $x \wedge y \notin H$. This means that $H \le H(x \wedge y)^{-1}x < Hx$. in set $R(G : H)$ of right cosets of H. Since $R(G : H)$ is totally ordered thus we have that $H(x \wedge y)^{-1}x \le Hx_\alpha$ for some $\alpha \in A$. But then there exists $y_1 \in H$ such that $e \le (x \wedge y)^{-1}x \le y_1 x_\alpha \in M$. Thus $(x \wedge y)^{-1}x \in M$ and $x \in M$.□

A *distributive radical* $D(G)$ of l-group G is an intersection of all closed prime subgroups of G.

LEMMA 5. *If G is an l-group, then*

$$D(G) = \bigcap_{\gamma \in \Gamma} M_\gamma,$$

where $\{M_\gamma | \gamma \in \Gamma\}$ is a set of all minimal primes subgroups of G.

The proof of Lemma is straightforward and we leave the details to the reader.□

LEMMA 6. *Let G be an l-group and $D(G)$ be the distributive* *radical of* G. *Then* $D(G) =$

$$= \{g \in G \mid |g| \text{ is not first member of any distributive pair } (|g|, g^*)\}.$$

Proof. Let $g \in G$ be the first member of the distributive pair $(|g|, g^*)$. Let M be the value of g^* and $\bar{M}$ be the closure of M. If $|g| \in \bar{M}$, then there is a subset $\{g_\alpha | \alpha \in A\}$ of positive elements of M such that $|g| = \bigvee_{\alpha \in A} g_\alpha$. Then $e < g^* \leq g_\alpha$ for some $\alpha \in A$ and $g^* \in M$, a contradiction. Therefore, $|g| \notin \bar{M}$ and $|g| \notin D(G) = \bigcap_{\gamma \in \Gamma} \bar{M}_\gamma$.

Conversely, Let $g \notin D(G)$, $g > e$. Then $g \notin M$ for some closed prime subgroup M. Let coset Mg cover M in the lattice $R(G : M)$ of right cosets of M. By Lemma 2, there is an element $a \in G$ such that $a > e$ and $a \leq x$ for all $x \in Mg$, $x > e$. Let $g = \bigvee_{\alpha \in A} g_\alpha$, $g_\alpha \geq e$ for all $\alpha \in A$. Since M is closed, then $g_\zeta \notin M$ for some $\zeta \in A$. Hence $Mg_\zeta = Mg$ and $a \leq g_\zeta$. So, the pair (g, a) is distributive.

Let coset Mg do not cover M in $R(G : M)$ that is $M < Mh < Mg$ for some $h \in G$, $h > e$. Let a be the strictly positive lower bound of the set of strictly positive elements of coset Mh. Let $g = \bigvee_{\alpha \in A} g_\alpha$, $g_\alpha > e$ for all $\alpha \in A$. Then by Lemma 1, $Mg = M(\bigvee_{\alpha \in A} g_\alpha) = \bigvee_{\alpha \in A} Mg_\alpha$. Since M is a prime subgroup of G, then $Mh < Mg_\alpha$ for some $\alpha \in A$. Hence $Mh = M(h \wedge g_\alpha)$ and $a < h \wedge g_\alpha \leq g_\alpha$. This means that (g, a) is a distributive pair.□

THEOREM 2. (Byrd and Lloyd [1].) *An l-group* G *is completely distributive if and only if* $D(G) = E$. *The quotient group* $G/D(G)$ *of any l-group* G *by its distributive radical* $D(G)$ *is a completely distributive l-group.*

The proof of this theorem follows immediately from Lemma 5 and Lemma 6.□

PROPOSITION 3. (Byrd and Lloyd [1].) *Let* G *be an l-group. Then*

(4) $H \cap D(G) = D(H)$ *for each convex l-subgroup* H *of* G;

(5) $D(D(G)) = D(G)$;

(6) $D(\overset{c}{\bigvee_{\alpha\in A}} H_\alpha) = \overset{c}{\bigvee_{\alpha\in A}} D(H_\alpha)$, $D(\overset{c}{\bigwedge_{\alpha\in A}} H_\alpha) = \overset{c}{\bigwedge_{\alpha\in A}} D(H_\alpha)$ *for each family* $\{H_\alpha | \alpha \in A\}$ *of convex l-subgroups of* G;

(7) $D(\bar{\prod_{\alpha\in A}} G_\alpha) = \bar{\prod_{\alpha\in A}} D(G_\alpha)$, $D(\prod G_\alpha) = \prod_{\alpha\in A} D(G_\alpha)$ *for each family* $\{G_\alpha | \alpha \in A\}$ *of l-groups.*

Proof. Properties (4),(7) are easy consequences of Lemma 6. The properties (5),(6) follow from (4).□

4⁰. The following class of completely distributive l-groups was discovered by Lloyd [1].

THEOREM 3. (Lloyd [1].) *For each totally ordered set* Ω, *the lattice-ordered group* $A(\Omega)$ *of all order permutations of* Ω *is completely distributive.*

Proof. It is enough to prove that each positive element $g \in A(\Omega)$ is the first element in some distributive pair (g, g^*). If element g is indecomposable into a join in $A(\Omega)$ (equivalently, the equality $g = \bigvee_{\iota\in I} g_i$ implies $g = g_\iota$ for some $\iota \in I$), then the pair (g, g) is a distributive pair in $A(\Omega)$. Now, let $g = \bigvee_{\iota\in I} g_\iota$ where $\{g_\iota | \iota \in I\}$ is a subset of strictly positive elements of G and $g_\iota \neq g$ for all $\iota \in I$. Without loss of generality we can assume that g is a bump and $o(g, \zeta) = \mathrm{supp}(g) = \{\xi \in \Omega | \zeta g^n \leq \xi \leq \zeta g^{n+1}\}$ for some integer n where $\zeta \in \Omega$ is such that $\zeta g \neq \zeta$. Now we divide the proof of the theorem into several steps.

Step 1. *If* $A(\Omega)$*-orbit* $\zeta A(\Omega) = \{\zeta f | f \in A(\Omega)\}$ *has empty intersection with open interval* $(\zeta, \zeta g) = \{\xi \in \Omega | \zeta < \xi < \zeta g\}$, *then there is an element* $h \in A(\Omega)$, $h > e$ *such that*

(8) $\mathrm{supp}(h) = \{\xi \in \Omega | \xi h \neq \xi\}$ *is contained in the open interval* $(\zeta, \zeta g)$ *and the pair* (g, h) *is the distributive pair for each element* h.

Let $s \in A(\Omega)$ such that $e < s < g$. Then there is an element $\eta \in o(g, \zeta), \eta < \eta s$ and $\zeta g^n < \eta < \zeta g^{n+1}$ for some positive integer n. Then $\zeta g^n \leq \zeta s g^n \leq \zeta g^{n+1}$. Since $\zeta A(\Omega) \cap (\zeta, \zeta g) = \emptyset$, then $\zeta g^n = \zeta s g^n$ or $\zeta s g^n = \zeta g^{n+1}$. Let $\zeta s g^n = \zeta g^n$. Then $\zeta s g^{n+1} = (\zeta s g^n) g = \zeta g^{n+1}$ and the order permutation $\bar{s}$ of totally ordered set Ω defined by the rule:

$$\xi\bar{s} = \begin{cases} \xi s & \text{if} \quad \xi \in o(s,\eta), \\ \xi & \text{if} \quad \xi \notin o(s,\eta), \end{cases}$$

is clearly positive. By definition, $\mathrm{supp}(\bar{s}) \subseteq (\zeta g^n, \zeta g^{n+1})$. Let $\zeta h = g^{-n}\bar{s}g^n$, then $h > e$ and $\mathrm{supp}(h) \subseteq (\zeta, \zeta g)$. Now let $\zeta g^n s = \zeta g^{n+1}$. Then $\zeta g^m s = \zeta g^{m+1}$ for all integers m. As $s \neq g$, then the permutation gs^{-1} is non-trivial on some open interval $(\zeta g^k, \zeta g^{k+1})$. Let $\bar{h}$ be an order permutation of Ω defined by the rule

$$\xi\bar{h} = \begin{cases} \xi g s^{-1}, & \text{if} \quad \zeta g^k < \xi < \zeta g^{k+1}, \\ \xi, & \text{if} \quad \xi \notin (\zeta g^k, \zeta g^{k+1}). \end{cases}$$

It is clear that $\bar{h}$ is strictly positive and $\mathrm{supp}(\bar{h}) \subseteq (\zeta g^k, \zeta g^{k+1})$. Then $h = g^{-k}\bar{h}g^k > e$ and $\mathrm{supp}(h) \subseteq (\zeta, \zeta g)$. Therefore it is proved that in Step 1 there exists an element $h \in A(\Omega)$ with property (8). We claim that (g, h) is a distributive pair. Indeed, let $g = \bigvee_{\iota \in I} g_\iota$, $g_\iota > e$. If $\zeta g_\iota = \zeta$ for all $\iota \in I$, then $\xi h^{-1} g \geq \xi g \geq \xi g_\iota$ for all $\xi \notin (\zeta, \zeta g)$, $\iota \in I$ and $\xi h^{-1} g > \zeta g$ for all $\zeta < \xi < \zeta g$. Hence $\xi h^{-1} g \geq \xi g_\iota$ for all $\xi \in \Omega$ and $\iota \in I$. Now assume that $\xi g_\iota > \zeta g$ for some $\iota \in I$ and some $\xi \in (\zeta, \zeta g)$. Then $\xi > \zeta g g_\iota^{-1}$. Since $\zeta g g_\iota^{-1} \in o(g, \zeta)$, $o(g, \zeta) \cap (\zeta, \zeta g) = \emptyset$ and $\xi > \zeta$ then $\xi > \zeta g$, contradicts the choice $\xi \in (\zeta, \zeta g)$. So, $h^{-1} g \geq g_\iota$ for all $\iota \in I$ contradicts the inequality $g = \bigvee_{\iota \in I} g_\iota$. Therefore there is an index $\kappa \in I$ such that $\zeta g_\kappa = \zeta$. Then $h < g_\kappa$ and the pair (g, h) is distributive.

Step 2. *Let $k \in A(\Omega)$, $e < k < g$ and $\zeta < \zeta k < \zeta g$. Then either (g, k) is a distributive pair of $A(\Omega)$ or there is an element $f \in A\zeta(\Omega)$, $e < f < g$ such that* $\mathrm{supp}(f)$ *is bounded in totally ordered set* $o(g, \zeta)$.

Let us assume, on the contrary that (g, k) is not a distributive pair of $A(\Omega)$. Then there is a family $\{g_\iota | \iota \in I\}$ of positive elements of $A(\Omega)$ such that $g = \bigvee_{\iota \in I} g_\iota$ and the inequality $k \leq g_\iota$ does not hold for all $\iota \in I$. Therefore there is g_ι $(\iota \in I)$ and $\alpha, \beta \in o(g, \zeta)$ such that $\alpha g_\iota < \alpha k$, $\beta g_\iota > \beta k$. Without loss of generality we can assume that $\alpha < \beta$. If there is an element $\gamma \in o(g, \zeta)$ such that $\gamma < \alpha$ and $\gamma g_\iota \geq \gamma k$

then $o(kg_\iota^{-1}, \alpha) \subseteq (\gamma, \beta)$ since $\alpha kg_\iota^{-1} > \alpha$, $\gamma kg_\iota^{-1} \leq \gamma$, $\beta kg_\iota^{-1} < \beta$. Now we define the order permutation f by the rule:

$$\xi f = \begin{cases} \xi kg_\iota & \text{if} \quad \xi \in o(kg_\iota^{-1}, \alpha), \\ \xi & \text{if} \quad \xi \notin o(kg_\iota^{-1}, \alpha). \end{cases}$$

It is clear that $f < g$ and $\mathrm{supp}(f)$ is bounded in the totally ordered set $o(g, \zeta)$. Let $\xi g_\iota < \xi k$ for all $\xi \in o(g, x)$, $\xi < \alpha$ and let $\zeta g^p < \alpha \leq \zeta g^{p+1}$, $\zeta g^m \leq \beta < \zeta g^{m+1}$ for some integers p, m. Now we set $r = g^{-m-1+p} kg_\iota^{-1} g^{m+1-p}$, $q = g_\iota k^{-1}$. Then $\beta r = \beta g^{-m-1+p} kg_\iota^{-1} g^{m+1-p} > \beta$ since $bg^{-m-1+p} \in (\zeta g^{p-1}, \zeta g^p)$. Therefore, $\beta g^{-m-1+p} < \alpha$ and $\beta r > \beta$. So $\beta(r \wedge q) > \beta$.

Similar arguments show that $\alpha(r \wedge q) < \alpha$ and $\beta g^{m+1-p}(r \wedge q) < \beta g^{m+1-p}$. Thus $o(r \wedge q, \beta) \subseteq (\alpha, \beta g^{m+1-p})$. Let the order permutation f_1 be defined by the rule:

$$\xi f_1 = \begin{cases} \xi(r \wedge q), & \text{if} \quad \xi \in o(r \wedge q, \beta), \\ \xi, & \text{if} \quad \xi \notin o(r \wedge q, \beta). \end{cases}$$

Then $e < f_1 < g$ and $\mathrm{supp}(f_1)$ is bounded in the totally ordered set $o(g, \zeta)$.

Step 3. *If there is an element $f \in A(\Omega)$ such that* $\mathrm{supp}(f)$ *is bounded in the totally ordered set $o(g, \zeta)$ and $e < f < g$, then there exists $s \in A(\Omega)$ such that*

$$e < s < g, \tag{9}$$

$$\mathrm{supp}(s) \subseteq (\zeta, \zeta g^2), \tag{10}$$

$$\mathrm{supp}(s) < (\mathrm{supp}(s))g. \tag{11}$$

Without loss of generality let f be a bump and $\mathrm{supp}(f) = o(f, \alpha)$. Let $\zeta g^p \leq \mathrm{supp}(f) \leq \zeta g^m$ for some integers p, m where p is the greatest and m is the least integers with this property. Let $f_1 = g^p f g^{-p}$ and $n = m - p$. Then $e < f_1 < g$ and $\mathrm{supp}(f_1) \subseteq (\zeta, \zeta g)$. If $n = 1$, then $s = f_1$ has properties (9)-(11). Let $n > 1$ and there is an element $\beta \in \mathrm{supp}(f_1)$ such that $\zeta < \beta < \zeta g$, $\beta g^{n-1} \in \mathrm{supp}(f_1)$. Then we set $f_2 = g^{n-1}f_1g^{-n+1}$ and $s = f_1 \wedge f_2$. It is easy to see that the order permutation s has properties (9)-(11). If $\xi g^{n-1} \notin \mathrm{supp}(f_1)$ for all $\xi \in (\zeta, \zeta g) \cap \mathrm{supp}(f_1)$. In this case $f_2 = g^{n-2}f_1g^{-n+2}$ has properties (9)-(11), too.

Step 4. In order to complete the proof of the theorem for any $g > e$, $g \in A(\Omega)$ we need to construct an element $s \in A(\Omega)$ with properties (9)-(11). Let $\xi \in o(g, \zeta)$ and $\xi < \xi s$. Since $\xi \in \mathrm{supp}(s)$, then $\xi g \notin \mathrm{supp}(s)$. Then $\xi g s^{-n} = \xi g > \xi$ for all positive integers n and therefore $\xi g > \xi s^n > \xi$. If s is the first member of some distributive pair in G, then g is also the first member of some distributive pair in G. In fact, otherwise case by Steps 1-3 there is an element $g^* \in A(W)$ for which $e < g^* < s$, $\mathrm{supp}(g^*) \subseteq (\xi, \xi s^2)$ and (g, g^*) is a distributive pair in $A(\Omega)$. Indeed, let $g^{**} = s^{-2}g^*s^2$. Then g^* and g^{**} have the following properties:$e < g^* < g$, $e < g^{**} < g$, $\mathrm{supp}(g^*) \cup \mathrm{supp}(g^{**}) \subseteq (\xi, \xi g)$, $\mathrm{supp}(g^*) < \mathrm{supp}(g^{**})$. Let $g = \bigvee_{\iota \in I} g_\iota$ and $g_\iota > e$ for all $\iota \in I$. If $\xi g_\iota < \mathrm{supp}(g^{**})$ for all $\iota \in I$, then $g(g^{**})^{-1} > g_\alpha$ for all $\iota \in I$. This contradicts $g = \bigvee_{\iota \in I} g_\iota$. Hence there exists $\iota \in I$ such that $\xi g_\iota > \eta$ for some $\eta \in \mathrm{supp}(g^{**})$ and $g^* \leq g_\iota$. Then (g, g^*) is a distributive pair in $A(\Omega)$.□

COROLLARY 1. *Each l-group G can be isomorphically embedded into a completely distributive l-group.*□

COROLLARY 2. *A closed convex l-subgroup of l-group $A(\Omega)$ of all order permutations of the totally ordered set Ω is a completely distributive l-group.*□

5^0. Of course, there are l-groups which are not completely distributive.

EXAMPLE 1. (Lloyd [1].) Let $G = \mathbf{ZP}(\mathbf{R})$ be an l-group from Example 4.7.3. More exactly $G = \{g \in A(\mathbf{R}) | (\mathbf{x} + \mathbf{n})\mathbf{g} = (\mathbf{x})\mathbf{g} + \mathbf{n}$ for some positive integer $n = n(g)$ and $x \in \mathbf{R}\}$. It was proved in Chapter 4, Section 7 that G is a simple l-group. We claim that G is not completely distributive. Let $h \in G$, $h > e$ and $f \in G$ is defined by the rule:$(x)f = x + 1$ for all $x \in \mathbf{R}$. By definition of G, the subset $\mathrm{sup}(h)$ is unbounded in $\mathbf{R}$. Let $\{x_n | n = 1, 2 \ldots\}$ be a sequence of real numbers such that $x_n \in \mathrm{supp}(h)$ and $n + 2 \leq x_n$. For every integer n let t_n be an integer for which $x_n \leq t_n$. Now we define the order permutation g'_n of the closed interval $[0, t_n]$ by the rule:

$$(x)g'_n = \begin{cases} (2^n+1)x, & \text{if} \quad 0 \leq x \leq \frac{1}{2^n}, \\ x+1, & \text{if} \quad \frac{1}{2^n} \leq x \leq n, \\ \frac{x+n+2}{2}, & \text{if} \quad n \leq x \leq n+2, \\ x, & \text{if} \quad n+2 \leq x \leq t_n. \end{cases}$$

For every integer n the order permutation g'_n can be extended to order permutation g_n of $\mathbf{R}$ for which $g_n f^{t_n} = f^{t_n} g_n$. It is clear that $f = \bigvee_{n=1}^{\infty} g_n$ and for each g_n is not valid $g_n > h$ since $x_n g_n = x_n$ and $x_n h > x_n$. Hence f cannot be the first element in any distributive pair in G.□

Some generalizations of complete distributivity were investigated by Jakubik [2]. For further results concerning complete distributive l-groups of order permutations of totally ordered sets see McCleary [2,4] and some other authors.

8.3 Compactness in l-Groups

1⁰. An element a of lattice L is *compact* if

(1) for each family $\{g_\alpha | \alpha \in A\}$ of elements of L such that $a \leq \bigvee_{\alpha \in A} g_\alpha$ there is a finite set of indexes $\alpha_1, \ldots, \alpha_n$ for which $a \leq \bigvee_{i=1}^{n} g_{\alpha_i}$.

If each element of the complete lattice L is compact, then the lattice L is called *compact.* If each element of complete lattice L is the supremum of some set of compact elements of L, then L is called *compactly generated.*

Similarly, the lattice-ordered group G is called compactly generated if each element of G is the supremum of some subset of compact elements of G. If each element $g \in G$ is compact, then G is called *compact by elements.* The investigations of the compactness in l-groups was initiated by F. Šik [3].

PROPOSITION 1. (Šik [3].) *Each element of a compactly generated l-group is compact.*

Proof. Let G be a compactly generated l-group and a be a compact element of G. Let $x \in G$ and $x \leq \bigvee_{\alpha \in A} x_\alpha$ where $x_\alpha \in G$ for all $\alpha \in A$. Then by Proposition 8.1.1, $a \leq (\bigvee_{\alpha \in A} x_\alpha)y$ for $y = x^{-1}a$. By compactness of the element a, it follows that there is a finite subset $\Gamma \subseteq A$ such that $a \leq \bigvee_{\alpha \in \Gamma} x_\alpha y$. Then $x = ay^{-1} \leq \bigvee_{\alpha \in \Gamma} x_\alpha$. Hence x is a compact element of G.□

PROPOSITION 2. (Šik [3].) *A lattice-ordered group G is compact by elements if and only if for each family $\{g_\alpha | \alpha \in A\}$ of elements of G such that there exists $\bigvee_{\alpha \in A} g_\alpha$ in G then there is a finite subset Γ of A for which $\bigvee_{\gamma \in \Gamma} g_\gamma = \bigvee_{\alpha \in A} g_\alpha$.*□

PROPOSITION 3. (Šik [3].) *A lattice-ordered group G is compact by elements if and only if for each family $\{g_\alpha | \alpha \in A\}$ of elements of G such that $\bigwedge_{\alpha \in A} g_\alpha = e$ there is a finite subset Γ of A for which $\bigwedge_{\gamma \in \Gamma} g_\gamma = e$.*□

2⁰. The characterization of compactly generated l-groups in terms of polars and prime subgroups was due to Šik [3] and Bigard, Keimel and Wolfenstein [1]. For this description we need the following notations. An element a of l-group G is called an *atom* if $a > e$ and for each $x \in G$, $e \leq x \leq a$ holds $x = e$ or $x = a$. If for every $g \in G$, $g >$ ethere is an atom a such that $e < a \leq g$, then G is called *discrete.*

THEOREM 1. *For l-group G the following statements are equivalent:*

(2) *G is compact by elements;*

(3) *G is discrete and each minimal prime subgroup is a polar of G;*

(4) *for each minimal prime subgroup convex l-subgroup M there is an atom $a \in G \setminus M$;*

(5) *each minimal prime subgroup M is closed in G and the totally ordered set $R(G : M)$ of right cosets of M is discrete.*

Proof. (2)⇒(3). Let G be a compact by elements l-group and $g \in G$, $g > e$. Let $\{g_\alpha | \alpha \in A\}$ be a maximal chain of elements containing in the interval

$$I = (e, g) = \{x \in G | e < x \leq g\}.$$

If $\bigwedge_{\alpha\in A} g_\alpha = e$, then $\bigwedge_{i=1}^{n} g_{\alpha_i} = e$ for some finite subset $\{\alpha_1, \ldots, \alpha_n\}$ of A by our assumption. Therefore, $g_{\alpha_i} = e$ for some α_i, a contradiction. Thus, $\bigwedge_{\alpha\in A} g_\alpha = a \neq e$, a is an atom in G and G is discrete.

Let $\{a_\alpha | \alpha \in A\}$ be the set of all atoms of G and M be a minimal prime subgroup of G. By Theorem 3.3.1, M contains the polar $x^\perp$ of each $x \notin M$, $x > e$. Then the greatest lower bound of each finite subset of the positive elements of $G \setminus M$ is not equal to e. Let us assume, on the contrary, that for each atom a_α $(\alpha \in A)$ there is $x_\alpha > e$, $x_\alpha \in G\setminus M$ such that $x_\alpha \wedge a_\alpha = e$. Then $\bigwedge_{x\in P(G)\setminus M} x = e$, a contradiction (here $P(G)$ is a positive cone of G as usual). Therefore there is an atom b for which $b \wedge x > e$ for all $x \in P(G) \setminus M$. Thus $b \leq x$ for all $x > e$, $x \in G \setminus M$,

$$M \supseteq \{x \in G|\ |x| \wedge b = e\}$$

and $b^\perp \subseteq M$. Conversely, if $x, y \notin b^\perp$, $e < x, y$, then $b \leq x$, $b \leq y$ and $b \leq x \wedge y$. By Theorem 3.3.1, $M = b^\perp$ is a prime subgroup.

(3)$\Rightarrow$(4). Let M be a minimal prime subgroup of G. By (3) M is a polar of G, $M = M^{\perp\perp}$ and $M^\perp \neq E$. Since G is discrete, then $M^\perp$ contains an atom.

(4)$\Rightarrow$(5). Let M be a minimal prime subgroup of G and a be an atom of G, $a \in G \setminus M$. It is easy to see that $M = a^\perp$. Let $x = \bigvee_{\alpha\in A} x_\alpha$, $x_\alpha \geq e$ and $x_\alpha \in a^\perp$.

Then by Proposition 8.1.1,

$$x \wedge a = (\bigvee_{\alpha\in A}) \wedge a = \bigvee_{\alpha\in A} (x_\alpha \wedge a) = e.$$

It is easy to see that the set $R(G : M)$ is homogeneous and Ma is an atom in $R(G : M)$. Therefore $R(G : M)$ is a discrete totally ordered set.

(5)$\Rightarrow$(2). Let $X = \{x_\alpha | \alpha \in A\}$ be a maximal subset of G such that $\bigwedge_{\alpha\in A} x_\alpha = e$ and $\bigwedge_{\gamma\in\Gamma} x_\gamma \neq e$ for each finite subset $\Gamma \subseteq A$. Let $H = \{x \in G|\ |x| \notin X\}$. It is clear that $H \neq \emptyset$ and $H \cap X = \emptyset$. We claim that G has the following property:

(6) $$\text{if } e < a < x \text{ and } x \in H, \text{ then } a \in H.$$

In fact, let $a \notin H$. Then $a \in X$. Let $X' = X \cup \{x\}$. By definition of X, there is a finite subset $\Gamma \subseteq A$ such that $x \wedge (\bigwedge_{\gamma\in\Gamma}) = e$. Then $a \wedge (\bigwedge_{\gamma\in\Gamma} x_\gamma) = e$ too. This is the contradiction. So, G has the property (6).

Let $a, b \in H$, $ab \neq e$. Then $|a|, |b| \in H$ and there is $c \in X$ such that $|a| \wedge c = |b| \wedge c = e$. Thus $|a| \cdot |b| \cdot |a| \wedge c = e$ and $|a| \cdot |b| \cdot |a| \in H$. By (6), $|ab| \in H$ and $ab \in H$. Hence H is a prime subgroup of G and $H \cap X = \emptyset$. Let M be a minimal prime subgroup of G containing in H. Then $M \cap X = \emptyset$, $Mx_\alpha \neq M$ for all $\alpha \in A$. By (5), M is closed. Hence by Lemma 8.2.1, the mapping $\varphi :\ x\varphi = Mx$ preserves infima and suprema. Thus

$$\bigwedge_{\alpha\in A} Mx_\alpha = \bigwedge_{\alpha\in A} (x_\alpha\varphi) = (\bigwedge_{\alpha\in A} x_\alpha)\varphi = M$$

contradicts the discreteness of $R(G : M)$.□

3⁰. THEOREM 2. (Šik [3].) *An l-group G is Archimedean and compact by elements if and only if G is isomorphic to the direct product of totally ordered infinite cyclic groups.*

Proof. It is clear that the direct product of totally ordered infinite cyclic groups is arfimedean and compact by elements.

Conversely, let G be an Archimedean compact by element l-group and M be a minimal prime subgroup of G. Then M is a polar of G and $M^\perp$ is an atom in the Boolean algebra $\mathcal{P}(G)$ of polars of G. As was mentioned in the proof of Theorem 1 the polar $M^\perp$ contains an atom a of an l-group G.

Now we claim that $M^\perp$ is l-isomorphic to additive group of integers and $G = M \times M^\perp$. Indeed, let $x \in G$, $x > e$. Since G is Archimedean, then there is the least positive integer n such that $a^n \leq x$, $a^{n+1} \not\leq x$. If $a^n = x$ or $a^{n+1} > x$ then $x \in M^\perp$. Let a^{n+1} is incomparable with x and $c = x \wedge a^{n+1} \geq e$. If $c = e$, then $x \in M = M^{\perp\perp}$. Otherwise $c > e$ and $e = ca^{-k} = xa^{-k} \wedge a^{n-k+1}$. Hence $xa^{-k} \in M$, $x \in M \times M^\perp$ and l-group G has a direct factor isomorphic to the totally ordered additive group of integers.

Let $\{G_\gamma | \gamma \in \Gamma\}$ be the set of all convex l-subgroups of G which are direct factors of G and are isomorphic to the totally ordered additive group of integers. By Theorem 3.1.1, subgroup H generated by all G_γ ($\gamma \in \Gamma$) is a convex l-subgroup of G and is isomorphic to the direct product of totally ordered additive groups of integers.

Moreover, $H^\perp = E$ and $H^{\perp\perp} = G$. Let φ be a map of G into the Cartesian product $\bar{G} = \overline{\prod_{\gamma\in\Gamma}} G_\gamma$ of o-groups G_γ ($\gamma \in \Gamma$) defined by the rule: if $x \in G$ and $x = x_\gamma y_\gamma$ where $x_\gamma \in G_\gamma$, $y_\gamma \in G_\gamma^\perp$, then $(x\varphi)(\gamma) = x_\gamma$. It is clear that φ is an isomorphism of l-group G into $\bar{G}$ and $a\varphi = a$ for all $a \in H$. If $G\varphi \neq \prod_{\gamma\in\Gamma} G_\gamma$, then there is an element $x \in G$ such that

$$x > e,\ x\varphi \in G\varphi \setminus \prod_{\gamma\in\Gamma} G_\gamma.$$

Thus $(x\varphi)(\delta) \neq e$ for some infinite subset $\Delta \subseteq \Gamma$. Thus

$$x\varphi = \bigvee_{\delta\in\Delta} (x\varphi)(\delta), \text{ or equivalently } x = \bigvee_{\delta\in\Delta} x_\delta \in G.$$

Since G is compact by elements, then $x = \bigvee_{\sigma\in\Sigma} x_\sigma$ for some finite subset $\Sigma \subseteq \Delta$. As the elements x_σ ($\sigma \in \Sigma$) are mutually orthogonal, then x is equal to the product of the elements $x_\sigma, \sigma \in \Sigma$. Therefore, $x \in G \setminus H$ contradicts the choice of x.□

Other facts about compactness in l-groups can be found in the works of Jakubik [1], Bigard,Keimel and Wolfenstein [1].

8.4 Descending Chain Condition for Positive Elements

l-Group G satisfies the *descending chain* condition (DCC) if every non-empty set of positive elements of G contains a minimal element. It is easy to see that the direct product of any family of totally ordered additive groups of integers satisfies (DCC). We now show that there are no other l-groups.

THEOREM 1. (Ward [1], Birkhoff [1].) *An l-group G satisfies the descending chain condition for positive elements if and only if G is* *isomorphic to the direct product of the family of totally ordered additive groups of integers.*

Proof. Let G be an l-group, G satisfy (DCC) and $g \in G$, $g > e$. Then by (DCC),$g \geq a_1$ for some atom a_1. If $g \neq a_1$, then $g_1 = ga_1^{-1} > e$. By above arguments, there is an atom a_2 such that $ga_1^{-1} \leq a_2$. If $ga_1^{-1} \neq a_2$, then $g_2 = g_1a_2^{-1} > e$. Let $g_k > e$. Then there is an atom g_{k+1} such that $g_k \geq a_{k+1}$. If $g_k \neq a_{k+1}$ then $g_{k+1} = g_k a_{k+1}^{-1} > e$. Thus we obtain a strictly decreasing sequence $g > g_1 > \cdots > g_k > \cdots > e$. By (DCC), $g_k = e$ for some positive integer k. Thus

$$g_k = g_{k-1}a_k^{-1} = g_{k-2}a_{k-1}^{-1}a_k^{-1} = \cdots = ga_1^{-1}a_2^{-1}\cdots a_k^{-1} = e$$

and $g = a_k a_{k-1} \cdots a_1$. Hence each element of G belongs to subgroup A generated by all atoms of G. Since any two different atoms are orthogonal they are permutable and subgroup A is isomorphic to the direct product $\prod_{\gamma\in\Gamma}(a_\gamma)$ of totally ordered infinite cyclic convex subgroups generated by all atoms a_γ $(\gamma \in \Gamma)$ of G.□

8.5 Lattice-Ordered Groups with Finite Orthogonal Rank

1⁰. A lattice-ordered group G has the *finite orthogonal rank* n if there are n mutually orthogonal elements $a_1, \ldots, a_n$ in G and each $n+1$ positive elements of G are not mutually orthogonal. Lattice-ordered groups with finite orthogonal rank were studied by Jaffard [6] and by Conrad [5]. In this section G is an l-group with the finite orthogonal rank n and $a_1, \ldots, a_n$ are mutually orthogonal strictly positive elements of G. For every a_i $(i = 1, 2, \ldots, n)$ let $A_i = \{a_1, \ldots, a_{i-1}, a_{i+1}, \ldots, a_n\}^\perp$ and A is the convex l-subgroup of G generated by all A_i, $1 \leq i \leq n$. It is clear that $a_i \in A_i$ for all $i = 1, 2, \ldots, n$.

LEMMA 1. *Each polar $A_i, 1 \leq i \leq n$, is a totally ordered subgroup of G. The convex l-subgroup A generated by all A_i, $1 \leq i \leq n$, is the direct product of totally ordered subgroups A_i, $1 \leq i \leq n$.*

Proof. Assume, on the contrary, that A_i is not an o-group for some i. Therefore there are $a_i', a_i'' \in A_i$ such that $a_i', a_i'' > e$ and $a_i' \wedge a_i'' = e$.

Then the subset $a_1, \ldots, a_{i-1}, a_i', a_i'', a_{i+1}, \ldots, a_n$ contains $n+1$ mutually orthogonal elements of G. This contradicts our assumption about G. The direct verification shows that A is the direct product of A_i, $1 \leq i \leq n$.□

LEMMA 2. *Each subset of n strictly positive elements $b_1, \ldots, b_n$ of G is contained in A. Moreover, for every b_i, $1 \leq i \leq n$, there is j, $1 \leq j \leq n$, such that $b_i \in A_j$.*

Proof. Let $b_1, \ldots, b_n$ be strictly positive mutually orthogonal elements of G. Assume, on the contrary, that there are two elements a_j, a_k such that $j \neq k$, $1 \leq j, k \leq n$, $b_i \wedge a_j \neq e$, $b_i \wedge a_k \neq e$. Then the elements $b_1, \ldots, b_{i-1}, b_i \wedge a_j, b_i \wedge a_k, b_{i+1}, \ldots, b_n$ are strictly positive and mutually orthogonal. This contradicts our assumption on G. Thus there is the unique j, $1 \leq j \leq n$, such that $b_i \wedge a_j \neq e$, $b_i \wedge a_s = e$ for all s, $s \neq j$, $1 \leq s \leq n$. Hence $b_i \in \{a_s | 1 \leq s \leq n,\ s \neq j\}^{\perp} = A_j$.□

COROLLARY 1. *The subgroup A is normal in G. Each inner automorphism of G induces the permutation on the set $\{A_1, \ldots, A_n.\}$*

Proof. Let φ be an automorphism of l-group G. Then the elements $a_1\varphi, \ldots, a_n\varphi$ are strictly positive and mutually orthogonal. Hence $a_1\varphi, \ldots, a_n\varphi \in A$. Now the proof follows from Lemma 2.□

LEMMA 3. *If $x \wedge c_i = e$, $c_i \in A_i$ and $c_i > e$, then $x \wedge y = e$ for each $y \in A_i$, $y > e$.*

Proof. Let $x \wedge y = y' > e$ for some $y \in A_i$. Since A_i is an o-group then $x \wedge z \geq y' > e$ for all $z \geq y'$, $z \in A_i$. By similar arguments, $x \wedge z = z$ for all $z \leq y'$. Then $e = x \wedge (c_i \wedge y') = c_i \wedge y' = \min\{c_i, y'\} > e$ in A_i, a contradiction.□

THEOREM 1. (Conrad [2].) *An l-group G has the finite orthogonal rank n if and only if the Boolean algebra $\mathcal{P}(G)$ of polars of G has the order 2^n.*

Proof. If $\mathcal{P}(G)$ has the order 2^n, then G has n minimal polars and therefore there are at least n of mutually orthogonal strictly positive elements in G. Assume that $b_1, \ldots, b_m$, $m > n$ are mutually orthogonal strictly positive elements of G. Then there are at least m different atoms in $\mathcal{P}(G)$. So $\mathcal{P}(G)$ has the order greater than 2^n.

Conversely, let G be an l-group with the finite orthogonal rank n, $a_1, \ldots, a_n$ be mutually orthogonal strictly positive elements of G and

$A_i = \{a_1, \ldots, a_{i-1}, a_{i+1}, \ldots, a_n\}^\perp$, $A = \prod_{i=1}^{n} A_i \subseteq G$. By Lemma 1 and Lemma 3, it follows that A_i is totally ordered for all i, $1 \leq i \leq n$. Then A_i is an atom in the Boolean algebra $\mathcal{P}(G)$ for all i, $1 \leq i \leq n$, and each polar of G contains at least one atom. Therefore, $\mathcal{P}(G)$ is the atomic Boolean algebra with n atoms. Thus $\mathcal{P}(G)$ is finite and has the order 2^n.□

So the class of l-groups with finite orthogonal rank is coincides with the class of l-groups with finite number of polars (see Chapter 3.5).

2^0. It is possible to characterize the class of l-groups with the finite number of polars in terms of breath of the lattice. The integer n is called a *breath* of the lattice L if n is the least positive integer such that any meet $x_1 \wedge x_2 \wedge \cdots \wedge x_m$, $m > n$, in the lattice L is always a meet of subset of n of x_i, $1 \leq i \leq m$. l-Group G has the breath n if the lattice G has the breath n.

THEOREM 2. (Jakubik [3].) *The orthogonal rank of l-group G is finite if and only if the breath of G is finite.*

We leave the proof of this theorem to reader.□

3^0. Let G be an l-group and P be a positive cone of G. We define the binary relation δ on P by the rule: $x \equiv y \pmod{\delta}$ if and only if $x^\perp = y^\perp$. Direct verification shows that δ is an equivalence relation on P. The equivalence class $x\delta$ of δ which contains $x \in P$ is called a *thread* of G and it is denoted by $x^\top$.

PROPOSITION 1. (Jaffard [1].) *Each thread $x^\top$ of l-group G is a convex subsemigroup of positive cone P of G.*

The proof is similar to that of Proposition 3.4.1.□

On the set $\mathcal{T}(G)$ of all threads of G we define the order relation $\leq$ by the rule: $x^\top \geq y^\top$ if and only if $x^\perp \subseteq y^\perp$. The interested reader can without difficulty obtain proofs of the following statements.

THEOREM 3. (Jaffard [2], Pierce R. [1].) *The map $\theta : x \to x^\top$ of the positive cone P of an l-group G onto $\mathcal{T}(G)$ of threads of G is an isomorphism of the lattice P and $x^\top \wedge y^\top = (xy)^\top$. So, $\mathcal{T}(G)$ is a distributive lattice.* □

THEOREM 4. (Kappos and Kehayopulos [1].) *The map ψ :*

$\mathcal{T}(G) \to \mathcal{P}(G)$ *of the lattice* $\mathcal{T}(G)$ *of threads of* G *into the Boolean lattice* $\mathcal{P}(G)$ *of polars of* G *where* $x^{\top}\psi = x^{\perp\perp}$ *is an isomorphic embedding of* $\mathcal{T}(G)$ *into* $\mathcal{P}(G)$ *and* $\mathcal{P}(G)$ *is the completionn of the sublattice* $\mathcal{T}(G) \cdot \psi$.□

COROLLARY 2. (Jaffard [2].) *If* $\mathcal{T}(G)$ *is finite, then* $\mathcal{T}(G)$ *is a Boolean algebra.*□

COROLLARY 3. (Conrad [2].) *l-Group* G *has the finite number of threads if and only if* G *has a finite orthogonal rank.*□

Chapter 9

Varieties of lattice-ordered groups

9.1 l-Varieties

1^0. We will regard lattice-ordered group G as an algebraic system of the signature $l = \{ \cdot , {}^{-1} , e , \vee , \wedge \}$. It was noted in Chapter 2 that the class of all l-groups $\mathcal{L}$ is defined in the signature l by the identities: $\{G; \cdot , {}^{-1} , e\}$ is a group;$\{G; \vee , \wedge\}$ is a lattice; for any $x, y, z, t \in G$ the equalities $x(y \vee z)t = xyt \vee xzt$, $\quad x(y \wedge z)t = xyt \wedge xzt$ are valid. Let us recall that an *identity of signature* l is a formula φ of the tight predicate calcules of the form $(\forall x_1) \ldots (\forall x_n)(w(x_1, \ldots, x_n) = e)$ where $w(x_1, \ldots, x_n) = w(X)$ is the term (or word) of the signature l on variables $x_1, \ldots, x_n$. As usual we omit quantifiers and therefore the identity φ has the form $w(X) = e$ where $w(X) = w(x_1, \ldots, x_n) = w(X)$ is the term of l.

Let Φ be some set of identities of the signature l, $\mathcal{M}$ be the class of all l-group G such that every identity φ, $\varphi \in \Phi$, is valid in G. Then the class $\mathcal{M}$ is a *variety of l-groups* or an *l-variety* for short. The set Φ is called the *basis of identities* of $\mathcal{M}$. We recall the fundamental theorem due to Birkhoff as shown in the book of Burris and Sankappanavar [1] (Theorem 2.11.9).

THEOREM 1. (Birkhoff) *The non-empty class $\mathcal{M}$ is an l-variety if and only if $\mathcal{M}$ is closed under*

(1) *the formation of l-subgroups;*
(2) *homomorphic images,*
(3) *Cartesian products.*

From Theorem 1 it follows that every l-variety is closed under ultraproducts. Now we will describe what it means for an l-groups to satisfy an identity. Let $X = \{x_1, x_2, \ldots\}$ be a countable set of variables or symbols. The identity has the form $w = e$ where $w = w(x_1, \ldots, x_n)$ is a term of the form

$$\bigvee_{i\in I}\bigwedge_{j\in J} w_{ij}(X) = \bigvee_{i\in I}\bigwedge_{j\in J}\prod_{k\in K} x_{ijk}^{\varepsilon_{ijk}},$$

where I, J, K are finite index sets, each x_{ijk} is either the element $x \in X$ or the signature symbol e; w_{ij} are words of the groups theoretic signature. A *substitution* for $w(X)$ in l-group G means the following: replace all occurences of each variables which occur in the term by some element of G and replace the signature symbol e by the identity element e of G. This gives us an element of G which we denote by $w(g) = w(g_1, \ldots, g_n)$. If $w(g) = e$ in G for all possible substitutions in G, we say that the identity $w(x_1, \ldots, x_n) = e$ is valid in the l-group G or G satisfies the identity $w(x_1, \ldots, x_n) = e$.

Since an identity involves only a finite number of variables, it is important to note that an l-group G belongs to an l-variety $\mathcal{M}$ iff every finitely generated l-subgroup of G lies in $\mathcal{M}$. Another very useful observation about identities in l-groups is that inequalities that hold universally in an l-group are equivalent to identities. Conversely, certain identities are equivalent to unequalities. Let $u(x_1, \ldots, x_n)$, $v(x_1, \ldots, x_n)$ be the terms of signature l (or, simply, l-group words). We say that l-group G satisfies the unequality

$$u(x_1, \ldots, x_n) \leq v(x_1, \ldots, x_n)$$

if and only if $u(g_1, \ldots, g_n) \leq v(g_1, \ldots, g_n)$ for all $g_1, \ldots, g_n \in G$. Then clearly G satisfies the unequality $u \leq v$ iff it satisfies the identity $(u \vee v)v^{-1} = e$. Consequently, we will use the inequalities between l-group words and variables freely when defining varieties by identities. Similar arguments show that identity $u = v$ is equivalent to $uv^{-1} = e$.

2⁰. A number of is l-varieties are defined by some set of identities in the signature of the group theory $\{\ \cdot\ ,\ ^{-1}\ ,\ e\ \}$. We call them the *group l-varieties* . For group l-varieties we use the following notation:

(4) $\mathcal{A}_l$ is the l-variety of all Abelian l-groups;

(5) $\mathcal{N}_k$ is the l-variety of all nilpotent l-groups of the class $\leq k$;

(6) $\mathcal{A}^{(k)}$ is the l-variety of all soluble groups of the soluble lenght $\leq k$;

(7) $\mathcal{A} \cdot \mathcal{A}_p$ is the l-variety of all l-groups defined by the identities $[x^p, y^p] = e,\ [[x, y] \cdot [u, v] = e$.

As usual, $\mathcal{E}$ denotes the trivial l-variety of all trivial l-groups E, $\mathcal{L}$ denotes the l-variety of all l-groups.

The following example shows the existence of non-abelian from the l-variety $\mathcal{A}\mathcal{A}_p$.

EXAMPLE 1. Let p be a positive integer and $G_p = \mathrm{gp}\{a_1, \ldots, a_p, b|$ $[a_i, a_j] = e,\ b^{-1}a_i b = a_{\bar{i}}\ (1 \leq i, j, \bar{i} \leq p,\ i \equiv \bar{i} \pmod{p})\}$. Then every element $g \in G_p$ can be represented uniquely in the form $g = b^k a_1^{n_1} \cdots a_p^{n_p}$. We define the order on G_p by the rule: $g \geq e$ iff $k > 0$ or $k = 0$ and $n_i \geq 0$ for all i, $1 \leq i \leq p$. It is readily verified that G_p is an l-group and $G_p \in \mathcal{A}\mathcal{A}_p \setminus \mathcal{A}$.

The set of all l-varieties **L** does not coincide with the set of all group l-varieties. There exists an l-variety $\mathcal{X}$ which can not be defined by the identities of signature $\{\ \cdot\ ,\ ^{-1}\ ,\ e\ \}$.

EXAMPLE 2. Let $\mathcal{X}$ be the l-variety defined by the identities $[x, y, z] = e$ and $[|[x, y]| \wedge |z|, t] = e$. It is easy to see that $G \in \mathcal{X}$ iff the ideal of G generated by the derived subgroup G' of G is central. It is not difficult to construct an example of a totally ordered nilpotent of class 2 group G such that a convex subgroup generated by G' is not central. Therefore, $\mathcal{X} \neq \mathcal{N}_2$. On the other hand, by Theorem 4.1.5, every torsion-free nilpotent group G has a total order such that centre Z of G is a convex subgroup of G. Now it is evident that every identity of the signature $\{\ \cdot\ ,\ ^{-1}\ ,\ e\ \}$ which is valid in all l-groups from $\mathcal{X}$ is valid in $\mathcal{N}_2$, too. Hence, l-variety $\mathcal{X}$ cannot be defined by the identities of the signature $\{\ \cdot\ ,\ ^{-1}\ ,\ e\ \}$.

Now we dwell on some very important l-varieties:

(8) $\mathcal{N}$ is the l-variety defined by the identity $|x| \cdot |y| \wedge |y^2| \cdot |x^2| = |x| \cdot |y|$;

(9) $\mathcal{R}$ is the l-variety defined by the identity $(x \wedge y^{-1}x^{-1}y) \vee e = e$;

(10) $\mathcal{W}a$ is the l-variety defined by the identity $x^{-1}|y|x|y|^{-2} \vee e = e$.

Every l-group $G \in \mathcal{N}$ is called a *normal-valued* l-group and $\mathcal{N}$ is called the l-variety of normal-valued l-groups. The properties of l-groups from $\mathcal{N}$ will be considered in the Section 2.

Every l-group $G \in \mathcal{R}$ is called a *representable* l-group and $\mathcal{R}$ is the l-variety of representable l-groups. As will be shown in Section 3 every $G \in \mathcal{R}$ is an l-subgroup of the Cartesian product of totally ordered groups.

Every l-group $G \in \mathcal{W}a$ is called a *weakly abelian* l-group and $\mathcal{W}a$ is called the l-variety of weakly abelian l-groups. This variety will be considered in Section 4.

Other examples of l-varieties will be defined later.

3^0. For any class $\mathcal{K}$ of l-groups let $H(\mathcal{K})$, $S(\mathcal{K})$, $P(\mathcal{K})$ denote the classes of l-groups which are homomorphic images, l-subgroups or Cartesian products respectevely, of elements of $\mathcal{K}$. Let $\mathrm{var}_l(\mathcal{K})$ be the l-variety generated by $\mathcal{K}$ (that is the smallest l-variety containing $\mathcal{K}$). Moreover, recall that l-group G is said to be *subdirectly irreducible* if it has the smallest non-trivial ideal. An l-group G is called *finitely subdirectly irreducible* if there do not exist non-trivial ideals H_1, H_2 such that $H_1 \cap H_2 = E$. For example, any totally ordered group is finitely subdirectly irreducible. By Theorem 4.4.1, it is easy to see that if l-group G has a transitive representation, then G is finitely subdirectly irreducible.

We say that an l-group G is *subcartesian product* of subdirectly irreducible l-groups G_α $(\alpha \in A)$ if G can be embedded in the Cartesian product $\overline{\prod\limits_{\alpha \in A}} G_\alpha$ in such a way that each projection is onto. From well-known results in universal algebra (see Burris and Sankappanavar [1], Theorem 2.8.6, 2.9.5) we have the following useful observations.

THEOREM 2. *For any class $\mathcal{K}$ of l-groups*

a) $\mathrm{var}_l(K) = HSP(\mathcal{K})$,

b) *every l-group is a subcartesian product of subdirectly irreducible l-groups,*

c) *if $\mathcal{K}$ is the class of subdirectly irreducible l-groups of the l-variety $\mathcal{V}$, then $\mathcal{V} = HSP(\mathcal{K})$.*

By Theorem 3.1.2, the lattice of ideals of any l-group is distributive. Therefore there is another description due to Jónsson, using ultraproducts, of l-variety generated by class $\mathcal{K}$. Let us recall that the family $\mathcal{F}$ of subsets of non-empty set I is a filter over I if $I \in \mathcal{F}$; $\emptyset \notin \mathcal{F}$; $A \cap B \in \mathcal{F}$ for $A, B \in \mathcal{F}$; $B \in \mathcal{F}$ for any $B \supseteq A$, $A \in \mathcal{F}$.

If for any subset $X \subseteq I$ holds $X \in \mathcal{F}$ or $I \setminus X \in \mathcal{F}$, then filter $\mathcal{F}$ is called an *ultrafilter* over I.

Let I be an arbitrary index set, $\{G_i | i \in I\}$ be the family of l-groups and $\mathcal{F}$ be an arbitrary ultrafilter over I. Then we define the equivalence relation $\equiv_{\mathcal{F}}$ on $\bar{G} = \bar{\prod\limits_{i\in I}} G_i$ by the rule:

$$f \equiv_{\mathcal{F}} g \text{ iff } \{i \in I | f(i) = g(i)\} \in \mathcal{F}.$$

It is easy to see that the set $\bar{G}/\mathcal{F}$ of equivalence classes of $\bar{G}$ is again an l-group. So we have the l-homomorphism μ: from $\bar{G}$ onto $\bar{G}/\mathcal{F}$. Direct calculations show that

$$\ker\mu = \{g \in \bar{\prod_{i\in I}} G_i | \{i \in I | g(i) = e\} \in \mathcal{F}\} = \bigvee_{K\in\mathcal{F}}^{c} T_K,$$

where $T_K = \{g \in \bar{G} | \{i \in I | g(i) = e\} \supseteq K\}$ and $\bigvee\limits_{K\in\mathcal{F}}^{c}$ is the supremum in the lattice of ideals of $\bar{G}$. The l-group $\bar{G}/\mathcal{F} = \bar{\prod\limits_{i\in I}} G_i/\mathcal{F}$ is called the *ultraproduct* of l-groups $\{G_i | i \in I\}$ by the ultrafilter $\mathcal{F}$.

LEMMA 1. (Jónsson [1].) *Let $\mathcal{W}$ be some family of subsets of non-empty set I such that*

(11) $I \in \mathcal{W}$,

(12) $K \in W$, *if* $J \in \mathcal{W}$, $J \subseteq K \subseteq I$,

(13) $K \in \mathcal{W}$ *or* $J \in \mathcal{W}$, *if* $K \cup J \in \mathcal{W}$.

Then there exsists an ultrafilter $\mathcal{F}$ over I with $\mathcal{F} \subseteq \mathcal{W}$.

Proof. If $\emptyset \in \mathcal{W}$, then $\mathcal{W}$ coincides with the Boolean algebra $\mathcal{B}(I)$ of subsets of I. So any ultrafilter $\mathcal{F}$ over I will do. If $\emptyset \notin \mathcal{W}$, then $\mathcal{B}(I)\setminus\mathcal{W}$ is a proper ideal in $\mathcal{B}(I)$. In fact, if $B \subset A$, $A \in \mathcal{B}(I)\setminus\mathcal{W}$, then $B \in \mathcal{B}(I)\setminus\mathcal{W}$. Let $A, B \in \mathcal{B}(I)\setminus\mathcal{W}$. Then $(A\cup B)\cup(I\setminus(A\cup B)) = I$ and so $A\cup B$ or $I\setminus(A\cup B)$ belongs to $\mathcal{W}$. If $A\cup B \in \mathcal{W}$, then by (13) $A \in \mathcal{W}$ or $B \in \mathcal{W}$, which is a contradiction. Therefore, $I\setminus(A\cup B) \in \mathcal{W}$ is a proper ideal. Now extend it to the maximal ideal and take the complementary ultrafilter.□

THEOREM 3. (Jónsson [1].) *Let $\mathcal{V} = \mathrm{var}_l(\mathcal{K})$ be an l-variety generated by some class $\mathcal{K}$ of l-groups. If A is a finitely subdirect irreducible l-group in $\mathcal{V}$, then $A \in HSP_U(\mathcal{K})$ where H, S, P_U are operators of l-homomorphic images,l-subgroups and ultraproducts, respectively. Hence,*

$$\mathcal{V} = \mathrm{var}_l(\mathcal{K}) = SPHSP_U(\mathcal{K}).$$

Proof. (Burris and Sankappanavar [1].) Let A be a nontrivial finitely subdirectly irreducible l-group in $\mathcal{V} = \mathrm{var}_l(\mathcal{K})$. Then for some choice of G_i $(i \in I)$ and for some l-subgroup $B \subseteq \overline{\prod_{i\in I}} G_i = \bar{G}$ there exists a homomorphism $\alpha : B \to \alpha$ which maps B onto A as $\mathcal{V} = HSP(\mathcal{K})$. Let $T = \ker\alpha$. Now for $J \subseteq I$ let $T_J = \{x \in \bar{G} | x(i) = e \text{ for } i \in J\}$. It is evident that T_J is an ideal in $\bar{G}$. Let $T_{JB} = T_J \cap B$ be the restriction of the ideal T_J on B and define $\mathcal{W} = \{J \subseteq I | T_{JB} \subseteq T\}$. Clearly, $I \in \mathcal{W}$, $\emptyset \notin \mathcal{W}$ (if $\emptyset \in \mathcal{W}$, then $T_\emptyset = \overline{\prod_{i\in I}} G_i$, $T_{\emptyset B} = B \cap T_\emptyset = B$ and $B/T = A = E$). If $J \in \mathcal{W}$ and $J \subseteq K \subseteq I$, then $T_{JB} \subseteq T$ and $T_{JB} \supseteq T_{KB}$. So $T_{KB} \subseteq T$.

Now suppose that $J \cup K \in \mathcal{W}$. Then $T_{(J\cup K)B} \subseteq T$. By definition, $T_{J\cup K} = T_J \overset{c}{\bigwedge} T_K$ and $T_{(J\cup K)B} = (T_J \overset{c}{\bigwedge} T_K) \overset{c}{\bigwedge} B = (T_J \cap B) \overset{c}{\bigwedge} (T_K \cap B) = T_{JB} \overset{c}{\bigwedge} T_{KB}$. By distributivity of the lattice of ideals of l-group B,

$$T = T \overset{c}{\bigvee} (T_{JB} \overset{c}{\bigwedge} T_{KB}) = (T \overset{c}{\bigvee} T_{JB}) \overset{c}{\bigwedge} (T \overset{c}{\bigvee} T_{KB}).$$

By Homomorphism theorem (see Theorem 2.3.3),

$$A \cong B/T = B/(T \overset{c}{\bigvee} T_{JB}) \overset{c}{\bigwedge} (T \overset{c}{\bigvee} T_{KB}).$$

Now we have from the fact that $B/T \cong A$ is finite subdirectly irreducible; $T = T \overset{c}{\bigvee} T_{JB}$ or $T = T \overset{c}{\bigvee} T_{KB}$. Hence, $T_{JB} \subseteq T$ or $T_{KB} \subseteq T$. So either J or K lies in $\mathcal{W}$. By Jónsson's lemma (Lemma 1), there is an ultrafilter $\mathcal{F}$ contained in $\mathcal{W}$. From the definition of $\mathcal{W}$ we have $T_{\mathcal{W}} = \overset{c}{\bigvee}_{K \in \mathcal{W}} T_K \cap B = T_{\mathcal{W}B} \subseteq T$.

Now let μ be a natural homomorphism from $\bar{G}$ onto $\bar{G}/\mathcal{W}$. Then let $\beta : B \rightarrow B\mu$ be the restriction of μ to B. As $\ker\beta = T_{\mathcal{W}} \cap B \subseteq T$ we have $A \cong B/T \cong (B/T_{\mathcal{W}} \cap B)/(T/T_{\mathcal{W}} \cap B)$. Now $B/T_{\mathcal{W}} \cap B \cong B\mu \subseteq \overline{\prod_{i \in I} G_i/\mathcal{W}}$ so $A \in HSP_U\mathcal{K}$. As every l-group is an l-subgroup of the Cartesian product of subdirectly irreducible l-groups we have $\text{var}_l(\mathcal{K}) = SPHSP_U(\mathcal{K})$.□

4⁰. It is easy to see that there are at most a continuum of l-varieties. Furthermore, the set **L** of all l-varieties is a complete lattice with the meet operations being the set-theoretical intersections and the join defined as follows: if $\{\mathcal{U}_\alpha | \alpha \in A\}$ is the set of l-varieties then $\overset{\mathbf{L}}{\bigvee}_{\alpha \in A} \mathcal{U}_\alpha = \bigcap_{\beta \in B} \mathcal{V}_\beta$ where $\{\mathcal{V}_\beta | \beta \in B\}$ is the set of all l-varieties such that $\mathcal{U}_\alpha \subseteq \mathcal{V}_\beta$ for all $\alpha \in A$. We use later the notations $\overset{\mathbf{L}}{\bigvee}$ and $\overset{\mathbf{L}}{\bigwedge}$ for the join and the meet respectively in the lattice **L** of all l-varieties.

It is usually quite a difficult problem to characterize more preisisely the join of two varieties of algebraic systems. Martinez [3] showed that the join of two l-varieties can be nicely described in the following way.

PROPOSITION 1. (Martinez [3].) *Let $\mathcal{U}$, $\mathcal{V}$ be l-varieties. Then an l-group G lies in $\mathcal{U} \overset{\mathbf{L}}{\bigvee} \mathcal{V}$ if and only if there exsist ideals U, V in G such that $G/U \in \mathcal{U}$, $G/V \in \mathcal{V}$ and $U \cap V = E$.*

Proof. Suppose that $G \in \mathcal{U} \cup \mathcal{V}$. Then by Theorem 2(1), there exist an l-group H with ideals A, B, C such that $H/A \in \mathcal{U}$, $H/B \in \mathcal{V}$, $A \cap B = E$, $H/C = G$. Since the lattice of ideals of H is distributive

(Theorem 3.2.2), then $AC \cap BC = (A \cap B)C = C$. If $U = AC/C$, $V = BC/C$, then $U \cap V = C, G/U \cong H/AC \in \mathcal{U}$, $G/V \cong H/BC \in \mathcal{V}$.

The converse statement is obvious.□

COROLLARY 1. *Let $\mathcal{U}, \mathcal{V}$ be l-varieties, G be a finitely irreducible l-group and $G \in \mathcal{U} \overset{\mathbf{L}}{\bigvee} \mathcal{V}$. Then $G \in \mathcal{U}$ or $G \in \mathcal{V}$.*□

5^0. Let $\mathcal{X}$ be an l-variety, $\mathcal{K}(\mathcal{X})$ be a class of groups which are embedded as subgroup into any l-group of $\mathcal{X}$. In other words $\mathcal{K}(\mathcal{X})$ is the class of all groups G which can be embedded into a group $H(G)$ which has a lattice order P such that l-group $(H(G), P)$ lies in $\mathcal{X}$. It is easy to see that $\mathcal{K}(\mathcal{X})$ does not coincide (in general) with the class of groups which can be turned into the l-groups from the l-variety $\mathcal{X}$. In fact, for the l-variety of all l-groups $\mathcal{L}$ the class $\mathcal{K}(\mathcal{L})$ is the class of all right-orderable groups. It is clear that there exist right-orderable groups which cannot be lattice-ordered. For example, a group from Example 2.2.1 has no nontrivial right order, but is right-orderable by Corollary 2 of Proposition 5.1.1.

THEOREM 4. *The class of groups $\mathcal{K}(\mathcal{X})$ is a quasivariety of groups for all l-variety $\mathcal{X}$.*

Proof. From standard results in the theory of universal algebras (see, for example, the book of Burris and Sankoppanavar [1], Theorem 5.2.25) it is known that the non-empty class $\mathcal{K}(\mathcal{X})$ is the quasivariety of groups iff it is closed under formation ultraproducts, subgroups, Cartesian products. The direct verification shows that this statements are valid for $\mathcal{K}(\mathcal{X})$.□

The following example shows that quasivariety of groups $\mathcal{K}(\mathcal{X})$ does not uniquely define l-variety $\mathcal{X}$.

EXAMPLE 3. As in Example 2 let $\mathcal{X}$ be the l-variety defined by the identities $[x, y, z] = e$ and $[|[x, y]| \wedge |z|, t] = e$ and $\mathcal{N}_2$ be the l-variety of all nilpotent l-groups of the class ≤ 2 defined by the identity $[x, y, z] = e$. Then, as was mentioned above, $\mathcal{N}_2 \neq \mathcal{X}$. On the other hand, every torsion-free nilpotent group G has a total order such that the centre Z of G is a convex subgroup in G. Therefore, $\mathcal{K}(\mathcal{X}) \neq \mathcal{K}(\mathcal{N}_2)$ and $\mathcal{K}(\mathcal{X})$ is a quasivariety of all torsion-free nilpotent groups of the class ≤ 2.

9.2 Normal-Valued l-Groups

1⁰. Let G be an l-group and $g \in G$, $g \neq e$. Then g defines the family $\{V_\alpha(g) | \alpha \in A\}$ of *values* of the element g i.e., a maximal convex l-subgroups of G with respect to not containing g. We call any convex l-subgroup $V_\alpha(g)$ ($\alpha \in A$) *regular*. Let $\bar{V}_\alpha(g)$ be a convex l-subgroup of G generated by $V_\alpha(g)$ and g. Each convex l-subgroup of G that properly contains $V_\alpha(g)$ contains g. Hence the convex l-subgroup $\bar{V}_\alpha(g) = \bigcap_{i \in I} M_i$ where $\{M_i | i \in I\}$ is the set of all convex l-subgroups of G such that $M_i \supset V_\alpha(g)$, $M_i \neq V_\alpha(g)$. Every convex l-subgroup $\bar{V}_\alpha(g)$ is called the *cover* of $V_\alpha(g)$ and there is no convex l-subgroup of G between $V_\alpha(g)$ and $\bar{V}_\alpha(g)$. The pair $V_\alpha(g) \prec \bar{V}_\alpha(g)$ is a jump in the lattice $\mathcal{C}(G)$ of all convex l-subgroups of G which is defined by element g. An l-group G is said to be *normal-valued* if $V_\alpha(g)$ is normal in its cover $\bar{V}_\alpha(g)$ for all jumps $V_\alpha(g) \prec \bar{V}_\alpha(g)$, $\alpha \in A$ from the lattice $\mathcal{C}(G)$.

Now we will have to characterize the class $\mathcal{N}$ of all normal-valued l-groups in the terms of identities of the signature l.

Let $\Gamma(G)$ be the set of all regular subgroups of G. The subset $\Delta \subset \Gamma(G)$ of regular subgroups of l-group G is said to be *plenary* if each element $g \in G$ has a value H in Δ and $K \in \Delta$ for every regular subgroup K of G such that $H \in \Delta$, $H \subset K$.

LEMMA 1. *Let G be an l-group and Δ be a plenary subset in $\Gamma(G)$. Then $x > g$ if and only if $Vx > V$ for all values V of x belonging to Δ.*

Proof. Since the condition is absolutely necessary we may suppose, by way of contradiction, that $g \wedge e < e$. Since $\bigcap_{M \in \Delta} M = E$, then $g^- = g \wedge e \notin H$ for some $H \in \Delta$. But then $|g^-|$ has value $V, V \supseteq H$. Since Δ is a root system (see Chapter 3 Section 2) in $\mathcal{C}(G)$, then $V \in \Delta$. Now V is also the value of g and $g^+ = g \vee e \in V$ since$|g^-| \wedge g^+ = e$ and V is a prime. Therefore, $V < Vg = Vg^+g^- < V$. This is the contradiction.□

LEMMA 2. *Let G be an l-group and the identity $|x| \cdot |y| \wedge |x|^2 |y|^2 = |x| \cdot |y|$ is valid in G. Then $A \cdot B = B \cdot A$ for all convex l-subgroups A, B of G and $A \cdot B$ is the least convex l-subgroup of G containing A and B.*

Proof. Let $g = ab$, $a \in A$, $b \in B$. Then $|a|^{-1} = a \wedge a^{-1}$, $|b|^{-1} =$

$b \wedge b^{-1}$ and $|a|^{-1}|b|^{-1} = ab \wedge a^{-1}b \wedge ab^{-1} \wedge a^{-1}b^{-1} \leq ab = g$

On the other hand, $ab \leq |a| \cdot |b|$. Hence $|a|^{-1}|b|^{-1} \leq g \leq |a| \cdot |b|$ and $|b|^{-2}|a|^{-2} \leq g \leq |b|^2|a|^2$. So $e \leq |b|^2 g|a|^2 \leq |b|^4|a|^4$. By Corollary 1 of Theorem 2.2.1, $|b|^2 g|a|^2 = b'a'$ where $e \leq b' \leq |b|^4$, $e \leq a' \leq |a|^4$. Therefore, $b' \in B$, $a' \in A$ and $g = |b|^{-2}b'a'|a|^{-2} \in BA$. Thus $AB = BA$ and AB is a subgroup of l-group G. By Theorem 3.1.1, the lattice $\mathcal{C}(G)$ of convex l-subgroups of G is a sublattice of the lattice of all subgroup of G therefore AB is a convex l-subgroup of G.□

LEMMA 3. *Let H be a convex l-subgroup of l-group G and $Hx > H$ for some $x \in G$. Then $Hx^{-1}hx < Hx$ for all $h \in H$.*

Proof. If $Hx > H$, then $H > Hx^{-1}$, $H = Hh > Hx^{-1}h$ and $Hhx = Hx > Hx^{-1}hx$ for all $h \in H$.□

LEMMA 4. *Let H be a regular l-subgroup of g and $\bar{H}$ be the cover of H. If for all $x \in \bar{H} \setminus H$, $x > e$ and for all $y \in G \setminus H$, $y > e$, there exists a positive integer n for which $Hx \leq Hy^n$, then H is normal value.*

Proof. It is clear that H is the value of x. Let $H_1 = H^x = x^{-1}Hx$. We wish to show that $H_1 = H$. Assume that $x > e$. Conjugation by x preserves inclusions and so H and H_1 are maximal subgroups in $\bar{H}$. Therefore, each cannot strictly contain the other. So, $H \neq H_1$ and there exist $z \in H$, $z > e$, such that $x^{-1}zx \notin H$. Then there is a positive integer n for which $Hx \leq Hx^{-1}z^n x$. This inequality contradicts Lemma 3. The case $x < e$ is similar. Since in general $x = x^+ \cdot x^-$, then $H_1 = H^x = H$.□

THEOREM 1. (Wolfenstein [1].) *The class $\mathcal{N}$ of normal-valued l-groups is an l-variety and is defined in the signature l by the identity*

$$|x| \cdot |y| \wedge |y|^2|x|^2 = |x| \cdot |y|.$$

Proof. Suppose G is a normal-valued l-group and $x, y \in G$, $x, y \geq e$. If $xy = e$, then $x = e = y$ and the identity holds for these x, y. Let $xy > e$ and H be a value of xy. Then $xy \in \bar{H}$. By our assumption, H is an ideal in $\bar{H}$. Therefore, $\bar{H}/H$ is an Archimedean totally ordered group and so $Hxy < Hy^2x^2$. Since H is an arbitrary value of xy, then by Lemma 1, $xy \leq y^2x^2$. So the identity $|x| \cdot |y| \wedge |y|^2|x|^2 = |x| \cdot |y|$ holds in G.

Conversely, let G be an l-group and the identity $|x| \cdot |y| \wedge |y|^2|x|^2 =$

$|x|\cdot|y|$ holds in G. Let H be a value of some element $g \in G$ and $\bar{H}$ be the cover of H. Let $x \in G\backslash H$, $x > e$, and $(x)_l = X$ be a convex l-subgroup of G generated by x. By Lemma 2, HX is a convex l-subgroup which consequently contains the cover $\bar{H}$. Therefore, $g = hy$ with $h \in H$ and $y \in X$. Then there exists a positive integer n for which $|y| \leq x^n$. So, $Hg = Hy \leq Hx^n$ and by Lemma 4, H is a normal subgroup in $\bar{H}$.□

The following example shows that $\mathcal{L} \neq \mathcal{N}$.

EXAMPLE 1. The l-group $A(\mathbf{R})$ of all order permutations of the real line is not normal-valued.

Indeed, let $f(x) = x + 1$ and

$$g(x) = \begin{cases} x, & \text{if } x \leq 1; \\ x^2 & \text{if } x > 1. \end{cases}$$

It is clear that $f, g \in A(\mathbf{R})$ and $f(g(x)) > g^2(f^2(x))$ for all $x > \frac{\sqrt{5}-1}{2}$ and therefore $A(\mathbf{R})$ is not normal-valued.

2^0. The variety $\mathcal{N}$ is the largest proper l-variety. This fact was proved by Ch.W.Holland [9].

LEMMA 5. *Let G be an l-group and $G \notin \mathcal{N}$. Then* $\mathrm{var}_l G$ *contains some l-group which is o-2-transitive on some totally ordered set.*

Proof. Let $G \notin \mathcal{N}$. Then there is a value H in G such that H is not normal in its cover $\bar{H}$. Let $M = \bigcap_{x \in \bar{H}} x^{-1}Hx$. Thus $K = \bar{H}/M \in \mathrm{var}_l G$ and natural action of $\bar{H}$ on $R(\bar{H} : H)$ induces a faithful representtation of $\bar{H}/M$ as transitive ordered permutation l-group $(\bar{H}/M, R(\bar{H} : H))$. Since $\bar{H}/M$ covers the stabilizer H/M of H and $H/M \neq M$ it follows from the Theorem of McCleary (see Theorem 4.5.1) that $(\bar{H}/M, R(\bar{H} : H))$ is an o-primitve ordered permutation l-group.

By Theorem 4.5.2, l-group $(\bar{H}/M, R(\bar{H} : H))$ is o-2-transitive or $\bar{H}/M$ contains an l-subgroup having o-2-transitive representations of some totally ordered set. Therefore, $\mathrm{var}_l G$ contains some o-2-transitive ordered permutation l-group.□

We say that an ordered permutation l-group (G, Ω) *mimics* an ordered permutation l-group (H, Λ) iff whenever $\lambda \in \Lambda$ and

$$\{w_r(x_1, \ldots, x_n) | 1 \leq r \leq k\}$$

is a finite set of words in the signature l in the variables $x_1, \ldots, x_n$ and $h_1, \ldots, h_n \in H$, then there exist $\alpha \in \Omega$ and $g_1, \ldots, g_n \in G$ such that $\lambda w_i(h_1, \ldots, h_n) < \lambda w_j(h_1, \ldots, h_n)$ iff $\alpha w_i(g_1, \ldots, g_n) < \alpha w_j(g_1, \ldots, g_n)$ for all $1 \leq i, j \leq k$.

We assert that (G, Ω) *mimics l-variety* $\mathcal{X}$ iff $G \in \mathcal{X}$ and (G, Ω) mimics all transitive ordered permutation l-groups $(H.\Lambda)$ with $H \in \mathcal{X}$.

LEMMA 6. *If* (G, Ω) *mimics the l-variety* $\mathcal{X}$, *then* $\mathrm{var}_l G = \mathcal{X}$.

Proof. It is clear that $\mathrm{var}_l G \subseteq \mathcal{X}$. Therefore, it is suffice to show that any identity which does not hold in $\mathcal{X}$ will not hold in G either. Let $w(x_1, \ldots, x_n = e)$ not hold in $\mathcal{X}$. Then for some subdirectly irreducible lattice $H \in \mathcal{X}$ and $h_1, \ldots, h_n \in H$ is valid $w(h_1, \ldots, h_n) \neq e$. By Corollary 3 of Theorem 4.4.1, l-group H has a faithful transitive representation by order permutations of totally ordered set Λ. Then for some $\lambda \in \Lambda$, $\lambda w(h_1, \ldots, h_n) \neq \lambda$. Since (G, Ω) mimics $\mathcal{X}$ there exists an $\alpha \in \Omega$ and $g_1, \ldots, g_n \in G$ such that $\alpha w(g_1, \ldots, g_n) \neq \alpha$. Hence $w(g_1, \ldots, g_n) \neq e$ in G, so the identity $w(x_1, \ldots, x_n) = e$ does not hold in G.□

LEMMA 7. *Let* (G, Ω) *be an 0-2-transitive ordered pemutation l-group and* $\mathcal{L}$ *be the l-variety of all l-groups. Then* (G, Ω) *mimics* $\mathcal{L}$.

Proof. (Reily [4].) Let (H, Λ) be any transitive ordered permutation l-group, $\lambda \in \Lambda$, $\{w_r(x_1, \ldots, x_n) | 1 \leq r \leq k\}$ be the finite set of words of the signature l and $h_1, \ldots, h_n \in H$. Every word $w_r(x_1, \ldots, x_n)$ can be written in the form $w_r(x_1, \ldots, x_n) = \bigvee_{i \in I} \bigwedge_{j \in J} w_{rij}(x_1, \ldots, x_n)$ where I, J are the finite set of indexes and each $w_{rij}(x_1, \ldots, x_n)$ is a word of the signature of the group theory. Let $\{u_t | t = 1, \ldots, m\}$ be the set of all initial segments of the words w_{rij} when written in a reduced form. Let $\lambda_0 = \lambda$, $\lambda_t = \lambda u_t(h_i, \ldots, h_n)$, $t = 1, 2, \ldots, m$. Since (G, Ω) is o-2-transitive, Ω is dense. Therefore we can find a set of elements $\{\alpha_0, \ldots, \alpha_m\} \subseteq \Omega$ such that the map:$\lambda_t \rightarrow \alpha_t$, $(t = 1, 2, \ldots, m)$ is an order isomorphism of the set $\{\lambda_0, \ldots, \lambda_m\}$ onto $\{\alpha_0, \ldots, \alpha_m\}$.

By Proposition 4.4.1, (G, Ω) is 0-$(m+1)^2$-transitive. It means that for each i we can find $g_i \in G$ such that $\lambda_s h_i \geq \lambda_t$ for $s, t \in \{0, \ldots, m\}$ iff $\alpha_s g_i \geq \alpha_t$. Thus g_i acting on $\{\alpha_0, \ldots, \alpha_m\}$ simply copies or mimics the action of the h_i on $\{\lambda_0, \ldots, \lambda_m\}$ and (G, Ω) mimics (H, Λ) and, therefore, the l-variety $\mathcal{L}$.□

THEOREM 2. (Holland [9].) *The l-variety $\mathcal{N}$ is the largest proper l-variety in the lattice of all l-varieties* $\mathbf{L}$.

Proof. Let $\mathcal{X}$ be an l-variety and $\mathcal{X} \not\subseteq \mathcal{N}$. Then there exists an l-group $G \in \mathcal{X} \backslash \mathcal{N}$. By Lemma 5, we can assume that G has a faithfui o-2-transitive representation by order permutations of some totally ordered set Ω. Then by Lemma 7, (G, Ω) mimics l-variety $\mathcal{L}$ and by Lemma 6 $\text{var}_l G = \mathcal{L}$. Therefore, $\mathcal{X} = \mathcal{L}$.□

COROLLARY 1. (Holland [9].) *Let $A(\mathbf{R})$ be the l-group of all order permutations of the real line. Then* $\text{var}_l A(\mathbf{R}) = \mathcal{L}$.□

3⁰. The l-variety $\mathcal{N}$ is ingeneously described in terms of ordered permutation l-groups.

PROPOSITION 1. *Let G be an l-group and Δ be a plenary subset in the set $\Gamma(G)$ of all regular convex l-subgroups of G. If every $V \in \Delta$ is normal in its cover $\bar{V}$, then G is a normal-valued l-group.*

Proof. Let $x, y > e$, $x, y \in G$ and $V \in \Delta$ be the value of xy. Then $Vxy < Vx^2y^2 = Vy^2x^2$ and by Lemma 1, $xy < y^2x^2$. Therefore, $G \in \mathcal{N}$ by Theorem 1.□

Let $g \in A(\Omega)$ and $\alpha \in \Omega$. Let us recall that the supporting interval of g or orbital $o(g, \alpha)$ is the convex subset of Ω swept out by t he images of α under powers of g:

$$o(g, \alpha) = \text{Conv}_{\Omega}(\alpha)(g) = \{\beta \in \Omega | \alpha g^n \leq \beta \leq \alpha g^m \text{ for some } n, m \in \mathbf{Z}\}.$$

We say that ordered permutation l-group (G, Ω) is *non-overlapping* if $o(g, \alpha)$ and $o(f, \alpha)$ are comparable with respect to inclusion for all $g, f \in G$, $\alpha \in \Omega$.

THEOREM 3. (Read [1].) *Let G be an l-group. Then the following are equivalent:*

(1) $G \in \mathcal{N}$;

(2) *for every faithful representation of l-group G by order permutations of totally ordered set Ω the ordered permutation l-group (G, Ω) is non-overlapping;*

(3) *there exists a faithfull representation of l-group G by order permutations of the totally ordered set Ω such that (G, Ω) is non-overlapping.*

Proof. (1)⇒(2). Let $G \in \mathcal{N}$ and (g, Ω) be a faithfull representation of G such that (G, Ω) is overlapping. Then there are $f, h \in G$, $\alpha \in \Omega$ such that $o(h, \alpha)$ and $o(f, \alpha)$ are incomparable. We can assume that $f, h > e$. Then there are integers m, n such that $\beta = \alpha f^n$, $\gamma = \alpha h^m$, $\beta < \alpha < \gamma$ and $\beta \in o(f, \alpha) \setminus o(h, \alpha)$, $\gamma \in o(h, \alpha) \setminus o(f, \alpha)$. Therefore, $\beta h^r < \alpha$, $\beta h^r f^s < \alpha f^s < \gamma$ in Ω for all $r, s \in \mathbf{Z}$. Let $H = (h)_l$, $F = (f)_l$ be the convex l-subgroups of G generated by elements h, f corresponingly. Then, by Lemma 2, $H \cdot F = F \cdot H$ is a convex l-subgroup generated by elements f and h. Hence for every $g \in H \cdot F$ there are integers $s, r \in \mathbf{Z}$ such that $g \leq h^r f^s$ and $\beta g \leq \beta h^r f^s < \gamma$. On the other hand, $f^{-n} h^m \in F \cdot H$ and $\beta f^{-n} h^m = \alpha h^m = \gamma$, a contradiction to our assumption.

Implication (2)⇒(3) is evident. (3)⇒(1). Let (G, Ω) be a non-overlapping l-group. For every $\alpha \in \Omega$ let G_α be a stabilizer l-subgroup on α. Then $\bigcap_{\alpha \in \Omega} G_\alpha = E$ and the set Δ of regular subgroup of G containing some G_α, $(\alpha \in \Omega)$, is a plenary subset in $\Gamma(G)$ by Propositon 3.3.3.

Let $V \in \Delta$ and $\bar{V}$ be the cover V. Then $V \supseteq G_\alpha$ for some $\alpha \in \Omega$. Let g, h be any positive elements of $\bar{V} \setminus V$. We claim that $o(g, \alpha) = o(h, \alpha)$. Conversely, let $o(g, \alpha) \subset o(h, \alpha)$. Then we consider the subset $S = \{f \in G | \alpha f^m \in o(g, \alpha)\}$. It is evident that S is a convex l-subgroup of G and $G_\alpha \subseteq S$. By Proposition 4.3.3, convex l-subgroups are prime. Hence $V \subseteq S \subseteq \bar{V}$ and $V \neq S$. This contradicts our assumption.

So, $o(g, \alpha) = o(h, \alpha)$ for all positive elements $g, h \in \bar{V} \setminus V$. Then there is a positive integer n such that $\alpha g \leq \alpha h^n$. Thus $\alpha g h^{-n} \leq \alpha = \alpha(g h^{-n} \vee e)$ and $g h^{-n} \vee e \in G_\alpha$, $g h^{-n} \vee e \in V$. Hence $V(g h^{-n} \vee e) = V g h^{-n} \vee V = V$, $V g h^{-n} \leq V$ and $V g \leq V h^n$. By Lemma 4, V is a normal subgroup in V.□

4⁰. The class of normal-valued l-groups is very close to the class of Conradian right-ordered groups (see Chapter 5).

THEOREM 4. (Kopytov [5].) *Let G be a normal-valued l-group and P be a positive cone of G. Then P is the intersection of some family of Conradian right orders.*

If G is a Conradian right-ordered group, then l-subgroup G^* of l-group $A(G)$ of totally ordered set $\{G, \leq\}$ generated by right regular representation of G is a normal-valued l-group.

Proof. We define a Conradian right order on the normal-valued l-group G. Let $\{H_\gamma | \gamma \in I\}$ be a set of all prime subgroups of l-group G. For every $\gamma \in I$ well order I by ordering $\prec^\gamma$ such that γ will be the least element I under this ordering $\prec^\gamma$:

$$\gamma_1 = \gamma \prec^\gamma \gamma_2 \prec^\gamma \cdots \prec^\gamma \gamma_\xi \prec^\gamma \cdots$$

where $\xi < \sigma$.

Now we define in l-group G by the ordering $\prec^\gamma$ the system of convex l-subgroup by the rule: $G_0^\gamma = G$, $G_\xi^\gamma = H_{\gamma_\xi} \cap H_{\gamma_{\xi+1}}$. It is clear that $G_{\xi+1}^\gamma$ is a prime subgroup in G_ξ^γ. This system of convex l-subgroup $\{G_\xi^\gamma | \xi < \sigma\}$ can be embedded into the maximal system $\mathcal{S}_\gamma(G)$ of convex l-subgroups of G totally ordered by inclusion. It is clear that $\mathcal{S}_\gamma(G)$ is closed under arbitrary unions and intersections of its members. But $G \in \mathcal{N}$, therefore the system $\mathcal{S}_\gamma(G)$ is subnormal with factors isomorphic to subgroup of the additive group of reals. As in Corollary 5.1.2 we define a right order P_γ on G by the rule: $x \in P_\gamma$ iff $x \in B \setminus A$ where $A \prec B$ is a jump in the system $\mathcal{S}_\gamma(G)$ and $Ax > A$ in Archimedean totally ordered group B/A. By definition the right order P_γ extends the lattice order P of G.

If $x \in G$ and $x \notin P$, then there is a maximal prime subgroup H_α $(\alpha \in I)$ such that $x \notin H_\alpha$ and $H_\alpha x < H_\alpha$. Therefore, $x \notin P_\alpha$ and $P = \bigcap_{\gamma \in I} P_\gamma$. The first part of the theorem is proved.

Let now P be a Conradian right order of a group G and $R : \gamma \to A(G)$ be a right regular representation of G on totally ordered set (under right order of G) $\{G :\leq\}$: $R(a)(x) = xa$ for $x \in \{G; \leq\}$, $a \in G$. Let G^* be an l-subgroup of $A(G)$ generated by $R(G)$. We claim that $G^* \in \mathcal{N}$. First, it is clear that $o(\bar{g}, x) = R(x)(o(x\bar{g}x^{-1}, e) = o(R(x)\bar{g}R(x^{-1}), e)x$ for every $\bar{g} \in G^*$, $x \in G$. Hence it is enough to prove that $o(\bar{g}, e)$ coinsides with some convex subgroup of right-ordered group G. Use induction on the lenght of $\bar{g} = \bigvee_{i \in I} \bigwedge_{j \in J} R(g_{ij})$ where $g_{ij} \in G$, I, J are finite sets of indexes. If $\bar{g} = R(g)$, then $o(\bar{g}, e) = \{y \in G | g^n \leq y \leq g^{n+1}$ for some integer $n\}$ and $o(\bar{g}, e)$ is a convex subgroup of right-ordered group G generated by the element g. Let $\bar{g} = \bar{h}_1 \vee \bar{h}_2$, $\bar{g}(e) \neq e$, $o(\bar{h}_1, e) = H_1$, $o(\bar{h}_2, e) = H_2$ where H_1, H_2 are convex subgroups in right-ordered group G and $H_1 \subseteq H_2$. We have three different cases: a) $\bar{h}_2(e) > e$; b) $\bar{h}_2(e) \leq e$, $\bar{h}_1(e) > (e)$; c) $\bar{h}_2(e) < e$, $\bar{h}_1(e) < e$.

a) In this case for $n > 0$

$$\bar{g}^n(e) = (\bar{h}_1 \vee \bar{h}_2)^n(e) = (\bigvee \bar{h}_1^{n_1} \bar{h}_2^{k_1} \cdots \bar{h}_1^{n_s} \bar{h}_2^{k_s})(e),$$

where $n_i, k_i \in \mathbf{Z}$, $n_i, ki \geq 0$ for $i = 1, 2, \ldots, s$ and $\sum_{i=1}^{s}(n_i + k_i) = n$. Then

$$\bar{g}^n(e) = \max\{(\bar{h}_1^{n_1} \bar{h}_2^{k_1} \cdots \bar{h}_1^{n_s} \bar{h}_2^{k_s})(e)\} \geq \bar{h}_2^n.$$

The sequence $\{\bar{h}_2^m(e) | m \geq 0,\ m \in \mathbf{Z}\}$ is unbounded above in the totally ordered set H_2, then the sequence $\{\bar{g}^n(e) | n \geq 0,\ n \in \mathbf{Z}\}$ is unbounded above in H_2, too. Hence $o(\bar{g}, e) \cap P = H_2 \cap P$.

If $n < 0$, then

$$\bar{g}^n(e) = (\bigwedge \bar{h}_1^{n_1} \bar{h}_2^{k_1} \cdots \bar{h}_1^{n_s} \bar{h}_2^{k_s})(e),$$

where $n_i, k_i \in \mathbf{Z}$, $n_i, k_i \leq 0$ and $\sum_{i=1}^{s}(n_i + k_i) = n$. Then $\bar{g}^n(e) \leq \bar{h}_2^n(e)$. The sequence $\{\bar{h}_2^m(e) | m \leq 0\}$ is unbounded below in the totally ordered set H_2, therefore the sequence $\{\bar{g}^n(e) | n \leq 0\}$ is unbounded below in H_2. Then $o(\bar{g}, e) \cap H_2 = H_2 \cap P^{-1}$ and $o(\bar{g}, e) = H_2$.

Cases b),c) are similar (in these cases $o(\bar{g}, e) = H_1$).

From the equalities $(\bar{h}_1 \wedge \bar{h}_2) = (\bar{h}_1^{-1} \vee \bar{h}_2^{-1})^{-1}$ and $o(\bar{g}, x) = o(\bar{g}^{-1}, x)$ it follows that $o(\bar{h}_1 \wedge \bar{h}_2, e)$ is equal to H_1 or H_2. This estabilishes the claim. Now the proof of the theorem follows immediately from Theorem 3.□

9.3 Representable l-Groups

We recall that l-group G is a subcartesian product of the totaly ordered groups G_i $(i \in I)$ if G is isomorphic to a l-subgroup in the Cartesian product $\overline{\prod_{i \in I}} G_i$ in such a way that each projection is onto. Obviously, l-group G is a subcartesian product of o-group if G possesses the family of normal prime subgroups whose intesection is an identity subgroup E of G. We will call such an l-group *representable.* So, every representable l-group is an l-subgroup of some Cartesian product of o-groups and therefore all representable l-groups are vector groups. Every homomorphism φ of l-group G into Cartesian product of o-groups is called a *realization.* The realization φ is *faithful* if φ is an l-isomorphism. Therefore, l-group G is representable iff G has a faithful realization.

The following results give another characterization of representable l-groups.

1⁰. Let us recall that element g of l-group G is regular if $|g| \wedge |x^{-1}gx| \neq e$ for all $x \in G$.

PROPOSITION 1. *The following properties are equivalent for l-group G:*

(1) *the identity $(x \wedge y^{-1}x^{-1}y) \vee e = e$ holds in G;*

(2) *every nonidentity element in G is regular.*

Proof. Let the identity $(x \wedge y^{-1}x^{-1}y) \vee e = e$ be valid in G and there are elements $a \in G$, $a > e$, and $c \in G$ such that $a \wedge c^{-1}ac = e$. Let $x = a(c^{-1}ac)^{-1}$, $b = c^{-1}ac$, $y = c^{-1}$. Then the direct verification shows that

$$e = (x \wedge y^{-1}x^{-1}y) \vee e = (ab^{-1} \wedge cba^{-1}c^{-1}) \vee e =$$
$$= (ab^{-1} \vee e) \wedge (aca^{-1}c^{-1} \vee e) = a(a \wedge c^{-1}ac)^{-1} \wedge (aca^{-1}c^{-1} \vee e) =$$
$$= a \wedge a(cac^{-1} \wedge a)^{-1} = a(e \wedge (cac^{-1} \wedge a)^{-1}) = a.$$

Conversely, let there be $x, y \in G$ such that $a = (x \wedge y^{-1}x^{-1}y) \vee e \neq e$. It is clear that x is incomparable with e. Let $b = yay^{-1} = (yxy^{-1} \wedge x^{-1}) \vee e$. Then

$$a \wedge yay^{-1} = ((x \wedge y^{-1}x^{-1}y) \vee e) \wedge ((yxy^{-1} \wedge x^{-1}) \vee e) =$$
$$= (x \vee e) \wedge (y^{-1}x^{-1}y \vee e) \wedge (yxy^{-1} \vee e) \wedge (x^{-1} \vee e) =$$
$$= (x \vee e) \wedge (x^{-1} \vee e) \wedge (y^{-1}x^{-1}y \vee e) \wedge (yxy^{-1} \vee e) = e.$$

This contradicts our assumption. □

THEOREM 1. (Lorenzen [2], Šik [2].) *The class $\mathcal{R}$ of all representable l-groups is the l-variety defined by the identity*

$$(x \wedge y^{-1}x^{-1}y) \vee e = e.$$

Proof. Let G be a representable l-group. Suppose that G is an l-subgroup of some Cartesian product $H = \overline{\prod\limits_{\alpha \in I}} G_\alpha$ of the family $\{G_\alpha | \alpha \in I\}$ of o-groups G_α. Let $x, y \in G$. Then $(x \wedge y^{-1}x^{-1}y)(\alpha) \leq e$ and $((x \wedge y^{-1}x^{-1}y) \vee e)(\alpha) = e$ for all $\alpha \in I$. Therefore, $(x \wedge y^{-1}x^{-1}y) \vee e = e$ for all $x, y \in G$.

Conversely, let the identity $(x \wedge y^{-1}x^{-1}y) \vee e = e$ hold in g. We can assume that G is a subcartesian product of subdirectly irreducible l-groups G_α, $\alpha \in I$. It is enough to prove that every G_α is a totally-ordered group. Suppose for reductio ad absurdum that there exists $a \in G_\alpha$ noncomparable with e. Then $a^+ = a \vee e > e$ and $(a^-)^{-1} = (a \wedge e)^{-1} = a^{-1} \vee e > e$ are orthogonal. Therefore, polars $A = (a^+)^\perp$ and $B = A^\perp$ are nontrivial convex l-subgroups of G_α and $A \cap B = E$. Suppose that there are $x \in A$ and $y \in G_\alpha$ such that $y^{-1}xy \notin A$. Then $c = a^+ \wedge y^{-1}xy| \neq e$ and

$$\begin{aligned} c \wedge ycy^{-1} &= (a^+ \wedge |y^{-1}xy|) \wedge (ya^+y^{-1} \wedge |x|) = \\ &= (a^+ \wedge |x|) \wedge (|y^{-1}xy| \wedge ya^+y^{-1}) = e \end{aligned}$$

as $x \in A$. Therefore, c is not regular element, which contradicts Proposition 1. So, $y^{-1}Ay = A$ and $y^{-1}By = B$ for all $y \in G_\alpha$. Then A and B are nontrivial ideals in G_α such that $A \cap B = E$ and G is not subdirectly irreducible. So, all l-groups G_α, $\alpha \in I$ are o-groups and G is a representable l-group.□

The interested reader can without difficulty obtain the proof of following theorem.

THEOREM 2. (Šik [2], Kontorovich and Kutyev [1].) *An l-group G is representable if and only if every polar in G is an ideal* .□

If G is an l-subgroup of the Cartesian product $\bar{G} = \overline{\prod\limits_{\alpha \in I}} G_\alpha$ of totally ordered groups G_α, $\alpha \in I$, then we can interpret the information about polars of G in terms of I, as follows. For $f \in \bar{G}$ let

$\mathrm{Supp}(f) = \{\alpha \in I | f(\alpha) \neq e\}$. If $X \subseteq \bar{G}$ let $\mathrm{Supp}(X) = \bigcup_{f \in X} \mathrm{Supp}(f)$. Then the map $\tau : H\tau = \mathrm{Supp}(H)$ from the set of polars $\mathcal{P}(G)$ of G into the set of all subsets of I is one-to-one and intersections preserving. Consequently, $\{\mathrm{Supp}(H) | H \in \mathcal{P}(G)\}$ can be made into a Boolean algebra under inclusion, which is isomorphic to $\mathcal{P}(G)$. Now we list some useful tests for the representability of l-groups.

PROPOSITION 2. *If every convex l-subgroup of l-group G is an ideal, then G is representable.*□

PROPOSITION 3. *Each Abelian l-group is representable.*□

It is clear that every representable l-group is an orderable group and the lattice order P of the representable l-group G is the intersection of all total orders P_α of G which are extensions of P. On the other hand, there are normal-valued l-groups which are not representable.

EXAMPLE 1. (See Example 7.6.1.) Let $G = \mathrm{gp}\{b, a_1, a_2 | a_1a_2 = a_2a_1,\ b^{-1}a_1b = a_2,\ b^{-1}a_2b = a_1\}$. If $P = \{x = b^k a_1^n a_2^m | k > 0 \text{ or} k = 0, n \geq 0, m \geq 0\}$, then P defines on G the lattice order. But G is not an orderable group.

On other hand, the l-group $D(\mathbf{R})$ of example 4.7.1 is an orderable group. But $D(\mathbf{R})$ is an o-2-transitive ordered permutation l-group and therefore, by Lemma 7, $D(\mathbf{R})$ cannot be normal-valued.

2^0. Let G be an l-group and $\bar{G} = \bar{\prod}_{\alpha \in A} G_\alpha$ be the Cartesian product of the family of o-groups $\{G_\alpha | \alpha \in A\}$. If φ is the realization of G into $\bar{G}$, then any factor G_α ($\alpha \in A$) is called a *component* of realization φ. For any $\alpha \in A$ let φ_α denote the *projection* $\varphi : G \to G$ defined by the rule: $g\varphi_\alpha = g\varphi(\alpha)$. It is evident that each projection φ_α of l-group G is a realization, too. For any subset $\Gamma \subseteq A$, let φ_Γ denote the projection of l-group G into the Cartesian product $\bar{\prod}_{\alpha \in \Gamma} G_\alpha$ where $g\varphi_\Gamma(\alpha) = g\varphi(\alpha)$ for all $\alpha \in \Gamma$.

A component G_β, $\beta \in A$, of the faithful realization φ is called *superfluous* if the projection $\varphi_{A \setminus \beta} : G \to \bar{\prod}_{\alpha \in A \setminus \beta} G_\alpha$ is the faithful realization too. Faithful realization φ of l-group G is called *irreducible* if every component of φ is not superfluous. There exist examples of the realizations of l-groups such that every component of realization is superfluous.

EXAMPLE 2. Let F be a non-commutative free group. Then the intersection of the lower central series

$$F = F_0 \supset F_1 \supset \cdots \supset F_n \supset F_{n+1} \supset \cdots$$

is E and all factors F_n/F_{n+1}, $n \in \mathbf{Z}$, are Abelian torsion-free groups. We order every factor F_n/F_{n+1} and define the order on F by the rule: $x > e$ iff $x \neq e$ and $xF_{n+1} > F_{n+1}$ in o-group F_n/F_{n+1} where $x \in F_n \setminus F_{n+1}$. The direct verification shows that this order on F is a total order and all subgroups F_n are convex normal subgroups in F. Then $H_n = F/F_n$ is a nilpotent o-group for every $n \in \mathbf{N}$. Let $\bar{F} = \bar{\prod\limits_{n\in\mathbf{N}}} H_n$ be the Cartesian product of totally ordered groups H_n, $n \in \mathbf{N}$. Since $\bigcap\limits_{n\in\mathbf{N}} H_n = E$, then the map $\varphi : F \to \bar{F}$ where $x\varphi(n) = xF_n$ is the faithful realization of o-group F and every component of φ is superfluous.

Now we consider l-groups which have irreducible realizations.

THEOREM 3. (Šik [2].) *Let G be a representable l-group. Then G has the faithful irreducible realization if and only if the Boolean algebra $\mathcal{P}(G)$ of polars of G is atomic.*

Proof. Let φ be the faithful irreducible realization of l-group G into the Cartesian product $\bar{G} = \bar{\prod\limits_{\alpha\in A}} G_\alpha$ of the family of totally ordered groups $\{G_\alpha | \alpha \in A\}$. Since every component G_α ($\alpha \in A$) is not superfluous, then the kernel K_β of the projection $\varphi_{A\setminus\beta}$ of G into $\bar{G}$ is not trivial. Moreover, the ideal $K'_\beta = \bigcap\limits_{\alpha\neq\beta} K_\alpha \neq E$ since otherwise the projection $\varphi_{A\setminus\beta}$ would be the faithful realisation of G into $\bar{G}$, a contradiction to irreducibility of φ. So $K'_\beta \neq E$ for all $\beta \in A$. Therefore, for every $\beta \in A$ there is an element $a_\beta > e$, $a_\beta \in K'_\beta$ such that $a_\beta\varphi(\alpha) = e$ for all $\alpha \neq \beta$ and $a_\beta\varphi(\beta) > e$. Let A_β be the smallest polar of G containing a_β. We claim that A_β is an atom in the Boolean algebra $\mathcal{P}(G)$ of polars of G. Suppose, on the contrary that there is polar $B \subset A_\beta, B \neq A_\beta$. Then

$B^\perp \supset A^\perp$, $B^\perp \neq A^\perp_\beta$. But $g \in A^\perp_\beta$ iff $g\varphi(\beta) = e$. If $B^\perp \supset A_\beta$, then there is $f \in B^\perp$ such that $f\varphi(\beta) \neq e$. Hence for every $\alpha \in A$ there is an element h_α, $h_\alpha \in B^\perp$ for which $h_\alpha\varphi(\alpha) \neq e$. Then $B = \beta^{\perp\perp} = E$ contradicts our assumption.

Let $X^\perp$ be a polar of G and A_X be the subset of A such that the polar A_β defined above is contained in $X^\perp$ for all $\beta \in A_X$. Now the direct verification shows that $X^\perp = \bigvee^{\mathcal{P}}_{\beta \in A_X} A_\beta$ (here $\bigvee^{\mathcal{P}}$ is supremum in the Boolean algebra $\mathcal{P}(G)$).

Conversely, let G be a representable l-group and Boolean algebra $\mathcal{P}(G)$ of polar of G be atomic. Let $\{A_\alpha | \alpha \in AA\}$ be the set of all atoms of $\mathcal{P}(G)$. It is clear that $A_\alpha = (x)^{\perp\perp}$ for any nonidentity element of A_α. Let now a_α be any nontrivial element of A_α. By Theorem 2, all A_α are ideals in G. Since A_α is an atom in $\mathcal{P}(G)$, then it is easy to see that A_α is a totally ordered l-subgroup of G and $G/A^\perp_\alpha$ is a totally ordered group, too.

If $\alpha, \beta \in A$, $\alpha \neq \beta$, then $A^\perp_\beta \supseteq A_\alpha$. Hence $\bigcap_{\alpha \in A \setminus \beta} A^\perp_\alpha$ and every component $G/A^\perp_\alpha$ of the realization φ of l-group G into $\overline{\prod_{\alpha \in A}} G/A^\perp_\alpha$ is not superfluous. This means that φ is a faithful realization.□

It follows from the proof of Theorem 3 that any faithful irreducible realization of a representable l-group G is uniquely determined by atoms of the Boolean algebra $\mathcal{P}(G)$ of polars of G. In other words, any two faithful irreducible realizations of G are isomorphic.

EXAMPLE 3. The lattice-ordered addidive group $C([0,1], \mathbf{R})$ of all continuous functions from $[0,1]$ into $\mathbf{R}$ is Abeleian. Hence $C([0,1], \mathbf{R})$ is a representable l-group. We claim that $C([0,1], \mathbf{R})$ has no faithful irreducible realizations. Indeed, let A be any polar of $C([0,1], \mathbf{R})$ and $I(A) = \{x \in [0,1] | a(x) = 0 \text{ for all } a \in A\}$. It is clear that $I(A)$ is closed in the natural topology on $[0,1]$. If $A \neq \{0\}$, then $I(A) \neq [0,1]$ and there is a closed subset $X \subset [0,1]$ such that $I(A) \subset X$. Let B be the set of all continuous functions $f \in C([0,1], \mathbf{R})$ such that $f(x) = 0$ for all $x \in X$. Since the natural topology of the closed interval $[0,1]$ is regular, it follows that $B \neq 0$, $B \subset A$ and $B \neq A$. Therefore, the Boolean algebra $\mathcal{P}(C([0,1], \mathbf{R}))$ is non-atomic. By Theorem 3, $C([0,1], \mathbf{R})$ has no faithful irreducible realizations.□

3^0. In Section 3 of Chapter 2 examples of vector l-groups were constructed for which not every embedding into the Cartesian product of o-groups is a realization. Therefore, the question concerning representability of arbitrary vector l-group arises.

Let G be an l-group and P be a lattice order of G. As usual, $\bigwedge_{\alpha\in A} x_\alpha$, $\bigvee_{\alpha\in A} x_\alpha$ are infima and suprema under P in G of the elements $\{x_\alpha | \alpha \in A,\ x_\alpha \in G\}$ where A is a finite index set. If P', P_ζ are other lattice orders of G, then $\overset{\prime}{\bigwedge_{\alpha\in A}} x_\alpha$, $\overset{\prime}{\bigvee_{\alpha\in A}} x_\alpha$, $\overset{\zeta}{\bigcap_{\alpha\in A}} x_\alpha$, $\overset{\zeta}{\bigvee_{\alpha\in A}} x_\alpha$ denote the infima and suprema in G of elements $\{x_\alpha | \alpha \in A\}$ under orders P', P_ζ accordingly.

LEMMA 1. (Conrad [5].) *Let G be a vector l-group, P be a lattice order on G and $\bigwedge_{1\le i\le n} \bigvee_{1\le j\le m} g_{ij} < e$ in G. Then there is a total order P' on G extending P such that $\overset{\prime}{\bigwedge_{1\le i\le n}} \overset{\prime}{\bigvee_{1\le j\le m}} g_{ij} <' e$.*

Proof. Since the l-group G is a vector l-group, then $P = \bigcap_{\alpha\in A} P_\alpha$ where $\{P_\alpha | \alpha \in A\}$ is a set of all total orders extending P (see Theorem 7.2.1). Let now the inequality $\bigwedge_{1\le i\le n} \bigvee_{1\le j\le m} g_{ij} < e$ be valid in G. Suppose, for reductio ad absurdum, that $\overset{\alpha}{\bigwedge_{1\le i\le n}} \overset{\alpha}{\bigvee_{1\le j\le m}} g_{ij} \ge<^\alpha e$. for all total orders P_α extending P. Then $\bigvee_{1\le j\le m} g_{ij} \in P_\alpha$ for all i, $1 \le i \le n$ and $\alpha \in A$. Hence $\bigvee_{1\le j\le m} g_{ij} \in \bigcap_{\alpha\in A} P_\alpha = P$, which contradicts the inequality $\bigwedge_{1\le i\le n} \bigvee_{1\le j\le m} g_{ij} < e$. So, there exists a total order P' extending P such that $\overset{\prime}{\bigwedge_{1\le i\le n}} \bigvee_{1\le j\le m} g_{ij} <' e$. Then there is k, $1 \le k \le m$, for which $h_k = \bigvee_{1\le j\le m} g_{kj} \in (P')^{-1} \setminus \{e\}$. Therefore, $g_{kj} \le h_k$ under lattice order

P for all j, $1 \leq j \leq m$. Hence $\overset{\prime}{\bigvee\limits_{1\leq j\leq m}} g_{kj} \leq' h_k < e$. Then

$$\overset{\prime}{\bigwedge\limits_{1\leq i\leq n}} \overset{\prime}{\bigvee\limits_{1\leq j\leq m}} g_{ij} \leq' \overset{\prime}{\bigvee\limits_{1\leq j\leq m}} g_{kj} < e$$

under P''.□

THEOREM 4. *Each vector lattice-ordered group is representable.*

Proof. Let G be a vector l-group and $P = \bigcap\limits_{\alpha\in A} P_\alpha$, where $\{P_\alpha | \alpha \in A\}$ is a set of all total orders extending P. Now we define an order homomorphism θ (as in Theorem7.2.1) of G into the Cartesian product $H = \overline{\prod\limits_{\alpha\in A}} G_\alpha$ of the totally ordered groups G'_α $\alpha \in A$, where G_α is group G totally ordered by order P_α and $g\theta(\alpha) = g$ for all $\alpha \in A$. Let $\bar{G}$ be an l-subgroup of l-group H generated by $G\theta$. Then each element x of $\bar{G}$ can be represented in the form $x = \overline{\bigwedge\limits_{1\leq i\leq n}} \overline{\bigvee\limits_{1\leq j\leq m}} g_{ij}\theta$ where $g_{ij} \in G$. By distributivity of the lattice operations $\bar{\vee}, \bar{\wedge}$ every element $x \in G$ can also be represented in the form $x = \overline{\bigvee\limits_{1\leq r\leq p}} \overline{\bigwedge\limits_{1\leq s\leq q}} g_{rs}\theta$ Here $\bar{\vee}, \bar{\wedge}, \bar{\leq}$ are lattice operations and order relation in $\bar{G}$ correspondingly.

Now we define a map φ of $\bar{G}$ into G by the rule: $(\overline{\bigwedge\limits_i} \overline{\bigvee\limits_j} g_{ij}\theta)\varphi = \bigwedge\limits_i \bigvee\limits_j g_{ij}$. We claim that the map φ is is well-defined. In fact, let $\bigwedge\limits_i \bigvee\limits_j g_{ij} < e$ in G then, by Lemma 1, there is a total order P' of G extending P such that $\overset{\prime}{\bigwedge\limits_i} \overset{\prime}{\bigvee\limits_j} g_{ij} <' e$. Therefore, let $\overline{\bigwedge\limits_i} \overline{\bigvee\limits_j} g_{ij}\theta \neq e$ in $\bar{G}$. If $\bigwedge\limits_i \bigvee\limits_j g_{ij}$ is incomparable with e in G, then $(\bigwedge\limits_i \bigvee\limits_j g_{ij}) \wedge e < e$ and $(\overline{\bigwedge\limits_i} \overline{\bigvee\limits_j} g_{ij}\theta)\bar{\wedge}\bar{<}e$ in $\bar{G}$. Hence $\overline{\bigwedge\limits_i} \overline{\bigvee\limits_j} g_{ij}\theta \neq e$ in $\bar{G}$. If $h = \overline{\bigwedge\limits_i} \overline{\bigvee\limits_j} g_{ij} > e$ in G, then $h = \bigvee\limits_r \bigwedge\limits_s g'_{rs}$. Then $h^{-1} = \bigwedge\limits_r \bigvee\limits_s g'^{-1}_{ij} < e$ and by similar arguments, $\overline{\bigwedge\limits_r} \overline{\bigvee\limits_s} g'^{-1}_{ij}\theta \neq e$ in $\bar{G}$. By distributivity of the lattice operations $\bar{\vee}, \bar{\wedge}$ in

G,

$$e \neq (\bar{\bigwedge_r}\bar{\bigvee_s}(g'_{rs})^{-1})^{-1} = (\bar{\bigvee_i}\bar{\bigwedge_j} g_{ij}^{-1})^{-1}.$$

Therefore, $\bar{\bigvee_i}\bar{\bigwedge_j} g_{ij}^{-1}\theta \neq e$ and $\bar{\bigwedge_i}\bar{\bigvee_j} g_{ij}\theta \neq e$. Thus the map θ is well-defined. The direct verification shows that φ if a homomorphism of $\bar{G}$ onto G. It is clear that $\bar{G}$ is representable. By Theorem 1, the class $\mathcal{R}$ of representable l-groups is closed under homomorphisms. Hence G is representable.□

4^0. The following test for representability was proposed by Byrd.

THEOREM 5 (Byrd [1].) *Each transitive representable l-group of order permutations of some totally ordered set is totally ordered.*

Proof. Let (G, Ω) be a transitive representable l-group. Assume on the contrary that G is not totally ordered and element g is noncomparable with e in the lattice order of g. Then there are $\alpha, \beta \in \Omega$ such that $\alpha g = \gamma > \alpha$ and $\beta g = \delta < \beta$. By transitivity of (G, Ω) there is an element $h \in G$ such that $\beta h = \alpha$. Then $\alpha(h^{-1}gh) = \beta(gh) = \delta h < \beta h = \alpha$ and $\alpha < \alpha(h^{-1}g^{-1}h)$. Therefore,

$$\alpha(g \wedge h^{-1}g^{-1}h) = \min\{\alpha g, \alpha(h^{-1}g^{-1}h)\} > \alpha$$

and $\alpha((g \wedge h^{-1}g^{-1}h) \vee e) > \alpha$. The last inequality is impossible because G is representable.□

The following two corollaries are similar to those of 1 and 3 of Theorem 4.3.1.

COROLLARY 1. (Byrd [1].) *A lattice-ordered group G is representable if and only if each minimal prime subgroup is normal.*□

COROLLARY 2. (Byrd [1].) *Each subdirectly irreducible representable l-group is totally ordered.*

We use now the notations from the proof of Theorem 9.1.3.

COROLLARY 3. (Conrad and McAlister [1].) *Let G_i $(i \in I)$ be a family of totally ordered groups and M be a minimal prime subgroup of Cartesian product $\bar{\prod_{i\in I}} G_i$. Then there exists an ultrafilter $\mathcal{F}$ such that $T_{\mathcal{F}} = M$.*

Proof. In fact, in terms of the proof of Jónsson's theorem (Theorem 9.1.3) we have: a) $B = \bar{\prod\limits_{i\in I}} G_i$; b) $\beta = \mu$ and $\ker\beta = \ker\mu = \bigvee\limits_{J\in\mathcal{F}}^{c} T_J$. By Corollary 1, the minimal prime subgroup M is an ideal in $\bar{\prod\limits_{i\in I}} G_i$ and $A = \bar{\prod\limits_{i\in i}} G_i/M$ is a totally ordered group. It is clear that every totally ordered group is finitely subdirectly irreducible. So, by repeating Jónsson's arguments, we have an ultrafilter $\mathcal{F}$ over I such that $T_{\mathcal{F}} \subseteq M$ and $\bar{\prod\limits_{i\in I}} G_i/M$ is a homomorphic image of $\bar{\prod\limits_{i\in I}} G_i/\mathcal{F} = \bar{\prod\limits_{i\in I}} G_i/T_{\mathcal{F}}$. But it easy to see that the ultraproduct $\bar{\prod\limits_{i\in I}} G_i/\mathcal{F}$ of the family of o-groups $\{G_i | i \in I\}$ is an o-group, too. From minimality of the prime ideal M it follows $M = T_{\mathcal{F}}$.□

COROLLARY 4. *Let M be a minimal prime subgroup of Cartesian product $G = \bar{\prod\limits_{i\in I}} G_i$ of the family of l-groups $\{G_i | i \in I\}$. Then there is an ultrafilter $\mathcal{F}$ over I such that $T_{\mathcal{F}} \subseteq \bigcap\limits_{g\in G} M^g$.*

Proof. Let $H = \bigcap\limits_{g\in G} M^g$ in G. By Theorem 4.4.1, $\bar{\prod\limits_{i\in I}} G_i/H$ has a transitive representation by order permutations of a totally ordered set. Therefore, G/H is finitely subdirectly irreducible and by arguments of Theorem 9.1.3, we have an ultrafilter $\mathcal{F}$ over I such that $T_{\mathcal{F}} \subseteq H$.□

9.4 Weakly Abelian l-Groups

1⁰. Let us recall that l-group G is *weakly abelian* if the identity

(1) $$x^{-1}|y|x|y|^2 \vee e = e$$

in G. It is clear that the identity (1) is equivalent to inequalitiy

(2) $$[x, |y|^{-1}] \le |y|,$$

or

(3) $$|[x, y]| \ll |x|$$

for all $x, y \in G$. ($|a| \ll |b|$ denotes $|a|^n \leq |b|$ for all positive integers n.)

The l-variety of all weakly abelian l-groups is denoted by $\mathcal{W}a$. We begin with some basic properties regarding weakly abelian l-groups.

PROPOSITION 1. *All convex l-subgroups of weakly abelian l-group G are normal.*□

Proof. Let g be a weakly abelian l-group, H be a convex l-subgroup of g and $a \in H$, $a > e$, $x \in G$. Then $e \leq |[a,x]| \leq a$ and $[a,x] \in H$. Consequently, $x^{-1}ax \in H$.□

PROPOSITION 2. *Each weakly abelian l-group is representable and is a subdirect product of totally ordered weakly abelian l-groups.*□

PROPOSITION 3. *Let G be a weakly abelian l-group and P be a lattice order on g. Then P is the intersection of some set of weakly abelian total orders of G.*□

PROPOSITION 4. *Group G admits weakly abelian lattice order if and only if G has a central series with torsion-free factors.*□

PROPOSITION 5. *If $A \prec B$ is a jump in the lattice of convex l-subgroups of g then $[G,B] \subseteq A$.*□

All these statements easily follow from the results of Section 2 of Chapter 6 and Section 2 of Chapter 8.

2^0. In Chapter 6 it was proved that each total order of a locally nilpotent group is weakly abelian. The following result due to Kopytov is even strouger.

THEOREM 1. (Kopytov [5].) *Each locally nilpotent l-group is weakly abelian.*

Proof. Since G is a weakly abelian l-group if and only if every finitely generated l-subgroup is weakly abelian we can assume that G if finitely generated as an l-group. Since G is a subcartesian product of subdirectly irreducible l-groups which must also be locallly nilpotent, we can assume that G is non-trivial finitely generated subdirectly irreducible locally nilpotent l-group generated by $a_1, \ldots, a_k$ and N is the smallest non-trivial ideal of G. Now we divide the proof of the theorem into several steps.

Step 1. *The centre Z_1 of l-group G is non-trivial.*

Let $A = \mathrm{gp}(a_1, \ldots, a_k)$ be a subgroup of G generated by $a_1, \ldots, a_k$.

Then A is a nilpotent group and each element $g \in G$ can be written in the form $g = \bigvee_{1 \le i \le n} \bigwedge_{1 \le j \le m} h_{ij}$ where $h_{ij} \in A$ vor all i, j, $1 \le i \le n$, $1 \le j \le m$. Let c be a nonidentity central elementt of A. Then

$$cg = c(\bigvee_{1 \le i \le n} \bigwedge_{1 \le j \le m} h_{ij}) = \bigvee_{1 \le i \le n} \bigwedge_{1 \le j \le m} ch_{ij} =$$

$$\bigvee_{1 \le i \le n} \bigwedge_{1 \le j \le m} h_{ij}c = (\bigvee_{1 \le i \le n} \bigwedge_{1 \le j \le m} h_{ij})c = gc$$

for all $g \in G$. Hence the centre Z_1 of G is non-trivial. Let $Z_1^0 = Z_1 \cap N$.

Step 2. *The centre Z_1 of group G is totally ordered.* Suppose for reducto ad absurdum that there is an element $z \in Z_1$ incomparable with e. Then $z^+ = z \vee e$ and $z_-^{-1} = z^{-1} \vee e$ are orthogognal elements of Z_1. Therefore, the polars $X_1 = (z^+)^\perp$ and $Z_2 = (z^+)^{\perp\perp}$ are non-trivial ideals in G. It is clear that $X_1 \cap X_2 = E$, which contradicts the subdirect irreducibility of G.

Step 3. *For every $g \in N$ there exists $z \in Z_1$ such that $|g| \le |z|$.* It is a well-known fact that in each nilpotent group every non-trivial normal subgroup has the non-trivial intersection with the centre of the group (see the book of D.Robinson [1], part 1, Lemma 2.16). Let $b \in N$, $b \neq e$ and H be a subgroup of G generated by b, A. Then H is nilpotent and $N \cap H \neq E$ is normal in H. Therefore, $(N \cap H) \cap Z(H) \neq E$ where $Z(H)$ denotes the centre of H. But $Z(H) \subseteq Z_1$ and so $Z_1^0 = Z_1 \cap N \neq E$. The convex l-subgroup K of l-group G generated by Z_1^0 is an ideal in G. Therefore $N \subset K$. On the other hand, $Z_1^0 \subseteq N$ and $K \subseteq N$. Thus, $K = N$.

Step 4. *The subgroup Z_1^0 is an Archimedean totally ordered group.* Assume, on the contrary that there is a convex subgroup C in Z_1^0 such that $C \neq E$, $C \neq Z_1^0$. Let $B = \{g \in G | \, |g| \le |z| \text{ for some } z \in C\}$. It is clear that B is non-trivial ideal of G. Then $B \cap Z_1^0 = C \neq Z_1^0$, a contradiction to the minimality of N.

Step 5. Let $Z_2^0 = \{a \in N | [a, x_a] \in Z_1^0 \setminus E \text{ for some } x_a \in G\}$. *Then every element of Z_2^0 is comparable with e.* If $a \in Z_2^0$, then $x_a^{-1} a x_a \neq a$ and $c = [a, x_a] \in Z_1^0$. We can assume that $c > e$ (in another case we can consider $c = [a, x_a^{-1}]$). Then $x_a^{-1} x_a = ac$ and $x_a^{-n} a x_a^n = ac^n$ for all integers n. Since $a \in N$, then by Step 3, there is $z \in Z_1^0$ for which $z^{-1} < a < z$. By Step 4, Z_1^0 is Archimedean. So, there exists an integer t such that $c^t > z$. Then $c^{-t} < a < c^t$ from which we get $ac^t > e$, that is $x^{-t}ax^t = ac^t > e$ so $a > e$.

Step 6. *The ideal N coincides with the subgroup Z_1^0.* Suppose not. Then there exist $b \in Z_2^0$, $x \in G$ such that $b > e$, $[b, x] \in Z_1^0$, $[b, x] > e$. By Step 4, there is an integer n such that $e < b < [b, x]^n$. Lattice-ordered group G is locally nilpotent. Therefore, by Theorem 6.5.2, the lattice order of G can be extended to the total order of G. Hence $e < b < [b, x]^n$ in this totall order. But all total orders on locally nilpotent groups are weakly abelian (see Proposition 6.2.1). Thus, inequalities $e < b < [b, x]^n$ are absurd, a contradiction.

Step 7. *Lattice-order group G is a lexicographic extension of ideal N i.e., $ax > e$ for all $a > e$, $a \in G \backslash N$ and $x \in N$.* Let $a > e$, $a \in G \backslash N$ and $z \in N$. Then $a \wedge z = z_1$ for some $z_1 \in N$. Thus $az_1^{-1} \wedge zz_1^{-1} = e$ and $az_1^{-1} \notin N$, $zz_1^{-1} \in N$. By Step 2, it follows that $zz_1^{-1} = e$, $z = z_1$ and $a \wedge z = z$, $z < a$.

As l-group G is the lexicographical extension of ideal N by l-group G/N, then by Proposition 2.3.9 the l-group G/N is totally ordered. Hence l-group G is the lexicographical extension of 0-group N by o-group G/N and G is an o-group, too. By Proposition 6.2.1, G is weakly abelian.□

COROLLARY 1. *All convex l-subgroups of locally nilpotent l-group are normal.*□

COROLLARY 2. *Each lattice order of the locally nilpotent lattice-ordered group G is the intersection of some set of total orders of G and is strongly isolated.*□

COROLLARY 3. *Each finitely generated l-subgroup of locally nilpotent l-group is nilpotent.*

Proof. Let G be a locally nilpotent l-group. Then G is an l-subgroup of the Cartesian product $\bar{\prod}_{i \in I} G_i$ of totally order groups G_i $(i \in I)$. By φ_i we denote a projection of $\bar{\prod}_{i \in I} G_i$ onto G_i. Let now A be a subgroup of G generated by $g_1, \ldots, g_n$ and B be an l-subgroup of G generated by $g_1, \dots, g_n$. Each element $g \in B$ can be represented in the form $g = \bigvee_{1 \le k \le p} \bigwedge_{1 \le m \le q} a_{km}$ where $a_{km} \in A$. Since all G_i, $(i \in I)$, are totally ordered, then $B\varphi_i = A\varphi_i$ for all $i \in I$. The subgroup A is nilpotent of class r. Then all $B\varphi_i$ are nilpotent groups of class $\le r$, too. Therefore, B is a nilpotent group of class r.□

Independently and by different methods, Hollister [1] also proved that every nilpotent l-group is representable.

Now we consider the similar problem for Engel l-groups. Group G is called an *Engel* group if for each pair of elements $x, y \in G$ there is an integer $t = t(x, y) > 0$ such that

$$(3) \qquad [x, \underbrace{y, \dots, y}_{t}] = e.$$

If identity (3) holds in group G for some positive integer t, then G is called a t-Engel group.

LEMMA 1. *Let G be an l-group and $x, y \in G$ such that $x > e$ and $x, x^y, \dots, x^{y^t}$, $(t \ge 1)$ are mutually orthogonal. Then in G*

$$[x, \underbrace{y, \dots,}_{t}] \ne e.$$

Proof. Evidently, each commutator $[x, \underbrace{y, \dots, y}_{t}]$ is represented as the product of the elements $x^{\pm 1}, (x^{\pm 1})^y, \dots, (x^{\pm 1})^{y^t}$ and the elements $x^{\pm 1}$ have only one occurence in this product. It is clear that elements $x^{\pm 1}, (x^{\pm 1})^y, \dots, (x^{\pm 1})^{y^t}$ are mutually commutative and linearly independent over $\mathbf{Z}$. Therefore, $[x, \underbrace{y, \dots, t}_{t}] \ne e$ in G.□

For any pair $x, y \in G$, $x > e$, let $M(x, y) = \{x \wedge y^{-i_1} x y^{i_1} \wedge \dots \wedge y^{-i_k} x y^{i_k} | i_1, \dots, i_k \in \mathbf{Z},\ k \ge 1\}$.

THEOREM 2. (Medvedev [12].) *Let G be an Engel l-group. Then G is representable if and only if for every $x, y \in G$, $x > e$ there exists a positive integer $t = t(x, y)$ such that the equality $[u, \underbrace{y, \dots, y}_{t}] = e$ holds in G for all $u \in M(x, y)$.*

Proof. If G be a totally ordered group, then $u = x \wedge y^{-i_1}xy^{i_1} \wedge \cdots \wedge y^{-i_k}xy^{i_k} = \min\{x, y^{-i_1 x y^{i_1}}, \ldots, y^{-i_k}x\eta^{i_k}\} = y^{-i_p}xy^{i_p}$ for some $i_p \in \mathbf{Z}$. Then

$$[u, \underbrace{y, \ldots, t}_{t}] = [y^{-i_p}xy^{i_p}, \underbrace{y, \ldots, y}_{t}] == y^{-i_p}[x, \underbrace{y, \ldots, y}_{t}]y^{i_p} = e$$

since $[x, \underbrace{y, \ldots, y}_{t}] = e$. Therefore, if G is an Engel representable l-group, then G is subcartesian product of totally ordered Engel group and by above arguments, the equality $[u, \underbrace{y, \ldots, y}_{t}] = e$ holds for all $u \in M(x, y)$.

Let now G be not representable and $[u, \underbrace{y, \ldots, y}_{t}] = e$ for all $u \in M(x, y)$, where $t = t(x, y)$. Then there are elements $x, y \in G$, $x > e$ such that $x \wedge x^y = e$ By Lemma 1, the set of mutually orthogonal elements $u, y^{-1}uy, \ldots, y^{-k}uy^k$ contain less or equal to $t = t(x, y)$ of different elements. Therefore, there exists an element $u_0 \in M(x, y)$, $u_0 > e$ for which the set of mutually orthogonal elements contains maximal number $k_0, k_0 \leq t = t(x, y)$ of different elements. Then for every positive integer n the element $w(n) = u_0 \wedge y^{-k_0}u_0y^{k_0} \wedge \cdots \wedge y^{-k_0 n}u_0 y^{k_0 n} \neq e$, $w(n) \in M(x, y)$ because otherwise for the element $w(n-1) = u_0 \wedge y^{-k_0}u_0y^{k_0} \wedge \cdots \wedge y^{-k_0(n-1)}u_0y^{k_0(n-1)} \neq e$, $w(n-1) \in M(x, y)$ the set of mutually orthogonal elements $w(n-1), y^{-1}w(n-1)y, \ldots, y^{-k_0}w(n-1)y^{k_0}$ contains $k_0 + 1$ different elements. Now we consider the subset $M(u_0, y^{k_0}) \subseteq M(x, y)$. Then for every $w(n) \in M(u_0, y^{k_0})$ the equality

$$[w(n), \underbrace{y^{k_0}, \ldots, y^{k_0}}_{t(u_0, y^{k_0})}] = e$$

is valid. Now we fix any positive integer $n_0 > k_0 t(u_0, y^{k_0})$, $n_0 > t(x, y)$ and take the least positive integer $t_0, t_0 \leq t(u_0, y^{k_0})$ such that

$$[w(n_0), \underbrace{y^{k_0}, \ldots, y^{k_0}}_{t_0}] = e \text{ and } [w(n_0), \underbrace{y^{k_0}, \ldots, y^{k_0}}_{t_0 - 1}] = z \neq e,$$

then the element z is the product of the elements

$$w(n_0^{\pm 1}), y^{-k_0}w(n_0)^{\pm 1}y^{k_0}, \ldots, y^{-k_0(t_0-1)}w(n_0)y^{k_0(t_0-1)}.$$

All these elements are mutually orthogonal to the elements

$$w(n_0^{\pm 1}), y^{-i}w(n_0)^{\pm 1}y^i, \ldots, y^{-k_0(t_0-1)-i}w(n_0)y^{k_0(t_0-1)+i},$$

$1 \leq i \leq k_0 - 1$. Hence the elements $|z|, y^{-1}|z|, \ldots, y^{-k_0+1}|z|y^{k_0-1}$ are mutually orthogonal, too, and $y^{-k_0}|z|y^{k_0} = |z|$. Let now $H = \mathrm{gp}(z, y)$ be a subgroup of G generated by z and y. It is easy to see that subgroup H is isomorphic to the group

$$G_{k_0} = \mathrm{gp}(a_1, \ldots, a_{k_0}, b \| [a_i, a_j] = e \text{ for } 1 \leq i, j \leq k_0, \\ b^{-1}a_p b = a_q \text{ where } p \equiv q \pmod{k_0}).$$

But the direct calculations show that group G_{k_0}, $k_0 \geq 2$, is not an Engel group.□

COROLLARY 1. *Let G be an n-Engel l-group. Then G is representable.*

3^0. In the theory of groups it is known that every variety of nilpotent groups of the class $\leq k$ is generated by some group with $k+1$ generators (see for example the book of H.Neumann [1], Theorem 35.11). The similar result is not true for l-varieties.

THEOREM 3. (Kopytov [7].) *Each l-variety $\mathcal{N}_k$ of all nilpotent l-groups of the class $\leq k$, $(k \geq 2)$, cannot be generated by any finitely generated l-group.*

Proof. For any l-group G and $x, y_i, z_i \in G$, $i \in \mathbf{N}$ let

$$c_1(x, y_1, z_1) = |[|x|, y_1]| \wedge |z_1|, c_{k+1}(x, y_1, z_1 \ldots, y_{k+1}, z_{k+1}) = \\ = c_1(c_k(x, y_1, z_1 \ldots, y_k, z_k), y_{k+1}, z_{k+1}).$$

It is evident that for each jump $A \prec B$ in the lattice $\mathcal{C}(G)$ of convex l-subgroups of the weakly abelian l-group G $c_1(x, y, z) \in A$ for all $x \in B$, $y, z \in G$ and the convex l-subgroup of G generated by the elements $c_1(x, y, z)$ where $x \in B$, $y, z \in G$, is contained in A.

Let now $\mathcal{X}$ be any l-variety of nilpotent l-groups of the class $\leq k$ generated by some l-group G with n generators $x_1, \ldots, x_n$. By Theorem 1, l-group G is the subcartesian product of totally ordered groups

$G_\alpha = \mathrm{gp}(x_1^\alpha, \ldots, x_n^\alpha)$, $\alpha \in I$, where every G_α is an l-homomorphic image of G. Therefore, $\mathcal{X}$ is generated by all totally ordered groups G_α, $\alpha \in I$. We claim that there is an identity which distingnishes l-varieties $\mathcal{X}$ and $\mathcal{N}_k$.

Each n-generated totally ordered group G_α (in the signature of the theory of groups $\{\ \cdot\ ,\ ^{-1}\ ,\ e\ \}$) is a nilpotent group of the class $\leq k$. Therefore, the special rank of G is bounded above by some positive integer $r = r(n,k)$. It is clear that $r = r(n,k)$ is less or equal to the special rank of free n-generated nilpotent group of class $\leq k$. Hence each totally ordered group G_α, $\alpha \in I$, has less or equal to $r(n,k)+1$ different convex l-subgroups. Thus the identity $c_r(x, y_1, z_1, \ldots, y_r, z_r) = e$ is valid in all totally ordered groups G_α and therefore holds in all l-groups of l-variety $\mathcal{X}$.

Now, let $H_i = \mathrm{gp}(a_i, b_i, c_i \mid [a_i, b_i] = c_i,\ [a_i, c_i] = [b_i, c_i] = e)$ for $i = 1, 2, \ldots, r+1$. Then each $x \in H_i$ has the unique normal form $x = a_i^{p_i} b_i^{q_i} c_i^{s_i}$ where $p, q, s \in \mathbf{Z}$. Let us totally order H_i by calling an element positive iff $p > 0$ or $p = 0, q > 0$ or $p = q = 0, s > 0$. Let $H = \overrightarrow{\prod\limits_{1 \leq i \leq r+1}} H_i$ be the lexicographical product of totally ordered groups H_i. Then for $x \in H$, $x = h_1 h_2 \cdots h_{r+1}$ holds $x > e$ iff $h_j > e$ in H_j where j is the least index i such that $h_i \neq e$. So, H is a nilpotent group of the class 2 and $H \in \mathcal{N}_k$ for all $k \geq 2$. Let M_i be a convex l-subgroup of H generated by $H_i, H_{i+1}, \ldots, H_{r+1}$. Then $H = M_1 \supset \cdots \supset M_i \supset \cdots \supset M_{r+1} = E$ and all M_i $(i = 1, 2, \ldots, r+1)$ are convex in the totally ordered group H. Direct calculations show that $M_i \supseteq C_i \supseteq M_{i+1}$, $C_i \neq M_i, M_{i+1}$ for every i, $1 \leq i \leq r$ where C_i is a convex subgroup of H generated by all values of the l-word $c_i(x, y_1, z_1, \ldots, y_i, z_i)$ when x, y_i, z_i pass through G. Therefore, the identity $c_r(x, y_1, z_1, \ldots, y_r, z_r) = e$ is not valid in the totally ordered group H and also in $\mathcal{N}_k$ for $k \geq 2$.□

9.5 Torsion and Radical Classes

1⁰. A non-empty class of l-groups $\mathcal{F}$ is a *radical class* of l-groups if it is closed with respect to

(K.1) convex l-subgroups;
(K.2) l-isomorphisms;
(K.3) for every l-group G, forming arbitrary joins of convex l-subgroups of G which belong to the class $\mathcal{F}$.

Therefore, for any radical class $\mathcal{F}$ and every l-group G there is the greatest convex l-subgroup $\mathcal{F}(G) \subseteq G$ such that $\mathcal{F}(G) \in \mathcal{F}$. The convex l-subgroup $\mathcal{F}(G)$ is called *$\mathcal{F}$-radical* of G. The radical class of l-groups $\mathcal{F}$ is called a *torsion class* if it is closed under l-homomorphisms.

PROPOSITION 1. *Each torsion class $\mathcal{F}$ is radical class.*□

PROPOSITION 2. *For every radical class $\mathcal{F}$ the $\mathcal{F}$-radical $\mathcal{F}(G)$ of G is an l-characteristic ideal in G, i.e., $\mathcal{F}(G)\varphi = \mathcal{F}(G)$ for all l-automorphisms φ of l-group G.*□

It is not difficult to construct an example of l-group G and torsion class $\mathcal{T}$ for which $\mathcal{T}$-radical $\mathcal{T}(G)$ is not a fully invariant l-subgroup of G, i.e., there is an l-isomorphism φ of G into G such that $\mathcal{T}(G)\varphi \neq \mathcal{T}(G)$.

PROPOSITION 3. (Martinez [4].) *Let $\mathcal{F}$ be a torsion class and G be an l-group. Then*
(1) *if A is a convex l-subgroup of G, then $\mathcal{F}(A) = \mathcal{F}(G) \cap A$;*
(2) *if φ is an l-homomorphism of l-group G onto l-group H, then $\mathcal{F}(G)\varphi = \mathcal{F}(H)$;*
(3) *$\mathcal{F}(\mathcal{F}(G)) = \mathcal{F}(G)$;*
(4) *if $\{A_\alpha | \alpha \in I\}$ is a family of convex l-subgroups of G, then $\mathcal{F}(\overset{c}{\bigvee_{i\in I}} A_\alpha) = \overset{c}{\bigvee_{\alpha\in I}} \mathcal{F}(A_\alpha)$.*

Proof. Properties (1)–(3) are clear. Property (4) follows from the property of the lattice of convex l-subgroups (Theorem 3.1.2).□

PROPOSITION 4. (Martinez [4].) *Let $\bar{\mathcal{F}}$ be an operator which compares for any l-group G an ideal $\mathcal{F}(G)$ with properties (1)–(2) of Proposition 3. Then the class $\mathcal{F}$ of l-groups G such that $\mathcal{F}(G) = G$ is a torsion class and for any l-group G the ideal $\mathcal{F}(G)$ is $\mathcal{F}$-radical of G.*

Proof follows from Proposition 3.□

2⁰. LEMMA 1. (Byrd and Lloyd [1], Martinez [4].) *The l-variety $\mathcal{N}$ of normal-valued l-groups is a torsion class.*

Proof. It is clear that $\mathcal{N}$ is closed under convex l-subgroups and l-homomorphisms. So we need to prove that for any l-group G arbitrary join (in the lattice $\mathcal{C}(G)$ of all convex l-subgroups of G) of normal-valued convex l-subgroups is a normal-valued l-group.

It is clear that in any l-group G the join of ascending series of normal-valued subgroups is also normal-valued. Therefore, it is enough to prove that if A, B are normal-valued convex l-subgroups, then $A \overset{c}{\bigvee} B$ is a normal-valued convex l-subgroup.

Let now $T \prec T^*$ be a jump in the lattice $\mathcal{C}(C)$ of convex l-subgroups $C = A \overset{c}{\bigvee} B$ and $t \in T^* \setminus T,\ t > e$. By Theorem 3.1.1, $t = a_1 b_1 \cdots a_n b_n$ where $a_i \in A,\ b_i \in B$ for $i = 1.2, \ldots n$. From $t \notin T$ it follows that $a_k \notin T$ or $b_k \notin T$ for some $k, 1 \leq k \leq n$. Let $a_k \notin T$. Then $A \cap T \prec A \cap T^*$ is a jump in the lattice $\mathcal{C}(A)$ of convex l-subgroups of A and $a_k \in (A \cap T^*) \setminus (A \cap T)$. By our assumption, $A \in \mathcal{N}$, hence $A \cap T$ is normal in $A \cap T^*$. Since $T = (T \cup A) \overset{c}{\bigvee} (T \cap B),\ T^* = (T^* \overset{c}{\bigvee} \cap B),\ T^* \cap A \neq (T \cap A)$, T^* is a minimal convex l-subgroup of G containing T and a_k then $T^* \cap B = T \cap B$.

Let now $a \in T^* \cap A,\ b \in T \cap B,\ a > e,\ b > e$. Then

$$c = a \wedge b \in (T^* \cap A) \cap (T \cap B) \subseteq T \cap A \cap B \subseteq T \cap A,$$
$$ac^{-1} \wedge bc^{-1} = e,\ (ac^{-1})^{-1} bc^{-1} ac^{-1} = bc^{-1},\ ca^{-1} bc^{-1} ac^{-1} = bc^{-1}$$

and $a^{-1} baa^{-1} ca = c^{-1} a,\ a^{-1} ba = c^{-1} ba^{-1} c^{-1} a$. The subgroup $T \cap A$ is normal in $T^* \cap A$, then $a^{-1} c^{-1} a \in T \cap A$. Similarly $aba^{-1} \in T$. Therefore, $a^{-1}(T \cap B)a \subseteq T$ for all $a \in A,\ a > e$. Since $T^* = (T^* \cap A) \overset{c}{\bigvee} (T \cap B)$ and $T \cap A$ is normal in $T^* \cap A$ it follows that T is normal in T^*.□

LEMMA 2. (Holland [9].) *Let G be a subdirectly irreducible normal-valued l-group generated by elements $g_1, \ldots, g_n$. Then G coinsides with the convex l-subgroup $(g_k)_l$ generated by one element g_k for some k, $1 \leq k \leq n$.*

Proof. Let C be the value of some element of the minimal ideal N, $N \neq E$ of G. Then $g_i \notin C$ for some i, $1 \leq i \leq n$. Let K be the largest number of the non-empty finite chain

$$\{M | C \subseteq M,\ M \text{ is value of some } g_i\}.$$

Then K is the value of g_k for some k, $1 \leq k \leq n$. K is normal in its cover K^*, as well. It is clear that $K^* = G$ and G/K is l-isomorphic to a subgroup of the Archimedean ordered group of reals. Moreover, G is a lexicographic extension of K. On the contrary, suppose that there are incomparable positive elements $h \in K$ and $g \in G \setminus K$. We can assume that $h \wedge g = e$ and $h > e$, $g > e$. Since $\bigcap_{x \in G} x^{-1}Cx$ is an ideal of G which does not contain the minimal ideal N, it must be that $\bigcap_{x \in G} x^{-1}Cx = E$. Hence there exists $x \in G$ such that $h \notin x^{-1}Cx$. But C is a prime l-subgroup, hence $x^{-1}Cx$ must be prime. As $g \wedge h = e$, $h \notin x^{-1}Cx$, then $g \in x^{-1}Cx$. But $C \subseteq K$ and is an ideal of G. Therefore, $x^{-1}Cx \subseteq K$ and $g \in K$, which contradicts the choice of g. This means that G is a lex-extension of K. Since $g_k \notin K$ and G/K is an Archimedean o-group, it follows that $G = (g_k)_l$.□

THEOREM 1. (Holland [14].) *Every l-variety is a torsion class.*

Proof. The first two properties in the definition of torsion class obviously hold for any l-variety. To verify the third property we assume that l-variety $\mathcal{X} \neq \mathcal{L}$; G is an l-group; $\{A_\alpha | \alpha \in \Lambda\}$ is the family of convex l-subgroups of G and each A_α satisfies the nontrivial identity $w(x_1, \ldots, x_n) = e$ of l-variety $\mathcal{X}$. Let now $A = \overset{c}{\bigvee_{\alpha \in \Lambda}}$ be the convex l-subgroup of G generated by all subgroups $A_{\alpha,\, \alpha\in\Lambda}$. By Theorem 1, A is a normal-valued l-group. Let $a_1, \ldots, a_n \in A$. Then $a_i = b_{i1}b_{i2} \cdots b_{im}$, $1 \leq i \leq n$, $1 \leq j \leq m$ where $b_{ij} \in \bigcup_{\alpha \in \Lambda} A_\alpha$. Let H be the l-subgroup of A generated by the elements b_{ij}, $1 \leq i$, $1 \leq j \leq m$. As an l-subgroup of normal-valued l-group H is also normal-valued.

By Theorem 9.1.2, H is the subcartesian product of subdirectly irreducible l-groups. Let $\bar{H}$ be any subdirectly irreducible factors of H and let $\bar{g} \in \bar{H}$ denote the image of $g \in H$ under the natural map from H onto $\bar{H}$. Then $\bar{H}$ is normal-valued and generated by $\{\bar{b}_{ij}, |1 \leq i \leq n,\ 1 \leq j \leq m\}$. Therefore, by Lemma 2, $\bar{H} = (b_{ks})_l$ for some k, s, $1 \leq k \leq n$, $1 \leq s \leq m$. Since $b_{ks} \in A_\alpha$ for some $\alpha \in \Lambda$ and since the image $\overline{H \cap A_\alpha}$ is a convex l-subgroup of $\bar{H}$, then $\bar{H} = \overline{H \cap A_\alpha}$. A_α satisfies the identity $w(x_1, \ldots, x_n) = e$, so does $\overline{A_\alpha \cap H} = \bar{H}$.

Finally, H is the subcartesian product of subdirectly irreducible factors, each for which satisfies to $w(x_1, \ldots, x_n) = e$ and therefore so does H. In particular, $w(a_1, \ldots, a_n) = e$ and A also satisfies identity $w(x_1, \ldots, x_n) = e$.□

3⁰. Now we give a list of popular torsion classes which are not l-varieties.

1) $\mathcal{Z}$, the class of direct products of the copies of integers.
2) $\mathcal{R}e$, the class of direct products of the copies of the subgroups of reals.
3) $\mathcal{D}a$, the class of all divisible Abelian l-groups.
4) $\mathcal{O}$, the class of the direct products of o-groups.
5) $\mathcal{V}_{\aleph}$, the class of l-groups in which each non-identity element exceed one having at most $\aleph$ different values (here $\aleph$ is an infinite cardinal)
6) $\mathcal{V}_f$, the class of *finite-valued l-groups.* An l-group belongs to $\mathcal{V}_f$ iff each non-identity element determines the finite number of values.

The proof that classes 1)–6) are torsion classes is straightforward and we leave the details for the reader.

PROPOSITION 5. (Medvedev.) *The class loc$\mathcal{N}$ of all locally nilpotent l-groups is a torsion class.*

Proof. It is obvious that the class $loc\mathcal{N}$ is closed under convex l-subgroups and l-homomorphisms. Therefore, we need only to prove that for any $G \in loc\mathcal{N}$ and for arbitrary family $\{C_\alpha | \alpha \in A\}$ of convex l-subgroups of G such that $c_\alpha \in loc\mathcal{N}$, $\overset{c}{\bigvee\limits_{\alpha \in A}} C_\alpha \in loc\mathcal{N}$. By Theorem 9.4.1, all C_α $(\alpha \in A)$ are contained in $\mathcal{W}a$-radical $\mathcal{W}a(G)$ of l-group G and $C = \overset{c}{\bigvee\limits_{\alpha \in A}} C_\alpha \subseteq \mathcal{W}a(G)$.

Let $K \subseteq C_{\alpha_1} \overset{c}{\vee} C_{\alpha_2} \overset{c}{\vee} \cdots \overset{c}{\vee} C_{\alpha_n}$ for some positive integer n. Convex l-subgroups C_{α_i} in weakly abelian l-group C are normal in C. Therefore, $C_{\alpha_1} \overset{c}{\vee} C_{\alpha_2} \overset{c}{\vee} \cdots \overset{c}{\vee} C_{\alpha_n} = C_{\alpha_1} \cdot C_{\alpha_2} \cdots C_{\alpha_n}$. It is a well-known group-theoretical fact that the product of locally nilpotent normal subgroups is locally nilpotent. Therefore, C is a locally nilpotent group. So K is a nilpotent subgroup of G and the class $loc\mathcal{N}$ of locally nilpotent l-groups is a torsion class.□

It is not hard to see that the set of torsion classes is a lattice when partially ordered by inclusion. By Theorem 1, this lattice contains the lattice **L** of l-varieties. However, J.Smith [1] has shown that the lattice of l-varieties is not a sublattice of the lattice of torsion classes.

Torsion and radical classes have abttracted considerable attention in their own right and the interested reader can find more information on this topic in the works by W.Ch.Holland and J.Martinez [1], J.Jakubik [2], J.Martinez [4],[5].

4^0. Let $\mathcal{X}$ be an l-variety and G be an arbitrary lattice ordered group and $\mathcal{X}(G)$ be $\mathcal{X}$-radical of G. The convex l-subgroups $\mathcal{X}(G)$ do have an interesting closure property which follows from the following result due to Bernau [3].

Let us recall that l-subgroup H of l-group G is closed in G if for all subsets $\{x_i|i \in I\}$ of H such that $a = \bigvee\limits_{i \in I} x_i$ exists in G we have $a \in H$. The *closure* $\bar{H}$ of H is the smallest l-subgroup of G containing H which is closed. Clearly $\bar{H}$ is the intersection of all convex l-subgroups of G containing H.

THEOREM 2. (Bernau [3]) *Let $\mathcal{X}$ be an l-variety and H be an l-subgroup of l-group G. If $H \in \mathcal{X}$, then so is $\bar{H}$.*

Proof. (Reily, private communication) Let K be the l-subgroup of G which is generated by the supreme of subsets of H. It will suffice to show that $K \in \mathcal{X}$. It is easily seen that any element of K can be written in the form

$$a = \bigvee_{i \in I} \bigwedge_{j \in J} h_{ij}, \tag{1}$$

where the index sets I and J are finite and where h_{ij} is the form

$$dh_{ij} = (\bigvee M_{ij1})^{\varepsilon_{ij1}} \cdots (\bigvee M_{ijn(ij)})^{\varepsilon_{ijn(ij)}} \tag{2}$$

for some $\varepsilon_{ijk} = \pm 1$ and some subsets M_{ijk} of H for which there exist suprema in G.

Let $w(x_1, \ldots, x_n) = e$ be any identity that holds in $\mathcal{X}$. If we can show that K satisfies this identity, then $K \in \mathcal{X}$. Consider the statement

$$S(k); \quad w(a_1, \ldots, a_n) = e \text{ if } a_1, \ldots, a_k \in K,\ a_{k+1}, \ldots, a_n \in H.$$

Then K satisfies the identity $w(x_1, \ldots, x_n) = e$ iff $S(n)$ holds and we will estabilish that $S(n)$ holds by induction. The statement $S(0)$ holds since this just means that $w(x_1, \ldots, x_n) = e$ holds in H. For the sake of notational simplicity, we will only estabilish that $S(1)$ holds, the proof of the general step $S(k) \rightarrow S(k+1)$ being entirely similar. So, let $a_1 = a$ be of the form described in (5)and (6) and $a_2, \ldots, a_n \in H$. For each i, j, k let m_{ijk} represent the supremum of the arbitrary finite subset of M_{ijk} and let $a_1^* = \bigvee_{i \in I} \bigwedge_{j \in J} m_{ij1}^{\varepsilon_{ij1}} \cdots m_{ijn(ij)}^{\varepsilon_{ijn(ij)}}$. Then $a_1^* \in G$ so that $w(a_1^*, a_2, , \ldots, a_n) = e$. As before, we may take w to be in the form

$$w(x_1, \ldots, x_n) = \bigvee_{\alpha \in A} \bigwedge_{\beta \in B} w_{\alpha\beta}(x_1, \ldots, x_n)$$

where $w_{\alpha\beta} = x_{\alpha\beta 1}^{\varepsilon 1} \cdots x_{\alpha\beta k}^{\varepsilon_k}$ $(x_{\alpha\beta k} \in \{x_1, \ldots, x_n\},\ \varepsilon_i = \pm 1)$.

Let $b_{\alpha\beta j} = a_{\alpha\beta j}$ if $x_{\alpha\beta j} \neq x_1$ and $b_{\alpha\beta j} = a_1^*$ if $x_{\alpha\beta j} = x_1$. Then $w(a_1^*, a_2, \ldots, a_n) = \bigvee_{\alpha \in A} \bigwedge_{\beta \in B} w_{\alpha\beta}^* = e$ where $w_{\alpha\beta}^* = b_{\alpha\beta 1}^{\varepsilon_1} \cdots b_{\alpha\beta k}^{\varepsilon_k}$. We now replace certain terms in $w_{\alpha\beta}^*$ by others, but always in such a way as to obtain a smaller elements of G than $w_{\alpha\beta}^*$.

If $x_{\alpha\beta\gamma} \neq x_1$, then the term $b_{\alpha\beta\gamma}^{\varepsilon_\gamma}$ is not changed. Now consider $b_{\alpha\beta\gamma}^{\varepsilon_\gamma}$ where $x_{\alpha\beta\gamma} = x_1$. Then $b_{\alpha\beta\gamma} = a_1^*$.

a) If $\varepsilon_\gamma = 1$ and $\varepsilon_{ijk} = -1$ or $\varepsilon_\gamma = -1$ and $\varepsilon_{ijk} = 1$ replace $m_{ijk}^{\varepsilon_{ijk}}$ in $b_{\alpha\beta\gamma}^{\varepsilon_\gamma}$ by $(\bigvee M_{ijk}^{\varepsilon_{ijk}})$.

b) If $\varepsilon_\beta = 1$ and $\varepsilon_{ijk} = 1$ or $\varepsilon_\beta = -1$ and $\varepsilon_{ijk} = -1$ replace $m_{ijk}^{\varepsilon_{ijk}}$ in $b_{\alpha\beta\gamma}^{\varepsilon_\gamma}$ by $g^{\varepsilon_{ijk}})$ where g is any element of M_{ijk}.

Denote the element so obtained from $b_{\alpha\beta j}$ by $c_{\alpha\beta j}$ and the element so obtained from $w_{\alpha\beta}^*$ by $u_{\alpha\beta}^*$. Then the changes (5)and (6) above ensure that $c_{\alpha\beta j}^{\varepsilon_j} \leq b_{\alpha\beta j}^{\varepsilon_j}$ and so $u_{\alpha\beta}^* \leq w_{\alpha\beta}^*$.

Now $w(a_1, \ldots, a_n) = \bigvee_{\alpha} \bigwedge_{\beta} w_{\alpha\beta}(a_1, \ldots, a_n)$ where

$$w_{\alpha\beta}(a_1, \ldots, a_n) = a_{\alpha\beta 1}^{\varepsilon_1} \cdots a_{\alpha\beta k}^{\varepsilon_k} = \bigvee u_{\alpha\beta}^*$$

and the join is taken over all the elements from M_{ijk} introduced in (2) above. This holds, since each $u_{\alpha\beta}^*$ only differs from $w_{\alpha\beta}$ in that whenever a_1 appears in $w_{\alpha\beta}(a_1, \ldots, a_n)$ and ε_j and ε_{ijk} are either both positive or both negative, then $\bigvee M_{ijk}$ is replaced by a single element of

M_{ijk}. But then taking joins over all choices of elements from M_{ijk} and employing the distributive law $\bigvee x(\bigvee y_\lambda)z = \bigvee xy_\lambda z$ (see Proposition 8.1.1), whenewer $\bigvee y_\lambda$ exists, achieves the desired result. Hence

$$w(a_1, \ldots, a_n) = \bigvee_\alpha \bigwedge_\beta w_{\alpha\beta}(a_1, \ldots, a_n) =$$
$$= \bigvee(\bigvee_\alpha \bigwedge_\beta u^*_{\alpha\beta}) \leq \bigvee(\bigvee_\alpha \bigwedge_\beta w^*_{\alpha\beta}) = \bigvee w(a^*_1, a_2, \ldots, a_n) = e.$$

Appling the same technique to w^{-1} we have that $w^{-1}(a_1, \ldots, a_n) \leq e$ and so $w(a_1, \ldots, a_n) = e.\square$

If H is a convex l-subgroup of G, then by Lemma 8.2.3, $\bar{H} = K$ and $\bar{H}$ is a convex l-subgroup of G, too. Therefore, the following result is true.

COROLLARY 1. *For any l-variety $\mathcal{X}$ and any l-group G $\mathcal{X}$-radical of G is closed in G.*$\square$

9.6 Finite Basis Problem

Each l-variety is completely determined by the set of identities that hold in the l-variety. Since there are only a countable infinite number of identities there can be at most $2^{\aleph_0}$ distinct l-varieties. Therefore, one of the central problems in the theory of varieties of l-groups is the following finite basis problem: does every l-variety have a finite basis of identities? Here we consider different solutions of this problem.

Olshanskii [1] showed that there is a continuum $\{\mathcal{X}_\alpha | \alpha \in A\}$ of the varieties of groups that are soluble of lenght 4. Vanghau-Lee [1] constructed also a continuum of the varieties of groups of lenght 3. Here we will apply these results to l-varieties.

THEOREM 1. (Kopytov and Medvedev [2].) *There exists a continuun of l-varieties which are deretmined by identities of the group theory.*

Proof. Let $F(X)$ be a free group with infinite countable set X of free generators and $\{\mathcal{X}_\alpha | \alpha \in A\}$ be any continuum of soluble varieties of groups. Let N_α be a fully invariant subgroup of free group $F(X)$ corresponding to $\mathcal{X}_\alpha$, $(\alpha \in A)$. Then for every $\alpha \in A$ the subgroup

$N''_\alpha = [[N_\alpha, N_\alpha], [N_\alpha, N_\alpha]]$ is also fully invariant and defines the $\mathcal{Y}_\alpha$ which is a product of the variety of metabelian groups and the variety $\mathcal{X}_\alpha$. But the semigroup of varieties of groups is free and has a cancellation property (see the book of H.Neumann [1], Theorem 23.3D). Therefore, all varieties $\mathcal{Y}_\alpha$ are distinct. Moreover, by Theorem 5.3.3, the groups $F(X)/N''_\alpha$ can be totally ordered and $F(X)/N''_\alpha \in \mathcal{Y}_\alpha \setminus \mathcal{Y}_\beta$ for all $\alpha \neq \beta$. These totally ordered groups $F(X)/N''_\alpha$ are distingnished by the identities of the signature of the groups theory. Therefore, the l-varieties generated by the o-group $F(X)/N''_\alpha$ are distinct and provide a continuum of l-varieties.□

COROLLARY 1. *There exists a representable soluble l-variety of lenght 6 which is not finitely based.*□

2⁰. Here we present Reily's solution of the finite basis problem. Reily in [2] presented the correspondence between varieties of groups and l-varieties.

Let $X = \{x_1, \ldots, x_n, \ldots,\}$ be a countable set of variables and let $X' = X \cup \{z\}$ where $z \notin X$. Let $F = F(X)$ be a free group with the set of free generators X which will be considered as a subset of F. For any l-variety $\mathcal{X}$ we define

$$r\mathcal{X} = \{u \in F | z^+ \wedge u^{-1}(z^-)^{-1}u = e \text{ is an identity in } \mathcal{X}\}$$

and for any l-group G we define

$$rG = \{u \in F | z^+ \wedge u^{-1}(z^-)^{-1}u = e \text{ is an identity in } G\}$$

Clearly, $rG = r\mathrm{var}_l(G)$ and $r\mathcal{X} = \bigcap_{G \in \mathcal{X}} rG$. We sometimes write $l(u) = z^+ \wedge u^{-1}(z^-)^{-1}u$.

Let φ be a mapping from X' into group G. Then the restriction of φ on X extends to the unique homomorphism F into G. Therefore, φ determines the unique mapping of $\{z\} \cup F$ into G which is a homomorphism on F and the restriction of which to X' is just φ. We denote this extension by φ, too.

THEOREM 2. (Reily [2].) *For any l-group G (or l-variety $\mathcal{X}$) rG (respectively $r\mathcal{X}$) is a fully invariant subgroup of F.*

Proof. Let $u, v \in rG$. Then $l(u) = e$, $l(v) = e$ are identities in l-group G. Hence for any mapping $\varphi : X' \to G$ with $z\varphi = g$, $u\varphi =$

h, $v\varphi = k$, $g^+ \wedge h^{-1}(g^-)^{-1}h = e$. Let $f = g^+ \wedge h^{-1}(g^-)^{-1}h$. Then $f^+ = g^+$ and $f^- = h^{-1}g^-h$. Since $l(v) = e$ is the identity in G, we have $f^+ \wedge k^{-1}(f^-)^{-1}k = e$. Equivalently, $g^+ \wedge (hk)^{-1}(g^-)^{-1}hk = e$, that is $z^+\varphi \wedge (uv\varphi)^{-1}(z^-\varphi)^{-1}(uv)\varphi = e$. Hence $l(uv) = e$ is the identity in G and $uv \in rG$. Furthemore, for any $g \in G$ is valid $(g^{-1})^+ \wedge u^{-1}((g^{-1})^-)^{-1}u = e$ or $(g^-)^{-1} \wedge u^{-1}g^+u = e$. Therefore, $(u^{-1})^{-1}(g^-)^{-1}u^{-1} \wedge g^+ = e$ and $u^{-1} \in rG$. Thus rG is a subgroup of F.

We claim now that rG is fully invariant. Let ψ be any endomorphism of F and let $\varphi' : X' \to G$ be any mapping. Define $\theta : X' \to G$ by $z\theta = z\varphi$, $x\theta = x\psi\varphi$ $(x \in X)$. Let $u \in rG$. Then $l(u) = e$ is an the identity in G. Therefore, $(z\theta)^+ \wedge (u\theta)^{-1}((z\theta)^-)^{-1}(u\theta) = e$ that is $(z\varphi)^+ \wedge (u\psi\varphi)^{-1}((z\varphi)^-)^{-1}(u\varphi\psi) = e$. Since φ has been chosen arbitrarily, it follows that $l(u\psi) = e$ is the identity in G and that $u\psi \in rG$. Therefore, rG is a fully invariant subgroup of F. The result for $r\mathcal{X}$ follows immediately. □

Theorem 2 associates a fully invariant subgroup of F with each l-variety. Conversely, each fully invariant subgroup of F may by used to define an l-variety as follows.

Let U be a fully invariant subgroup of F. Then $q\mathcal{R}U$ denotes the l-variety satisfying the identities $l(u) = e$ for all $u \in U$. It is clear that $q\mathcal{R}E = \mathcal{L}$ where $\mathcal{L}$ is the l-variety of all l-groups. It is clear also that $q\mathcal{R}F = \mathcal{R}$ where $\mathcal{R}$ is the l-variety of all representable l-groups. We will call an l-variety $\mathcal{X}$ *quasi-representable* if $\mathcal{X} = q\mathcal{R}U$ for some fully invariant subgroup U of F.

Let F be a free group totally ordered under any total order P and (a) be a totally ordered infinite cyclic group generated by element a. Let $H = (a)\mathrm{Wr}F$ be the wreath product of the groups (a) and F. Then H is semidirect product of the Cartesian product $\bar{A} = \overline{\prod_{f \in F}} (a_f)$ of the family $\{(a_f) | f \in F\}$ of infinite cyclic groups (a_f), $(f \in F)$, and the free group F. The group operation in H is defined by the rule: $f_1\bar{a}_1 \cdot f_2\bar{a}_2 = f_1 f_2 \bar{a}_1^{f_2}\bar{a}_2$ where $\bar{a}_1, \bar{a}_2 \in \bar{A}$, $f_1, f_2 \in F$ and $\bar{a}_1^{f_2}(g) = \bar{a}_1(gf_2)$ for all $g \in F$.

Now we define a lattice order on H by the rule: $x = f\bar{A} \geq e$ in H iff $f > e$ in F in the order P or $f = e$, $\bar{a}(g) \geq e$ in (a_g) for all $g \in F$.

Let now U be a fully invariant subgroup of group F. Then we define the subset L_U of $H = (a)\mathrm{Wr}F$ as follows

$$L_U = \{x\bar{a} | \bar{a}(g) = \bar{a}(h) \text{ if } Ug = Uh\}$$

LEMMA 1. *L_U is an l-subgroup of H.*

Proof. Let $f_1\bar{a}_1,\ f_2\bar{a}_2 \in L_U$. Then $f_1\bar{a}_1 \cdot f_2\bar{a}_2 = f_1f_2\bar{a}_1^{f_2}a_2$. Let $Ug = Uh$. Then $Ugf_2 = Uhf_2$ and so

$$(\bar{a}_1^{f_2}\bar{a}_2)(g) = \bar{a}_1(gf_2)\bar{a}_2(g) = \bar{a}_1(hf_2)\bar{a}_2(h) = \bar{a}_1^{f_2}\bar{a}_2(h).$$

Thus L_U is closed under products. By similar arguments, L_U is closed under inverses and is a subgroup. It can readily be verified that for any $f\bar{a} \in L_U$ is valid

$$f\bar{a} \vee e = \begin{cases} f\bar{a}, & \text{if } f > e \quad \text{in o-group } F; \\ e, & \text{if } f < e \quad \text{in o-group } F; \\ \bar{a} \vee e, & \text{if } f = e \quad \text{in o-group } F. \end{cases}$$

LEMMA 2. *Let U be a fully invariant subgroup of F. Then*
(1) L_U *satisfies the identity $l(u) = e$, if $u \in U$;*
(2) L_U *does not satisfy the identity $l(v) = e$ if $v \notin U$.*

Proof. (1) Let $u \in U$, $\varphi : X' \to L_U$ and $w = z\varphi = f\bar{b}$ where $\bar{b} \in \bar{A}$. If $f \neq e$ then, since F is an o-group, either $f > e$ or $f < e$. This means that either $w^- = e$ or $w^+ = e$ and therefore, $w^+ \wedge (u\varphi)^{-1}w^{-1}u\varphi = e$. Now suppoose that $w = \bar{b}$ and let $x = u\varphi = f\bar{a}$. Let $\pi : L_U \to F$, where $(f\bar{a})\pi = f$. Then the restriction $\varphi\pi_{|F}$ of $\varphi\pi$ on F is an endomorphism of F. Since U is fully invariant it follows that $f \in U$ and also $f^{-1} \in U$. We have

$$x^{-1}(w^-)^{-1}x = (f\bar{a})^{-1}\bar{b}^{-1}f\bar{a} = \bar{a}^{-1}f^{-1}(\bar{b}^-)^{-1}f\bar{a} = ((\bar{b}^-)^f)^{-1}.$$

Since $f^{-1} \in U$, then for any $g \in F$ is valid

$$((\bar{b}^-)^{-1})^f(g) = (\bar{b}^-)^{-1}((gfg^{-1})g) = (\bar{b}^-)^{-1}(g).$$

Thus $supp(((\bar{b}^-)^{-1})^f) = supp((\bar{b}^-)^{-1})$. Hence $w^+ \wedge x^{-1}(w^-)^{-1}x = \bar{b}^+ \wedge ((\bar{b}^-)^{-1})^f = e$. Therefore, $l(u) = e$ is the identity in L_U.

(2). Now let $v \in F \setminus U$. Let $\bar{a} \in \bar{A}$ be defined by the rule:

$$\bar{a}(f) = \begin{cases} a & \text{if } f \in U; \\ a^{-1} & \text{if } f \in U_v; \\ e & \text{otherwise }. \end{cases}$$

Then $w = \bar{a} \in L_U$. Define $\varphi : X' \to L_U$ by $z\varphi = w$, $x\varphi = x$ for all $x \in X$. Then $v\varphi = v$ and

$$((\bar{a}^-)^{-1})^v(e) = (\bar{a}^-)^{-1}(v) = (\bar{a}^-(v))^{-1} = e = \bar{a}^+(e).$$

Thus $\bar{a}^+ \wedge (\bar{a}^- \wedge^{-1} \neq e$ and so $(z\varphi)^+ \wedge (v\varphi)^{-1}((z\varphi)^-)^{-1}v\varphi \neq e$ i.e., $l(v) = e$ is not the identity in L_U.□

THEOREM 3. (Reily [2].) *Let $\mathcal{U}$, $\mathcal{V}$, $\mathcal{X}_\lambda$ $(\lambda \in \Lambda)$ be l-varieties and U, V, X_λ $(\lambda \in \Lambda)$ be fully invariant subgroups of free group F. Then*

$$q\mathcal{R} \textit{ is one-to-one }; \tag{3}$$
$$\textit{if } \mathcal{U} \subseteq \mathcal{V}, \textit{ then } r\mathcal{V} \subseteq r\mathcal{U}; \tag{4}$$
$$U \subseteq V \textit{ if and only if } q\mathcal{R}V \subseteq q\mathcal{R}U; \tag{5}$$
$$\mathcal{U} \subseteq q\mathcal{R}r\mathcal{U}; \tag{6}$$
$$U = rq\mathcal{R}U; \tag{7}$$
$$\bigcap_{\lambda\in\Lambda} r\mathcal{X}_\lambda = r(\bigwedge_{\lambda\in\Lambda}^{\mathbf{L}} \mathcal{X}_\lambda); \tag{8}$$
$$\bigcap_{\lambda\in\Lambda} q\mathcal{R}X_\lambda = q\mathcal{R}(\bigvee_{l\in\Lambda}^{\mathbf{L}})X_\lambda; \tag{9}$$
$$\bigvee_{\lambda\in\Lambda}^{\mathbf{L}} r\mathcal{X}_\alpha \subseteq r(\bigcap_{\lambda\in\Lambda} \mathcal{X}_\lambda); \tag{10}$$
$$\bigvee_{\lambda\in\Lambda}^{\mathbf{L}} q\mathcal{R}X_\lambda \subseteq q\mathcal{R}(\bigcap_{\lambda\in\Lambda} X_\lambda). \tag{11}$$

Proof. (3) Let U, V be distinct fully invariant subgroup of free group F. Let $v \in V \setminus U$. Then it follows from Lemma 2, that $L_U \in q\mathcal{R}U \setminus q\mathcal{R}V$. Hence $q\mathcal{R}$ is one-to-one. The direct part of (4),(5) and (6) are immediate from the definition of r and $q\mathcal{R}$.

(7) It is also follows from the definitions that $U \subseteq rq\mathcal{R}U$. Let $v \in F \setminus U$. By Lemma 2 $L_U \in q\mathcal{R}U$ and L_U does not satisfy $l(v) = e$. Hence $v \notin rq\mathcal{R}U$ and so we must have the equality.

(8) By (4), $r(\bigvee^{\mathbf{L}}_{\lambda\in\Lambda} \mathcal{X}_\lambda) \subseteq \bigcap_{\lambda\in\Lambda} r\mathcal{X}_\lambda$. Now, for any $\lambda \in \Lambda$, $r\mathcal{X}_\lambda \supseteq \bigcap_{\lambda\in\Lambda} r\mathcal{X}_\lambda$ and so by (5) and (6),

$$\mathcal{X}_\lambda \subseteq q\mathcal{R}r\mathcal{X}_\lambda \subseteq q\mathcal{R}(\bigcap_{\lambda\in\Lambda} r\mathcal{X}_\lambda).$$

hence $\bigvee^{\mathbf{L}}_{\lambda\in\Lambda} \mathcal{X}_\lambda \subseteq q\mathcal{R}(r)\mathcal{X}_\lambda$. Applyng r to both sides we obtain

$$r(\bigvee^{\mathbf{L}}_{\lambda\in\Lambda} \mathcal{X}_\lambda) \supseteq rq\mathcal{R}(\bigcap_{\lambda\in\Lambda} rX_\lambda).$$

Therefore, we have the equality.

For the converse implication in (5), now suppose that $q\mathcal{R}V \subseteq q\mathcal{R}U$. Then by (4), $rq\mathcal{R}U \subseteq rq\mathcal{R}V$ which by (7), is reduced to $U \subseteq V$.

(9) By (5), we have that $q\mathcal{R}(\bigvee^{\mathbf{R}}_{\lambda\in\Lambda} X_\lambda) \subseteq \bigcap_{\lambda\in\Lambda} q\mathcal{R}X_\lambda$. On the other hand, for each λ, $\bigcap_{\lambda\in\Lambda} q\mathcal{R}X_\lambda \subseteq q\mathcal{R}X_\lambda$ implies that $X_\lambda = rq\mathcal{R}X_\lambda \subseteq r(\bigcap_{\lambda\in\Lambda} q\mathcal{R}X_\lambda)$. Hence

$$\bigvee^{\mathbf{L}}_{l\in\Lambda} X_\lambda \subseteq r(q\mathcal{R} \text{ and } q\mathcal{R}(\bigvee^{\mathbf{L}}_{\lambda\in\Lambda} X_\lambda) \supseteq q\mathcal{R}r(\bigcap_{\lambda\in\Lambda} q\mathcal{R}X_\lambda) \supseteq \bigcap_{\lambda\in\Lambda} q\mathcal{R}X_\lambda.$$

Properties (10) and (11) follow from (4) and (5) immediately. □

Since the mapping $q\mathcal{R}$ is one-to-one and it is known that the set of all fully invariant subgroups of free group F is uncountable, then it follows that there is an uncountable family of quasi-representable l-varieties. It has been shown by S.Adijan [1] that there exists an uncountable family of pairwise incomparable varieties of groups. Then by Theorem 3(5), $q\mathcal{R}$ maps such varieties of groups onto incomparable l-varieties in the lattice **L** of all l-varieties.

COROLLARY 1. (Reily [2].) *There exists a continuum of pairwise incomparable quasi-representable l-varieties in the lattice* **L** *of l-varieties.*□

3⁰. The following ingenious method of establishing that the cardinality of the lattice of all l-varieties **L** is due to Feil [2].

Let $w_1 = w_1$, $w_2 = w_2$ be any two terms of the signature l. For positive integers p and q let $\mathcal{W}_{p/q}(w_1, w_2)$ be the representable l-variety defined by the identities:

$$(F)\quad \begin{cases} (F.1) & |w_1|^p \wedge |w_2|^q = |w_2|^q, \\ (F.2) & (x \wedge y^{-1}x^{-1}y) \vee e = e, \end{cases}$$

or in the equivalent form

$$(F)\quad \begin{cases} (F.3) & |w_1|^p \geq |w_2|^q, \\ (F.4) & x \wedge y^{-1}x^{-1}y \leq e. \end{cases}$$

For representable l-groups the inequalities $x^n \leq y^n$ and $x \leq y$ are equivalent for any positive integer n. Hence if $0 < \frac{p}{q} \leq \frac{m}{n}$ for any positive integers n, m, then $|w_1|^p \geq |w_2|^q$ implies $|w_1|^{mp} \geq |w_2|^{mq} \geq |w_2|^{np}$ and $|w_1|^m \geq |w_2|^n$. So $\mathcal{W}_{p/q}(w_1, w_2) \subseteq \mathcal{W}_{m/n}$. It follows that if $\frac{m}{n} = \frac{p}{q}$, then $\mathcal{W}_{m/n}(w_1, w_2) = \mathcal{W}_{p/q}(w_1, w_2)$ so these l-varieties are well defined.

Let now $w_1 = [|x| \vee |y|, |[|x| \vee |y|, |y|]|]$, $w_2 = [|x| \vee |y|, |y|]$ and $0 < \frac{p}{q} \leq 1$. We denote for these w_1, w_2 $\mathcal{W}_{p/q}(w_1, w_2) = \mathcal{W}_{p/q}$ and show that $\mathcal{W}_{p/q} \subset \mathcal{W}_{m/n}$, $\mathcal{W}_{p/q} \neq \mathcal{W}_{m/n}$ if $\frac{p}{q} < \frac{m}{n}$.

Let t be a real number, $0 < t < 1, \beta(t) = \frac{t}{t+1}$ and $T_\beta(t)$ be a semidirect product of the additive group of real numbers **R** and the infinite cyclic group (b) with group operation defined by the rule: for $x = b^r u$, $x = b^s v$ in $T_\beta(t)$ where $u, v \in \mathbf{R}$, $b^r, b^s \in (b)$ is valid $xy = b^r u \cdot b^s v = b^{r+s}(\beta(t)^s r + s)$.

The group $T_{\beta(t)}$ is an o-group under the lexicographical order: $x = b^r u > e$ iff $r > 0$ or $r = 0$, $u > 0$ in **R**.

LEMMA 3. *For* $0 < t \leq \frac{p}{q}$, $T_{\beta(t)} \in \mathcal{W}_{p/q}$.

Proof. We need only to verify the validility of identity (F.3) in totally ordered group $T_\beta(t)$. Let $x = b^r u, y = b^s v \in T_\beta(t), x > y \geq e$. If $[|x| \vee |y|, |y|] = e$, then the result is trivial. So assume that $[|x| \vee |y|, |y|] = a > e$ As $r \geq s \geq 0$, $r \neq$, then

$$|w_1| = |[|x| \vee |y|, |[|x| \vee |y|, |y|]|]| = |[x, a]| = |(1 - \beta(t)^{-r})a| \leq$$
$$|(1 - \beta(t)^{-1})a| \leq \tfrac{a}{t} \leq \tfrac{q}{p}.$$

Hence $p(\beta(t)^{-r}-1)a \geq qa$ and

$$|[|x|\vee|y|,|[|x|\vee|y|,|y|]|]|^p \geq |[|x|\vee|y|,|y|]|^q$$

for these $x, y \in T_{\beta(t)}$.□

LEMMA 4. *For* $0 < \frac{p}{q} < t \leq 1$ *is valid* $T_{\beta(t)} \notin \mathcal{W}_{p/q}$.

The proof is similar to that of Lemma 3.□

THEOREM 4. (Feil [2].) *The family of l-varieties*

$$\{\mathcal{W}_{p/q} |\ 0 < \frac{p}{q} \leq 1\}$$

is a chain of representable l-varieties and if $0 < \frac{m}{n} < \frac{p}{q}$, *then* $\mathcal{W}_{m/n} \subset \mathcal{W}_{p/q}$, $\mathcal{W}_{m/n} \neq \mathcal{W}_{p/q}$.

The proof follows immediately from Lemmas 3 - 4.□

Now for $a \in \mathbf{R} \setminus \mathbf{Q}$, $0 < a < 1$, define $\mathcal{W}_a = \bigcap\limits_{\frac{p}{q}>a} \mathcal{W}_{p/q}$. It follows that $\mathcal{W}_{p/q} \supseteq \mathcal{W}_a \supset \mathcal{W}_{m/n}$ for $\frac{p}{q} > a > \frac{m}{n}$. In fact $T_{\beta(t)} \in \mathcal{W}_a \setminus \mathcal{W}_{p/q}$ where $t = \frac{a}{a+1}$ and $\frac{p}{q} < a$. This now gives us a chain of l-varieties whose order type is that of the set of real numbers. Let $\mathcal{A}^2 \cap \mathcal{R}$ be the l-variety of representable l-groups that consists of l-groups that have an Abelian ideal with Abelian quotient l-group. Since $T_{\beta(t)}$ given above are in $\mathcal{A}^2 \cap \mathcal{R}$, then the following results is true.

COROLLARY 1. (Feil [2].) *There exsists a chain of representable l-varieties in* $\mathcal{A}^2$ *whose order type is that of continuum.*□

COROLLARY 2. (Feil [2].) *There exists a continuum of representable l-varieties in* $\mathcal{A}^2$ *which are not finitely based.*

Proof. It is clear that any l-variety which is the intersection of the infinite properly descending chain of the finitely based varieties cannot be finitely based. The l-varieties $\mathcal{W}a \cap \mathcal{A}^2$ with $a \in \mathbf{R} \setminus \mathbf{Q}$, $0 < a < 1$ have the required properties.□

4^0. In the theory of varieties of groups it is known that every nilpotent variety of groups has a finite basis of identities (see, for example, the book of H.Neumann [1]). Here we will establish the existence of l-varieties of nilpotent of class ≤ 2 which are not finitely based.

Let us consider the following $2n$ ($n \in \mathbf{N}$) terms of the signature l;

$$\begin{aligned} u_1 &= (|[x,y]| \vee t_1) \wedge |[x,y]|^2, & w_1 &= [|x| \vee |y|, u_1],\\ u_2 &= (|w_1| \vee t_2) \wedge |w_1|^2, & w_2 &= [|x| \vee |y|, u_2],\\ &\dots & &\dots\\ u_n &= (|w_{n-1}| \vee t_n) \wedge |w_{n-1}|^2, & w_n &= [|x| \vee |y|, u_n]. \end{aligned}$$

Now for every positive integer $k \geq 3$ we define the following identity

$$f_k = w_1^+ \wedge \cdots \wedge w_k^+ \wedge [u_1,u_2]^+ \wedge [u_2,u_3]^+ \wedge \cdots \wedge [u_{k-1},u_k]^+ \wedge [u_1,u_k]^+ = e$$

Let $\mathcal{U}_n$ be an l-variety defined by the identities $f_k = e$ ($k = 3, \ldots, n$) and $[x,y,z] = e$. It is evident that

$$\mathcal{U}_3 \supseteq \mathcal{U}_4 \supseteq \cdots \mathcal{U}_n \supseteq \mathcal{U}_{n+1} \supseteq \cdots$$

We claim that all inclusions are proper. Let G_n be a nilpotent group class 2 generated by the elements $b, a_0, a_1, \ldots, a_n$ with defining relations:$[a_0, a_i] = e$ for all i, $1 \leq i \leq n$. Now let us consider any total order on G_n such that the following inequalities are valid in G_n :

$$\begin{array}{ll}
(12) & b \gg a_0 \gg a_1 \gg \cdots \gg a_n > e;\\
(13) & [b, a_i] = c_{i+1} > e \text{ for all } i = 0, 1, \ldots, n;\\
(14) & c_i < a_i < c_i^2 \text{ for all } i = 1, 2, \ldots, n;\\
(15) & |[a_i, a_j]| \gg |[a_k, a_s]| \text{ where } 1 \leq i < j \leq n,\\
 & 1 \leq k < s \leq n \text{ if } i < k \text{ or } k = i \text{ and } j < s;\\
(16) & [a_i, a_{i+1}] > e \text{ for } i = 1, 2, \ldots, n-1;\\
(17) & [a_i, a_j] < e \text{ for } i = 1, 2, \ldots, n-2,\ i+2 \leq j \leq n;\\
(18) & [a_1, a_n] > e,\ [a_1, a_j] < e \text{ for } j = 3, 4, \ldots, n-1;\\
(19) & c_{n+1} \gg |[a_i, a_j]| \text{ for } 1 \leq i < j \leq n.
\end{array}$$

Here for $x, y > e$ we denote $x \gg y$ if $x > y^m$ for all $m \in \mathbf{N}$.

It is clear that the total order on G_n with properties (12)-(19) exists and it is not unique. Subgroup C generated by elements $c_{n+1}, [a_i, a_j]$, ($1 \leq i < j \leq n$), is convex under every such total order of G_n. Each element $x \in G_n$ can be uniquely represented in the form

$$x = b^p a_0^{p_0} (c_1^{p_1} a_1^{q_1}) \cdots (c_n^{p_n} a_n^{q_n}) c_{n+1}^{p_{n+1}} \prod_{1 \leq i < j \leq n} [a_i, a_j]^{p_{ij}}$$

where $p, p_0, \ldots, p_{n+1}, q_1, \ldots, q_n, p_{ij} \in \mathbf{Z}$ for $1 \leq i < j \leq n$.

From (12)-(19) it follows that $x > e$ in totally ordered group G_n iff the first nontrivial factor among

$$b^p, a_0^{p_0}, (c_1^{p_1} a_1^{q_1}), \ldots, (c_n^{p_n} a_n^{q_n}), c_{n+1}^{p_{n+1}}, [a_i, a_j]^{p_{ij}} \ (1 \leq i < j \leq n)$$

is positive in o-group G_n.

Now, let $x = b$, $y = a_0$, $t_1 = a_1, \ldots, t_n = a_n$ Then the direct calculation shows that

$w_1 = c_2 > e, \ldots, w_n = c_{n+1} > e$; $[u_1, u_2] = [a_1, a_2] = e$, $[u_2, u_3] = [a_2, a_3] > e, \ldots, [u_{n-1}, u_n] = [a_{n-1}, a_n] > e$, $[u_1, u_n] = [a_1, a_n] > e$

Therefore, $f_n(b, a_0, a_1, \ldots, a_n) \neq e$ in G_n. Now we claim that all identities $f_k(x, y, t_1, \ldots, t_k) = e$ hold in G_n for $k \neq n$. Suppose, for reductio ad absurdum, that there are $x, y, t_1, \ldots, t_k \in G_n$ such that $w_1 > e, \ldots, w_k > e$, $[u_1, u_2] > e, \ldots, [u_{k-1}, u_k] > e$, $[u_1, u_k] > e$. We consider now two cases.

Case a). Let

$$|x| \vee |y| = b^p a_0^{p_0} (c_1^{p_1} a_1^{q_1}) \cdots (c_n^{p_n} a_n^{q_n}) c_{n+1}^{p_{n+1}} \prod_{1 \leq i < j \leq n} [a_i, a_j]^{p_{ij}}$$

Then $[x, y] \in C$. Therefore, $u_1, u_2 \in C$ and $[u_1, u_2] = e$. This contradics our assumption.

Case b). Let

$$|x| \vee |y| = b^p a_0^{p_0} (c_1^{p_1} a_1^{q_1}) \cdots (c_n^{p_n} a_n^{q_n}) c_{n+1}^{p_{n+1}} \prod_{1 \leq i < j \leq n} [a_i, a_j]^{p_{ij}}$$

Then

$$u_1 = (c_1^{p_{11}} a_1^{q_{11}}) \cdots (c_n^{p_{n1}} a_n^{q_{n1}}) c_{n+1}^{p_{n+1\,1}} \prod_{1 \leq i < j \leq n} [a_i, a_j]^{p_{ij1}}$$

$$\cdots$$

$$u_k = (c_1^{p_{1k}} a_1^{q_{1k}}) \cdots (c_n^{p_{nk}} a_n^{q_{nk}}) c_{n+1}^{p_{n+1\,k}} \prod_{1 \leq i < j \leq n} [a_i, a_j]^{p_{ijk}}$$

Let $(c_{i(1)}^{p_{i(1)1}} a_{i(1)}^{q_{i(1)1}}), \ldots, (c_{i(k)}^{p_{i(k)k}} a_{i(k)}^{q_{i(k)k}})$ be the first non-trivial factors of the elements u_j $(1 \leq j \leq k)$ with non-zero exponents $q_{i(1)1}, \ldots, q_{i(k)k}$. If all factors $c_i^{p_{ij}} a_i^{q_{ij}}$ of the element u_j $(1 \leq j \leq k)$ contain generators

a_i $(1 \le i \le n)$ with zero exponents, then the element u_j is central in G_n and therefore, $[u_{j-1}, u_j] = e = [u_j, u_{j+1}]$, a contradiction. From property (13), $i(1) < i(2) < \cdots < i(k)$ and from properties (12)-(19) of the total order on G_n and from the inequalities $w_1 = [|x| \vee |y|, u_1] > e, \ldots, w_k = [|x| \vee |y|, u_k] > e$, it follows that $q_{i(1)1} > 0, \ldots, q_{i(k)k} > 0$. From the inequalities $[u_1, u_2] > e$, $[u_2, u_3] > e, \ldots, [u_{k-1}, u_k] > e$ it follows that $i(1), i(2), \ldots, i(k)$ is a sequental subset of the totally ordered set $\{1, 2, \ldots, n\}$. If $i(1) \neq 1$, then $[a_{i(1)}, a_{i(1)+k}] < e$ and $[u_1, u_k] < e$, a contradiction. Therefore, $i(1) = 1$. In this case $k < n$ and $[a_1, a_k] < e$. Hence $[u_1, u_k] < e$. This means that the identity $f_k(x, y, t_1, \ldots, t_k) = e$ holds in G_n. If $m > n$, then for all $x, y, t_1, \ldots, t_m \in G_n$ the element u_m lies in C and $[u_1, u_m] = e$. Therefore, the identity $f_m = e$ $(m > n)$ holds in o-group G_n.

THEOREM 5. (Medvedev [11].)

(20) *The l-variety* $\mathcal{U} = \bigcap_{n \in \mathbf{N}} \mathcal{U}_n$ *of nilpotent l-groups of nilpotent class* ≤ 2 *is not finitely based.*

(21) *The set of identities* $\{f_n = e | n \in \mathbf{N}\}$ *independent.*

(22) *The lattice of nilpotent l-varieties of the nilpotent class* ≤ 2 *has uncountable cardinality.*□

COROLLARY 1. *The l-variety* $\mathcal{U} = \bigcap_{n \in \mathbf{N}} \mathcal{U}_n$ *has an infinite axiomatic rank.*

Proof. Suppose, on the contrary, that the axiomatic rank of $\mathcal{U}$ is equal to s, $s < \infty$. As was mentioned in the proof of Theorem 9.4.3 that every nilpotent o-group with s-generators has less or equal to $\varphi(s)$ of convex subgroups, where $\varphi : \mathbf{N} \to \mathbf{N}$ is some function on $\mathbf{N}$. Therefore, in all nilpotent o-groups with s generators the identity

$$f_n = e \ (n > \varphi(s))$$

is valid. Every nilpotent l-group K with s generators is a subcartesian product of nilpotent 0-groups with s generators. Therefore, all identities $f_n = e$ $(n > \varphi(s))$ are valid in K, too. If the axiomatic rank of $\mathcal{U}$ is equal to s, then $G \in \mathcal{U}$ iff all s-generated subgroups of G belongs to $\mathcal{U}$. Therefore, $G \in \mathcal{U}$ iff the identities $f_3 = e, \ldots, f_{\varphi(s)} = e$ are valid in G. This means that $f_3 = e, \ldots, f_{\varphi(s)} = e$ is the finite basis of identities of $\mathcal{U}$. This is the contradiction.□

Other examples of nilpotent l-varieties which are not finitely based were constructed by S.A.Gurchenkov [5],[6].

Chapter 10

Free l-groups

10.1 Description of Free l-groups

1^0. From well-known results in universal algebra each variety or quasi-variety $\mathcal{K}$ of algebraic systems has a free algebraic system F with the set $X = \{x_i | i \in I\}$ of generators such that if $\hat{\varphi}$ is a map of X into an algebraic system $G \in \mathcal{K}$, then there is a homomorphism φ from F into G extending $\hat{\varphi}$ (see the book of Burris and Sankappanavar [1], Theorem 11.4). The set X is called in this case the set of free generators. Therefore, each l-variety $\mathcal{X}$ has a *free l-group* in $\mathcal{X}$. If $\mathcal{X} = \mathcal{L}$, then a free l-group in $\mathcal{L}$ is a called *free l-group.* The cardinality of the set of free generators of a free l-group in l-variety is called a *rank* of a free l-group. It is clear that a free l-group is a free l-group over trivially ordered free group (see Section 1 of Chapter 7). Let $F_l(n)$ be a free l-group with free generators $x_1, \ldots, x_n$ and $\Phi = \{w_i(X) = w_i(x_1, \ldots, x_n) | i \in I\}$ be a set of elements of $F_l(n)$, i.e., the set of words of the signature l (l-words) in the variables $x_1, \ldots, x_n$. Let G be an l-group. Then the convex l-subgroup H generated by all substitutions of $w_i(x_1, \ldots, x_n)$ in G is called a *verbal* l-subgroup of G. A convex l-subgroup H of G is called a *fully invariant* l-subgroup of G if $H\varphi \subseteq H$ for each l-homomorphism φ of G into G.

PROPOSITION 1. *Each l-variety is uniquely determined by its free l-group of the countable infinite rank.*□

PROPOSITION 2. *Each verbal l-subgroup of l-group G is a con-*

vex fully invariant l-subgroup. Each convex fully invariant l-subgroup of free l-group F is a verbal l-subgroup of F.□

PROPOSITION 3. *A free l-group $F_{\mathcal{X}}$ of rank $\aleph$ in l-variety $\mathcal{X}$ with the basis of identities $\Phi = \{w_i(X) = e | i \in I\}$ is l-isomorphic to the quotient group $F/H(\Phi)$ of free l-group F of rank $\aleph$ by verbal l-subgroup $H(\Phi)$ defined by the set Φ.*□

2^0. It is known (M.Hall [1]) that free group has a solvable word problem. The answer to the same problem for the free l-group is not quite so obvoius. Let us recall that each element x of free l-group F with the set $X = \{x_i | i \in I\}$ of free generators can be represented in the form $x = \bigvee_{\alpha\in A} \bigwedge_{\beta\in B} w_{\alpha\beta}(X)$ where $w_{\alpha\beta}$ are elements of free group F_0 with the set X of free generators.

THEOREM 1. (Holland and McCleary [2].) *The word problem for a free l-group F with recursive set of free generators X is solvable.*

Proof. We will construct the decision procedure for deciding in finite number of steps whether the element x of free l-group F is the identity element or not. Since each word $x \in F$ contains only a finite subset of free generators we can assume that the set $X = \{x_1, \ldots, x_r\}$ of free generators is finite. Let the word x be represented in the form

$$x = \bigvee_{\alpha\in A} \bigwedge_{\beta\in B} w_{\alpha\beta}(x_1, \ldots, x_r) = \bigvee_{\alpha\in A} \bigwedge_{\beta\in B} w_{\alpha\beta}(X) \tag{1}$$

where A, B are finite sets of indexes, A contains $n(A)$ elements and B contains $n(B)$ elements, $w_{\alpha\beta}(X)$ are reduced words of the signature of the group theory of the lenght $n(\alpha, \beta)$ correspondingly. Let W be the set of all initial segments of the words $w_{\alpha\beta}$ $(\alpha \in A, \beta \in B)$.

Let us consider the following algoritmic problem. Is there a finite totally ordered set S containing less than or exactly

$$2n \cdot n(A) \cdot n(B) \cdot \sum_{\alpha\in A, \beta\in B} n_{\alpha\beta} = 2k$$

elements with subset T, $T \subseteq S$ such that T contains less than or exactly k elements and the mappings $\bar{x}_i : T \to S$ and $\bar{x}_i^{-1} : T \to S$, $1 \leq i \leq r$, for which the following properties are valid?

(2) If $a < b$ in T, then $a\bar{x}_i < b\bar{x}_i$, $a\bar{x}_i^{-1} < b\bar{x}_i^{-1}$.
(3) If $a \in T$, $b \in T$, $b = a\bar{x}_i$ (or $b = a\bar{x}_i^{-1}$), then $b\bar{x}_i^{-1} = a$ ($b\bar{x}_i = a$, correspondingly).
(4) There is an element $\bar{t} \in T$ such that the element $\bar{t} \cdot w(\bar{x}_1, \ldots, \bar{x}_r)$ is well defined in S for every $w(x_1, \ldots, x_n) \in W$ and $\max\limits_{\alpha \in A} \min\limits_{\beta \in B} \bar{t} \cdot w(\bar{x}_1, \ldots, \bar{x}_r) \neq \bar{t}$.

It is clear that this algoritmical problem is solvable.

Now we claim that the word problem for the l-word x (see (1)) is equivalent to this algoritmical problem. Let $x \neq e$ in F. Then by Holland's embedding theorem (Theorem 4.1), we identify F with some l-group of order permutations of totally ordered set M. Since $x \neq e$, then $\bar{t} \cdot x \neq \bar{t}$ for some $\bar{t} \in M$. Then we set

$$T = \{\bar{t} \cdot w | w \in W\}, \quad S = \{t \cdot x_i,\ t \cdot x_i^{-1} | t \in T,\ i = 1, 2, \ldots, r\}$$

and set $t \cdot \bar{x}_i = t \cdot x_i$, $t\bar{x}_i^{-1} = t \cdot x_i^{-1}$ for $t \in T$, $i = 1, 2, \ldots, r$. The direct verification shows that the cardinality of S is less than or exactly $2k$ and the cardinality of T is less than or equal k and properties (2)–(4) hold.

Conversely, let the totally ordered sets T, S ($T \subseteq S$) and values $t \cdot \bar{x}_i, t\bar{x}_i^{-1}$ ($1 \leq i \leq r$) be defined and properties (2)–(4) be valid. Let θ be an order isomorphism of the totally ordered set S into naturally ordered open interval $(a, b) \subseteq [0, 1]$ of reals. Now we identify the totally ordered set S with its image $S\theta$. Let $t_1 < t_2, \ldots, t_m$ ($m \leq k$) be all elements of T and $t_0 = 0$, $t_{m+1} = 1$. We define for every integer p, $1 \leq p \leq r$, the order authomorphism y_p of $[0, 1]$ by the rule:

$$t \cdot y_p = \frac{t_{i+1} \cdot \bar{x}_p - t_i \bar{x}_p}{t_{i+1} - t_i}(t - t_i) + t_i \cdot \bar{x}_p$$

for $t \in [t_i, t_{i+1}]$ $i = 0, 1, \ldots, m$. Then $\bar{t} \cdot \bigvee\limits_{\alpha \in A} \bigwedge\limits_{\beta \in B} w_{\alpha\beta}(y_1, \ldots, y_r) \neq \bar{t}$ in l-group $A(\mathbf{R})$ and $\bigvee\limits_{\alpha \in A} \bigwedge\limits_{\beta \in B} w_{\alpha\beta}(x_1, \ldots, x_r) \neq e$ in F.□

3^0. A very succesful approach to the study of free l-groups is to use ordered permutations groups. In this Section, following Kopytov [8], we construct one special embedding of free l-group F into l-group $A(\mathbf{R})$ of order permutations of the reals.

Let $D^* = D^*(\mathbf{R})$ be the set of all order permutations of R such that for each $f \in D^*$ there is a well-ordered ascending under the natural order of $\mathbf{R}$ subset $A_f = \{a_0 < a_1 < \cdots < a_\xi < \cdots < a_{\sigma(f)} \leq \infty\}$ of the set $\mathbf{R}\vec{\cup}\infty$ such that

(5) $(xf) = x$ for all $x \in \mathbf{R}$, $x < a_0$;

(6) f is a monotone linear function on the segment $[a_\xi, a_\xi + 1]$ for each $\xi < \sigma(f)$.

Here $\sigma(f)$ is some ordinal determined by f and $a_\tau = \sup\limits_{\xi<\tau} a_\xi$ for all limits ordinal $\tau < \sigma(f)$.

Each $a_\xi \in A_f$ is called a *break point* of f. It is clear that D^* is an l-group under natural pointwise order: $f \leq g$ in D^* iff $xf \leq xg$ for all $x \in \mathbf{R}$. Let D_l^* denote this l-group. It is clear that D_l^* contains Dlab's group $D_l(\mathbf{R})$ (see Section 7 of Chapter 4).

THEOREM 2. (Glass [2], Kopytov [8], McCleary [7].) *The free l-group F of rank r ($r \geq 2$) has a faithful transitive representation by order permutations.*

Proof. We will prove this theorem for finite rank r ($r \geq 2$). For infinite countable rank thus is done similarly. Let $X = \{x_1, \ldots, x_r\}$ be a set of free generators of F. F is countable and we enumerate its nonidentity elements $w_1, \ldots, w_n, \ldots$. Let

$$(7) \qquad w_n(x_1, \ldots, x_r) = w_n(X) = \bigvee_{\alpha\in A} \bigwedge_{\beta\in B} w_{\alpha\beta}(x_1, \ldots, x_r) \neq e$$

be a word of F. Then there is a representation θ_n of F by order automorphisms of some totally ordered set $\bar{T}$ such that $\bar{t}_n \cdot w_n(X)\theta_n \neq \bar{t}_n$ for some $\bar{t}_n \in \bar{T}$. Let $T_n = \{\bar{t}_n \cdot w(X)\theta_n | w \in W\}$ where W is a set of all initial segments of words $w_{\alpha\beta}$ $\alpha \in A, \beta \in B$. Let $S_n$be a set $\{t \cdot x_i\theta_n | t \in T_n,\ i = 1, 2, \ldots, r\}$ and a_n, b_n be the least and greatest elements of S_n, correspondingly.

Now we define totally ordered sets $T_n(2), S_n(2)$ for which $T_n(2) \subset S_n(2)$ in the following way. Let us consider the left side of S_n. If $a_n \in T_n$, then $a_n \cdot x_i\theta_n \geq a_n$ and three different cases arise.

Case 1.1. Let $a_n \in T_n$, $a_n \cdot x_i\theta_n = a'_n$ for all $i = 1, 2, \ldots, r$ and $a_n < a'_n$. Then we set $T_n(1) = T_n$, $S_n(1) = S_n$ and $a_n = u_n$, $a'_n = c_n$, $t^{(i)} = t \cdot x_i\theta_n$ for all i, $1 \leq i \leq r$ and $t \in T_n(1)$.

Case 1.2. Let $a_n \in T_n$, $a_n \cdot x_i\theta_n > a_n$ for all $i = 1, 2, \dots, r$ and $a_n \cdot x_i\theta_n \neq a_n \cdot x_i\theta_n$ for some i, j $(1 \leq i, j \leq r)$. Then we set $S_n(1) = \{u_n\} \vec{\cup} S_n$ where $u_n < s$ for all $s \in S_n$, $T_n(1) = \{u_n\} \vec{\cup} T_n$, $t^{(i)} = t \cdot x_i\theta_n$ for all i, $1 \leq i \leq r$, $t \in T$ and $a_n = c_n$.

Case 1.3. Let $a_n \in T_n$, $a_n \cdot x_p\theta_n = a_n$, $a_n < a_n \cdot x_q\theta_n$ for some p, q, $1 \leq p, q \leq r$. Then we set

$$S_n(1) = \{u_n = a_{n3}, a_{n2}, a_{n1}\} \cup S_n$$

where $a_{n3} < a_{n2} < a_{n1} < s$ for all $s \in S_n$,

$$T_n(1) = T_n(1) = \{u_n = a_{n3}, a_{n2}, a_{n1}\} \cup T_n$$

and $t^{(i)} = t \cdot x_i\theta_n$, $u_n^i = a_{n2} = c_n$,

$$a_{n2}^{(i)} = \begin{cases} a_n & \text{for } a_n \cdot x_i\theta_n > a_n, \\ a_{n1} & \text{for } a_n \cdot x_i\theta_n = a_n. \end{cases}$$

Let now $a_n \notin T_n$.

1.4. We set

$$S_n(1) = \{a_{n7}, \dots, a_{n1}\} \cup S_n,$$

$a_{ni+1} < a_{ni}$ for all $i, 1 \leq i \leq 6$, $u_n = a_{n7}$ and $a_{ni} < s$ for all $i = 1.2, \dots, 7$, $s \in S_n$;

$$T_n(1) = \{a_{n7}, \dots, a_{n1}\} \cup T_n,$$

$t^{(i)} = t \cdot x_i\theta_n$ for all $t \in T_n$ and

$$a_n^i = \begin{cases} a_{n1} & \text{for } i = 1, \\ a_{n3} & \text{for } i > 1; \end{cases} \qquad a_{n3}^i = \begin{cases} a_{n2} & \text{for } i = 1, \\ a_{n4} & \text{for } i > 1; \end{cases}$$

$$a_{n6}^i = \begin{cases} a_{n3} & \text{for } i = 1, \\ a_{n5} & \text{for } i > 1; \end{cases} \qquad u_n^i = a_{n7}^i = a_{n6} = c_n.$$

So we have constructed finite totally ordered sets $S_n(1), T_n(1)$, $T_n(1) \subset S_n(1)$ and elements $t^{(i)} \in S_n(1)$ for $1 \leq i \leq r$, $t \in T_n(1)$.

Now we consider the right side of $S_n(1)$.

2.1. Let $b_n \notin T_n(1)$. Then we set

$$S_n(2) = S_n \cup \{v_n\},$$

$s < v_n$ for all $s \in S_n(1)$ and

$$T_n(2) = \{b_n\} \cup T_n(1),$$

$b_n^{(i)} = v_n, \; d_n = b_n.$

2.2. Let $b_n \in T_n(1)$. Then we fix some index $i(1), \; 1 \leq i(1) \leq r$ for which $b_n \cdot x_{i(1)}\theta_n \leq b_n \cdot x_i\theta_n$ for all $i = 1, 2, \ldots, r$ and $b_n \cdot x_{i(1)}\theta_n < b_n$. In this case we set

$$S_n(2) = S_n(1) \cup \{b_{n1}, b_{n2}, b_{n3}\},$$

where $s < b_{n1} < b_{n2} < b_{n3}$ for all $s \in S_n(1)$ and

$$T_n(2) = T_n(1) \cup \{b_{n1}, b_{n2}\}.$$

Let $b_{n2}^i = b_{n3}$,

$$b_{n1}^i = \begin{cases} b_{n2} & \text{for } i \neq i(1), \\ b_n & \text{for } i = i(1), \end{cases}$$

and $b_{n2} = d_n, \; b_{n3} = v_n$.

Thus finite totally ordered sets $S_n(2), \; T_n(2), T_n(2) \subset S_n(2)$ for any positive integer n have been constructed. For each $i, 1 \leq i \leq r$, $t \in T_n(2)$ there are elements $t^{(i)} \in S_n(2)$ such that $t_1 < t_2$ in $T_n(2)$ implies $t_1^i < t_2^i$.

Now we define o-embeddings γ_n of $S_n(2)$ into the segment $[n, n+\frac{4}{3}]$ of the real line with the following properties: $u_n\gamma_n = n, \; c_n\gamma_n = n + \frac{1}{3}, \; d_n\gamma_n = n+1, \; v_n\gamma_n = n + \frac{4}{3}$. For each $i, \; o = 1, 2, \ldots, r$ we define the function $f_i \in D_l^*$ for which $A_{f_i} = \{0\} \cup \bigcup\limits_{n=1}^{\infty} T_n(2)\gamma_n, \; x \cdot f_i = t^{(i)}\gamma_n$ for $x \in A_{f_i}, \; x \neq 0, \; x = t\gamma_n, t \in T_n(2)$.

From the definition of f_i it follows that the positive integer $n \in \{2, 3, \ldots, \}$ is the image of $u_n \in S_n(2)$ under γ_n and it is the image of d_{n-1} under γ_{n-1}. Since $u_n^{(i)} = c_n, \; d_n^{(i)} = v_{n-1}$, then

$$u_n^{(i)}\gamma_n = n + \frac{1}{3} = (n-1) + \frac{4}{3} = d_{n-1}^{(i)}\gamma_{n-1}.$$

Therefore, f_i is well-defined on the set A_{f_i}. It is clear that

$$t\gamma_n \cdot f_i = (t \cdot x_i\theta_n)\gamma_n \tag{8}$$

for all $t \in T_n$. Now we define f_i on the real line $\mathbf{R}$ by the rule

$$x \cdot f_i = \begin{cases} x & \text{for} \quad x \leq 0; \\ \frac{4}{3} & \text{for} \quad 0 \leq x \leq 1; \\ \frac{t_1^{(i)}\gamma_n - t_2^{(i)}\gamma_n}{t_1\gamma_n - t_2\gamma_n}(x - t_2\gamma_n) + t_2^{(i)}\gamma_n & \text{for} \quad t_1\gamma_n \leq x \leq t_2\gamma_n; \end{cases}$$

where are neighbouring elements of $T_n(2)$.

By definition, $f_i \in D_l^*$ for all $i = 1, 2, \ldots, r$. Now we determine the action φ of F in $A(\mathbf{R})$ if we set $x_i\varphi = f_i$.

We claim that φ is a representantion of F. In fact, let

$$w = w_n = w_n(x_1, \ldots, x_r) = \bigvee_{\alpha \in A} \bigwedge_{\beta \in B} w_{\alpha\beta}(x_1, \ldots, x_r) \neq e$$

in F, then $\bar{t}_n\gamma_n \cdot w_n\varphi =$

$$= \bigvee_{\alpha \in A} \bigwedge_{\beta \in B} \bar{t}_n\gamma_n \cdot w_{\alpha\beta}(f_1, \ldots, f_r) = \max_{\alpha \in A} \min_{\beta \in B}\{\bar{t}_n\gamma_n \cdot w_{\alpha\beta}(f_1, \ldots, f_r)\}.$$

By (8),

$$\begin{aligned} \bar{t}_n\gamma_n \cdot w_n\varphi &= \max_{\alpha \in A} \min_{\beta \in B}\{\bar{t}_n \cdot w_{\alpha\beta}(x_1\theta_n, \ldots, x_r\theta_n)\gamma_n\} = \\ &= (\max_{\alpha \in A} \min_{\beta \in B}\{\bar{t}_n w_{\alpha\beta}(x_1, \ldots, x_r)\theta_n\}\gamma_n = \\ &= (\bar{t}_n(\bigvee_{\alpha \in A} \bigwedge_{\beta \in B} w_{\alpha\beta}(x_1, \ldots, x_r))\theta_n)\gamma_n. \end{aligned}$$

Since

$$\bar{t}_n(\bigvee_{\alpha \in A} \bigwedge_{\beta \in B} w_{\alpha\beta}(x_1, \ldots, x_r))\theta_n = \bar{t}_n \cdot w_n\theta_n \neq \bar{t}_n,$$

then φ is a faithful representation F in $A(\mathbf{R})$.

In order to complete the proof of the theorem we need only to prove that the orbit of $F\varphi$ containing $\bar{t}_n\gamma_n$ possesses all positive integers n. In fact, the orbit $\bar{t}_n\gamma_n \cdot F\varphi$ of l-group $F\varphi$ containing $\bar{t}_n\gamma_n$ possesses the subset $S_n\gamma_n$. In cases 1.1 and 1.2 $a_n \in T_n$, $a_n = u_n$, $u_n\gamma_n = n$. Therefore, $n \in \bar{t}_n\gamma_n \cdot F\varphi$. In case 1.3 $a_n \cdot x_i\theta_n > a_n$ and $n \cdot f_i^2 = u_n\gamma_n \cdot f_i^2 = u_n^i\gamma_n \cdot f_i = a_{n2}\gamma_n \cdot f_i = a_n\gamma_n \in \bar{t}_n\gamma_n \cdot F\varphi$. Hence $n \in \bar{t}_n\gamma_n \cdot F\varphi$.

In case 1.4 $a_n \in S_n$, $n \cdot f_1^2 f_2^{-1} = u_n\gamma_n \cdot f_1^2 f_2^{-1} = u_n^{(1)}\gamma_n \cdot (f_1 f_2^{-1}) = a_{n3}\gamma_n f_2^{-1} = a_n^{(2)}\gamma_n \cdot f_2^{-1} = a_n\gamma_n \in \bar{t}_n\gamma_n \cdot F\varphi$. Therefore, $n \in \bar{t}_n\gamma_n \cdot F\varphi$.

In case 2.1 we have $n = v_n\gamma_n = b_n^{(i)}\gamma_n$. Since $b_n \in S_n$, then $n \in \bar{t}_n\gamma_n \cdot F\varphi$. In case 2.2 for $i \neq i(1)$ holds $n = b_n\gamma_n \cdot f_{i(1)}^{-1} f_i^2$ and $n \in \bar{t}_n\gamma_n \cdot F\varphi$.

Thus, the orbit $\bar{t}_1\gamma_1 \cdot F\varphi$ contains all positive integers n and therefore $\bar{t}_n\gamma_n \in \bar{t}_1\gamma_1 F\varphi$.□

COROLLARY 1. (Kopytov [8].) *The lattice order of the free l-group F can be extended to the total order of F.*

Proof. It is clear that the pointwise lattice order of D_l^* can be extended to the total order on D_l^*. (see Section 7 of chapter 4.) Since F is not representable for $r > 1$, then the lattice order of F is not the intersection of total orders of F.□

COROLLARY 2. (Kopytov [8].) *The free l-group F is a group with the unique roots and has a soluble infrainvariant system of subgroups with Abelian torsion-free factors.*□

Let F be a free l-group of countable rank. By modifying the representation φ of Theorem 2, we can construct the representation $\hat{\varphi}$ of F into D_l^* such that supp$\hat{\varphi} \subseteq [0,1]$ for all $g \in F$ and $\mathbf{Q} \cap [0,1] \cdot F\hat{\varphi} \subseteq \mathbf{Q} \cap [0,1]$ where $\mathbf{Q}$ is the set of rational numbers.

10.2 Approximability of Free l-Groups

1⁰. Here we describe free l-groups in arbitrary l-varieties.

Let $F_0 = F_0(X)$ be a free group with free generators $X = \{x_i | i \in I\}$ with the trivial partial order. Let $\{\leq^\lambda \mid \lambda \in \Lambda\}$ be the set of all right orders on F_0. For each $\lambda \in \Lambda$, let F_0^λ be the ro-group obtained when F_0 is endowed with the right order $\leq^\lambda$ and P_0^λ be the positive cone of F_0^λ. By R_λ we denote the right regular represention of F_0^λ by order permutations of the totally ordered set $\{F_0^\lambda; \leq^\lambda\}$ i.e.,

$$R_\lambda : F_0^\lambda \to A(\{F_0^\lambda; \leq^\lambda\}) = A(F_0^\lambda) = A_\lambda,$$

$$xR_\lambda(g) = xg \text{ for } x \in \{F_0^\lambda; \leq^\lambda\}, \quad g \in F_0^\lambda.$$

Let B_λ be an l-subgroup of l-group A_λ generated by $R_\lambda(G)$, $B = \bar{\prod\limits_{\lambda \in \Lambda}} B_\lambda$ be the Cartesian product of l-groups B_λ and φ be a mapping F_0 into B for which $(x\varphi)(\lambda) = R\lambda(x)$ for all $x \in F_0, \lambda \in \Lambda$. Then the l-subgroup of B generated by $F_0\varphi$ is a free l-group with free generators $\{x_i\varphi | i \in I\}$ (see Section 1 of chapter 7). This construction of the free l-group was due to P.Conrad [5] for the l-variety $\mathcal{L}$ of all l-groups, for the l-variety $\mathcal{R}$ of representable l-groups and to E.Weinberg [2] for the l-variety $\mathcal{A}$ of all Abelian l-groups. Conrad's construction of the free l-group mentioned above is the direct generalization of Weinberg's construction of free Abelian l-group. It is interesting to note that a free l-group with one generator is l-isomorphic to the direct product of two totally ordered infinite cyclic groups (G.Birkhoff [3]). These constructions have been extended by Kopytov [6] to arbitrary l-varieties.

Let G be a right-ordered group and H be a convex subgroup of G. Then the right regular represention R of group G by order permutations of the totally ordered set $\{G; \leq\}$ induces the representation of G by order permutations of the totally ordered set $R(G : H)$ of right cosets G by H. This representation is also denoted by R and is called the right regular representation of G by order permutations of $R(G : H)$ corresponding to right order $\leq$ on G.

LEMMA 1. *Let $X = \{x_i | i \in I\}$ be the set of free generators of free l-group F in an l-variety $\mathcal{X}$ and F_0 be a subgroup of F generated by X. Then F_0 is a free group in the quasivariety $\mathcal{K}(\mathcal{X})$ of groups which can be embedded in l-groups of $\mathcal{X}$ and X is a set of free generators for F_0.*

Proof. Let G be any group in $\mathcal{K}(\mathcal{X})$ and $Y = \{y_i | i \in I\}$ be a subset of G. We claim that any mapping $\varphi : x_i \to y_i$ can be exteded to the homomorphism from F_0 into G. Since $G \in \mathcal{K}(\mathcal{X})$, there exists an l-group H in $\mathcal{X}$ such that G is a subgroup of H. The l-group F is free in $\mathcal{X}$. Therefore, the mapping $\varphi : x_i \to y_i$ $(i \in I)$ can be extended to the homomorphism from F into H. Then the restriction of this homomorphism onto F_0 is a homomorphism of group F_0 into G.□

It is clear that the elements $\{x_i | i \in I\}$ are incomparable with e in F. Moreover, the lattice order of F induces the trivial order on the subgroup F_0.

Now we extend the notation from the beginning of this section to arbitrary l-varieties. Later in this section $F = F_{\mathcal{X}}$ denotes a free l-group in l-variety $\mathcal{X}$ with the set of free generators $X = \{x_i | i \in I\}$ and F_0 denotes a free group in the quasivariety $\mathcal{K}(\mathcal{X})$. We assume that F_0 is a subgroup of F generated by X. Let $\{P_\lambda | L \in \Lambda\}$ be the set of all right orders on F_0. For each $\lambda \in \Lambda$ let F_0^λ be the ro-group F_0 with right order $\leq^\lambda$ and positive cone P_0^λ. If H is a convex subgroup of F_0^λ, then R_λ is the right regular representation of the group F_0^λ by order permutations of $R(F_0^\lambda : H)$ where $Hx \cdot R_\lambda(g) = Hxg$ for $x, g \in F_0^\lambda$. Let $A_\lambda(H)$ be an l-subgroup of l-group $A(\{(G : H); \leq^\lambda\})$ generated by $R_\lambda(F_0^\lambda)$. For every $\lambda \in \Lambda$ let H_0^λ be the least convex subgroup of F_0^λ such that $A_\lambda(H) \in \mathcal{X}$. It is clear that H_0^λ exsists in F_0^λ and is unique. We write now $A_\lambda(H_0^\lambda) = A_\lambda$. Let F^* be the Cartesian product $\bar{\prod\limits_{\lambda \in \Lambda}} A_\lambda$ of l-groups A_λ, $\lambda \in \Lambda$ and $\bar{F}$ be the l-subgroup of F^* generated by the elements $\{\bar{x}_i | i \in I\}$ such that $\bar{x}_i(\lambda) = R_\lambda(x_i)$ for all $i \in I$, $\lambda \in \Lambda$.

THEOREM 1. (Kopytov [6].) *For each l-variety $\mathcal{X}$ the l-group $\bar{F}$ is l-isomorphic to the free l-group $F = F_{\mathcal{X}}$ of l-varietiy $\mathcal{X}$ and elements $\{\bar{x}_i | i \in I\}$ are free generators for $\bar{F}$.*

Proof. Let $\{H^\gamma | \gamma \in \Gamma\}$ be the set of all minimal primes l-subgroups of the free l-group $F = F_{\mathcal{X}}(X)$ of $\mathcal{X}$. All sets $R(F : H^\gamma)$ are totally ordered and by Theorem 4.1.1, the l-group F is approximated by l-groups F/N_γ where N_γ is the kernel of representation $R_{H^\gamma} : H^\gamma x \cdot R_{H^\gamma}(g) = H^\gamma xg$ of F in $A(R(F : H_\gamma)$. It is clear that $F/N_\gamma \in \mathcal{X}$ and $F \cdot R_{H^\gamma}$ is an l-subgroup of $A(R(F : H^\gamma))$ generated by $\{x_i \cdot R_{H^\gamma} | x_i \in F_0,\ i \in I\}$.

Let P be any right order on F. Now we define a new right order P^γ for every $\gamma \in \Gamma$ by the rule: $x \leq^\gamma y$ iff $H^\gamma x < H^\gamma y$ in the totally ordered set $R(F : H^\gamma)$ of right cosets F by H^γ or $H^\gamma x = H^\gamma y$ and $x \leq y$ in right order P. The direct verification shows that $\leq^\gamma$ is a total order on the set F. Let $x, y, z \in F$ and $x \leq^\gamma y$. If $H_\gamma x < H^\gamma y$ in $R(F : H^\gamma)$, then by Proposition 3.3.2, $H^\gamma xz < H^\gamma yz$ in $R(F : H^\gamma)$ and $xz \leq yz$. If $H^\gamma x = H^\gamma y$, then $x \leq y$ in P. Thus $H^\gamma xz = H^\gamma yz$, $xz \leq yz$ under P and $xz \leq^\gamma yz$. Therefore, $\leq^\gamma$ is a right order on F. Let P^γ be a positive cone ro-group $F^\gamma = \{F; \leq^\gamma\}$.

The subgroup H^γ is a convex subgroup in the ro-group F^γ. Moreover, the total order on $R(F : H^\gamma)$ induced by right order $\leq^\gamma$ of F^γ coincides with the total order induced on the $R(F : H_\gamma)$ by the lattice order of F. This means that we can consider the right regular representation R_γ of l-group F by order permutations of $R(F : H^\gamma)$ corresponding the right order P^γ on F^γ.

Let x be an element of l-group F and $x = \bigvee_{k\in K}\bigwedge_{j\in J} w_{kj}$ where K, J are finite index sets and $w_{kj} \in F_0$. Since the map $x \to H^\gamma x$ is a homomorphism of lattice F onto totally ordered set $R(F : H^\gamma)$ (see Proposition 3.3.1), then $H^\gamma x = H^\gamma(\bigvee_{k\in K}\bigwedge_{j\in J} w_{kj}) = \bigvee_{k\in K}\bigwedge_{j\in J} H^\gamma w_{kj} = h^\gamma w_{\bar{k}\bar{j}}$ for some $\bar{k} \in K$, $\bar{j} \in J$. Hence, each coset $H^\gamma x$ contains an element of the subgroup F_0. Let $H_0^\gamma = H^\gamma \cap F_0$ and $R(F_0^\gamma : H_0^\gamma)$ be the totally ordered set of right cosets of $F_\wedge^\gamma$ of F_0^γ by H_0^γ under induced right order $P^\gamma \cap F_0$ of F_0. The mapping θ defined by the rule

$$H_0^\gamma x \cdot \theta = H^\gamma x \text{ for } x \in F_0$$

is an order homomorphism of the totally ordered set $R(F_0^\gamma : H_0^\gamma)$ into $R(F : H^\gamma)$. In fact, θ maps $R(F_0^\gamma : H_0^\gamma)$ onto $R(F : H^\gamma)$. If $H_0^\gamma x < H_0^\gamma y$ and $x, y \in F_0$, then $h <^\gamma yx^{-1}$ for all $h \in H_0^\gamma$. Since H^γ is a convex subgroup of F in the right order P^γ, $h <^\gamma yx^{-1}$ for all $h \in H^\gamma$. Therefore, $H^\gamma x < H^\gamma y$ in $R(F : H^\gamma)$ and θ is an order isomorphism.

For each free generator x_i of F_0 and right regular representation R_γ of the right-ordered group F_0^γ by order permutations of the totally ordered set $R(F_0^\gamma : H_0^\gamma)$,

$$(H_0^\gamma x \cdot R_\gamma(x_i))\theta = (H_0^\gamma x x_i)\theta = H^\gamma x x_i = H^\gamma x \cdot R_\gamma(x_i).$$

Therefore, the l-subgroup of the l-group $A(R(F_0))^\gamma : H_0^\gamma))$ generated by order permutations $R_\gamma(x_i)$ is l-isomorphic to the l-group $R_\gamma(F)$ and the mapping $\varphi_0^\gamma :\ F \to A(R(F_0^\gamma : H_0^\gamma))$ for which

$$(\bigvee_{k\in K}\bigwedge_{j\in J} w_{kj})\varphi_0^\gamma = \bigvee_{k\in K}\bigwedge_{j\in J} R_\gamma(w_{kj})$$

is a representation of l-group F by order permutations of the totally ordered set $R(F_0^\gamma : H_0^\gamma)$ with the kernel $N_\gamma = \ker R_\gamma$.

Since the set of all right orders $\{P_0^\gamma = P^\gamma \cap F_0 | \gamma \in \Gamma\}$ is contained in the set $\{P_0^\lambda | \lambda \in \Lambda\}$ of all right orders of the group F_0, the mapping $x_i \to \bar{x}_i$ can be extended to the l-isomorphism of l-group F onto l-subgroup $\bar{F}$ of l-group F^*.□

COROLLARY 1. (Kopytov [6].) *Let a free l-group F of l-variety $\mathcal{X}$ have a transitive faithfull representation by order permutations of* *totally ordered set. Then there is the right order P_0 on the free group F_0 of the quasivariety $\mathcal{K}(\mathcal{X})$ and convex subgroup H_0 of ro-group F_0 with the order P_0 such that the the mapping $x_i \to R(x_i)$ can be extended to an l-isomorphism from l-group F into the l-group $A(R(F_0 : H_0))$.*□

COROLLARY 2. (Kopytov[6].) *There exists some right order P_0 of free group F_0 with free generators $X = \{x_1, \ldots, x_r\}$, $(r \geq 2)$, for which the mapping $x_i \to R(x_i)$, $1 \leq i \leq r$, extends to the faithful representation of free l-group F with set X of free generators by order permutations of the totally ordered set (F_0, P_0).*□

2^0. For some l-varieties $\mathcal{X}$ the construction of a free l-groups in $\mathcal{X}$ is not so complicated as in Theorem 1. Let $\mathcal{R}$ be the l-variety of all representable l-groups, and $F = F_{\mathcal{R}}(X)$ be an l-group free in $\mathcal{R}$ with the set of free generators X. Let F_0 be a free group with the set X of free generators and F_0 be a subgroup of F. By Corollary 1 of Theorem 9.3.5, each minimal prime subgroup H^λ of F is an ideal. Hence the quotient group $\bar{F}_0 = F_0 H^\lambda / H^\lambda$ is totally ordered and F_0/H_0^λ is an o-group under the induced total order $\bar{P}_o^\lambda$ where $H_0^\lambda = H^\lambda \cap F_0^\lambda$. Let P_0^λ be any total order on F_0^λ for which the subgroup H^λ is convex and P_0^λ induces the total order $\bar{P}_0^\lambda$ on F_0/H_0^λ. Let F_0^λ be group F_0 totally ordered by P_0^λ. It is not difficult to see that the kernel of the right regular representation R_λ of o-group F_0^λ by order permutations of the totally ordered set $R(F_0^\lambda : H_0^\lambda)$ is contained in the kernel of the right regular representation of F_0^λ by order permutations of the totally ordered set $R(F_0^\lambda : E) = F_0^\lambda$. Therefore the following result is true.

THEOREM 2. (Conrad [5].) *Let $\{P_0^\lambda | \lambda \in \Lambda\}$ be the set of all total orders of free group F_0 with the set of free generators $X = \{x_i | i \in I\}$, F_0^λ be a free group F_0 totally ordered by P_0^λ and $F^* = \overline{\prod_{\lambda \in \Lambda}} F_0^\lambda$. Then the mapping φ for which $x_i \to \bar{x}_i$, $\bar{x}_i(\lambda) = R_\lambda(x_i)$, $i \in I$, can be extended to an l-isomorphism of a free representable l-group F into the l-group F^*.*□

Let l-variety $\mathcal{X}$ satisfy the following property:

(1) $\mathcal{X} \subseteq \mathcal{R}$ and o-group G is contained in $\mathcal{X}$, then G is o-isomorphic to a quotient group of o-group F_0^λ where F_0^λ is a totally ordered free group in $\mathcal{K}(\mathcal{X})$ and $F_0^\lambda \in \mathcal{X}$.

THEOREM 3. (Conrad [5].) *Let $\mathcal{X}$ be an l-variety with property (1), $F_0 = F_0(X)$ be a free in $\mathcal{K}(\mathcal{X})$ group with the set $X = \{x_i | i \in I\}$ of free generators. Let $\{P_0^\lambda | \lambda \in \Lambda\}$ be a set of all total orders of F_0 such that $F_0^\lambda = (F_0; P_0^\lambda) \in \mathcal{X}$ for all $\lambda \in \Lambda$,* $F^* = \overline{\prod_{\lambda\in\Lambda}} F_0^\lambda$.

Then the mapping φ of the free generators x_i, $i \in I$, of the l-group F free in $\mathcal{X}$ into F^* such that $x_i\varphi(\lambda) = R_\lambda(x_i)$ can be extended to an l-isomorphism of F into F^*.□

We leave the details of the proof to the reader. Each l-variety defined by identities of the signature of the group theory has property (1). The l-variety $\mathcal{W}a$ of all weakly abelian l-groups has this property as well, but of course, there are l-varieties without property (1).

EXAMPLE 1. Let $\mathcal{X}$ be an l-variety defined by the identities:

$$|[[[x_1, y_1], [z_1, t_1]], [u_1, v_1]]| \wedge |[[[x_2, y_2], z_2], [[t_2, u_2], v_2]| = e,$$
$$(x \wedge y^{-1}x^{-1}y) \vee e = e.$$

It is clear that l-variety $\mathcal{X}$ does not coincide with the l-variety of all representable l-groups which are extensions of nilpotent groups of class 2 by Abelian groups and with l-variety of representable l-groups which are extensions of Abelian groups by nilpotent groups of class 2.

It can be readily verified that free group F_0 in the quasivariety $\mathcal{K}(\mathcal{X})$ is isomorphic to a free group in the variety of groups defined by the identity

$$[[[[x_1, y_1], [z_1, t_1]], [u_1, v_1]], [[[x_2, y_2], z_2], [[t_2, u_2], v_2]] = e.$$

It is clear that $(F_0; P) \notin \mathcal{X}$ for each total order P on F_0.

Let now $\mathcal{X}$ be an l-variety and $\mathcal{X} \in \mathcal{N}$. Then for the construction of free l-group in $\mathcal{X}$ we need to consider on the free in $\mathcal{K}(\mathcal{X})$ group F_0 the Conradian right orders only, i.e., the right orders on F_0 such that the system of convex subgroups are soluble (see Theorem 8.2.4).

10.3 Direct Indecomposability of Free l-Groups

It is well-known in the theory of groups that each subgroup of a free group F is free, too. For free l-groups the situation is more complicated. For example, in free l-group F with r $(r \geq 2)$ free generators there are non-identity orthogonal elements. Therefore, F contains l-subgroups which are direct products of non-trivial l-subgroups. It is also known that each free Abelian group is a direct product of infinite cyclic groups. The same result is not valid for free Abelian l-groups.

1°. THEOREM 1. (Weinberg [2], Chisamiev [1].) *A free Abelian l-group is approximated by totally ordered additive groups of integers.*

Proof. (Glass, Holland, McCleary [1].) We use in this theorem the additive notation for the group operation. Without loss of generality we can assume that $F = F_{\mathcal{A}}$ is finitely generated. Let $X = \{x_1, \ldots, x_n\}$ be the set of free (in $\mathcal{A}$) generators of F, and $w = w(x_1, \ldots, x_n) \in F$, $w \neq e$. Then there is an Abelian o-group G generated by elements $a_1, \ldots, a_n$ for which $w(a_1, \ldots, a_n) \neq e$. Let

$$w(a_1, \ldots, a_n) = \bigvee_{\alpha \in A} \bigwedge_{\beta \in B} w_{ij}(a_1, \ldots, a_n) > e$$

in G. Here $w_{\alpha\beta}(a_1, \ldots, a_n)$ are reduced words of the Abelian groups theory signature from the variables $a_1, \ldots, a_n$, A, B are finite sets of indeces. Thus

$$w_{\bar{\alpha}\beta}(a_1, \ldots, a_n) > e \tag{1}$$

for some $\bar{\alpha} \in A$ and all $\beta \in B$. Since G is a finitely generated Abelian l-group, G is a subgroup of the antilexicographic sum $R_1 \overleftarrow{+} \cdots \overleftarrow{+} R_r$ of copies of the reals. Then the typical element of the set $D = \{w_{\bar{\alpha}\beta} | \beta \in B\}$ is $d = (d_1, \ldots, d_r)$, $d_1 \in R_1, \ldots, d_r \in R_r$ and $d_j > 0$ where $j = \max\{k | d_k \neq 0,\ 1 \leq k \leq r\}$. Let now

$$\frac{1}{c_i} = \min\{|d_i| \mid d_i \neq 0\} \text{ or } c_i = 0,\ 1 \leq i \leq r.$$

Let $k_1 = 1$, and for $2 \leq i \leq r$ let

$$k_i > \max_{d \in D} |d_1 k_1 c_1 d_2 k_2 c_2 + \cdots + d_{i-1} k_{i-1} c|.$$

Now define $\psi : R_1 \overleftarrow{+} \cdots \overleftarrow{+} R_r \to \mathbf{R}$ by the rule $(b_1, \ldots, b_r)\psi = \sum_{t=1}^{r} b_i k_i c_i$ where $b_i \in R_i$ $(i = 1, 2, \ldots, r)$. Then ψ is a homomorphism of the direct sum $R_1 + R_2 + \cdots R_r$ into $\mathbf{R}$ and ψ preserves the signs of $D = \{w_{\bar{\alpha}\beta} | \beta \in B\}$. For if $d = (d_1, \ldots, d_r)$ is an element of D with $d_j > 0$, then

$$\begin{aligned} d_j k_j c_j > d_j c_j |d_1 k_1 c_1 + \cdots + d_{j-1} k_{j-1} c_{j-1}| \geq \\ \geq |d_1 k_1 c_1 + \cdots + d_{j-1} k_{j-1} c_{j-1}| \end{aligned}$$

which implies $d\psi < 0$ for all $d \in D$ in $\mathbf{R}$.

Hence we may maintain that G is a subgroup of R and $a_1, \ldots, a_n$ are reals. The subgroup of G generated by $a_1, \ldots, a_n$ is a free Abelian group with the basis $u_1, \ldots, u_m$, say. As before, let

$$D = \{w_{\bar{\alpha}\beta}(a_1, \ldots, a_n) | \beta \in B.\}$$

Each $d \in D$ has the form $d = \sum_{i=1}^{m} f_i(d) u_i > 0$ where each $f_i(d)$ is an integer. Thus $(u_1, \ldots, u_m)$ is the solution of the system of inequalities $\sum_{i=1}^{m} f_i(d) z_i > 0$. By continuity, this system must also have a solution in rational numbers and therefore must have a solution in integers $(k_1, \ldots, k_m)$. That is, $\sum_{i=1}^{m} f_i(d) k_i > 0$. Let $\varphi : G \to \mathbf{Z}$ be a homomorphism of G into $\mathbf{Z}$ defined by the rule

$$(\sum_{i=1}^{m} f_i(d) k_i u_i)\varphi = \sum_{i=1}^{m} f_i(d) k_i.$$

Then $w_{\bar{\alpha}\beta}(k_1, \ldots, k_m) > 0$ in $\mathbf{Z}$ for all $\beta \in B$. This means that there is an l-homomorphism $\tau : F \to \mathbf{Z}$ for which $w(x_1, \ldots, x_n)\tau > 0$ in $\mathbf{Z}$.□

COROLLARY 1. (Weinberg [2].) *Each free Abelian l-group is Archimedean.*

Proof. The class $\mathcal{A}rch$ of all Archimedean l-groups is closed under Cartesian products and subgroups and the totally order group $\mathbf{Z}$ of integers is contained in $\mathcal{A}rch$.□

COROLLARY 2. (Weinberg [2].) *The smallest non-trivial l-variety is the l-variety $\mathcal{A}$ of all Abelian l-groups.*

Proof. Every non-trivial l-variety $\mathcal{X}$ contains the totally ordered infinite cyclic group. Therefore the free Abelian l-group with the countable set of free generators is contained in $\mathcal{X}$.□

2⁰. LEMMA 1. (Bernau [2].) *Let F be a free Abelian l-group with the set of free generators $X = \{x_i | i \in I\}$ and F_0 be a subgroup of F generated by X. Then for each finite set $\{a_1, \ldots, a_n\}$ of non-identity elements in F_0 and for each $u, v \in F$ such that $u \wedge v = e$ and $uv = a_1^+ \cdots a_n^+$ is valid $u = e$ or $v = e$.*

Proof. It is clear that for any l-homomorphism $\varphi : F \to \mathbf{R}$, $u\varphi = 0$ or $v\varphi = 0$. Let Φ be the family of all l-homomorphisms from F into $\mathbf{R}$ such that $(uv)\varphi \neq 0$. We claim that if $u\varphi = 0$ for some $\varphi \in \Phi$, then $u\psi = 0$ for all $\psi \in \Phi$. Assume, on the contrary, that $u\psi \neq 0$ for some $\psi \in \Phi$.

Let $\theta_t = t\varphi + (1-t)\psi$ be a map from F_0 into R. If $t \in [0,1]$, then θ_t is a homomorphism from F_0 into $\mathbf{R}$. Then there exists an l-homomorphism from F into $\mathbf{R}$ extending θ_t. We denote it by θ_t, too. It is clear that $(uv)\theta > 0$. Therefore, $\theta \in \Phi$. Let

$$U = \{t | t \in [0,1],\ u\theta_t = 0\}, \quad V = \{t | t \in [0,1],\ v\theta_t = 0\}.$$

Since $u\theta_1 = u\varphi = 0$ and $v\theta_0 = v\psi = 0$, then $U \neq \emptyset$, $V \neq \emptyset$. It is clear that $U \cup V = [0,1]$. Since θ_t is a counttinuous function from $[0,1]$ into $\mathbf{R}$ and the subsets U, v are closed in the natural topology of the segment $[0,1]$, then $U \cap V \neq \emptyset$. Therefore there is $\bar{t} \in U \cap V$. Then $(uv)\theta_{\bar{t}} = u\theta_{\bar{t}} + v\theta_{\bar{t}} = 0$ contradicts the choice $\theta_t \in \Phi$ for all $t \in [0,1]$.□

THEOREM 2. (Bernau [2].) *A free Abelian l-group $F = F_{\mathcal{A}}(r)$ with r free generators is decomposed into direct product of l-groups if and only if $r = 1$.*

Proof. If $r = 1$, then F is a free Abelian l-group with one free generator x. Then $F = (x^+) \times (x^-)$ where (x^+), (x^-) are totally ordered infinite cyclic groups generated by x^+ and x^- correspodingly.

Let F be a free Abelian l-group with the set $X = \{x_i | i \in I\}$ of free generators and $F = U \times V$ where convex l-subgroups U, V are nontrivial. Let F_0 be an subgroup of F generated by X and $a = uv$ where $u \in U$, $v \in V$. Then $a^+ = u^+ \cdot v^+$ and $u^+ \wedge v^+ = e$. By Lemma 1, $u^+ = e$ or $v^+ = e$. Therefore, $a^+ \in U$ or $a^+ \in V$. By similar arguments, $a^- \in U$ or $a^- \in V$. Let us assume that $a^+ \in U$. If $a^- \in U$, then $|a| \in U$, $a \in U$. By Lemma 1, the polar $(a)^\perp$ is an identity subgroup E. Hence $V \subseteq (a)^\perp = E$ and $V = E$, a contradiction. So we can assert $a^+ \in U$, $a^- \in V$ or $a^- \in U$, $a^+ \in V$ for every $a \in F_0$.

Let a, b be rationally independent elements of F_0. Without loss of generality, we can suppose that $a^+, b^+ \in U$, $a^-, b^- \in V$. From results of Chapter 6, there is a total order P on F_0 for which $a \in P$, $b \in P^{-1}$. The identical isomorphism of F_0 onto totally ordered group $(F_0; P)$ can be extended to the l-homomorphism ψ from free Abelian l-group F onto $(F_0; P)$. Then $a^+\psi = a$, $a^-\psi = b^+\psi = e$, $b^-\psi = b$. Hence in the o-group $(F_0; P)$ holds $e = (a^+ \wedge (b^-)^{-1})\psi = a^+\psi \wedge (b^-)^{-1}\psi = a \wedge b^{-1}$. This contradicts our assumption. Therefore the subgroup F_0 does not contain two rationally independed elements and F_0 is an infinite cyclic group.□

THEOREM 3. (Medvedev [4].) *Let $\mathcal{X}$ be an arbitrary non-trivial l-variety and F be a free l-group in l-variety $\mathcal{X}$. Then F is decompossed into the direct product of l-groups if and onli if F has one generator.*

Proof. Let F be a free in $\mathcal{X}$ l-group with the set $X = \{x_i | i \in I\}$ of free generators. If the cardinality $|I|$ of I is 1, then F is a direct product of two infinite cyclic o-groups.

Now, let $\mathcal{X}$ be the l-variety $\mathcal{L}$ of all l-groups and $|I| > 1$. By Theorem 10.1.2, the l-group F has a faithful transitive representation and by Corollary 4 of Theorem 4.3.1, F is finitely subdirectly irreducible. Therefore it is indecomposable into direct product.

If $\mathcal{X}$ is the l-variety $\mathcal{A}$ of all Abelian l-groups the result follows from Theorem 2.

So we can assert that $\mathcal{X} \neq \mathcal{L}, \mathcal{A}, \mathcal{E}$ where $\mathcal{E}$ is a trivial l-variety. Hence by Theorem 9.2.2 and Corollary 2 of Theorem 1, $\mathcal{A} \subset \mathcal{X} \subseteq \mathcal{N}$. Let K be an ideal of F generated by the derived subgroup $[F, F]$. Then F/K is l-isomorphic to the free Abelian l-group $F_{\mathcal{A}}$ with the set X of free generators. Assume, on the contrary, that $F = A \times B$. Then

$F/K = \bar{A} \times \bar{B}$ where $\bar{A}, \bar{B}$ are images of A, B under the natural l-homomorphism of F onto $F_{\mathcal{A}}$. Since $F_{\mathcal{A}}$ is indecomposable into direct product, then $A \subseteq K$ or $B \subseteq K$. Suppose that $A \subseteq K$ and $|I| = r$ is a finite number. By Theorem 9.1.2, F is a subcartesian product of subdirectly irreducible finitely generated l-groups G_λ $(\lambda \in \Lambda)$. Let $\varphi_\lambda : F \to G_\lambda$ be an l-homomorphism from F onto G_λ. Then $F\varphi_\lambda = A\varphi_\lambda \times B\varphi_\lambda = G_\lambda$ and $A\varphi_\lambda = E$ or $B\varphi_\lambda = E$ in G_λ. Let $B\varphi_\lambda = E$. Then $A\varphi_\lambda = G_\lambda = K\varphi_\lambda$ and therefore G_λ coinsides with the convex l-subgroup of G_λ generated by the derived subgroup $[G_\lambda, G_\lambda]$. Since $G_\lambda \in \mathcal{N}$ is finitely generated, then by Lemma 9.5.2, G_λ coinsides with the convex l-subgroup $(g)_l$ generated by one element $g \in G_\lambda$. Then the value C of g in G_λ is the ideal of G_λ and G_λ/C is an Abelian l-group. Hence the convex l-subgroup of G_λ generated by the derived subgroup $[G_\lambda, G_\lambda]$ is a proper ideal of G_λ. A contradiction to our assumption $B\varphi_\lambda = E$. Thus $A\varphi_\lambda = E$ for all $\lambda \in \Lambda$ and $A = E$ in F, or equivalently, F is indecomposable into direct product.

Let now $|I|$ be an infinite cardinal and $F = A \times B$ where $A \neq E, B \neq E$. Then there are $a \neq e$, $a \in A$, $b \neq e$, $b \in B$ such that $a = v(x_1, \ldots, x_r)$, $b = w(x_1, \ldots, x_r)$, where v, w are l-words. Let $\bar{F}$ be a free in $\mathcal{X}$ l-group with the free generators $\{\bar{x}_1, \ldots, \bar{x}_r\}$. Then the mapping φ of $X = \{x_i | i \in I\}$ into $\bar{F}$ is defined by the rule :

$$x_i\varphi = \begin{cases} \bar{x}_i & \text{if } 1 \leq i \leq r, \\ e & \text{if i¿r}, \end{cases}$$

can be extended to the l-homomorphism from F onto $\bar{F}$ and

$$F\varphi = (A \times B)\varphi = A\varphi \times B\varphi, \quad A\varphi \neq E, \quad B\varphi \neq E.$$

A contradiction to the direct indecomposibility of finitely generated free in $\mathcal{X}$ l-group with finite set of free generators.□

The centralizers of free generators of a free in $\mathcal{L}$ l-group F were investigated by A.K.Arora and S.H.McCleary [1]. By the use of o-2-transitive representation of F, S.H.McCleary [7] proved that F has no basic elements and is not completely distributive. In [8] S.H.McCleary described the root systems of prime subgroups of free l-groups with finite number of free generators and proved that these systems are isomorphic to each other.

Chapter 11

The semigroup of l-varieties

The set of all l-varieties $\mathbf{L}$ is a semigroup where, for any two l-varieties $\mathcal{X}, \mathcal{Y}$ their *product* is

$$\mathcal{X} \cdot \mathcal{Y} = \{G \in \mathcal{L} | G \text{ has an ideal } A \text{ such that } A \in \mathcal{X} \text{ and } G/A \in \mathcal{Y}\}.$$

It is easily verified that $\mathcal{X} \cdot \mathcal{Y}$ is an l-variety and the operation of multiplication of l-varieties is associative. It is evident that l-variety $\mathcal{L}$ of all l-groups is zero and l-variety $\mathcal{E} = \{E\}$ is the unit of the semigroup of l-varieties. One of the powerful and very useful tools for the investigation of the semigroup of l-varieties $\mathbf{L}$ is the wreath product of ordered permutation l-groups. This notion was introduced by W.Ch.Holland and S.H.McCleary [1].

11.1 The Wreath Product

Here we use notations and terminology of Chapter 4. We first define the *wreath product* of two ordered permutation l-groups. Let (G, Ω), (H, T) be transitive ordered permutation l-groups. For each $\tau \in T$ let G_τ be an l-group identitically isomorphic to G. An element g of the Cartesian product $\overline{\prod\limits_{\tau \in T}} G_\tau = \bar{G}$ will be regarded as a function with domain of definition T and range of values G. It is will be denoted by

$$g = \{g_\tau | \tau \in T\} = \{g_\tau\}.$$

Here $g_\tau \in G$ for each $g \in \bar{G}$, $\tau \in T$.

Let $\Lambda = \Omega \overleftarrow{\times} T$ and

$$W = \{(\{g_\tau\}, h) | h \in H,\ g_\tau \in G \text{ for all } \tau \in T\}.$$

Define binary operation $\cdot$ and unary operation $^{-1}$ on W by the rules:

$$(1) \qquad (\{f_\tau\}, h_1) \cdot (\{g_\tau\}, h_2) = (\{a_\tau\}, h_1 h_2),$$

where $a_\tau = f_\tau \cdot g_{\tau h_1}$ $(\tau \in T)$;

$$(2) \qquad (\{g_\tau\}, h)^{-1} = (\{g'_\tau\}, h^{-1}),$$

where $g'_\tau = g^{-1}_{\tau h^{-1}}$ $(\tau \in T)$.
For $(\alpha, \sigma) \in \Lambda$ where $\alpha \in \Omega$, $\sigma \in T$ let

$$(3) \qquad (\alpha, \sigma) \cdot (\{g_\tau\}, h) = (\alpha g_\sigma, \sigma h).$$

The direct verificaton shows that $(\{g_\tau\}, h)$ is an order permutation of the totally ordered set $\Lambda = \Omega \overleftarrow{\times} T$.

LEMMA 1. (Holland and McCleary [1].) *Let (G, Ω), (H, T) be transitive ordered permutation l-groups. Then (W, Λ) is a transitive ordered permutation l-group with the identity $e = (e_\tau, e)$, where $e_\tau = e$ in G_τ, $(\tau \in T)$.*

Proof. The easy verification shows that (W, Λ) is an ordered permutation group with the identity $e = (e_\tau, e)$, where $e_\tau = e$ in G_τ $(\tau \in T)$. Let $(\alpha, \sigma), (\alpha', \sigma') \in \Lambda$. By transitivity (G, Ω) and (H, T) there are $g \in G$, $h \in H$ such that $\alpha g = \alpha'$, $\sigma h = \sigma'$. Let $w = (\{g_\tau\}, h) \in W$ where $g_\sigma = g$ and $g_\tau = e$ for all $\tau \in T$, $\tau \neq \sigma$. Then $(\alpha, \sigma)w = (\alpha, \sigma)(\{g_\tau\}, h) = (\alpha g_\sigma, \sigma h) = (\alpha', \sigma')$. Hence (W, Λ) is transitive.

Let us recall that $w \geq e$ in (W, Λ) iff $\lambda w = (\alpha, \sigma)w \geq (\alpha, \sigma)$ for all $\lambda = (\alpha, \sigma) \in \Lambda$. Now we define the operations $\vee$ and $\wedge$ on W by the rules:

$$(4) \qquad w \vee e = (\{g_\tau\}, h) \vee e = (\{f_\tau\}, h \vee e)$$

where

$$f_\tau = \begin{cases} g_\tau & \text{if } \tau h > \tau, \\ e & \text{if } \tau h < \tau, \\ g_\tau \vee e_\tau & \text{if } \tau h = \tau; \end{cases}$$

$$w \wedge e = (\{g_\tau\}, h) \wedge e = (\{f_\tau\}, h \wedge e) \tag{5}$$

where

$$f_\tau = \begin{cases} g_\tau & \text{if } \tau h < \tau, \\ e & \text{if } \tau h > \tau, \\ g_\tau \wedge e_\tau & \text{if } \tau h = \tau. \end{cases}$$

The direct verification shows that the operations $\vee$ and $\wedge$ are well-defined on W and

$$\lambda(w \vee e) = \max\{\lambda w, \lambda\}, \quad \lambda(w \wedge e) = \min\{\lambda w, \lambda\}$$

for all $\lambda \in \Lambda$. Therefore (W, Λ) is an ordered permutation l-group.□

The ordered permutation l-group (G, Λ) is called the *wreath product* of (G, Ω) and (H, T) and is written as

$$(W, \Lambda) = (G, \Omega)\mathrm{Wr}(H, T).$$

The subset of W of all elements $w = (\{g_\tau\}, h)$ such that $g_\tau = e$ for all but a finite number of $\tau \in T$ is called the *restricted wreath product* of (G, Ω) and (H, T) and it is denoted by $(G, \Omega)\mathrm{wr}(H, T)$. We will also write $G\mathrm{Wr}(H, T)$ and $G\mathrm{wr}(H, T)$ if we wish to emphasize the l-group part, since W is independent of the particular representation H.

If $w = (\{g_\tau\}, h) \in (G, \Omega)\mathrm{Wr}(H, T)$, then h is called the *global component* of w and $\{g_\tau | \tau \in T\}$ the set of *local components* of w. Note that $G\mathrm{Wr}(H, T)$ and $G\mathrm{wr}(H, T)$ are splitting extensions of the *base subgroup* $\overline{\prod_{\tau \in T}} G_\tau$ and $\prod_{\tau \in T} G_\tau$ respectively, by H. The *natural congruence* on $(W, \Lambda) = (G, \Omega)\mathrm{Wr}(H, T)$ is given by the rule $\lambda_1 = (\alpha_1, \tau_1) \equiv (\alpha_2, \tau_2) = \lambda_2$ iff $\tau_1 = \tau_2$ in T. The class of $\lambda = (\alpha, \tau)$ under this congruence will be denoted by $\bar{\lambda}$. Then $L_\equiv = \overline{\prod_{\tau \in T}} G_\tau, \quad \Lambda/\equiv$ is o-isomorphic to T and ordered permutation l-groups $(W/L_\equiv, \Lambda_\equiv)$, (H, T) are isomorphic.

Let (G, Ω) be a transitive ordered permutation l-group and ζ be a any convex G-congruence on W, $\varepsilon \leq \zeta \leq \upsilon$ where ε, υ are the least and the greatest convex G-congruences on Ω correspondingly. Then (ε, ζ)-component of (G, Ω) is equal to $(\hat{G}_{(\Delta)}, \Delta/\varepsilon) = (\hat{G}_{(\Delta)}, \Delta)$ where Δ is an arbitrary congruence class of the G-congruence ζ and (ζ, υ)-component of (G, Ω) is equal to $(G/L_\zeta, \Omega/\zeta)$ (see Section 5 of Chapter 4).

THEOREM 1. (Holland and McCleary [1].) *Let* (G, Ω) *be a transitive ordered permutation l-group,* ζ *be a convex G-congruence on* Ω *and* Δ *be any congruence class of* ζ. *Then* (G, Ω) *is l-isomorphic to l-subgroup of* $(W, \Lambda) = (\hat{G}_{(\Delta)}, \Delta)\mathrm{Wr}(G/L(\zeta), \Omega/\zeta)$.

Proof. (Glass [6].) For each class A of convex G-congruence ζ let $c_A \in \gamma$ such that $\Delta c_A = A$, with $c_\Delta = e$. The elements c_A, $A \in \{\alpha \cdot \zeta | \alpha \in \Omega\}$ are called the *coordinate maps* and will remain fixed throughout the proof. Now we define the map φ; $\Omega \to \Lambda = \Delta \overleftarrow{\times} \Omega/\zeta$ by the rule: $\alpha \cdot \varphi = (\alpha c_{\alpha \cdot \zeta}^{-1}, \alpha \cdot \zeta)$. It is clear that φ maps Ω onto Λ. Let $\alpha \neq \beta$ and $\alpha \cdot \varphi \beta \cdot \varphi$. Then $\alpha\zeta = \beta\zeta$ and $\alpha c_{\alpha\zeta}^{-1} = \beta c_{\alpha\zeta}^{-1}$, $\alpha = \beta$. This contradicts our assumption. Thus φ is one-to-one. If $\alpha < \beta$, then $\alpha\zeta \leq \beta\zeta$. If $\alpha\zeta < \beta\zeta$, then clearly $\alpha\varphi < \beta\varphi$. Otherwise, $\alpha\zeta = \beta\zeta$ and $\alpha\varphi = (\alpha c_{\alpha\zeta}^{-1}) < (\beta c_{\alpha\zeta}^{-1}, \beta\zeta)$. Therefore φ preserves order.

Let $g \to \bar{g}$ be the natural map G onto $\bar{G} = G/L(\zeta)$, and for each class A of the convex G-congruence ζ, let $\hat{g}_A = c_A g c_{Ag}^{-1}|_\Delta \in \hat{G}_{(\Delta)}$. Now we define the map $\psi : G \to W$ by the rule $d\psi = (\{\hat{g}_A\}, \bar{g})$. We claim that (ψ, φ) is an l-isomorphic embedding (G, Ω) into (W, Λ).

If $f < g$, then $\alpha\zeta f \leq \alpha\zeta g$ for all $\alpha \in \Omega$. Therefore, $\bar{f} \leq \bar{g}$ in $G/L(\zeta)$. Let $A = \alpha\zeta$ and $A\bar{f} = A\bar{g}$, then $\alpha\hat{f}_A = \alpha c_A f c_{Af}^{-1} = \alpha c_A f c_{Ag}^{-1} \leq \alpha c_A g c_{Ag}^{-1} = \alpha\hat{g}_A$ for each $\alpha \in A$. Moreover, if $\beta f < \beta g$, either $\beta\zeta\bar{f} < \beta\zeta g$ or $\beta\hat{f}_{\beta\zeta} < \beta\hat{f}_{\beta\zeta}$ by above arguments.

So ψ preserves order and is one-to-one. Let $h \in G$, then $\hat{fh}_A = c_A f h c_{Afh}^{-1} = (c_A f c_{Af}^{-1})(c_{Af} h c_{Afh}^{-1}) = \hat{f}_A \cdot \hat{h}_{Af}$ and

$$(fh)\psi = (\{\hat{fh}_A\}, \bar{fh}) = (\{\hat{f}_A, f\}) \cdot (\{\hat{h}_A\}, \bar{h}).$$

Hence ψ is anisomorphism of group G into W. It is clear that $(f \vee e)\psi = f\psi \vee e$. So ψ is an l-isomorphism.

Let now $\alpha \in \Omega$, $g \in G$. Then $\alpha g \cdot \varphi = (\alpha c_{\alpha\zeta}^{-1}, \alpha\zeta)(\{\hat{g}_A\}, \bar{g}) = (\alpha\varphi)(g\psi)$ and the pair (ψ, φ) defines the l-isomorphism of ordered permutation l-group (G, Ω) into (W, Λ).□

LEMMA 2. (Associativity of wreath products). *Let* $(G, \Omega), (H, T), (F, \Gamma)$ *be ordered permutation l-groups. Then the ordered permutation l-groups* $((G, \Omega)\mathrm{Wr}(H, T))\mathrm{Wr}(F, \Gamma)$ *and* $(G, \Omega)\mathrm{Wr}((H.T)\mathrm{Wr}(F, \Gamma))$ *are l-isomorphic.*

The proof follows immediately from the definitions.□

Now we define the wreath product of an arbitrary family of ordered permutation l-groups. Let K be a totally ordered set and (G_k, Ω_k) be an ordered permutation l-group for each $k \in K$. Let $\Omega^+ = \bar{\prod}_{k \in K} \Omega_k$ be a Cartesian product of totally ordered sets Ω_k and let 0 be any fixed point of Ω^+. For each $\alpha \in \Omega^+$ let $\sigma(\alpha) = \{k \in K | \alpha(k) \neq 0(k)\}$. Let $\hat{\Omega} = \{\alpha \in \Omega^+ | \sigma(\alpha)$ be an inversely well-ordered subset of $K\}$. For $\alpha, \beta \in \hat{\Omega}$, $\alpha \neq \beta$ we denote

$$K(\alpha, \beta) = \{k \in K | \alpha(k) \neq \beta(k)\} \subseteq \sigma(\alpha) \cup \sigma(\beta).$$

Then $K(\alpha, \beta)$ is an inversely well-ordered subset of K and there is the greatest element $\bar{k}$ in $K(\alpha, \beta)$ for which $\alpha(\bar{k}) \neq \beta(\bar{k})$. We order $\hat{\Omega}$ lexicographically from above: $\alpha < \beta$ in $\hat{\Omega}$ iff $\alpha(\bar{k}) < \beta(\bar{k})$ in Ω_k. This makes $\hat{\Omega}$ a totally ordered set.

Now for each $k \in K$ we define equivalence relations $\equiv^k$ and $\equiv_k$ on $\hat{\Omega}$ by the rules:

$$\alpha \equiv^k \beta \text{ iff } \alpha_{k'} = \beta_{k'} \text{ for all } k' > k,$$
$$\alpha \equiv_k \beta \text{ iff } \alpha_{k'} = \beta_{k'} \text{ for all } k' \geq k.$$

It is clear that $\alpha \equiv^k$ and $\alpha \equiv_k$ are convex subsets in $\hat{\Omega}$ for all $\alpha \in \hat{\Omega}$, $k \in K$. If $\alpha, \beta \in \hat{\Omega}$, $\alpha \neq \beta$ and $\bar{k}$ is the greatest element in $K(\alpha, \beta)$, then $\alpha \equiv^k \beta$ and $\alpha \equiv_k \beta$ for all $k > \bar{k}$. Let $W_1 = \{g \in A(\hat{\Omega}) | \equiv^k$ and $\equiv_k$ be a convex (g)-congruence for each $k \in K\}$. Clearly, the totally ordered set $\hat{\Omega}/ \equiv^k / \equiv_k$ (ordered in the natural way) is isomorphic to the totally ordered set Ω_k. Now for each $k \in K$ and $\alpha \in \hat{\Omega}$ let $\alpha^k \in \bar{\prod}_{k'>k} \Omega_{k'}$ where $\alpha^k(k') = \alpha(k')$ for $k' > k$. Note that $\alpha^k = \beta^k$ iff $\alpha \equiv^k \beta$. Let $g \in W_1$, $\alpha \in \hat{\Omega}$ and $k \in K$. Then g determines the mapping $g_{k,\alpha} : \Omega_k \to \Omega_k$ defined by the rule:$\xi g_{k,\alpha} = \alpha' g(k)$ where $\xi \in \Omega_k$, $\alpha' \equiv^k \alpha$ and $\alpha'(k) = \xi$.

LEMMA 3. *The mapping $g_{k,\alpha}$ is an order permutation of Ω_k for each $k \in K$, $\alpha \in \hat{\Omega}$ and $g \in W_1$.*

Proof. Let $\alpha \equiv^k \beta$ and $\beta(k) = \xi$, then $\alpha' \equiv_k \beta$. Since $g \in W_1$, then $\beta g \equiv_k \alpha' g$. Therefore, $\beta g(k) = \alpha' g(\mathrm{k})$ and $g_{k,\alpha}$ is well-defined. The easy verification shows that $g_{k,\alpha}$ is one-to-one and preserves order.□

Let $\{(G_k, \Omega_k)|k \in K\}$ be a family of ordered permutation l-groups. We define the wreath product $\mathrm{Wr}\{(G_k, \Omega_k)|k \in K\}$ to be $(W, \hat{\Omega})$ where W is the set of all order permutations $g \in W_1$ such that $g_{k,\alpha} \in G_k$ for all $k \in K,\ \alpha \in \hat{\Omega}$. Elements of W may be represented as $K \times \hat{\Omega}$ matrices $\|g_{k,\alpha}\|$ with entries $g_{k,\alpha} \in G_k$ whose "rows" satisfy the condition $g_{k,\alpha} = g_{k,\beta}$ if $\alpha \equiv^k \beta$ or, equivalently, $\alpha^k = \beta^k$, and whose "column" have an inversely well-ordered set of non-identity elements. The multiplication on these matrices is defined by the rule: $\|f_{k,\alpha}\| \cdot \|g_{k,\alpha}\| = \|h_{k,\alpha}\|$ where $h_{k,\alpha} = f_{k,\alpha} g_{k,\alpha f}$ and $\delta g_{k,\alpha}(k) = \delta(k) g_{k,\delta},\ \delta \in \hat{\Omega},\ k \in K$.

LEMMA 4. *The wreath product* $(W, \hat{\Omega}) = \mathrm{Wr}\{(G_k, \Omega_k)|k \in K\}$ *of transitive ordered permutation l-groups* (G_k, Ω_k), $(k \in K)$ *where* K *is a totally ordered set, is a transitive ordered permutation l-group.*

Proof. Let $\alpha, \beta \in \hat{\Omega}$. For each $k \in K$ there is $g_k \in G_k$ such that $\alpha_k g_k = \beta_k$ where $g_k = e$ for all $k \notin K(\alpha, \beta)$. Now we define $g \in A(\hat{\Omega})$ by the rule : $\gamma g(k) = \gamma_k g_k$ where $\gamma \in \hat{\Omega},\ k \in K$. It is clear that $\sigma(\gamma g) \subseteq \sigma(\gamma) \cup K(\alpha, \beta)$. Therefore, g is an order permutation of $\hat{\Omega},\ g \in W$ and W is a transitive ordered permutation group on $\hat{\Omega}$.

The direct verification shows that the pointwise supremum of g and e is equal to $h = \|h_{k,\alpha}\|$ where

$$h_{k,\alpha} = \begin{cases} g_{k,\alpha} & \text{if} \quad \alpha^k g > \alpha^k, \\ e & \text{if} \quad \alpha^k g < \alpha^k, \\ g_{k,\alpha} \vee e & \text{if} \quad \alpha^k g = \alpha^k. \end{cases}$$

Clearly, $h \in W$. Therefore, $g \vee e \in W$ and W is an l-group variety.□

We can now define the *restricted wreath product* $\mathrm{wr}\{(G_k . \Omega_k|k \in K)\}$ as the l-subgroup of $W = \mathrm{Wr}\{(G_k, \Omega_k)|k \in K\}$ consisting of those matrices $\|g_{k,\alpha}\|$ in which all entries are e except for finitely many k, and for each k, $g_{k,\alpha} = e$ in G_k except at most finitely many $\alpha \equiv^k$-classes.

Our definition of the wreath product depends on the choice of the reference point $0 \in \hat{\Omega}$. Now we will prove that if each (G_k, Ω_k) is transitive, then the wreath product $\mathrm{Wr}\{(G_k, \Omega_k)|k \in K\}$ does not depend on the choice of) $\in \hat{\Omega}$.

LEMMA 5. *Let* (G_k, Ω_k) *be a transitive ordered permutation l-*

group for all $k \in K$. Let $0' \in \Omega^+ = \bar{\prod}_{k\in K} \Omega_k$ and $(W', \hat{\Omega}')$ be the wreath product of $\{(G_k, \Omega_k) | k \in K\}$ with reference point $0'$. Then $(W, \hat{\Omega})$ and $(W', \hat{\Omega}')$ are l-isomorphic as ordered permutation l-groups.

Proof. (Glass [6].) For each $k \in K$ there is $f_k \in G_k$ such that $0kf_k = 0'(k)$. Define $\varphi: \Omega^+ \to \Omega^+$ by the rule: $\alpha\varphi(k) = \alpha(k)f_k$ ($\alpha \in \Omega^+$). Now φ is an isomorphism of totally ordered set $\hat{\Omega}$ and $\hat{\Omega}'$. Now define $\psi: W_1 \to W_1'$ by the rule: $g\psi = \varphi^{-1}g\varphi$. Clearly $\alpha\varphi \cdot g\psi = (\alpha g)\varphi$. Hence (ψ, φ) is an l-isomorphism of $(W, \hat{\Omega})$ onto $(W', \hat{\Omega}')$.□

THEOREM 2. (Holland and McCleary [1].) *Let (G, Ω) be a transitive ordered permutation l-group, $K = K(G, \Omega)$ and $\{(G_k \Omega_k) | k \in K\}$ be the set of all o-primitive components of (G, Ω). Then there is l-isomorphism of (G, Ω) into $(W, \hat{\Omega}) = \mathrm{Wr}\{(G_k, \Omega_k) | k \in K\}$ as ordered permutation l-groups.*

The proof is straightforward (see the book of A.M.W.Glass [6], Theorem 5.9).□

11.2 Wreath Products and Extensions

1⁰. We denote by $(\mathbf{Z}, \mathbf{Z})$ and $(\mathbf{R}, \mathbf{R})$ the ordered permutation l-groups of totally ordered groups $\mathbf{Z}$ and $\mathbf{R}$ of integers and reals which act as right regular represetions on the totally ordered sets $\mathbf{Z}$ and $\mathbf{R}$ correspondingly. In this Section we say, that $(\mathbf{Z}, \mathbf{Z})$ and $(\mathbf{R}, \mathbf{R})$ are right regular represention of totally ordered groups $\mathbf{Z}$ and $\mathbf{R}$.

THEOREM 1. (Read [2].) Let (G, Ω) be a transitive ordered permutation l-group and $G \in \mathcal{N}$. Then (G, Ω) is l-isomorphic to an l-subgroup of wreath product $(W, \hat{\Omega}) = \mathrm{Wr}\{(R_k, R_k | k \in K)\}$ where each (R_k, R_k) is the right regular representation of reals and K is some totally ordered set.

Proof. By Theorems 4.5.2 and 9.2.2, each transitive o-primitive normal valued l-group is l-isomorphic to an l-subgroup of $(\mathbf{R}, \mathbf{R})$. By Theorem 11.1.2, (G, Ω) is l-isomorphic to an l-subgroup of $\mathrm{Wr}\{(G_k, \Omega_k) | k \in K\}$ where (G_k, Ω_k) are primitive components of (G, Ω) and K is the set of all covering pairs of G-congruences on Ω. Then each (G_k, Ω_k) is an l-subgroup of the ordered permutation l-group (R_k, R_k) where (R_k, R_k)

is isomorphic to $(\mathbf{R}, \mathbf{R})$. Therefore, (G, Ω) is l-isomorphic to an l-subgroup of the ordered permutation l-group $\mathrm{Wr}\{(R_k, R_k) | k \in K\}$.□

It is evident that $\mathrm{Wr}\{(R_k, R_k) | k \in K\}$ is not a totally ordered group for every totally ordered set K which contains more than one element.

At the beginning of Chapter 11 we introduced the concept of the product $\mathcal{X} \cdot \mathcal{Y}$ of two l-varieties $\mathcal{X}$ and $\mathcal{Y}$.

LEMMA 1. *Let $\mathcal{X}$, $\mathcal{Y}$ be l-varieties and (G, Ω), (H, T) be transitive ordered permutation l-groups such that $G \in \mathcal{X}$, $H \in \mathcal{Y}$ and $(W, \Lambda) = (G, \Omega)\mathrm{Wr}(H, T)$. Then $W \in \mathcal{X} \cdot \mathcal{Y}$.*

Proof. Let $\bar{G} = \overline{\prod_{\tau \in T}} G_\tau$ be the base subgroup of (W, Λ). Then $\bar{G} \in \mathcal{X}$, $\bar{G}$ is an ideal of W and $W/\bar{G} \in \mathcal{Y}$.□

Let

$$\mathrm{Wr}^n(\mathbf{R}, \mathbf{R}) = \underbrace{(\mathbf{R}, \mathbf{R})\mathrm{Wr}(\mathbf{R}, \mathbf{R})\mathrm{Wr} \cdots \mathrm{Wr}(\mathbf{R}, \mathbf{R})}_{\mathbf{n}},$$

$$\mathrm{Wr}^n(\mathbf{Z}, \mathbf{Z}) = \underbrace{(\mathbf{Z}, \mathbf{Z})\mathrm{Wr}(\mathbf{Z}, \mathbf{Z})\mathrm{Wr} \cdots \mathrm{Wr}(\mathbf{Z}, \mathbf{Z})}_{\mathbf{n}},$$

$$\mathrm{wr}^n(\mathbf{R}, \mathbf{R}) = \underbrace{(\mathbf{R}, \mathbf{R})\mathrm{wr}(\mathbf{R}, \mathbf{R})\mathrm{wr} \cdots \mathrm{wr}(\mathbf{R}, \mathbf{R})}_{\mathbf{n}},$$

$$\mathrm{wr}^n(\mathbf{Z}, \mathbf{Z}) = \underbrace{(\mathbf{Z}, \mathbf{Z})\mathrm{wr}(\mathbf{Z}, \mathbf{Z})\mathrm{wr} \cdots \mathrm{wr}(\mathbf{Z}, \mathbf{Z})}_{\mathbf{n}}.$$

By Lemma 11.1.2, $\mathrm{Wr}^n(\mathbf{Z}, \mathbf{Z})$ and $\mathrm{Wr}^n(\mathbf{Z}, \mathbf{Z})$ are well-defined. An easy induction, using Lemma 1 yields

COROLLARY 1. $\mathrm{Wr}^n(\mathbf{R}, \mathbf{R}) \in \mathcal{A}^n$ *for all positive integer n.*□

LEMMA 2. (Martinez [3].) $\mathcal{N}^2 = \mathcal{N}$.

Proof. Let $G \in \mathcal{N}^2$. Then there is an ideal A of G such that $A, G/A \in \mathcal{N}$. Let V ba any value in G. If $A \subset V$ then V/A is a value in G/A. Therefore, V/A is normal in its cover $\bar{V}/A$ and V is normal in its cover $\bar{V}$. If $V \not\subseteq A$, then by Lemma 7.6.2, $A \cap V$ is a value in A. Since $A \in \mathcal{N}$, then $A \cap V$ is normal in its cover. Therefore, V is always normal in its cover, and $G \in \mathcal{N}$.□

THEOREM 2. (Glass,Holland and McCleary [1].)

$$\mathcal{N} = \bigvee_{n\in\mathbf{N}}^{\mathbf{L}} \mathcal{A}^n.$$

Proof. It is obvious that $\mathcal{A}^n \subseteq \mathcal{N}$ for all $n \in \mathbf{N}$. Hence $\mathcal{N} \subseteq \bigvee_{n\in\mathbf{N}}^{\mathbf{L}} \mathcal{A}^n$. Now let $e = w(x_1,\ldots,x_n) = \bigvee_{i\in I}\bigwedge_{j\in J} w_{ij}(x_1,\ldots,x_n)$ be a identity which fails in $\mathcal{N}$ and hence in some subdirectly irreducible l-group $G \in \mathcal{N}$. (Here $w_{ij}(x_1,\ldots,x_n)$ are elements of free group witht free generators $x_1,\ldots,x_n$, I, J are finite index sets.) We must show that $w(x_1,\ldots,x_n) = e$ is violated in some $\mathcal{A}^m$, $m \in \mathbf{N}$.

By Corollary 3 of Theorem 4.3.1, G has a faithful transitive representation by order permutations of some totally ordered set Ω. By Theorem 1, we can assume that (G,Ω) is an l-subgroup of the wreath product $(W,\hat{\Omega}) = \mathrm{Wr}\{(R_k,R_k)|k \in K\}$ where (R_k,R_k) is the right regular representation of the reals. Hence the identity $w = e$ fails in $\mathrm{Wr}\{(R_k,R_k)|k \in K\}$. Therefore for some $g_1,\ldots,g_n \in W$, $\lambda \in \hat{\Omega}_k$ is valid $\lambda(\bigvee_{i\in I}\bigwedge_{j\in J} w_{ij}(g_1,\ldots,g_n)) \neq \lambda$. Consider the finite set $\Lambda = \{\lambda\} \cup \{\lambda u(g_1,\ldots,g_n)|u(x_1,\ldots,x_n)$ is an initial segment of $w_{ij}(x_1,\ldots,x_n)\}$.

For each pair (λ_1,λ_2) of distinct elements of Λ there is unique $k \in K$ such that $\lambda_1(k) \neq \lambda_2(k)$ but $\lambda_1(k') = \lambda_2(k')$ for all $k' > k$. Let Δ be the set of all such k. It is clear that Δ is a finite set and let m be the number of distinct elements of Δ.

Let $\overleftarrow{\Omega} = \overleftarrow{\prod_{\delta\in\Delta}} R_\delta$ and let π denote the projection of $\hat{\Omega}$ onto $\overleftarrow{\Omega}$. We note that π is one-to-one and order preserving on Λ and for $\xi \in \Lambda$ let $\bar{\xi} = \xi\pi$. For each $g \in \{g_1,\ldots,g_n\}$ define $\bar{g} = \|g_{k,\xi}\| \in \mathrm{Wr}\{(R_\delta,R_\delta)|\delta \in \Delta\}$ where

$$\bar{g}_{k,\bar{\mu}} = \begin{cases} g_{k,\xi} & \text{if there exist } \xi,\xi^* \in \Lambda \text{ such that } \bar{\xi} \equiv^k \bar{\mu} \text{ in } \overleftarrow{\Omega} \text{ and} \\ & \xi g \equiv^k \xi^{ast} \text{ in } \overleftarrow{\Omega}, \\ e & \text{otherwise.} \end{cases}$$

Then $\bar{g}_{k,\bar{\mu}}$ is well defined and if g,h are inverses, then $\bar{g},\bar{h}$ will be , too.

Now consider any $g_t \in \{g_1,\ldots,g_n\}$ and suppose that $w_{ij} = ux_i$.

Then $\xi = \lambda u(g_1, \ldots, g_n) \in \Lambda$. Let $\delta \in \Delta$. Then

$$(\bar{\xi}\bar{g})(k) = \bar{\xi}(k)g(k, \bar{\xi}) = \xi(k)g_{k,\xi} = \bar{\xi g}(k)$$

since both ξ and $\xi g = \lambda u(g_1, \ldots, g_n)g_t \in \Lambda$. Thus $\bar{\xi}\bar{g} = \bar{\xi g}$ and for each i, j,

$$\bar{\lambda} w_{ij}(\bar{g}_1, \ldots, \bar{g}_n) = \overline{\lambda w_{ij}(g_1, \ldots, g_n)}.$$

Since π is one-to-one and order preserving on Λ and these suprema and infima are being taken in the totally ordered sets, then

$$\bar{\lambda} \bigvee_{i \in I} \bigwedge_{j \in J} w_{ij}(g_1, \ldots, g_n) \neq \bar{\lambda}.$$

Thus the identity $w(x_1, \ldots, x_n) = e$ is not valid in $\bar{W}, \overleftarrow{\Omega} = \mathrm{Wr}^m(R_k, R_k)$. By Corollary 1 of Theorem 1, $\mathrm{Wr}^m(R_k, R_k) \in \mathcal{A}^m$, then

$$w(x_1, \ldots, x_n) = e$$

does not hold in $\mathcal{A}^m$.□

COROLLARY 1. *If $\mathcal{V}$ is an l-variety and $\mathcal{A} \subseteq \mathcal{V} \subseteq \mathcal{N}$, then*

$$\mathcal{N} = \bigvee_{n \in \mathbf{N}}^{\mathbf{L}} \mathcal{V}^n. \square$$

COROLLARY 2. *The only idempotents in the semigroup of the l-varieties* **L** *are $\mathcal{E}, \mathcal{N}$ and $\mathcal{L}$.*□

2^0. Let $\mathcal{R}$ be the l-variety of all representable l-groups. Following Feil [3] for any l-varieties $\mathcal{U}, \mathcal{V} \subseteq \mathcal{R}$ let their *product in* $\mathcal{R}$ be $\mathcal{U} * \mathcal{V} = \mathcal{U} \cdot \mathcal{V} \cap \mathcal{R}$. In connection with Theorem 2 it is an interesting question whether $\mathcal{R} = \bigvee_{n \in \mathbf{N}}^{\mathbf{L}} \mathcal{A}^{*n}$ where $\mathcal{A}^{*2} = \mathcal{A} * \mathcal{A}$, $\mathcal{A}^{*n+1} = \mathcal{A}^{*n} * \mathcal{A}$. T.Feil in [3] showed that this hypothesis is not true. Let $\mathcal{B}$ be an l-variety defined by the identities :

(1) $|[[x|, |[|x| \vee |y|, |x|]|]|^n \wedge |x| = |[|x|, |[|x| \vee |y|, |x|]|]|^n$, $(n \in \mathbf{N})$.
(2) $x(y^{-1}x^{-1}y) \vee e = e$.

We first show that for each positive integer n is valid $\mathcal{A}^{*n} \subseteq \mathcal{B}$. It is enough to prove that each totally ordered group of $\mathcal{A}^{*n}$ satisfies the identities (1)-(2).

So, let G be a non-trivial totally ordered group in $\mathcal{A}^{*n}$. We first show that G contains a finite solvable system of normal convex subgroups with torsion-free factors. Let N_1 be the unique maximal Abelian normal convex subgroup of G. (N_1 is the Abelian radical of G.) Let N_2 be the preimage in G of Abelian radical of G/N_1. We continue in this manner. Since $G \in \mathcal{A}^{*n}$, then $N_n = G$. So

$$E = N_0 \subseteq N_1 \subseteq \cdots \subseteq N_n = G$$

is the solvable series of convex normal subgroup in G with $N_{i+1}/N_i \in \mathcal{A}$. Some inclusions, of course, may not be proper.

Now, let $e \leq b \leq a$ with $a, b \in G$. If $b = e$, then $|[b, |[a,b]|]| = e = b$ so the identities Σ hold. Therefore we assume that $e < b$. Choose i such that $b \in N_i \setminus N_{i-1}$. Since N_i is normal in G, then $[a,b] \in N_i$. Since N_i/N_{i-1} is abelian, then $N_{i-1}[b, |[a,b]|] = N_{i-1}$. Hence $|[b, |[a,b]|]| \ll b$. So we have that $\mathcal{A}^{*n} \subseteq \mathcal{B}$ for each positive integer n. It follows that $\bigvee^{\mathbf{L}}_{n\in\mathbf{N}} \mathcal{A}^n \subseteq \mathcal{B}$.

To show that $\mathcal{B} \neq \mathcal{R}$, we give the example of a totally ordered group that does not satisfy identities (1)-(2).

We consider the Clifford's example of simple totally ordered group (see A.H.Clifford [1].). Let

$$C = \mathrm{gp}\{g(r) | r \in \mathbf{Q} | g(s)^{-1} g(r) g(s) = g(\frac{r+s}{2}) \text{ for } s > r\}.$$

This allows us to write each non-identity element g in the unique normal form

$$g = g(r_1)^{n_1} g(r_2)^{n_2} \cdots g(r_k)^{n_k}$$

where $n_i \neq 0$ for all i, $1 \leq i \leq k$ and $r_1 < r_2 < \cdots < r_k$. We say $g > e$ iff $n_k > 0$. This makes this group totally ordered. The direct verification shows that group C does not satisfy identity (1) for $x = g_0$, $y = g_1$.

T.Feil in [3] has shown that

$$(\bigvee^{\mathbf{L}}_{n\in\mathbf{N}} \mathcal{A}^{*n}) * (\bigvee^{\mathbf{L}}_{n\in\mathbf{N}} \mathcal{A}^{*n}) \not\subseteq \mathcal{B}$$

and therefore, $(\bigvee^{\mathbf{L}}_{n\in\mathbf{N}} \mathcal{A}^{*n}) \not\subseteq \mathcal{B}$ is not idempotent in the semigroup of l-varieties of representable l-varieties.

3⁰. Now we need some technical results on wreath products of transitive ordered permutation l-groups.

PROPOSITION 1. (Glass, Holland and McCleary [1].) *Let (G,Ω) and (H,T) be transitive ordered permutation l-groups. Then*

$$\mathrm{var}_l\{(H,T)\mathrm{wr}(G,\Omega)\} = \mathrm{var}_l\{(H,T)\mathrm{Wr}(G,\Omega)\}.$$

Proof. Obviously, the first l-variety is contained in the second. Conversely, suppose that the identity

$$w(x_1,\ldots,x_n) = \bigvee_{i\in I}\bigwedge_{j\in J} w_{ij}(x_1,\ldots,x_n) = e$$

fails in $(H,T)\mathrm{Wr}(G,\Omega)$. Then for some $g_1,\ldots,g_n \in (H,T)\mathrm{Wr}(G,\Omega)$, $\lambda \in \Lambda = T \overleftarrow{\times} \Omega$,

$$\lambda \neq \lambda w(g_1,\ldots,g_n) = \lambda \bigvee_{i\in I}\bigwedge_{j\in J} w_{ij}(g_1,\ldots,_n). \tag{3}$$

For each $g_i = (\{h_\alpha, g\}) \in (H,T)\mathrm{Wr}(G,\Omega)$, $i = 1,\ldots,n$ only finitely many components h_α, $\alpha \ni \Omega$, are involved in inequality (3). Thus it is possible to violate the identity $w(x_1,\ldots,x_n) = e$ in $(H,T)\mathrm{wr}(G,\Omega)$.□

Let us recall that $\bar{\Omega}$ is the Dedekind completion of Ω. As usual, we identify (G,Ω) with $(G,\bar{\Omega})$ in $A(\bar{\Omega})$. Let (G,Ω) be an order permutation l-group and C be a convex l-subgroup of G. Let Σ be the set of all fixed points of C in $\bar{\Omega}$ and C^0 be the pointwise stabilizer of Σ:

$$\Sigma = \{\sigma \in \bar{\Omega} | \sigma g = \sigma \text{ for all } g \in C\}$$
$$C^0 = C_\Sigma = \{g \in G | \sigma g = \sigma \text{ for all } \sigma \in \Sigma\}$$

It is clear that sets of fixed points of C and C^0 are the same. Let C^* be a closure of C^0 in G, i.e C^* is the least closed convex l-subgroup of G containing C^0.

PROPOSITION 2. (Glass [6].) $C^* = C^0$. *Moreover,* $C = C^0$ *if and only if* C *is closed.*

Proof. It is clear that $C \subseteq C^0$, $C \subseteq C^*$ and C^0 is closed in G. Then $C^* \subseteq C^0$. If $C^* \neq C^0$, then the convex closure Δ^*_β of the orbit βC^* in Ω is properly containend in the convex closure Δ^0_β of the orbit βC^0 in Ω : $\Delta^*_\beta \subseteq \Delta^0_\beta$ for all $\beta \in \Omega$.

If $\Delta^*_\beta \neq \Delta^0_\beta$ for some $\beta \in \Omega$, then $\bar{\sigma} = \sup_{\bar{\Omega}} \Delta^*_\beta$ is contained in the convex closure $\bar{\Delta}^0_\beta$ of the orbit βC^0 in $\bar{\Omega}$. Then $\bar{\sigma}C^* = \bar{\sigma}$ and $\beta \leq \bar{\sigma} < \beta h$ for some $h \in C^0$. This contradicts the choice of C^0.

Let $e \leq g \in C^0$ and $\alpha \in \Omega$. There exists $f_\alpha \in C^*$, $e \leq f_\alpha$ such that $\alpha g \leq \alpha f_\alpha$. Since $e \leq g \wedge f_\alpha \leq f_\alpha$, then $g \wedge f_\alpha \in C^*$. Now, $g = \bigvee_{\alpha \in \Omega} (g \wedge f_\alpha) \in C^*$ as C^* is closed. Therefore, $C^* = C^0$.□

PROPOSITION 3. (Glass [6].) *Let (G, Ω) be a transitive ordered permutation l-group and H be a closed ideal of G. Then there is a convex G-congruence ζ on Ω for which $H = L_\zeta$ and whose classes are H-orbits.*

Proof. Since H is an ideal, then $\zeta : \ \alpha \equiv \beta \pmod{\zeta}$ iff there exist $h, g \in H$ such that $\alpha f \leq \beta \leq \alpha h$ in Ω is a convex G-congruence on Ω and the lazy subgroup L_ζ of ζ:

$$L_\zeta = \{g \in G | \Delta g = \Delta \text{ for all } \zeta\text{-classes } \Delta \} = \bigcap_{\Delta} G_{(\Delta)}$$

is an ideal of G. By Proposition 2, $L_\zeta = H^0 = H$.□

Now, by Theorem 9.1.1, the ordered permutation l-group (G, Ω) is l-isomorphic to the l-subgroup of $(W, \Lambda) = (\hat{G}_{(\Delta)}, \Delta)\mathrm{Wr}(G/L_\zeta, \Omega/\zeta) = (\hat{G}_{(\Delta)}, \Delta)\mathrm{Wr}(G/H, \Omega/\zeta)$ where Δ is any congruence class of ζ. Here $\hat{G}_{(\Delta)}$ is isomorphic to $G_{(\Delta)}/G_\Delta = \bar{\gamma}_\Delta$. Let $H_{(\Delta)}$ be a restriction of H onto Δ. Then $(\bar{H}_{(\Delta)}, \Delta)$ is a transitive ordered permutation l-group of totally ordered set Δ, where $\bar{H}_{(\Delta)} = H_{(\Delta)}G_\Delta/G_\Delta$. Thus $\bar{H}_{(\Delta)} \subseteq \bar{G}_{(\Delta)}$ and we can consider the closure $\bar{H}^*_{(\Delta)}$ in $\bar{G}_{(\Delta)}$. By Proposition 3, $\bar{H}^*_{(\Delta)}$ defines a convex $\bar{G}_{(\Delta)}$-congruence on Δ. By transitivity $(\bar{H}_{(\Delta)}, \Delta)$ on Δ, the congruence defined by $\bar{H}^*_{(\Delta)}$ is trivial. Therefore, $\bar{H}^*_{(\Delta)} = \bar{G}_{(\Delta)}$ and since $\mathrm{var}_l \bar{H}^*_{(\Delta)} = \mathrm{var}_l \bar{H}_{(\Delta)}$, then the following result is true.

PROPOSITION 4. (Glass, Holland and McCleary [1].) *Let $\mathcal{X}$ be any l-variety, (G, Ω) be a transitive ordered permutation l-group and $H = \mathcal{X}(G)$ be an $\mathcal{X}$-radical of G. Then (G, Ω) is l-isomorphic to an l-subgroup of $(W, \Lambda) = (\bar{H}^*_{(\Delta)}, \Delta)\mathrm{Wr}(G/H, \Omega/\zeta)$ where $\bar{H}^*_{(\Delta)} \in \mathcal{X}$.*

11.3 Wreath Products and the Semigroup of l-Varieties

We say that the class $\mathcal{G} = \{(G_s, \Omega_i)|s \in S\}$ of ordered permutation l-groups *mimics* an l-variety $\mathcal{X}$ if

(1) each l-group G_s belongs to $\mathcal{X}$;

(2) whenever (H, Λ) with $H \in \mathcal{X}$ is a transitive ordered permutation l-group, $\lambda \in \Lambda$, $\{w_p(x_1, \ldots, x_n)|p \in P\}$ a finite number of l-words, and elements $h_1, \ldots, h_n \in H$, then there exist $(G_s, \Omega_s) \in \mathcal{G}$, $\alpha \in \Omega_s$ such that $\lambda w_p(h_1, \ldots, h_n) < \lambda w_q(h_1, \ldots, h_n)$ in T iff $\alpha w_p(g_1, \ldots, g_n) < \alpha w_q(g_1, \ldots, g_n)o$ in Ω_s.

For example, the class of all transitive ordered permutation l-groups (G, Ω) with $\gamma \in \mathcal{X}$ mimics $\mathcal{X}$. We also note that if class $\{(G_s, \Omega_s)|s \in S\}$ mimics $\mathcal{X}$, then $\text{var}_l\{G_s|s \in S\} = \mathcal{X}$ since any identity $w = e$ which fails in $\mathcal{X}$ must fail in the subdirectly irreducible l-group $H \in \mathcal{X}$. Then H must have a faithful transitive representation (H, Λ) by Corollary 3 of Theorem 6.3.1. Therefore the identity $w(x_1, \ldots, x_n) = e$ fails in some G_s.

THEOREM 1. (Glass, Holland and McCleary [1].) *Let* $\{U_s|s \in S\}$ *be a family of l-groups,* $\{G_t|t \in T\}$ *be a family of order permutation l-groups and* $\mathcal{U} = \text{var}_l\{U_s|s \in S\}$. *If* $\{(G_t, \Omega_t)|t \in T\}$ *mimics l-variety* $\mathcal{V}$, *then* $\text{var}_l\{U_s\text{wr}(G_t, \Omega_t)|s \in S,\ t \in T\} = \mathcal{U} \cdot \mathcal{V}$.

Proof. By Lemma 9.2.1,

$$\text{var}_l\{U_s\text{wr}(G_t, \Omega_t)|s \in S,\ t \in T\} \subseteq \mathcal{U} \cdot \mathcal{V}.$$

For the converse inclusion, let us assume that $w(X) = w(x_1, \ldots, x_n) = e$ is an identity which fails in some l-group of $\mathcal{U} \cdot \mathcal{V}$. Then this identity fails in some l-group $F \in \mathcal{U} \cdot \mathcal{V}$ which has a faithful transitive representation (F, Λ). Thus the identity

$$w(x_1, \ldots, x_n) = e$$

fails in $(U, \Delta)\text{Wr}(F/\mathcal{U}(F), \Gamma)$ where (U, Δ) is a transitive ordered permutation l-group, $U \in \mathcal{U}$ and $\mathcal{U}(F)$ is the $\mathcal{U}$-radical of F. We apply

Proposition 9.2.4 to embed (F, Λ) into $(W, \Lambda) = (U, \Delta)\mathrm{Wr}(F/\mathcal{U}(F), \Gamma)$, identifying the two sets denoted by Λ.

Since identity $w(X) = e$ fails in $\mathcal{V}$, then it fails in some G_t since $\mathrm{var}_l\{G_t | t \in T\} = \mathcal{V}$. In this case, the identity fails in any $U_s\mathrm{wr}(G_t, \Omega_t)$.

Henceforth, we can assume that $w(X) = e$ holds in $\mathcal{V}$ but fails in $(U, \Delta)\mathrm{Wr}(F/\mathcal{U}(F), \Gamma)$. Let

$$w(x_1, \ldots, x_n) = \bigvee_{i \in I} \bigwedge_{j \in J} w_{ij}.$$

Here $w_{ij} = w_{ij}(x_1, \ldots, x_n) = x_{ij1}x_{ij2} \cdots w_{ijk}$ are reduced group words of variables $x_1, \ldots, x_n$, $x_{ijk} \in X \cup X^{-1}$, $1 \leq k \leq k(i,j)$ for all i, j, k.

Then for some $h_1, \ldots, h_n \in W$,

$$e \neq w(h_1, \ldots, h_n) = \bigvee_{i \in I} \bigwedge_{j \in J} w_{ij}(h_1, \ldots, h_n).$$

and so there exists a point $\lambda \in \Lambda$ such that $\lambda \neq \lambda w(h_1, \ldots, h_n)$. Since $w(X) = e$ holds in $F/\mathcal{U}(F)$, $\lambda\zeta = \lambda w(\bar{h}_1, \ldots, \bar{h}_n)\zeta$ where ζ is the natural congruence on $(U, \Delta)\mathrm{Wr}(F/\mathcal{U}(F), \Gamma)$. Then $\bar{\lambda} = \bar{\lambda} w(\bar{h}_1, \ldots, \bar{h}_n)$ where $\bar{\lambda} = \lambda\zeta$ is the class of λ under congruence ζ and $\bar{h}_1, \ldots, \bar{h}_n$ are images of $h_1, \ldots, h_n$ in $\bar{F} = F/\mathcal{U}(F)$.

Let $w_{ij}^q = w_{ij}^q(x_1, \ldots, x_n) = x_{ij1}x_{ij2} \cdots w_{ijq}$, $(q \leq k(i,j))$, be the initial part of the group word w_{ij}. Now consider the set Φ of words $\{w_{ij} | i \in I, j \in J\} \cup \{w_{ij}^r \cdot (w_{sp}^q)^{-1} | i \in I,\ j \in J,\ r \leq k(i,j),\ q \leq k(s,p)\}$. Since $\{(G_t, \Omega_t)\}$ mimics $\mathcal{V}$, there are $(G, \Omega) \in \{(G_t, \Omega_t)\}$, $\alpha \in \Omega$ and $g_1, \ldots, g_n \in G$ such that $\bar{\lambda} u(\bar{h}_1, \ldots, \bar{h}_n) < \lambda v(\bar{h}_1, \ldots, \bar{h}_n)$ iff $\alpha u(g_1, \ldots, g_n) < \alpha v(g_1, \ldots, g_n)$ for all $u, v \in \Phi$. Thus this forces several results:

(3) $\bar{\lambda} w_{ij}(\bar{h}_1, \ldots, \bar{h}_n) < \bar{\lambda} w_{sp}(\bar{h}_1, \ldots, \bar{h}_n)$ iff $\alpha w_{ij}(g_1, \ldots, g_n) < \alpha w_{sp}(g_1, \ldots, g_n)$,
$\bar{\lambda} w_{ij}(\bar{h}_1, \ldots, h_n) < \bar{\lambda}$ iff $\alpha w_{ij}(\bar{h}_1, \ldots, \bar{h}_n) < \alpha w_{ij}(g_1, \ldots, g_n) < \alpha$,
$\bar{\lambda} < \bar{\lambda} w_{sp}(\bar{h}_1, \ldots, \bar{h}_n)$ iff $\alpha < \alpha w_{sp}(g_1, \ldots, g_n)$,

(4) $\bar{\lambda} u(\bar{h}_1, \ldots, \bar{h}_n) = \bar{\lambda} v(\bar{h}_1, \ldots, \bar{h}_n)$ iff $\alpha u(g_1, \ldots, g_n) = \alpha v(g_1, \ldots, g_n)$.

Recall that $\bar{\lambda} = \bar{\lambda} w(\bar{h}_1, \ldots, \bar{h}_n) = \bigvee_{i \in I} \bigwedge_{j \in J} \bar{\lambda} w_{ij}(\bar{h}_1, \ldots, \bar{h}_n)$ the suprema in infima being taken in the totally ordered set $\bar{\Lambda}$ (which is order isomorphic to Γ).

Let $I(0)$ be the non-empty set $\{i \in I \mid \bigwedge_{j\in J} \bar{\lambda} w_{ij}(\bar{h}_1,\ldots,\bar{h}_n) = \bar{\lambda}\}$; for all other i, is valid $\bar{\lambda} \bigwedge_{j\in J} w_{ij} < \lambda$. For each $i \in I(0)$ let $J(i)$ be the non-empty set $J(i) = \{j \in J|\bar{\lambda} w_{ij}(\bar{h}_1,\ldots,\bar{h}_n) = \bar{\lambda}\}$; for all other j, $\bar{\lambda} < \bar{\lambda} w_{ij}(\bar{h}_1,\ldots,\bar{h}_n)$. Let

$$w'(x_1,\ldots,x_n) = \bigvee_{i\in I(0)} \bigwedge_{j\in J(i)} w_{ij}(x_1,\ldots,x_n).$$

Then $\bar{\lambda} w'(\bar{h}_1,\ldots,\bar{h}_n) = \bar{\lambda} w(\bar{h}_1,\ldots,\bar{h}_n)$. Now we obtain a new word

$$w''(y_{ijk|i\in I, J\in J, 1\le k\le k(i,j)})$$

which is constructed from $w'(x_1,\ldots,x_n)$ by partitioning the various occurances x_{ijk} of each variable according to where $\bar{\lambda}$ has arrived in $\bar{\Lambda}$ when x_{ijk} is reached during the application of $w_{ij}(\bar{h}_1,\ldots,\bar{h}_n)$ to $\bar{\lambda}$. More precisely, we replace each x_{ijk} by a new variable y_{ijk} in such a way that

(5) $y_{ijk} = y_{spt}$ iff $x_{ijk} = x_{spt}$ (as elements of $X \cup X^{-1}$) and $\bar{\lambda} w_{ij}^{(k-1)}(\bar{h}_1,\ldots,\bar{h}_n) = \bar{\lambda} w_{spt}^{(k-1)}(\bar{h}_1,\ldots,\bar{h}_n)$,

(6) $y_{ijk}^{\pm 1} = y_{spt}^{\mp 1}$ iff $x_{ijk}^{\pm 1} = x_{spt}^{\mp 1}$ and $\bar{\lambda} w_{ij}^{k-1}(\bar{h}_1,\ldots,\bar{h}_n) = \; = \bar{\lambda} w_{sp}^{t}(\bar{h}_1,\ldots,\bar{h}_n)$ (equivalently, $\bar{w}_{ij}^{(k)}(\bar{h}_1,\ldots,\bar{h}_n) = \bar{\lambda} w_{sp}^{t-1}(\bar{h}_1,\ldots,\bar{h}_n)$).

(7) $y_{ijk} = e$ iff $x_{ijk} = e$.

We claim that the identity $w''(y_1,\ldots,y_n) = e$ fails in $\mathcal{U}$.

Write $h_{ijk} = (\hat{f}_{ijk}, \bar{h}_{ijk})$, where $\hat{f}_{ijk} : \bar{\Lambda} \to U$ and $\bar{h}_{ijk}$ denotes the action of h_{ijk} on $\bar{\Lambda}$. Write $w'(h_1,\ldots,h_n) = (f,e)$. Since $\lambda \ne \lambda w(h_1,\ldots,h_n) = \lambda w'(h_1,\ldots,h_n) = \bigvee_{i\in I(0)} \bigwedge_{j\in J(i)} w_{ij}(h_1,\ldots,h_n)$ and since $\bar{\lambda} \ne \bar{\lambda} w_{ij}(\bar{h}_1,\ldots,h_n) = \bar{\lambda}$ for all w_{ij}'s involved in $w'(x_1,\ldots,x_n)$ we have

$$e \ne \bigvee_{i\in I(0)} \bigwedge_{j\in J(i)} \prod_{1\le k\le k(i,j)} \hat{f}_{ijk}(\lambda w_{ij}^{(k-1)}(h,\ldots,h_n)).$$

Hence the identity

$$w'' = w''(y_{ijk}|i \in I(0),\ j \in J(i),\ 1 \le k \le k(i,j)) = e$$

fails in U, for when $y_{ijk}^{\mp 1} = y_{spt}^{\pm 1}$, i.e., $x_{ijk}^{\mp 1} = x_{spt}^{\pm 1}$ and

$$\bar{\lambda} w_{ij}^{(k-1)}(\bar{h}_1, \ldots, \bar{h}_n) = \bar{\lambda} w_{sp}^{(t)}(\bar{h}_1, \ldots, \bar{h}_n),$$

then $\hat{f}_{ijk}(\bar{\lambda} w_{ij}^{(k-1)}(\bar{h}_1, \ldots, \bar{h}_n))$ and $\hat{f}_{spt}(\bar{\lambda} w_{sp}^{(t)}(\bar{h}_1, \ldots, \bar{h}_n))$ are inverses in U. Hence $w'' = e$ fails in U. Since $\text{var}_l U_s = \mathcal{U}$, the identity $w'' = e$ also fails in some $U \in \{U_s\}$ via some substitution $y_{ijk} \to u_{ijk}$. Thus

$$e \neq w''(u_{ijk} | i \in I, j \in J, 1 \leq k \leq k(i,j)) = \bigvee_{i \in I(0)} \bigwedge_{j \in J(i)} u_{ij1} u_{ij2} \cdots u_{ijk(i,j)}.$$

Now we define the elements m_{ijk} in $U\mathrm{Wr}(G, \Omega)$ by the rule: $m_{ijk} = (\hat{b}_{ijk}, g_{ijk})$. Here $\hat{b}_{ijk} : \Omega \in U$ is defined as follows:

(8) whenever $x_{ijk} = x_{spt}$ and $w_{sp}(x_1, \ldots, x_n)$ is involved in $w'(x_1, \ldots, x_n)$, $\hat{b}_{ijk}(\alpha w_{sp}^{(t-1)}(g_1, \ldots, g_n)) = u_{spt}$;

(9) whenever $x_{ijk}^{\mp 1} = x_{spt}^{\pm 1}$ and w_{sp} is involved in $w'(x_1, \ldots, x_n)$, $\hat{b}_{ijk}(\alpha w_{sp}^{(t)}(g_1, \ldots, g_n)) = u_{spt}^{-1}$,

(10) all other components $\hat{b}_{ijk}(\alpha) = e$, $\alpha \in \Omega$.

In checking that that $\hat{b}_{ijk}$ is well defined, the only difficulty is when x_{spt} satisfies (8) and $x_{s'p't'}$ satisfies (9), and

$$\alpha w_{sp}^{(t-1)}(g_1, \ldots, g_n) = \alpha w_{s'p'}^{t'}(g_1, \ldots, g_n).$$

Then

$$\bar{\lambda} w_{sp}^{(t-1)}(\bar{h}_1, \ldots, \bar{h}_n) = \bar{\lambda} w_{s'p'}^{t'}(\bar{h}_1, \ldots, h_n) = \bar{\lambda} w_{s'p'}^{(t')}\bar{(}h_1, \ldots, \bar{h}_n)$$

by (4), and since $x_{spt}^{\mp 1} = x_{s'p't'}^{\pm 1}$ we have $y_{spt}^{\mp 1} = y_{s'p't'}^{\pm 1}$. Thus u_{spt} and $u_{s'p't'}$ are inverses and elements $\hat{b}_{ijk}$ are well-defined.

Now is valid $w(m_{ijk} | i \in I,\ j \in J,\ 1 \leq k \leq k(i,j)) = (\hat{b}, e)$ in $U\mathrm{Wr}(G, \Omega)$ where by (3),

$$\hat{b}(\alpha) = \bigvee_{i \in I(0)} \bigwedge_{j \in J(i)} \prod_{1 \leq k \leq k(i,j)} \hat{b}_{ijk}(\alpha w_{ij}^{(k-1)}(g_1, \ldots, g_n)) =$$
$$w''(u_{ijk} | i \in I(0),\ j \in J(i),\ 1 \leq k \leq k(i,j)) \neq e.$$

Hence the identity $w = e$ fails in $U\mathrm{Wr}(G, \Omega)$ and in $U\mathrm{wr}(G, \Omega)$.□

COROLLARY 1. *Let* $\mathrm{var}_l\{U_s|s \in S\} = \mathcal{U}$ *and* $\{(G_t, \Omega_t)|t \in T\}$ *be the class of all transitive ordered permutation l-groups for which* $G_t \in \mathcal{V}$. *Then* $\mathrm{var}_l\{U_s \mathrm{wr}(G_t, \Omega_t)\} = \mathcal{U} \cdot \mathcal{V}$.□

THEOREM 2. (Glass, Holland and McCleary [1].) *The right regular represention* $(\mathbf{Z}, \mathbf{Z})$ *of the o-groups of integers mimics the l-variety of Abelian l-groups* $\mathcal{A}$.

Proof. Let (H, Λ) be a transitive Abelian ordered permutation l-group. Then H is an o-group and (H, Λ) is a regular representation of the totally ordered group, we may henceforth omit mentioning of the totally ordered sets and discuss only $\mathbf{Z}$ and H. Without loss of generality we can assume that H is a finitely generated o-group. Now arguments are similar to the proof of Theorem 8.6.1.□

COROLLARY 1. $\mathrm{var}_l\mathbf{Z} = \mathcal{A}$.□

Let $Z_i = \mathbf{Z}$ for $i = 1.2, \ldots, n$. We denote

$$\mathrm{wr}^n(\mathbf{Z}, \mathbf{Z}) = \underbrace{(\mathbf{Z}, \mathbf{Z})\mathrm{wr}(\mathbf{Z}, \mathbf{Z})\mathrm{wr} \cdots \mathrm{wr}(\mathbf{Z}, \mathbf{Z})}_{n} = \mathrm{wr}^n(\mathbf{Z}, \mathbf{Z}),$$

$$\mathrm{wr}\{(Z_i, Z_i)|i \in \mathbf{N}\} = \mathrm{wr}^\infty(\mathbf{Z}, \mathbf{Z}).$$

COROLLARY 2. $\mathrm{var}_l(\mathrm{wr}^n(\mathbf{Z}, \mathbf{Z})) = \mathcal{A}^n$.□

COROLLARY 3. $\mathrm{var}_l(\mathrm{wr}^\infty(\mathbf{Z}, \mathbf{Z})) = \mathcal{N}$.

Proof. In fact, $\mathrm{wr}^n(\mathbf{Z}, \mathbf{Z}) \in \mathcal{N}$ and $\mathrm{wr}^n(\mathbf{Z}, \mathbf{Z}) \subset \mathrm{wr}^\infty(\mathbf{Z}, \mathbf{Z})$ for all $n \in \mathbf{N}$. By Corollary 2 of Theorem 2, $\mathcal{A}^n \subset \mathrm{var}_l(\mathrm{wr}^\infty(\mathbf{Z}, \mathbf{Z}))$ and by Theorem 11.2.2, $\mathcal{N} = \bigvee^{\mathbf{L}}_{n \in \mathbf{N}} \mathcal{A}^n \subseteq \mathrm{var}_l(\mathrm{wr}^\infty(\mathbf{Z}, \mathbf{Z})) \subseteq \mathcal{N}$.□

COROLLARY 4. $\mathcal{N}$ *and* $\mathcal{L}$ *are the only l-varieties closed under wreath products, even above (below) by* $(\mathbf{Z}, \mathbf{Z})$. *In particular, if* $\mathcal{V} \neq \mathcal{N}, \mathcal{L}$ *then* $\mathcal{V} \cdot \mathcal{A} \nsubseteq \mathcal{V}$ *and* $\mathcal{A} \cdot \mathcal{V} \nsubseteq \mathcal{V}$.□

We will say that the l-variety $\mathcal{V}$ has *dimension* n ($\dim \mathcal{V} = n$) if $\mathcal{A}^n \subseteq \mathcal{V}$ and $\mathcal{A}^{n+1} \nsubseteq \mathcal{V}$. By Corollary 2 of Theorem 2, this is equivalent to $\mathrm{wr}^n(\mathbf{Z}, \mathbf{Z}) \in \mathcal{V}$ and $\mathrm{wr}^{n+1}(\mathbf{Z}, \mathbf{Z}) \notin \mathcal{V}$. We also say $\mathcal{E}$ has dimension 0 and $\mathcal{N}$ and $\mathcal{L}$ have dimension ∞. By Theorem 11.2.2, every proper l-subvariety of $\mathcal{N}$ has finite dimension.

LEMMA 1. (Glass, Holland and McCleary [1].) *Let* H *be an l-group and* $(G.\Omega)$ *a transitive ordered permutation l-group. Then*

(11) *every ideal N of $H\mathrm{wr}(G,\Omega)$ is related by inclusion to* $\prod_{\alpha\in\Omega} H_\alpha$;

(12) *if* $N \subseteq \prod_{\alpha\in\Omega} H_\alpha$, *then* $N = \prod_{\alpha\in\Omega}(H_\alpha \cap N)$, $H_\alpha \cap N$ *and* $H_\beta \cap N$ *are isomorphic for all* $\alpha, \beta \in \Omega$;

(13) *if* $N \supseteq \prod_{\alpha\in\Omega}$, *then* $N = H\mathrm{wr}(L,\Omega)$ *where L is some ideal of G.*

Proof. Let N be an ideal with $N \not\subseteq \prod_{\alpha\in\Omega} H_\alpha$. Then there exists $(\hat{h}, g) \in N$ with $g > e$. Hence there exists $\beta \in \Omega$ for which $\beta < \beta g$. Then $H_\beta \subseteq N$ and therefore, $H_\alpha \in N$ for all $\alpha \in \Omega$. Since all factors are in N, then $\prod_{\alpha\in\Omega} H_\alpha \subseteq N$. The proof of other statements are left to the reader.

COROLLARY 1. *The only ideals of* $\mathrm{wr}^n(\mathbf{Z},\mathbf{Z})$ *are kernels of the natural l-homomorphisms* $\varphi_k :\ \mathrm{wr}^n(\mathbf{Z},\mathbf{Z}) \to \mathrm{wr}^k(\mathbf{Z},\mathbf{Z})$ *where* $k \leq n$.□

PROPOSITION 1. *Let* $\mathcal{U},\mathcal{V}$ *be l-varieties and* $\dim\mathcal{U} = n$, $\dim\mathcal{V} = m$ *then* $\dim\mathcal{U}\cdot\mathcal{V} = n + m$. *Hence* $\dim\mathcal{A}^n = n$ *for all* $n \in \mathbf{N}$.

Proof. It is clear that $\mathrm{wr}^{n+m}(\mathbf{Z},\mathbf{Z}) = (\mathrm{wr}^n(\mathbf{Z},\mathbf{Z}))\mathrm{wr}(\mathrm{wr}^m(\mathbf{Z},\mathbf{Z})) \in \mathcal{U}\cdot\mathcal{V}$. Now suppose that $\mathrm{wr}^i(\mathbf{Z},\mathbf{Z}) \in \mathcal{U}\cdot\mathcal{V}$. Then by Lemma 2, the $\mathcal{U}$-radical $\mathcal{U}(\mathrm{wr}^i(\mathbf{Z},\mathbf{Z}))$ is the kernel of the natural l-homomorphism and so $\mathrm{wr}^i(\mathbf{Z},\mathbf{Z})/\mathcal{U}(\mathrm{wr}^i(\mathbf{Z},\mathbf{Z}))$ is isomorphic to $\mathrm{wr}^j(\mathbf{Z},\mathbf{Z})$ for some j and $\mathrm{wr}^j(\mathbf{Z},\mathbf{Z}) \in \mathcal{V}$. Therefore $j \leq n$. Moreover, $\mathcal{U}(\mathrm{wr}^i(\mathbf{Z},\mathbf{Z})) = \prod_{\gamma\in\Gamma}(\mathrm{wr}^{i-j}(\mathbf{Z},\mathbf{Z}))_\gamma$, so $i - j \leq m$. Hence $i \leq m + n$.□

COROLLARY 1. *If* $\mathcal{U},\mathcal{V}$ *are proper l-subvarieties of* $\mathcal{N}$, *so is* $\mathcal{U}\cdot\mathcal{V}$.□

COROLLARY 2. *For any l-varieties* $\mathcal{U}$ *and* $\mathcal{V}$, $\dim(\mathcal{U} \overset{\mathbf{L}}{\bigvee}) = \max\{\dim\mathcal{U}, \dim\mathcal{V}\}$.

If $\mathcal{U}$ *and* $\mathcal{V}$ *are proper l-subvarieties of* $\mathcal{N}$, *so is* $\mathcal{U} \overset{\mathbf{L}}{\bigvee} \mathcal{V}$.

Proof. By Corollary 1 of Lemma 2, $\mathrm{wr}^n(\mathbf{Z},\mathbf{Z})$ is subdirectly irreducible. By Corollary 1 of Proposition 9.1.1,

$$\mathrm{wr}^n(\mathbf{Z},\mathbf{Z}) \in \mathcal{U} \quad \text{or} \quad \mathrm{wr}^n(\mathbf{Z},\mathbf{Z}) \in \mathcal{V}.\square$$

COROLLARY 3. $\mathcal{A}^n$ *is finitely irreducible for all* $n \in \mathbf{N}$.□

LEMMA 2. *Let $\mathcal{U}_i, \mathcal{V}_i$ be l-varieties ($i = 1, 2$) with $\mathcal{U}_1 \cdot \mathcal{V}_1 \subseteq \mathcal{U}_2 \cdot \mathcal{V}_2$. If $\mathcal{U}_1 \not\subseteq \mathcal{U}_2$, then $\mathcal{A} \cdot \mathcal{V}_1 \subseteq \mathcal{V}_2$ and if $\mathcal{V}_1 \not\subseteq \mathcal{V}_2$ then $\mathcal{U}_1 \cdot \mathcal{A} \subseteq \mathcal{U}_2$.*

Proof. Let (G, Ω) be a transitive ordered permutation l-group which $G \in \mathcal{V}_1$. Let $H \in \mathcal{U}_1 \setminus \mathcal{U}_2$. Then $H\mathrm{wr}(G, \Omega) \in \mathcal{U}_1 \cdot \mathcal{V}_1 \subseteq \mathcal{U}_2 \cdot \mathcal{V}_2$. Since $H \notin \mathcal{U}_2$, $\mathcal{U}_2(H\mathrm{wr}(G, \Omega)) \not\supseteq \prod_{\alpha \in \Omega} H_\alpha$. It follows from Lemma 2 that $\mathcal{U}_2(H\mathrm{wr}(G, \Omega))$ is properly contained in $\prod_{\alpha \in \Omega} H_\alpha$. Now

$$H\mathrm{wr}(G, \Omega)/\mathcal{U}_2(H\mathrm{wr}(G, \Omega)) \in \mathcal{V}_2.$$

Hence $\mathbf{Z}\mathrm{wr}(G, \Omega) \subseteq (H/\mathcal{U}_2(H))\mathrm{wr}(G, \Omega)$ wich is isomorphic to $H\mathrm{wr}(G, \Omega)/\mathcal{U}_2(H\mathrm{wr}(G, \Omega)) \in \mathcal{V}_2$. Therefore,

$$\mathcal{V}_2 \supseteq \mathrm{var}_l\{\mathbf{Z}\mathrm{wr}(G, \Omega) \mid G \in \mathcal{V}_1\} = \mathcal{A} \cdot \mathcal{V}_1$$

by Corollary 1 of Theorem 1. The proofs of other statements are similar.□

COROLLARY 1. *Let $\mathcal{U}, \mathcal{V}$ and $\mathcal{W}$ be l-varieties and $\mathcal{V} \neq \mathcal{N}, \mathcal{L}$. If $\mathcal{U} \cdot \mathcal{V} \subseteq \mathcal{W} \cdot \mathcal{V}$, then $\mathcal{U} \subset \mathcal{W}$; and if $\mathcal{V} \cdot \mathcal{U} \subseteq \mathcal{V} \cdot \mathcal{W}$, then $\mathcal{U} \subseteq \mathcal{W}$. If $\mathcal{U} \cdot \mathcal{V} = \mathcal{W} \cdot \mathcal{V}$, then $\mathcal{U} = \mathcal{W}$; and if $\mathcal{V} \cdot \mathcal{U} = \mathcal{V} \cdot \mathcal{W}$, then $\mathcal{U} = \mathcal{W}$.*□

Now let n be a positive integer and $calV$ an l-variety of dimension at least n. We define $C_n(\mathcal{V}) = \{H \mid H\mathrm{wr}(G, \Omega) \in \mathcal{V}$ for all transitive (G, Ω) with $G \in \mathcal{A}^n\}$. The direct verification shows that $C_n(\mathcal{V})$ is non-empty, closed under l-subgroups, l-homomorphic images and Cartesian products. Hence $C_n(\mathcal{V})$ is an l-variety. By definition $C_n(\mathcal{V}) \cdot \mathcal{A}^n \subseteq \mathcal{V}$.

We will say that l-variety $\mathcal{W} \neq \mathcal{E}$ is *indecomposable* if

$$\mathcal{U} \cdot \mathcal{V} = \mathcal{W}$$

implies $\mathcal{U} = \mathcal{E}$ or $\mathcal{V} = \mathcal{E}$. We are now ready to prove the main theorem of this chapter.

THEOREM 3. (Glass, Holland and McCleary [1].) *Every proper l-variety different from $\mathcal{N}$ is uniquely the product of indecomposable l-varieties, i.e., the proper l-varieties other than $\mathcal{N}$ form a free semigroup on the set of indecomposable l-varieties as free generators.*

Proof. If $\mathcal{V}$ is a proper l-subvariety of $\mathcal{N}$, then $\dim \mathcal{V} = n$ for some finite n. From easy induction on n, it follows that $\mathcal{V}$ is a product of indecomposable l-varieties.

Now let $\mathcal{U}_i, \mathcal{V}_i$ $(i = 1, 2)$ be l-varieties, $\mathcal{U}_1$ and $\mathcal{U}_2$ be indecomposable, and $\mathcal{U}_1 \cdot \mathcal{V}_1 = \mathcal{U}_{\in} \cdot \mathcal{V}_2$. If $\mathcal{U}_1$ and $\mathcal{U}_2$ are incomporable, then by Lemma 3, $\mathcal{A} \cdot (\mathcal{A} \cdot \mathcal{V}_1) \subseteq \mathcal{A} \cdot \mathcal{V}_2 \subseteq \mathcal{V}_1$, violating Corollary 4 of Theorem 2. Without loss of generality, we can assume that $\mathcal{U}_2 \subseteq \mathcal{U}_1$, and hence $\mathcal{V}_1 \subseteq \mathcal{V}_2$. Let $\dim \mathcal{V}_1 = n$ and $\mathcal{U} = C_n(\mathcal{V}_2) \cap \mathcal{U}_1$. We will show that $\mathcal{U}_1 = \mathcal{U}_2 \cdot \mathcal{U}$. First, $\mathcal{U}_2 \cdot \mathcal{U} \cdot \mathcal{A}^n \subseteq \mathcal{U}_2 \cdot \mathcal{V}_2 = \mathcal{U}_1 \cdot \mathcal{V}_1$, and since $\mathcal{A}^{n+1} \not\subseteq \mathcal{V}_\infty$, it follows from Lemma 3 that $\mathcal{U}_2 \cdot \mathcal{U} \subseteq \mathcal{U}_1$. For the converse inclusion, let $H \in \mathcal{U}_1 \setminus \mathcal{U}_2$ and let $G \in \mathcal{A}^n$, (G, Ω) be a transitive ordered permutation l-group. Then $H\mathrm{wr}(G, \Omega) \in \mathcal{U}_1 \cdot \mathcal{V}_1 = \mathcal{U}_2 \cdot \mathcal{V}_2$, and since $H \notin \mathcal{U}_2$, by Lemma 2, $\mathcal{U}_2(H\mathrm{wr}(G, \Omega)) = \prod_{\alpha\in\Omega} \mathcal{U}_2(H)_\alpha$. Moreover, $(H/\mathcal{U}_2(H))\mathrm{wr}(G, \Omega)$ is isomorphic to $(H\mathrm{wr}(G, \Omega))/\mathcal{U}_2(H\mathrm{wr}(G, \Omega)) \in \mathcal{V}_2$. It follows that $H/\mathcal{U}_2(H) \in C_n(\mathcal{V}_2) \cap \mathcal{U}_1 = \mathcal{U}$. Hence $H \in \mathcal{U}_2 \cdot \mathcal{U}$. Therefore, $\mathcal{U}_1 = \mathcal{U}_2 \cdot \mathcal{U}$. Since $\mathcal{U}_1$ is indecomposable, $\mathcal{U}_1 = \mathcal{U}_2$. By Corollary 1 of Lemma 3, $\mathcal{V}_1 = \mathcal{V}_2$. The theorem now follows by induction, since $\dim \mathcal{V}_1 < \dim(\mathcal{U}_1 \cdot \mathcal{V}_1)$.□

As it was mentioned earlier, every l-variety of dimention one is indecomposable. There are other indecomposable l-varieties of arbitrary large dimension. Let $\mathcal{X}$ be any l-variety of dimention one and $\mathcal{Y}$ any l-variety incomparable to $\mathcal{X}$. Then we claim that $\mathcal{X} \overset{\mathbf{L}}{\bigvee} \mathcal{Y}$ is indecomposable. Suppose, on the contrary, that $\mathcal{X} \overset{\mathbf{L}}{\bigvee} \mathcal{Y} = \mathcal{U} \cdot \mathcal{V}$ with $\mathcal{U}$ and $\mathcal{V}$ non-trivial l-varieties. Let $\dim \mathcal{U} = i$, $\dim \mathcal{V} = j$ and $n = \dim \mathcal{Y}$. Then $i + j = n$. Let H be any non-trivial subdirectly irreducible l-group of $\mathcal{U}$ and (G, Ω) be any non-trivial transitive ordered permutation l-group, $G \in \mathcal{V}$. Then $H\mathrm{wr}(G, \Omega) \in \mathcal{U} \cdot \mathcal{V} = \mathcal{X} \overset{\mathbf{L}}{\bigvee} \mathcal{Y}$ and is subdirectly irreducible. Hence $H\mathrm{wr}(G, \Omega) \in \mathcal{X}$ or $H\mathrm{wr}(G, \Omega) \in \mathcal{Y}$. Since $\dim \mathcal{X} = 1$, $H\mathrm{wr}(G, \Omega) \in \mathcal{Y}$. Then $\mathcal{U} \cdot \mathcal{V} \subseteq \mathcal{Y}$ and $\mathcal{X} \subseteq \mathcal{X} \overset{\mathbf{L}}{\bigvee} \mathcal{Y} = \mathcal{U} \cdot \mathcal{V} \subseteq \mathcal{Y}$. This is the contradiction.

11.4 The Order on the Semigroup of l-Varieties

The aim of this Section is to investigate the connections between lattice operations $\overset{\mathbf{L}}{\bigvee}$, $\overset{\mathbf{L}}{\bigwedge}$ in the lattice of all l-varieties and the multiplication in the semigroup of l-varieties.

LEMMA 1. (Glass, Holland and McCleary [1].) *Let G be a subdirectly irreducible l-group and N be a closed ideal of G. Then G/N is finitely subdirectly irreducible.*

Proof. By Corollary 3 of Theorem 4.2.1, G has a faithfull transitive representation (G, Ω). The orbits of N on Ω determine a convex G-congruence $\equiv$ on Ω. Now the arguments similar to the proof of Proposition 11.2.3 show that the naturally induced representation $(G/N, \Omega/\equiv)$ is faithfull and transitive. Therefore, G/N cannot have two disjoint ideals.□

THEOREM 1. (Holland, Glass and McCleary [1].)

$$\mathcal{U}\cdot(\mathcal{V}_1 \overset{\mathbf{L}}{\bigvee} \mathcal{V}_2) = \mathcal{U}\cdot\mathcal{V}_1 \overset{\mathbf{L}}{\bigvee} \mathcal{U}\cdot\mathcal{V}_2, \tag{1}$$

$$(\overset{\mathbf{L}}{\bigvee_{i\in I}} \mathcal{U}_i)\cdot\mathcal{V} = \overset{\mathbf{L}}{\bigvee_{i\in I}} \mathcal{U}_i\cdot\mathcal{V}, \tag{2}$$

$$\mathcal{U}\cdot(\overset{\mathbf{L}}{\bigwedge_{i\in I}} \mathcal{V}_i) = \overset{\mathbf{L}}{\bigwedge_{i\in I}} \mathcal{U}\cdot\mathcal{V}_i, \tag{3}$$

$$(\bigwedge_{i\in I})^{\mathbf{L}}\mathcal{U}_i\cdot\mathcal{V} = \overset{\mathbf{L}}{\bigwedge_{i\in I}} \mathcal{U}_i\cdot\mathcal{V}. \tag{4}$$

Proof. (1). It is clear that $\mathcal{U}\cdot(\mathcal{V}_1 \overset{\mathbf{L}}{\bigvee} V_2) \supseteq \mathcal{U}\cdot\mathcal{V}_1 \overset{\mathbf{L}}{\bigvee} \mathcal{U}\cdot\mathcal{V}_{\in}$. Let G be a subdirectly irreducible l-group of the left side. By Theorem 9.5.2, $\mathcal{U}$-radical $\mathcal{U}(G)$ of G is closed. By Lemma 1, $G/\mathcal{U}(G)$ is finitely subdirectly irreducible and belongs to $\mathcal{V}_1 \overset{\mathbf{L}}{\bigvee} \mathcal{V}_2$. Therefore, $G/\mathcal{U}(G) \in \mathcal{V}_1$ or $G/\mathcal{U}(G) \in \mathcal{V}_{21}$ by Proposition 9.1.1. In either case $G \in \mathcal{U}\cdot\mathcal{V}_1 \overset{\mathbf{L}}{\bigvee} \mathcal{U}\cdot\mathcal{V}_2$.

(2). Clearly, the right side is contained in the left one. Let $\mathcal{G}$ be the class of all transitive ordered permutation l-groups (G, Ω) such that

$G \in \mathcal{V}$. Clearly $\mathrm{var}_l\{\bigcup_{i\in I} \mathcal{U}_i\} = \overset{\mathbf{L}}{\bigvee_{i\in I}} \mathcal{U}_i$. Therefore by Theorem 11.3.1,

$$\mathrm{var}_l\{U\mathrm{wr}(G,\Omega) | U \in \bigcup_{i\in I} U_i,\ (G,\Omega) \in \mathcal{G}\} = (\overset{\mathbf{L}}{\bigvee_{i\in I}} \mathcal{U}_i) \cdot \mathcal{V}.$$

Obviously, each $U\mathrm{wr}(G,\Omega) \in \overset{\mathbf{L}}{\bigvee_{i\in I}} \mathcal{U}_i \cdot \mathcal{V}$, so $(\overset{\mathbf{L}}{\bigvee_{i\in I}}) \cdot \mathcal{V} \subseteq \overset{\mathbf{L}}{\bigvee_{i\in I}} \cdot \mathcal{V}$.

(3). It is clear that the left side is contained in the right one. Let $G \in \overset{\mathbf{L}}{\bigwedge_{i\in I}} \mathcal{U} \cdot \mathcal{V}_i$. Thus $G/\mathcal{U}(G) \in \mathcal{V}_i$ for all $i \in I$ and $G \in \mathcal{U} \cdot (\overset{\mathbf{L}}{\bigwedge_{i\in I}} \mathcal{V}_i)$.

(4). Let $G \in \mathcal{U}_i \cdot \mathcal{V}$ for all $i \in I$. Then $G/\mathcal{U}_i \in \mathcal{V}$ for all $i \in I$. Hence $G/\bigcap_{i\in I} \mathcal{U}_i(G) \subseteq \bar{\prod_{i\in I}} G/\mathcal{U}_i(G) \in \mathcal{V}$. But $\bigcap_{i\in I} \mathcal{U}_i(G) = (\overset{\mathbf{L}}{\bigwedge_{i\in I}} \mathcal{U}_i(G))$ and therefore $G \in (\overset{\mathbf{L}}{\bigwedge_{i\in I}} \mathcal{U}_i) \cdot \mathcal{V}$. The converse inclusion is evident.□

From Theorem 1 it follows that the semigroup of proper l-varieties other than N forms a *lattice ordered semigroup* under inclusion, meaning that it is a lattice, the multiplication on either side preserves order, and finite distributive laws hold for suprema and infima. Moreover, this semigroup is *strong*, meaning that it satisfies the cancellation laws; and it is *positively ordered*, meaning that always $\mathcal{U} \subseteq \mathcal{U} \cdot \mathcal{V}$ and $\mathcal{U} \subseteq \mathcal{V} \cdot \mathcal{U}$. Writing $\mathcal{U} \equiv \mathcal{V}$ iff $\dim\mathcal{U} = \dim\mathcal{V}$ we obtain a congruence whose classes are convex sublattices. The quotient semigroup via this congruence is isomorphic to the semigroup of positive integers.

THEOREM 2. (Medvedev [12].) *The lattice ordered semigroup* **L** *of l-varieties does not satisfy the identity*

$$\mathcal{U} \cdot (\overset{\mathbf{L}}{\bigwedge_{i\in I}} \mathcal{V}_i) = \overset{\mathbf{L}}{\bigwedge_{i\in I}} \mathcal{U} \cdot \mathcal{V}_i$$

for infinite index set I.

Proof. Let β be a positive real number, $1 \neq \beta$ and $T_\beta = R_{\beta(n)}(b_\beta)$ be an o-group defined in $\mathbf{3^0}$ Section 6 of Chapter 9. Each element x of T_β has the unique representation in the form $x = b_\beta^n r$ where $b_\beta^n \in$

(b_β), $r \in R_\beta$ and R_β is isomorphic to the totally ordered additive group $\mathbf{R}$ of reals, (b_β) is an infinite cyclic group.

The group operation in T_β is defined by the rule

$$x \cdot y = (b_\beta^n r)(b_\beta^m s) = b_\beta^{n+m}(\beta^m r + s)$$

for $x = b_\beta^n r,\ y = b_\beta^m s$ in T_β. The group T_β is ordered lexicographically: $x = b_b^n r \geq e$ iff $n > 0$ or $n = 0,\ r > 0$ in natural total order of the additive group $\mathbf{R}$ of reals. Let N_0 be a nilpotent group of the class 2 with has the presentation

$$N_0 = \mathrm{gp}(a, b, c |\ [a, c] = [bc] = e,\ [a, b] = c)$$

totally ordered by the rule: $e \leq a^m b^n c^k$ iff $m > 0$ or $m = 0, n >$ or $m = n = 0, k > 0$. Let us consider positive real numbers $\beta(n) = \frac{n}{n+1}$, $(n \in \mathbf{N})$ and o-groups $T_{\beta(n)}$. Direct verification shows that following identity

$$t(n) \qquad (|x| \vee |y|)^{-1}(|[x, y]|^{n+1})(|x| \vee |y|) \leq |[x, y]|$$

is valid in $T_{\beta(n)}$. This identity is violated in o-group N_0 if $x = a,\ y = b$. Since the identity $(|x| \vee |y|)^{-1}(|[x, y]|^{n+1})(|x| \vee |y|) = |[x, y]|$ holds in N_0, $\mathrm{var}_l T_{\beta(n)} \overset{\mathbf{L}}{\bigwedge} \mathrm{var}_l N_0 = \mathcal{A}$.

Now we need the following statement.

LEMMA 2.

$$(5) \qquad \mathrm{var}_l N_0 \subseteq \overset{\mathbf{L}}{\bigvee_{1 \leq n \leq \infty}} \mathrm{var}_l T_{\beta(n)},$$

$$(6) \qquad \mathrm{var}_l N_0 \overset{\mathbf{L}}{\bigwedge} \mathrm{var}_l T_{\beta(n)} = \mathcal{A} \textit{ for all } \beta.$$

Proof. Let $G = \bar{\prod}_{n \in \mathbf{N}} T_{\beta(n)}$ be the Cartesian product of o-groups $T_{\beta(n)}$. It is evident that the subgroup $K = \bar{\prod}_{n \in \mathbf{N}} R_{\beta(n)}$, is an ideal in G and every element of K can be represented as an infinite sequence $k = (a_1, \ldots, a_n, \ldots)$.

Let G_0 be the l-subgroup of G generated by the element

$$g = (b_{\beta(1)}, b_{\beta(2)}, \ldots, b_{\beta(n)}, \ldots)$$

and the ideal K. Now consider the set

$$H = \{k = (a_1, , \ldots, a_n, \ldots) \in K | \lim_{n \to \infty} a_n = 0\}.$$

The easy verification shows that H is an ideal in l-group G_0. Now consider in G_0 the elements g and $c_1 = (2, 3, \ldots, n+1, \ldots)$. Then

$$c_2 = [g, c_1] = g^{-1}c_1^{-1}gc_1 =$$
$$= g^{-1} \cdot (-2, , \ldots, -n-1, \ldots) \cdot g \cdot (2, \ldots, n+1, \ldots) =$$
$$= (-1, -2, \ldots, -n, \ldots) \cdot (2, 3, \ldots, n+1, \ldots) = (1, 1, \ldots, 1, \ldots).$$

In the same way

$$c_3 = [g, c_2] = (\frac{1}{2}, \ldots, \frac{1}{n+1}, \ldots).$$

From these equalities it is easy to see that $c_3 \in H$. So in the quotient group $\bar{G} = G_0/H$ the following relations are valid:

$$\bar{g} \gg \bar{c}_1 \gg \bar{c}_2 > e,$$
$$[\bar{g}, \bar{c}_1] = \bar{c}_2, \quad [\bar{g}, \bar{c}_2] = [\bar{c}_1, \bar{c}_2] = e.$$

So the l-subgroup $\bar{N} = 1 - \text{gp}(\bar{g}, \bar{c}_1, \bar{c}_2)$ generated by elements $\bar{g}, \bar{c}_1, \bar{c}_2$ is isomorphic to the o-group N_0.□

Now we continue the proof of the theorem. It is evident that each l-group $G \in \text{var}_l T_{\beta(n)}$ has the following property:

for all $, y \in G$ is valid $|[x, y]| \neq e$ iff $[|x| \vee |y|, |[x, y]|] \neq e$.

It is clear also that

$$|[|x| \vee |y|, |[x, y]|]| = |(|x| \vee |y|)^{-1}|[x, y]|((|x| \vee |y|)|[x, y]|| \leq |[x, y]|$$

as $\beta(n) = \frac{n}{n+1} < 1$ for every n.

On the other hand, $(|x| \vee |y|)^{-(n+1)}|[x, y]|((|x| \vee |y|) \geq |[x, y]|^{-n}$ and therefore, $|[|x| \vee |y|, |[x, y]|]|^{n+1} \geq |[x, y]|$. Thus the elements $|[|x| \vee |y|, |[x, y]|]|$ and $|[x, y]|$ are Archimedean equivalent in G for all $x, y \in G$. Hence the sets of fixed points of the elements $|[|x| \vee |y|, |[x, y]|]|$ and

$|[x,y]|$ in each transitive representation of G by order permutations of totally ordered set Ω are the same.

Now we consider the class of l-groups $\{\mathbf{Z}\mathrm{wr}(G,\Omega)\}$ where $\mathbf{Z}$ is the totally ordered additive group of integers and (G,Ω) is a transitive ordered permutation l-group and $G \in \bigcup_{n\in\mathbf{N}} \mathrm{var}_l T_{\beta(n)}$. By Theorem 9.3.1, the class $\{\mathbf{Z}\mathrm{wr}(G,\Omega)\}$ generates the l-variety $\bigvee_{n=1}^{\infty} \mathcal{A}\cdot\mathrm{var}_l T_{\beta(n)}$. We claim that the following identity

$$(7)\quad (|[[u,v],[u_1,v_1]]| \wedge |[|x| \vee |y|, |[x,y]|]|) - 1|[[u,v],[u_1,v_1]]|)^{|[x,y]|} =$$
$$= (|[[u,v],[u_1,v_1]]| \wedge |[|x| \vee |y|, |[x,y]|]|)^{-1} \cdot |[[u,v],[u_1,v_1]]|$$

is valid in each l-group of the class $\{\mathbf{Z}\mathrm{wr}(G,\Omega)\}$.

Let $\mathbf{Z}\mathrm{wr}(G,\Omega)$ be any l-group of this class. We identify the l-group G with the subgroup of $\{\mathbf{Z}\mathrm{wr}(G,\Omega)\}$: $x = (\{e_\tau\}, x)$ where $x \in G$, 0_τ are identically elements of groups Z_τ and every Z_τ is isomorphic to the totally ordered additive group of integers. Let $w = (\{h_\tau\}, g) \in \mathbf{Z}\mathrm{wr}(G,\Omega)$. Then $w \geq e$ iff $g \geq e$ in G anf $h_\tau \geq 0_\tau$ $(h_\tau \in Z_\tau)$ for all $\tau \in \Omega$ such that $\tau g = \tau$. Let now $g > e$ in G, $f = (\{h_\tau\}, e)$ where $h_\tau = 0$ for all $\tau \in \Omega$ with $\tau g = \tau$. Then $g = (\{0_\tau\}, g) \gg$f in $\mathbf{Z}\mathrm{wr}(G,\Omega)$. Therefore for each elements $d = (\{h'_\tau\}, e) \geq e$, $f = (\{h_\tau\}, e)$, $g = (\{0_\tau\}, g) \geq e$ in $\mathbf{Z}\mathrm{wr}(G,\Omega)$ holds $(fg \wedge d)^{-1}d = (\{h''_\tau\}, e)$ where $g''_\tau = 0_\tau$ for all $\tau \in \Omega$ such that $\tau g \neq \tau$.

Let the elements g and g' have the same set of fixed points in Ω (as permutations of (G,Ω)). Then it is clear that $g'^{-1}((fg \wedge d)^{-1})g' = (fd \wedge d)^{-1}d$ where $g' = (\{0_\tau\}, g)$.

If $[x,y] = e$ or $[[u,v],[u_1,v_1]] = e$, then the identity (7) is valid. Let us assume that $[x,y] \neq e$, $[[u,v],[u_1,v_1]] \neq e$ in $\mathbf{Z}\mathrm{wr}(G,\Omega)$. Since group G is metabelian, then $|[u,v],[u_1,v_1]| = f = (\{h_\tau\}, e)$ where $h_\tau \in Z_\tau$. If $|[x,y]| = d = (\{h'_\tau\}, e)$, then $|[|x|, |y|, |[x,y]|]| = (\{h''_\tau\}, e) = \hat{d}$. As the elements $f, d, \hat{d}$ are mutually commutative in $\mathbf{Z}\mathrm{wr}(G,\Omega)$ then the identity (7) clearly holds in $\{\mathbf{Z}\mathrm{wr}(G,\Omega)\}$ in this case. Let now $|[x,y]| = d|[x_0,y_0]| = dg$ where $|[x_0,y_0]| = g \in G$, $g > e$ in G and $d = (\{\hat{h}_\tau\}, e)$. Then $|[|x| \vee |y|, |[x,y]|]| = d_1|[|x_0| \vee |y_0|, |[x_0,y_0]|]| = d_1 g_1$ where by above arguments, $g_1 > e$ in G and $d_1 = (\{\hat{h}'_\tau, e\})$. As well, the elements g and g_1 have the same set of fixed points in each transitive representation of (G,Ω).

Therefore, as it was mentioned above, $(dg)^{-1}((f \wedge d_1g_1)^{-1}f)(dg) = (f \wedge d_1g_1)^{-1}f$ and identity (7) holds in $\{\mathbf{Z}\text{wr}(G,\Omega)\}$. Thus identity (7) is valid in $\overset{\mathbf{L}}{\bigvee\limits_{1\leq n\leq\infty}} \mathcal{A}\cdot \text{var}_l T_{\beta(n)}$.

Now let (N_0, N_0) be the right regular representation of the o-group N_0 and $w = \mathbf{Z}\text{wr}(N_0, N_0)$. Each element $w \in \mathbf{Z}\text{wr}(n_0, N_0)$ has the unique representation in the form $w = (\{h_g\}, f)$ where $f \in N_0$ and $h_g \in \mathbf{Z}$ for all $g \in N_0$. Define the total order P on $\{\mathbf{Z}\text{wr}(N_0, N_0\}$ by the rule: $w = (\{h_g\}, f > e)$ iff $f > e$ in o-group N_0 or $f = e$ and $h_{\bar{g}} > 0$ in $Z_{\bar{g}}$ where $\bar{g} = \max\{g \in N_0 | h_g \neq 0 \text{ in } Z_g\}$. Let $(\mathbf{Z}\text{wr}(N_0, N_0; P)$ be the group $\mathbf{Z}\text{wr}(N_0, N_0$ totally ordered by P.

Let $d = (\{h_g\}, e)$ where $h_a = 1$ and $h_g = 0$ for all $g \neq a$, and $d' = (\{h'_g\}, e)$ where $h'_e = 1$ and $h'_g = 0$ for all $g \neq e$. Let

$$u = u_1 = (\{0_g\}, a),\ v = (\{0_g\}, b),$$
$$v_1 = (\{0_g\}, b)d,\ x = (\{0_g\}, a)d',\ y = (\{o_g\}, b).$$

Then the direct verification shows that identity (7) is violated in o-group $(\mathbf{Z}\text{wr}(N_0, N_0; P)$ for these values of variables u, u_1, v, v_1, x, y.

Therefore, $(\{\mathbf{Z}\text{wr}(N_0, N_0\}; P) \notin \overset{\mathbf{L}}{\bigvee\limits_{1\leq n\leq\infty}} \mathcal{A}\cdot \text{var}_l T_{\beta(n)}$. By Lemma 2,

$$(\mathbf{Z}\text{wr}(N_0, N_0; P) \in \mathcal{A}\cdot (\overset{\mathbf{L}}{\bigvee_{1\leq n\leq\infty}} \text{var}_l T_{\beta(n)}).\square$$

Chapter 12

The lattice of l-varieties

12.1 Fundamental Properties

As it was mentioned in Section 1 of Chapter 9, the set of all l-varieties **L**, is a complete lattice where, for any l-varieties $\mathcal{X}, \mathcal{Y}$, the greatest lower bound, or meet, of $\mathcal{X}, \mathcal{Y}$ is their set-theoretic intersection and their least upper bound, or join, is the intersection of all l-varieties containig both $\mathcal{X}$ and $\mathcal{Y}$; more formally

$$\mathcal{X} \overset{\mathbf{L}}{\bigwedge} \mathcal{Y} = \mathcal{X} \cap \mathcal{Y}, \quad \mathcal{X} \overset{\mathbf{L}}{\bigvee} \mathcal{Y} = \bigcap \{\mathcal{W} \mid \mathcal{X} \cup \mathcal{Y} \subseteq \mathcal{W}.$$

From the standard results in universal algebra (see for example the book of P.Cohn [2]) we have the following observations.

THEOREM 1. *The lattice L of all l-varieties is antiisomorphic to the lattice of all fully invariant ideals of the free l-group of countable rank.* □

THEOREM 2. *The lattice L of all l-varieties is distributive and, moreover, dually Browerian.*

Proof. By Theorem 3.1.2, the lattice of ideals of arbitrary l-group is dually Browerian. The direct verification shows that the product and the intersection of fully invariant ideals is also a fully invariant ideal and the result follows. □

Now a natural hypothesis on the complete distributivity of lattice L arises.

PROPOSITION 1. *The lattice of all l-varieties L is not Browerian and, therefore, completely distributive.*

Proof. By Lemma 11.4.2, there are l-varieties $\mathrm{var}_l T_{\beta(n)}, \mathrm{var}_l N_0$ such that $\overset{\mathbf{L}}{\bigvee}_{n\in\mathbf{N}} \mathrm{var}_l T_{\beta(n)} \supseteq \mathrm{var}_l N_0$ and $\mathrm{var}_l T_{\beta(n)} \overset{\mathbf{L}}{\bigwedge} \mathrm{var}_l N_0 = \mathcal{A}_l$. Therefore,

$$(\overset{\mathbf{L}}{\bigvee_{n\in\mathbf{N}}} T_{\beta(n)}) \overset{\mathbf{L}}{\bigwedge} \mathrm{var}_l N_0 = \mathrm{var}_l N_0 \neq \overset{\mathbf{L}}{\bigvee_{n\in\mathbf{N}}} (\mathrm{var}_l T_{\beta(n)} \overset{\mathbf{L}}{\bigwedge} \mathrm{var}_l N_0) = \mathcal{A}_l. \square$$

Since all these l-varieties are representable, then we have proved that the lattice of all representable l-varieties $\mathbf{L}_{\mathcal{R}}$ is not Browerian and, therefore, completely distributive.

12.2 Identities for Product and Join of l-Varieties

1^0. For any l-varieties $\mathcal{X}$ and $\mathcal{Y}$ it is possible to obtain the basis of identities for $\mathcal{X} \overset{\mathbf{L}}{\bigvee} \mathcal{Y}$ and $\mathcal{X} \overset{\mathbf{L}}{\bigwedge} \mathcal{Y}$ from the bases of identities for $\mathcal{X}$ and $\mathcal{Y}$, respectively.

Let $X = \{x_1, x_2, \ldots, x_n, \ldots\}$ and $w(X) = w(x_{i_1}, x_{i_2}, \ldots, x_{i_k})$ be the term of signature $l = \{\cdot, ^{-1}, e, \vee, \wedge\}$; using finitely many x_i's where, as before, $x_{ijk} \in X \cup \{e\}$ and $x_{ijk} = x_{i'j'k'}$ was possible without the subscripts being equal, now $x_p \in X$ and $x_p = x_q$ only if $p = q$. When we write expressions licke $w(y_1 \wedge v(X), y_2 \wedge v(X), \ldots)$, it should be understood that y_i's are elements of X not used in $v(X)$.

PROPOSITION 1 . *Let the l-variety $\mathcal{X}$ be defined by the identities $\Phi = \{v_i(x_{i1}, \ldots, x_{ik}) = e \mid i \in I, k \in N\}$ and the l-variety $\mathcal{Y}$ be defined by the identities $\Psi = \{u_j(x_{j1}, \ldots, x_{jr}) = e \mid j \in J, r \in R\}$ where the sets of indces I, J, K, R may be finite or infinite and*

$$x_{i1}, \ldots, x_{ik}, x_{j1}, \ldots, x_{jr} \in X = \{x_1, x_2, \ldots, x_n, \ldots\}.$$

Then:

(1) *l-variety* $\mathcal{X} \overset{\mathbf{L}}{\bigvee} \mathcal{Y}$ *is defined by the identities* $\{|v_i(x_{i1},\dots,x_{ik})| \wedge |u_j(y_{j1},\dots,y_{jr})| = e \mid i \in I, j \in J; k \in K, r \in R\}$,

(2) *l-variety* $\mathcal{X} \overset{\mathbf{L}}{\bigwedge} \mathcal{Y}$ *is defined by the identities* $\{\Phi \cup \Psi\}$.

Proof. (1) By Proposition 9.1.1, these identities are valid in $\mathcal{X} \overset{\mathbf{L}}{\bigvee} \mathcal{Y}$. Let (G, Ω) be a transitive ordered permutation group such that $G \notin \mathcal{X} \overset{\mathbf{L}}{\bigvee} \mathcal{Y}$. Then $G \notin \mathcal{X}, \mathcal{Y}$. Thus, there are elements α, β in Ω, and elements $g_{i1},\dots,g_{ik},h_{j1},\dots,h_{jr}$ in G, $(i \in I, j \in J, k \in K, r \in R)$ such that $\alpha|v_i(g_{i1},\dots g_{ik})| > \alpha$ and $\beta|u_j(h_{j1},\dots,h_{jr})| > e$. By transitivity of (G,Ω), there is $g_0 \in G$ such that $\alpha g_0 = \beta$. Then

$$\alpha|u_j(h_{j1},\dots,h_{jr})|^{g_0^{-1}} = \alpha g_0|u_j(h_{j1},\dots,h_{jr})|g_0^{-1} = \\ \beta|u_j(h_{j1},\dots,h_{jr})|g_0^{-1} > \beta g_0^{-1} = \alpha.$$

It is clear that

$$|u_j(h_{j1},\dots,h_{jr})|^{g_0^{-1}} = |u_j(h_{j1}^{g_0^{-1}},\dots,h_{jr}^{g_0^{-1}})|.$$

Then the identity $|v_i(x_{i1},\dots,x_{ik})| \wedge |u_j(y_{j1},\dots,y_{jr})| = e$ is violated in l-group G where $x_{i1} = g_{i1},\dots,x_{ik} = g_{ik}, y_{j1} = h_{j1}^{g_0^{-1}},\dots,y_{jr} = h_{jr}^{g_0^{-1}}$ since

$$\alpha(|v_i(g_{i1},\dots,g_{ir})| \wedge |u_j(h_{j1}^{g_0^{-1}},\dots,h_{jr}^{g_0^{-1}})|) = \\ = \min\{\alpha|v_i(g_{i1},\dots,g_{il})|, \alpha|u_j(h_{j1}^{g_0^{-1}},\dots,h_{jr}^{g_0^{-1}})| > \alpha.\square$$

COROLLARY 1. *The set of l-varieties with finite basis of identities is a sublattice in the lattice of l-varieties* **L**.□

2⁰. Now we will present the basis of identities for the product $\mathcal{X} \cdot \mathcal{Y}$ from the bases of $\mathcal{X}$ and $\mathcal{Y}$. Let the l-variety $\mathcal{X}$ be defined by the identities $\{v_i(X) \mid i \in I\}$ and l-variety $\mathcal{Y}$ be defined by the identities $\{u_j(X) \mid j \in J\}$ where the sets of indeces I and J are finite or infinite. Recall that l-group $G \in \mathcal{X} \cdot \mathcal{Y}$ if and only if $G/\mathcal{X}(G) \in \mathcal{Y}$, which is equivalent to saying that $v_i(g_1,\dots,g_n) \in \mathcal{X}(G)$ for each $i \in I$ and each substitution $X \to \mathbf{g}$ in G. We write now $v(\mathbf{g})$ for the value of the l-word $v(x_1, \dots, x_n)$ in G when we replace the variables $x_1, \dots, x_n$ by the elements $g_1, \dots, g_n$ for short. Clearly, this will be the case iff $u_j(\mathbf{h}) = e$ for all $j \in J$ and each substitution $X \to \mathbf{h}$ in $H(i, \mathbf{g})$. But the elements h of $H(i, \mathbf{g})$, are just those of the form $h = (f \wedge |v_i(\mathbf{g})|^n) \vee |v_i(\mathbf{g})|^{-n}$ for some positive integer n, and $f \in G$. Thus, the requirement that $u_j(\mathbf{h}) = e$ says

$$u_j((f_1 \wedge |v_i(\mathbf{g})|^{n(1)}) \vee |v_i(\mathbf{g})|^{-n(1)}, (f_2 \wedge |v_i(\mathbf{g})|^{n(2)}) \vee |v_i(\mathbf{g})|^{-n(2)}, \dots)$$

or simpler

$$u_j((f_1 \wedge |v_i(\mathbf{g})|^{n}) \vee |v_i(\mathbf{g})|^{-n}, (f_2 \wedge |v_i(\mathbf{g})|^{n}) \vee |v_i(\mathbf{g})|^{-n}, \dots) = e,$$

where $n = \max\{n(1), n(2), \dots\}$. This proves the following theorem.

THEOREM 1. (Glass, Holland and McCleary [1].) *Let the l - variety $\mathcal{X}$ be defined by the identities $\{u_j(X) = e \mid j \in J\}$ and the l-variety $\mathcal{Y}$ be defined by the identities $\{v_i(X) = e \mid i \in I\}$. Then $\mathcal{X} \cdot \mathcal{Y}$ is defined by the identities*

$$u_j((y_1 \wedge |v_i(X)|^{n}) \vee |v_i(X)|^{-n}, (y_2 \wedge |v_i(X)|^{n}) \vee |v_i(X)|^{-n}, \dots) = e$$

for all choices of $i \in I, j \in J$ and positive integers n.□

3⁰. It easy to see that an l-variety $\mathcal{X}$ is defined by a finite set of identities if and only if it is defined by a single identity. When the left l-variety ia Abelian, the following result is true.

THEOREM 2. (Glass, Holland and McCleary [1].) *Let the identity $v(X) = e$ define the l-variety $\mathcal{X}$. Then the identity*

$$[|y| \wedge |v(X)|, |z| \wedge |v(X)|] = e$$

defines the l-variety $\mathcal{A} \cdot \mathcal{X}$.

Proof. It is clear that this identity holds in $\mathcal{A}\mathcal{X}$. Conversely, let G be an l-group which satisfies the commutator identity in the theorem. Let $X \to \mathbf{g}$ be a substitution in G. Then all elements of the set $T = \{t \in G \mid e \leq |t| \leq |v(\mathbf{g})|\}$ commute with one another. If $e \leq h \leq |v(\mathbf{g})|^n$, then $h = h_1 h_2 \cdots h_n$ where for each $i, e \leq h_i \leq |v(\mathbf{g})|$, by the Riesz property

(Theorem 2.2.1). From this it follows that the convex l-subgroup of G generated by $v(\mathbf{g})$ is Abelian, and therefore it is contained in $\mathcal{A}(G)$. Thus, $G/\mathcal{A}(G) \in \mathcal{X}$ and $G \in \mathcal{A}\mathcal{X}$.□

COROLLARY 1. *Each l-variety $\mathcal{A}^n$ is defined by the single identity $c_n(X) = e$ where $c_1(X) = [x_1, x_2] = e$ and for $n > e$, $c_n(X) = [|y| \wedge |c_{n-1}(X)|, |z| \wedge |c_{n-1}(X)|]$.*□

It should also be noted that we do not know if $\mathcal{X}\mathcal{Y}$ can be defined by a single identity if $\mathcal{X}$ and $\mathcal{Y}$ can. Such a result (M.Litvinova [1]) is valid if $\mathcal{X}$ is an l-variety $\mathcal{L}_p$ defined by the identity $[x^p, y^p] = e$, (p is a prime number).

12.3 The Small l-Varieties

We will show that the set of all l-varieties contained in the l-variety $\mathcal{L}_p$ defined by the identity $[x^p, y^p] = e$ (where p is a prime number) is a well-ordered totally ordered set and each l-variety contained in $\mathcal{L}_p$ can be defined by a single identity. All these remarkable results were obtained by S.A.Gurchenkov [3], [6], C.Holland and N.Reily [1], C.Holland, A.Mekler and N.Reily [1]. Here we present Gurchenkov's approach [3].

Let P be a finite set of non-negative integers $\{0, 1, \dots, p-1\}$ and $M_n = \{(\sigma_1, \sigma_2, \dots, \sigma_n) \mid \sigma_i \in P,\ i = 1, 2, \dots, n\}$. For any l-group G, and $x, y \in G$ let

$$d_0(x,y) = x \wedge x^{-1}y^p x \wedge \dots x^{-(p-1)}y^p x^{p-1};$$
$$d_i(x,y) = x^{-i} d_0(x,y) x^i,\ (i = 1, 2, \dots p-1;)$$
$$c_i(x,y) = \bigvee_{j \in (P \setminus \{i\})} (d_i(x,y) \wedge d_j(x,y)),\ (i \in P);$$
$$a_i(x,y) = d_i(x,y) c_i^{-1}(x,y),\ (i \in P).$$

LEMMA 1. *In any l-group $G \in \mathcal{L}_p$ the following equations hold*

a) $$c_i(x,y) = c_j(x,y),\ (i, j \in P,);$$
b) $$a_i(x,y) \wedge a_j(x,y) = e,\ (i, j \in P,\ i \neq j);$$
c) $$x^{-1} a_i(x,y) x = a_{i+1}(x,y),\ (i \in P,\ i \neq p-1);$$
d) $$x^{-1} a_{p-1}(x,y) x = a_0(x,y).$$

Proof. Straightforward. □

LEMMA 2. (Scrimger [1].) *Let $G \in \mathcal{L}_p \setminus \mathcal{A}$. Then there are $x, y \in G$ such that $a_i(x, y) > e$ for all $i \in P$.*

Proof. Since in each orderable group the equality $[x^n, y^n] = e$ is equivalent to $[x, y] = e$ for all integers n, then $\mathcal{L}_p \wedge \mathcal{R} = \mathcal{A}$. Hence, G is a non-representable l-group, and there is a minimal prime subgroup $A \leq G$ such that $A \neq A^x = x^{-1}Ax$ for some $x \in G$. Then all prime subgroups $A, x^{-1}Ax, \ldots, x^{-(p-1)}Ax^{p-1}$ are distinct and $x^{-p}A^p = A$. Let $b \in A \setminus x^{-i}Ax^i$ and $y = b_1b_2 \cdots b_{p-1}$. We claim that $d_0(x, y) \neq c_0(x, y)$. Assume, on the contrary, that $d_0(x, y) = c_0(x, y)$. Then

$$d_0(x, y) = x \wedge x^{-1}y^px \wedge \ldots \wedge x^{-(p-1)}y^px^p \in x^{-j}Ax^j \text{ for all } j \in P.$$

As A is prime and $a \notin A$, then $x^{-t}y^px^t \in A$ for some $t \in \{1, 2, \ldots, p-1\}$ and $y^p \in x^{-p+t}Ax^{p-t}$. Since $e \leq b_{p-t} \leq y^p \in x^{-p+t}Ax^{p-t}$, it follows that $b_{p-t} \in x^{-p+t}Ax^{p-t}$. This contradicts the choice $b_{p-t} \in A \setminus x^{-p+t}Ax^{p-t}$.□

LEMMA 3. *Let $G \in \mathcal{L}_p$ and H be a convex l-subgroup of G such that $x^{-i}Hx^i \cap H = E$ for some $x \in G$ and $i \in 1, 2, \ldots, p-1$. Then H is Abelian.*

Proof. Suppose, on the contrary, that H is not Abelian. Then by Lemma 2, there are $x_1, y_1 \in H$ such that $a_i(x_1, y_1) > e$, $i \in P$ and by Lemma 1,

$$x_1^{-i}a_0(x_1, y_1)x_1^i \wedge a_0(x_1, y_1) = e, \quad i \in \{1, 2, \ldots, p-1\}.$$

Let $b = a_0^p(x_1, y_1)$. Since $x_1 \in H, x^{-i}bx^i \in x^{-i}Hx^i$ and $x^{-i}Hx^i \cap H = E$, then $[x_1, x^{-i}bx^i] = e$. Therefore, $b = (xx_1)^{-p}b(xx_1)^p = (x^px_1)^{-1}bx^px_1 = x_1^{-1}bx_1$, which contradicts the choice $x_1, y_1 \in H$. □

Let X be a non-empty subset of l-group G and A be a subgroup of G. Then $A^{<G>}$ is the least gp(X)-invariant subgroup of G containing A, or, simply, gp(X)-invariant subgroup of G generated by A. Let $l-$gp(X) be an l-subgroup of G generated X and $(X)_l$ be a convex l-subgroup generated X.

LEMMA 4.$\mathcal{L}_p \subseteq \mathcal{A}^2$.

Proof. By Lemmas 1-3 and Theorem 9.4.2, in each l-group $G \in \mathcal{L}_p$ there is a tower of ideals

$$E = A_0 \leq A_1 \leq A_2 \leq \ldots \leq A_i \leq \ldots \leq G,$$

where $A_{i+1}/A_i = \mathcal{A}(G/A_i)$ and $\mathcal{A}(G/A_i)$ denote $\mathcal{A}_l$-radical of G/A_i. Let $G \in \mathcal{L}_p \setminus \mathcal{A}^2$. Then $A_0 \neq A_1 \neq A_2 \neq A_3$. By Lemma 2, there are elements $x, y \in A_3/A_1$, such that $e < a_i(\bar{x}, \bar{y}), a_i(\bar{x}, \bar{y}) \wedge a_j(\bar{x}, \bar{y}) = e$ $(i \neq j, i, j \in P)$ in A_3/A_1. Let x, y be positive preimages of $\bar{x}, \bar{y}$ in A_3. Then for x and the convex l-subgroup $H = (a_0(x, y))_l$ is valid $H \vee H^{x^i} = E$ for $i \in \{1, 2, \ldots, p-1\}$. Hence, by Lemma 3, the convex l-subgroup H is Abelian, and so $H \subseteq \mathcal{A}(G)$,which contradicts the choice $a_0(x, y) \in A_3 \setminus A_1, A_1 = \mathcal{A}(G)$. □

LEMMA 5. *The identity* $[|[x, y]| \wedge |z|, u^p] = e$ *is valid in the l-variety* $\mathcal{L}_p$.

Proof. By Lemma 4, $([G, G])_l \subseteq \mathcal{A}(G)$. So it is enough to prove that $[a, u^p] = e$ for all $u \in G$ and $a \in \mathcal{A}(G)$. Let

$$a \wedge u^{-p} a u^p = z_1, \quad a \vee u^{-p} a u^p = z_2.$$

Since $a, u^{-p} a u^p \in \mathcal{A}(G)$, then $(a \wedge u^{-p} a u^p)^p = a^p \wedge u^{-p} a^p u^p = z_1^p$ and $z_2^p = a^p \vee u^p a^p u^p = a^p$. As $\mathcal{A}(G)$ is a group with unique roots, then $z_1 = a, z_2 = a$ and $a = u^{-p} a u^p$. □

Let A be an l-group and T be a subgroup of the group of all l-automorphisms of A. l-Group A is called *finitely subdirectly T-irreducible* iff there do not exist non-trivial T-invariant convex l-subgroups A_1, A_2 of A such that $A_1 \cap A_2 = E$.

LEMMA 6. *Let T be an elementary Abelian p-group and A be a finitely subdirectly T-irreducible Abelian l-group. Then A has a non-trivial totally ordered convex l-subgroup.*

Proof. By Theorem 7.2.3, an Abelian l-group A is embedded into SP-hull A^{SP}. Without loss of generality, we can assume that $A \leq A^{SP}$.

Then, as it was proved in Section 2 of Chapter 7 the following properties hold in A^{SP}:

(1) each polar of A^{SP} is a direct factor of A^{SP},

(2) for each $e < h, h \in A$ there are $h_1, h_2 \in A^{SP}$ such that $e < h_1 \leq h \leq h_2$,

(3) each l-automorphism g of A is uniquely extended to l-automorphism g^* of A^{SP}.

Let T^* be the subgroup of the group of all l-automorphisms generated by $g_1^*, \ldots, g_n^*$, where $g_1, \ldots g_n$ is a basis of T. As by (3), $a^{w(g_1^*, \ldots, g_n^*)} =$

$a^{w(g_1,\dots,g_n)}$ for all $a \in A$ and $w(g_1^*,\dots,g_n^*) \in T$, then the l-group $A^{SP}$is also T^*-finitely subdirectly irreducible. It is clear that for each convex l-subgroup $B \subseteq A^{SP}$ and $x, y \in T^*$,

$$x^p, [x,y] \in \mathrm{N}_{T^*}(B), \tag{4}$$

where $\mathrm{N}_{T^*}(T)$ is a normalizer of B into T^*. It is clear that for each non-trivial totally ordered convex l-subgroup C of A^{SP} the subgroup $A \cap C$ is also a non-trivial totally ordered convex l-subgroup of A. So it is enough to prove the existence of non-trivial totally ordered convex l-subgroup in A^{SP}. Assume, on the contrary, that there does not exist a non-trivial totally ordered convex l-subgroup in A^{SP}. We claim that for each positive integer m there is a polar $A_m \neq E$ in A^{SP} and finite subset $\{x_1,\dots,x_m\}$ of T^* such that

$$A_m \cap A_m^{x_1^{\sigma_1} x_2^{\sigma_2} \dots x_m^{\sigma_m}} = E \tag{5}$$

if $\sum\limits_{i=1}^{m} \sigma_i \neq 0$, where $(\sigma_1,\dots,\sigma_m) \in M_m$. We use induction on m. If each polar in A^{SP} is T^*-invariant, then it contradicts the T^*-finitely subdirectly irreducibility of A^{SP}. Let $A_{0,1} \neq E$ be a polar in A^{SP} and $x \in T^*$ such that $A_{0,1} \neq A_{0,1}^{x_1}$. As $x_1^p \in \mathrm{N}_{T^*}(A_{0,1})$, then the inclusions $A_{0,1} \subseteq A_{0,1}^{x_1}$ or $A_{0,1} \supseteq A_{0,1}^{x_1}$ are impossible. Hence,

$$A_{0,1} = \bar{A}_{1,1} \times (A_{0,1} \cap A_{0,1}^{x_1})$$

for some polar $\bar{A}_{0,1}$ in A^{SP}. The easy inductive arguments on i, $1 \leq i \leq p-1$ show the existence of polars $A_{i,1}(\neq E)$ such that $A_{1,1} = \bar{A}_{1,1}$ and $x_1^{-i} A_{i,1} x_1^i \cap A_{i,1} = E$. Let the polar A_{m-1} and the subset $\{x_1,\dots,x_m\}$ of T^* satisfy (5). Since A^{SP} is T^*-finitely subdirectly irreducible, then for each polar $B \neq E$ in A_{m-1} and for some $x \in T^*$,

$$B^{<x_1,\dots,x_{m-1}>} \neq x^{-1} B^{<x_1,\dots,x_{m-1}>} x.$$

It is clear that $x \in T^* \backslash \mathrm{gp}(x_1,\dots,x_{m-1})$. By (1), there is a polar $B_1 \neq E$ for which

$$B^{<x_1,\dots,x_{m-1}>} = \bar{B}_1 \times (B^{<x_1,\dots,x_{m-1}>} \cap (B^{<x_1,\dots x_{m-1}>})^x)$$

Let $B_1 = \bar{B}_1 \cap B$. By (4), the polar $\bar{B}_1$ is gp $< x_1, \ldots, x_{m-1} >$-invariant and, therefore, $B_1 \neq E$. Suppose that there is a polar $B_{i-1} \neq E$ in B such that

$$(6) \qquad B_{i-1}^{<x_1,\ldots,x_{m-1}>} \bigcap (B_{i-1}^{<x_1,\ldots,x_{m-1}>})^{x_m^k} = E,$$

where $1 \leq k \leq i-1, \quad 1 \leq i \leq p-1$.

Let $B^{<x_1,\ldots,x_{m-1}>} = (B^{<x_1,\ldots,x_{m-1}>})^{x_m^i}$. Since the greatest common divisor of i and p is equal to 1, then $1 = is + pt$ for some integers s and t. Thus,

$$x_m^{-(is+pt)} B^{<x_1,\ldots,x_{m-1}>} x^{(is+pm)} = B^{<x_1,\ldots,x_{m-1}>},$$

which cotradicts our assumption. Therefore,

$$B_{i-1}^{<x_1,\ldots,x_{m-1}>} = \bar{B}_i \times (B_{i-1}^{<x_1,\ldots,x_{m-1}>} \cap (B_{i-1}^{<x_1,\ldots x_{m-1}>})^{x^i}).$$

Then, for $B_i = \bar{B}_i \cap B_{i-1}$,

$$B_i^{<x_1,\ldots,x_{m-1}>} \cap (B_i^{<x_1,\ldots,x_{m-1}>})^{x_m^k} = E,$$

where $1 \leq k \leq i$.

Let $A_m = B_{p-1}$. Then for A_m, (5) holds for all k, $1 \leq k \leq p-1$.

Let $e \neq a \in A \cap A_m, m > |T|$, then, by (5), $a \wedge a^{x_1^{\sigma_1} \ldots x_m^{\sigma_m}} = e$ if $\sum_{i=1}^{m} \sigma_i \neq 0, (\sigma_1, \ldots, \sigma_m) \in M_n$.

By definition of T^*, $\quad x_s = w_s(g_1^*, \ldots, g_n^*)$, where g^* are extensions of g. Let $b_s = w_s(g_1, \ldots, g_n) \in T$. Thus

$$a \wedge a^{b_1^{\sigma_1} \ldots b_m^{\sigma_m}} = e \text{ if } \sum_{i=1}^{m} \sigma_i \neq 0, \ (\sigma_1, \ldots, \sigma_m) \in M_m.$$

Since $m > |T|$, then $b_k = b_j$ for some $k, j \in \{1, 2, \ldots, m\}$, $k \neq j$, contradicts $a^p = a^p \wedge (a^p)^{b_k b_j^{p-1}}$.□

Let $\mathcal{M}$ be a set of all finitely subdirectly irreducible l-groups of l-variety $\mathcal{L}_p$ such that the $|G/C_G(\mathcal{A}(G)| < \infty$, where $C_G(\mathcal{A}(G))$ is a centralizer of $\mathcal{A}(G)$ in G. From the definition given above and Lemmas 4, 5, 6 we get

LEMMA 7. *Let $G \in \mathcal{M}$ and $G \neq E$. Then there is a non-trivial convex totally ordered l-subgroup $A \subseteq G$ and $g_1, \dots, g_s \in G$, $(s < \infty)$, such that*

(7) $A \cap (g_1^{\sigma_1} \dots g_s^{\sigma_s})^{-1} A (g_1^{\sigma_1} \dots g_s^{\sigma_s}) = E$ *if* $\sum_{i=1}^{s} \sigma_i \neq 0$, $(\sigma_1, \dots, \sigma_s) \in M_s \neq 0$,

(8) *each element $u \in G$ can be uniquely represented in the form $u = g_1^{\sigma_1} \cdots g_s^{\sigma_s} g_u$ where $(\sigma_1, \dots, \sigma_s) \in M_s$ and $g_u \in C_G(A)$.* □

LEMMA 8. *The orthogonal rank of each $G \in \mathcal{M}$ is finite and is equal to some power of the prime number p.*

Proof. Let convex totally ordered l-subgroup A and the elements $g_1, g_2, \dots, g_s \in G$ satisfy hypotheses (7), (8) of Lemma 7. Let us assume that the orthogonal rank $o(G) = n$ of G is greater than p^s and $a_1, \dots, a_n$ are strictly positive mutually orthogonal elements of G. Since the convex l-subgroup $(a_i)_l \cap A^{<g_1, \dots, g_s>}$ is totally ordered and $n > p^s$, then

$$(a_i)_l \cap A^{<g_1, \dots, g_s>} = e$$

for some $i \in \{1, 2, \dots, n\}$. This equality contradicts the choice $G \in \mathcal{M}$ and hypothesis (8) of Lemma 7. □

In view of Theorem 3.5.1, we see that each l-group G of finite orthogonal rank $o(G)$ can be constructed from totally ordered groups using direct products and lexicographic extensions. Let $k(G)$, $G \in \mathcal{M}$, be a non-negative integer defined by the rule:

$$k(G) = \begin{cases} \max\{k(G_1), k(G_2)\} & \text{if } G = G_1 \times G_2 \text{ and } G_1, G_2 \neq E; \\ k(G) = 1 + k(G_1) & \text{if } G \text{ is a lexicographic extension of an Abelian l-group } G_1 \text{ by an o-group } G/G_1 \text{ where } G/G_1, G_1 \neq G, E \text{ and } G_1 = G_2 \times G_3 \text{ for some } G_2, G_3 \neq E; \\ k(G) = 0 & \text{if G is totally ordered.} \end{cases}$$

It is clear that $k(G)$ is well defined for each $G \in \mathcal{M}$.

LEMMA 9. *Let A be a maximal convex totally ordered l-subgroup of $G, G \in \mathcal{M}$. Then $G/A^{<G>} \in \mathcal{M}$.*

Proof. By Theorem 3.1.1, $A^{<G>}$ is finitely subdirectly irreducible. Suppose, on the contrary, that there are ideals $\bar{H}_1, \bar{H}_2$ in the quotient l-group $G/A^{<G>}$ such that $\bar{H}_1 \cap \bar{H}_2 = E$. Hence there are idels H_1, H_2 in G such that

$$H_1, H_2 \supseteq A^{<G>} \text{ and } H_1 \cap H_2 = A^{<G>}.$$

By Theorem 3.5.1 and maximality of totally ordered subgroup A, the inequalities $x \gg y$ hold for each $x \in G \setminus A^{<G>}$, $x > e$ and $y \in (x)_l \cap A^{<G>}$, $y > e$. Hence $x_1 \gg x_1 \wedge x_2$ and $x_2 \gg x_1 \wedge x_2$ for all $x_1 \in H_1 \setminus A^{<G>}, x_1 > e$ and $x_2 \in H_2 \setminus A^{<G>}, x_2 > e$. Thus, $x_1 \wedge x_2 = e$ and $H_1 \cap H_2 = E$ contradicts $G \in \mathcal{M}$.

Let A_1 be the inverse image of the Abelian radical $\mathcal{A}(\bar{G})$ of the quotient l-group $\bar{G} = G/A^{<G>}$. Then $[x, y] \in A^{<G>}$ for all $x, y \in A_1$. In fact, $x^g = xd$ for some $d \in ([G,G])_l \subseteq \mathcal{A}(G)$. Since $g \in C_G(\mathcal{A}(G))$, $x \in A_1$ then $g^{-p}xg^p = xd^p$ and $d^p = [x, g^p] \in A^{<G>}$. As $A^{<G>}$ is an isolated subgroup in G, then $d \in A^{<G>}$. Therefore,

$$C_G(\mathcal{A}(G))/A^{<G>} \subseteq C_{\bar{G}}(\mathcal{A}\bar{G}) \text{ and } \mid \bar{G}/C_{\bar{G}}(\mathcal{A}(\bar{G})) \mid < \infty. \ \square.$$

LEMMA 10. *Each l-group $G \in \mathcal{M}$ has a finite series of convex l-subgroups*

$$E = G_{(0)} \leq G_{(1)} \leq \ldots \leq G_{(k)} \leq G_{(k+1)} = G, (k = k(G))$$

such that:

(9) *for each $g \in G$ holds $G^g_{(i)} = G_{(i)}$ or $G^g_{(i)} \cap G_{(i)} = E$, $i \in \{0, 1, \ldots, k\}$;*

(10) *for each $i \in \{0, 1, \ldots, k\}$ there is $u \in G$ for which $G^u_{(i)} \cap G_i = E$;*

(11) *each convex l-subgroup $G_{(i+1)}$ is a lexicographic extension of the convex l-subgroup $G_{(i+1)} \cap G^{<G>}_{(i)}$ by the totally ordered group $G_{(i+1)}/G^{<G>}_{(i)}$, $i \in \{0, 1, \ldots, k\}$;*

(12) *$\bar{G}_i = G/G^{<G>}_{(i)} \in M$ for all $i \in \{0, 1, \ldots, k\}$;*

(13) *the quotient l-group $G_{(i+1)}/G_{(i+1)} \cap G^{<G>}_{(i)}$ is a maximal convex totally ordered subgroup of the l-group $\bar{G}_i = G/G^{<G>}_{(i)}$, $i \in \{0, 1, \ldots, k\}$.*

Proof. Let $G_{(1)}$ be a maximal convex o-subgroup of G. The direct verification shows that properties (9)-(13) are valid in $G_{(1)}$. Now let us assume that convex l-subgroups $E = G_0 < G_1 < \ldots < G_j$ are defined.

Also assume that for all $i \in \{0, 1, \ldots, j-1\}$ properties (9)-(13) are valid in G_j and for all $i \in \{0, 1, \ldots, j\}$ properties (9), (10), (12) hold in G_j. Then the l-subgroup $G_{(j)}/G_{(j)} \cap G^{<G>}_{(j-1)}$ is a maximal convex o-subgroup in $\bar{G}_{j-1} = G/G^{<G>}_{(j-1)}$. Therefore, there is a convex l-subgroup A of G, $A \not\subseteq G^{<G>}_{(j)}$ which is a lexicographic extension of $A \cap G^{<G>}_j$ by non-trivial totally ordered group $A/A \cap G^{<G>}_{(j)}$. By (9), $G^u_{(j)} \subseteq A$ for some $u \in G$. Without loss of generality, we can assert that $u = e$.

From the proof of Theorem 3.5.1. it follows that the set of all convex l-subgroups $\{A_\alpha \mid \alpha \in J\}$ of G which are lexicographic extensions of convex l-subgroups $A_\alpha \cap G^{<G>}_{(j)}$ by o-groups $A_\alpha/A_\alpha \cap G_{(j)}$ and for which $A_\alpha \cap G^{<G>}_{(j)} = A \cap G^{<G>}_{(j)}, (\alpha \in J)$ is totally ordered under inclusion. Let $G_{(j+1)} = (A_\alpha \mid \alpha \in J)_l$ be a convex l-subgroup of G generated by $\{A\alpha \mid \alpha \in J\}$. It is clear that properties (11), (13) hold for $i = j$. We claim that properties (9),(10),(12) hold in $G_{(j+1)}$ too. Let $g \in G$ and $G^g_{(j+1)} \neq G_{(j+1)}, G^g_{(j+1)} \cap G_{(j+1)} \neq E$. Let $x \in G^g_{(j+1)} \cap G_{(j+1)}, x > e$. If $x \notin G^{<G>}_{(j)}$, then

a) $$G^g_{(j+1)} \cap G^{<G>}_{(j)} = G_{(j+1)} \cap G^{<G>}_{(j)}.$$

In l-group $G/G^{<G>}_{(j)}$ the convex o-subgroup $G_{(j+1)}/G_{(j+1)} \cap G^{<G>}_{(j)}$ is centralized by the image $\bar{g}$ of g under the natural l-homomorphism G onto $G/G^{<G>}_{(j)}$. Hence

b) $$(G^g_{(j+1)}, G^{<G>}_{(j)})_l = (G_{(j+1)}, G^{<G>}_{(j)})_l.$$

It is clear that equalities a), b) contradict the distributivity of the lattice of convex l-subgroups of G. Therefore, $G^g_{(j+1)} \cap G_{(j+1)} \subseteq G^{<G>}_{(j)}$. Then $x_1 \gg x_1 \wedge x_2, x_2 \gg x_1 \wedge x_2$ for all $x_1 \in G^g_{(j+1)} \setminus G^{<G>}_{(j)}, x_1 > e$ and $x_2 \in G_{(j+1)} \setminus G^{<G>}_{(j)}, x_2 > e$. From these inequalities it follows that $G^g_{(j+1)} \cap G_{(j+1)} = E$ and (9) is valid in $G_{(j+1)}$.

Now suppose that $G^u_{(j+1)} = G_{(j+1)}$ for all $u \in G$. By our assumption, $G/G^{<G>}_j \in \mathcal{M}$ and $G_{(j+1)}/G_{(j+1)} \cap G^{<G>}_{(j)}$ is a maximal central convex o-subgroup in $\bar{G}_j$. Then $G = G_{(j+1)}, k(G) = j$. Hence 2) holds in $G_{(j+1)}$.

From Lemma 3 and property (10) it follows that the convex l-subgroup $G^{<G>}_{(j+1)}$ is Abelian. Therefore, $G^{<G>}_{(j+1)} \subseteq C_G(\mathcal{A}(G))$. Since

$\bar{G}_{j+1} = \bar{G}_j/(G^{<G>}_{(j+1)}/G^{<G>}_j)$, then property (12) follows by the direct application of Lemma 7 to the l-group $\bar{G}_j$ which is believed to have property (12).

Let $G \in \mathcal{M}$. Then easy inductive arguments show that $C_G(\mathcal{A}(G)) = C_G(G_{(i)})$ for all $i = 1, 2, \ldots, k(G)$.

LEMMA 11. *Let $G \in \mathcal{M}, a \in C_G(G^{<G>}_{(k)}), k = k(G)$, $u_1, \ldots u_n \in G$. Then there is $b \in G^{<G>}_{(k)}$ such that*

$$[ab, u_i] = e \text{ for all } i \in \{1, 2, \ldots, n\}. \tag{14}$$

Proof. Let $x, y \in C_G(\mathcal{A}(G))$. Then l-gp$(x, y)$ is a nilpotent group of the nilpotence class 2. By Theorem 9.4.1, l-gp(x, y) is representable. Since $\mathcal{L}_p \cap \mathcal{R} = \mathcal{A}$, then l-gp$(x, y)$ is Abelian, and, therefore, $[x, y] = e$ for all $x, y \in C_G(\mathcal{A}(G))$. In view of Lemma 7, we can assume that $\{u_1, \ldots, u_n\} = \{g_1, \ldots g_s\}$. Now we use induction on the orthogonal rank $o(G)$. Let $o(G) = 1$. Then G is an o-group, and, therefore, Abelian. Thus the result is obvious in this case. Suppose that the result holds for all l-groups $H \in \mathcal{M}$ for which $o(H) < o(G)$. Therefore we can assert that there is an element $b_0 \in G^{<G>}_{(k)}$ such that $d_i = [ab_0, g_i] \in G^{<G>}_{(1)}$ for all $i \in \{1, 2, \ldots, s\}$. By Lemmas 4,5 and commutator identities, the following equalities hold in G

$$d_i^{1+g_i+\ldots+g_i^{p-1}} = d_i d_i^{g_i} \cdots d^{g_i^{p-1}} = e, \; i \in \{1, 2, \ldots, s\}; \tag{15}$$

$$d_i^{g_j} d_j = d_i d_j^{g_i}, \; i, j \in \{1, 2, \ldots, s\}. \tag{16}$$

Each element $d_i \in G^{<G>}_1, i \in \{1, 2, \ldots, s\}$ can be uniquely represented in the form

$$d_i = \prod_{\sigma \in M_s} (g_1^{\sigma_1} \cdots g_s^{\sigma_s})^{-1}(b_{i,\sigma})(g_1^{\sigma_1} \cdots g_s^{\sigma_s}) \tag{17}$$

where $b_{i,\sigma} \in G_{(1)}, \sigma = (\sigma_1, \ldots, \sigma_s) \in M_s$. We will find the element $d \in B^{<G>}$ where $B = 1 - \text{gp}(b_{i,\sigma} \mid i \in \{1, 2, \ldots, s\}, \sigma \in M_s)$ such that

$$d(d^{-1})^{g_i} = d_i \tag{18}$$

for all $i \in \{1, 2, \ldots, s\}$. It is clear that for $b = b_0 d$ the result holds.

Without loss of generality, we can assume that finitely generated o-group B is an infinite cyclic group (f) generated by f. Then equalities (17) can be rewritten in the form

(17.1) $$d_i = \prod_{\sigma \in M_s} (g_1^{\sigma_1} \cdots g_s^{\sigma_s})^{-1}(f^{n_{i,\sigma}})(g_1^{\sigma_1} \cdots g_s^{\sigma_s}),$$

where $\sigma = (\sigma_1, \ldots, \sigma_s) \in M_s$, $i \in \{1, 2, \ldots, s\}$, $n_{i,\sigma} \in \mathbf{Z}$. Suppose that d has the form

(17.2) $$d = \prod_{\sigma \in M_s} (g_1^{\sigma_1} \cdots g_s^{\sigma_s})^{-1}(f^{x_\sigma})(g_1^{\sigma_1} \cdots g_s^{\sigma_s}), \ \sigma \in M_s$$

and x_σ are unknowns. By rewriting (18), we obtain the system of $p^s s$ equations

(18.1) $$x_\sigma - x_{\sigma(i)-1} = n_{i,\sigma}, \ i \in \{1, 2, \ldots, s\}, \ \sigma \in M_s.$$

From (15),(16),

(15.1) $$\sum_{j=0}^{p-1} n_{i,\sigma(i)+j} = 0,$$

(16.1) $$n_{i,\sigma j} + n_{j,\sigma} = n_{i,\sigma} + n_{j,\sigma(i)-1},$$

where $i, j \in \{1, 2, \ldots, s\}$, $\sigma \in M_s$, and $\sigma(i) + j$ denotes s-tuple $(\sigma_1, \ldots, \sigma_{i-1}, (\sigma_i + j) \pmod p, \sigma_{i+1}, \ldots, \sigma_s)$.

Now we claim that the system of equalities (18.1) has a solution in integers. Use induction on s. The result is trivially true for $s = 1$. From the equation $n_{s,\sigma(j)-1} + n_{j,\sigma} = n_{j,\sigma(s)-1} + n_{s,\sigma}$ follows that the equation $x_\sigma - x_{\sigma(j)-1} = n_{j,\sigma}$ is the consequence of the equations

$$\begin{aligned} x_{\sigma(j)-1} - x_{(\sigma(j)-1)(s)-1} &= n_{s,\sigma(j)-1}, \\ x_{\sigma(s)-1} - x_{(\sigma(s)-1)(j)-1} &= n_{j,\sigma(s)-1}, \\ x_\sigma - x_{\sigma(s)-1} &= n_{s,\sigma}. \end{aligned}$$

Hence the system (18.1) is equivalent to the system

(18.2) $$x_\sigma - x_{\sigma(s)-1} = n_{s,\sigma},$$

(18.3) $$x_\sigma - x_{\sigma(j)-1} = n_{j,\sigma},$$

where $j = 1, \ldots, s-1$, $\sigma \in M_s$, $\sigma_s = p - 1$.

From the equation $\sum_{j=0}^{p-1} n_{s,\sigma(s)+j} = 0$ it follows that the equation $x_\sigma - x_{\sigma(s)-1} = n_{s,\sigma}$ is a consequence of the equations

$$\begin{aligned} x_{\sigma(s)-1} - x_{\sigma-2} &= n_{s,\sigma(s)-1} \\ \cdots & \quad \cdots \\ x_{\sigma(s)-(p-1)} - x_\sigma &= n_{s,\sigma-(p-1)}. \end{aligned}$$

Therefore, system (18.2) is equivalent to the system

$$x_\sigma - x_{\sigma(s)-1} = n_{s,\sigma}, \ (\sigma \in M_s, \ 0 \le \sigma_s < p-1). \tag{18.4}$$

Thus, system (18.1) is equivalent to systems (18.3), (18.4). By assumption, system (18.3) has an integral solution. We can rewrite system (18.4) in the form

$$\begin{aligned} x_\sigma = x_{\sigma(s)-1} + n_{s,\sigma}, & \quad \sigma \in M_s, \ \sigma_s = 0, \\ x_\sigma - x_{\sigma(s)-1} = n_{s,\sigma}, & \quad \sigma \in M_s, \ 0 < \sigma_s < p-1. \end{aligned} \tag{18.5}$$

Since the matrix of the left side of system (18.5) has the determinant $\triangle = 1$ it follows that system (18.5) has the integral solution. □

Similar arguments show the fulfilment of the following statements.

LEMMA 12. *Let* $G \in \mathcal{M}$, $0 \le i \le k(G) - 1$, $a \in G_{(i+1)} \setminus G_{(i)}^{<G>}$ *and* $u_1, \dots, u_n$ *be elements of normalizer* $N_G(G_{(i+1)})$ *in* G. *Then there is* $b \in G_{(i)}^{<G>} \cap G_{(i+1)}$ *such that* $[ab, u_j] = e$, $j \in \{1, 2, \dots, n\}$. □

LEMMA 13. *Let* $G \in M$ *and* $u_1, \dots, u_n \in G$. *Then there are* $c_1, \dots, c_n \in G_k^{<G>}$, $(k = k(G))$ *such that* $[u_i c_i, u_j c_j] = e$ *for all* $i, j \in \{1, 2, \dots, n\}$.

Proof. Use induction on $o(G)$. Let $o(G) = p$. Then, by Lemma 7, $u_i = g^{n_i} x_i$ for some $g \in G$ and $x_i \in C_G(\mathcal{A}(G)) = C_G(G_{(1)})$ and $n_i \in \{0, 1, \dots, p-1\}$. By Lemma 11, there are $b_i \in G_{(k)}^{<G>}, i \in \{1, 2, \dots, n\}$. Therefore, $[u_i b_i, u_j b_j] = [g^{n_i} x_i b_i, g^{n_j} x_j b_j] = e$. Let us assume that result holds for $H \in \mathcal{M}$, $o(H) < p^s$ and $o(G) = p^s$. As it was mentioned in the proof of Lemma 11, we can suppose that $\{u_1, \dots, u_n\} = \{g_1, \dots, g_s\}$. Then the result holds for the l-group $\bar{G}_1 = G / G_{(1)}^{<G>}$, and therefore,

$$[g_i, g_j] \in G_{(1)}^{<G>} \text{ for all } i, j \in \{1, 2, \dots s\}. \tag{19}$$

Let $A = 1 - \text{gp}(g_2, \ldots, g_s, G_{(1)}^{<G>})$ and

$$A_i = (G_{(1)} \times g_1^{-1} G_{(1)} g_1 \times \ldots \times g_1^{-(i-1)} G_{(1)} g_1^{i-1} \times \\ \times g_1^{-(i+1)} G_{(1)} g_1^{i+1} \times \ldots \times g_1^{-(p-1)} G_{(1)} g_1^{p-1})^A \subseteq A.$$

We claim that $k(\bar{A}) = 1$ for $\bar{A} = A/A_i$. Assume, on the contrary, that $k(\bar{A}_i) > 1$. Then by Lemma 10, there is $u_i \in \bar{A}_i$ such that

$$a) \qquad (\bar{A}_i)_{(2)}^u \cap (\bar{A}_i)_{(2)} = E.$$

From (19), it follows that $\bar{A}_i/(G_{(1)}^{<G>}/A_i) = A/G_{(1)}^{<A>}$ is Abelian, which contradiction a). Thus, $k(\bar{A}_i) = 1$, $i \in \{0, 1, \ldots p-1\}$. Therefore, by our assumption, there are the elements

$$\{c_{ij} \mid c_{ij} \in G_{(1)}^{<G>},\ j = 2, 3, \ldots, s\},\ i \in \{0, 1, \ldots, p-1\}$$

such that

$$b) \qquad [g_j c_{ij}, g_t c_{it}] \in A_i,\ j, t \in \{2, \ldots, s\}.$$

Let $x_\sigma = (g_1^{\sigma_1} \cdots g_s^{\sigma_s})^{-1}(\tilde{x}_\sigma)(g_1^{\sigma_1} \cdots g_s^{\sigma_s})$ be the projection of x onto convex l-subgroup

$$(g_1^{\sigma_1} \cdots g_s^{\sigma_s})^{-1} G_{(1)} (g_1^{\sigma_1} \cdots g_s^{\sigma_s}),\ \sigma = (\sigma_1, \ldots, \sigma_s) \in M_s.$$

Let $d_{t,j} = [g_t, g_j]$, $t, j \in \{2, \ldots, s\}$ and $\sigma_1 = i$. Then

$$(\tilde{d}_{ij})_\sigma = (\tilde{c}_{i,t})_\sigma (\tilde{c}_{i,t}^{-1})_{\sigma(j)-1} (\tilde{c}_{i,j})_{\sigma(t)-1} (\tilde{c}_{i,j}^{-1})_\sigma,$$

where $t, j \in \{2, 3, \ldots, s\}$; $i \in \{0, 1, \ldots, p-1\}$. Therefore, for $c_j = \prod_{\sigma \in M_s} (c_{\sigma_1, j})_\sigma$, $j = 2, 3, \ldots, s$,

$$d_{t,j} = [g_t, g_j] = (c_t^{-1} \cdot g_j^{-1} c_t g_j \cdot g_t^{-1} c_j^{-1} g_t \cdot c_j)^{-1},\ t, j \in \{2, \ldots, s\}.$$

Thus, $[g_t c_t, g_j c_j] = e$, $c_t, c_j, [g_1, g_j] \in G_{(1)}^{<G>}$. Hence we may assert that

(20) $\quad [g_t, g_j] = e, \qquad t, j = 2, 3, \ldots, s,$

(21) $\quad [g_1, g_j] \in G_1^{<G>}, \quad j = 2, 3, \ldots, s.$

Now let $D = 1 - \mathrm{gp}(g_1, g_2, \ldots, g_s, G_{(1)}^{<G>})$. It is clear that $D \in \mathcal{M}$ and $g_i^p \in C_D(\mathcal{A}(D))$ for all $i \in \{1, 2, \ldots, s\}$. Therefore, $k(D) = 1$ and $D_{(k)}^{<D>} = G_{(1)}^{<G>}$. By Lemma 11, there are $b_i \in G_{(1)}^{<G>}$, $i \in \{2, \ldots, s\}$ for which $[g_i^p b_i, g_j] = e$, $j \in \{1, 2, \ldots, s\}$. In view of (20), (21), we have

(22) $\qquad [b_i, g_j] = e, \quad j \in \{2, \ldots, s\}.$

Let a_i be the element of $G_{(1)}^{<G>}$ defined by the rule:

$$(a_i)_\sigma = \begin{cases} (b_i)_\sigma & \text{for } \sigma_i = 0, \ (\sigma \in M_s) \\ e & \text{for } 1 \leq \sigma_i \leq p-1, \ (\sigma \in M_s). \end{cases}$$

Then (22) implies

(23) $\qquad [a_i, g_j] = e, \qquad j = 2, 3, \ldots, i-1, i+1, \ldots, s,$

(24) $\qquad a_i^{1+g_i+\cdots+g_i^{p-1}} = b_i, \quad i = 2, 3, \ldots, s.$

Similarly, (20), (22), (23), (24) imply

(25) $\qquad [g_t a_t, g_j a_j] = e, \quad t, j = 2, 3, \ldots, s,$

(26) $\qquad [g_1, (g_j a_j)^p] = e, \quad j = 2, 3, \ldots, s.$

Since $a_2, a_3, \ldots, a_s \in G_{(1)}^{<G>}$, then (21) implies

(27) $\qquad [g_1, g_j a_j] \in G_{(1)}^{<G>}, \quad j = 2, 3, \ldots, s.$

Let now $a_1 = g_1^{-1}(\bigvee_{\sigma \in M_s} g_1^{u_\sigma})$, where $\sigma = (\sigma_1, \ldots, \sigma_s) \in M_s$ and $u_\sigma = g_1^{\sigma_1 (g_2 a_2)^{\sigma_2} \cdots (g_s a_s)^{\sigma_s}}$. Then (27) implies $a_1 \in G_{(1)}^{<G>}$ and from (25), (26) it follows that $(g_1 a_1)^{g_j a_j} = g_1 a_1$, $j = 2, 3, \ldots, s$. □

Let $G = A \nabla B$ denote that an l-group is an antilexicographically ordered semidirect product of lattice-ordered group A and o-group B.

LEMMA 14. *Let G be a finitely generated l-group and $G \in \mathcal{M}$. Then G is the antilexicographically ordered semidirect product $G_{(k)}^{<G>} \nabla A$ of ideal $G_{(k)}^{<G>}$ and some totally ordered Abelian group A.*

Proof. Let $e < \bar{y}_1, \ldots, e < \bar{y}_n$ be a basis of the finitely generated Abelian group $\bar{G} = G/G_k^{<G>}$ and $y_1, \ldots, y_n$ be positive preimages of $\bar{y}_1, \ldots, \bar{y}_n$ in G. By Lemma 13, there are $c_1, \ldots, c_n \in G_{(k)}^{<G>}$ such that for $u_i = y_i c_i, (i = 1, 2, \ldots, n)$, $[u_i, u_j] = e$ for all $i, j \in \{1, 2, \ldots, n\}$. It is clear that the subgroup $A = \mathrm{gp}(u_1, \ldots, u_n)$ is a finitely generated Abelian group and $A \cap G_{(k)}^{<G>} = E$. We claim A is an o-group. In fact, $G/G_{(k)}^{<G>}$ is an o-group. Hence for each $g \neq e, g \in A$ there is $z \in G_{(k)}^{<G>}$ for which $gz > e$ or $gz < e$. Let $gz > e$. As G is the lexicographic extension of the ideal $G_{(k)}^{<G>}$ by o-group $G/G_{(k)}^{<G>}$, then $gz \gg y$ for all $y \in G_{(k)}^{<G>}$. Hence $gz \gg z$ and $g > e$. □

LEMMA 15. *Let G be a finitely generated l-group, $G \in \mathcal{M}$ and B be a finitely generated o-group. Then $G \overleftarrow{\times} B \in \mathrm{var}_l(G)$.*

Proof. By Lemma 14, $G = G_k^{<G>} \nabla A$ where $A = \mathrm{gp}(u_1, \ldots, u_n)$ is a free Abelian o-group with the set of free generators $u_1, \ldots, u_n$. Let $g = u_1^p \cdots u_n^p$. Then $g \in C(G)$.

Let $\{G_i \mid i \in I\}$ be a family of l-groups such that $G_i \cong G$ for all $i \in I$ and $\bar{\prod}_{i \in I} G_i$, $\prod_{i \in I} G_i$ be Cartesian and direct products correspondingly. Let a and a_k be the elements of $\bar{\prod}_{i \in I} G_i$ defined by the rules: $a(i) = g^i$, $a_x(i) = x$ for all $i \in I$. Let $\bar{a}$ and $\bar{a}_x$ be the images of a, a_x in the quotient group $\bar{\prod}_{i \in I} G_i / \prod_{i \in I} G_i$. It is easy to see that $1-\mathrm{gp}(\bar{a}_x \mid x \in G) \cong G$, $a \in C(G^*)$ and $a \gg w$ for all $w \in 1-\mathrm{gp}(\bar{a}_x \mid x \in G)$. Therefore, the l-subgroup $G^* = 1-\mathrm{gp}(\bar{a}_x, \bar{a} \mid x \in G) = 1-\mathrm{gp}(\bar{a}_x \mid x \in G) \overleftarrow{\times} (a)$. Without loss of generality we can assume that $G^* = G \overleftarrow{\times} (a)$. By Corollary 1 of Theorem 11.3.1, $\mathrm{var}_l(\mathbf{Z}) = \mathrm{q}_l(\mathbf{Z}) = \mathcal{A}$ and, therefore, each finitely generated Abelian o-group B is o-isomorphic to some o-subgroup of the reduced product $\bar{\prod}_{\beta \in J} Z_\beta / \mathcal{D}$ for some index set J and a filter $\mathcal{D}$ on J, where $Z_\beta \cong \mathcal{Z}$ and $\mathcal{Z}$ is a totally ordered additive group of integers. Let $a_1 > e, \ldots, a_n > e$, where $a_1, \ldots a_n \in \bar{\prod}_{\beta \in J} Z_\beta$ are preimages of some basis $\bar{a}_1, \ldots, \bar{a}_n$ of B. Now consider the l-subgroup G^{**} of the reduced products $\bar{\prod}_{\beta \in J} G_\beta^* / \mathcal{D}$ generated by the elements $\{\bar{b}_1, \ldots, \bar{b}_m, \bar{a}_x \mid x \in G\}$ where $\{\bar{b}_i (i \in \{1, 2, \ldots, m\}),\ a_x,\ (x \in G)\}$ are images of the elements $\{b_1, \ldots, b_m, a_x \mid x \in G\}$ of $\bar{\prod}_{\beta \in J} G_\beta^*$ such that $b_i(\alpha) = \bar{a}^{a_i(\alpha)}$, $i \in \{1, 2, \ldots, m\}, a_x(\alpha) = x,\ x \in G,\ \alpha \in J$. It is clear that $G^{**} = G \overleftarrow{\times} B$.□

LEMMA 16. *Let G be a finitely generated l-group, $G \in \mathcal{M}$ and $o(G) = p^s$. Let $g_1, \ldots, g_s \in G$ such that $G_{(1)} \cap G_{(1)}^{g_1^{\sigma_1} \cdots g_s^{\sigma_s}} = E$ for all*

$\sigma = (\sigma_1, \ldots, \sigma_s) \in M_s$ *with* $\sum_{i=1}^{s} \sigma_i \neq 0$ *and* $[g_i, g_j] = e$ *for all* $i, j \in \{1, 2, \ldots, s\}$. *Then*

$$\mathrm{var}_l(G) = \mathrm{var}_l(1 - \mathrm{gp}(g_1, \ldots, g_s, G_{(k)}^{<G>})).$$

Proof. It is easy to see that $G^* = 1 - \mathrm{gp}(g_1, \ldots, g_s, G_{(k)}^{<G>})$ is l-isomorphic to $G_{(k)}^{<G>} \nabla \mathrm{gp}(g_1, \ldots, g_s)$ where $\{g_1, \ldots, g_s\}$ is a basis of $\mathrm{gp}(g_1, \ldots, g_s)$. By Lemma 14, $G = G_{(k)}^{<G>} \nabla \mathrm{gp}(u_1, \ldots, u_n)$, where $u_1 > e, \ldots, u_n > e$ is the basis of finitely generated Abelian o-group $\mathrm{gp}(u_1, \ldots, u_n)$. Let $B = \mathrm{gp}(b_1, \ldots, b_n)$ be a finitely generated Abelian o-group which is l-isomorphic to $\mathrm{gp}(u_1, \ldots, u_n)$ under the map $b_i \to u_i$, $i \in \{1, \ldots, n\}$. Then for each $i \in \{1, 2, \ldots, n\}$, $u_i = y_i x_i$ for some

$$y_i \in \text{l-gp}\,(g_1, \ldots, g_n),\ x_i \in C_G(\mathcal{A}(G)).$$

Hence the l-subgroup $1 - \mathrm{gp}(b_1 y_1, \ldots, b_n y_n, G_{(k)})^{<G>})$ of the l-group $G^* \overleftarrow{\times} B$ is l-isomorphic to l-group G. By Lemma 15,

$$G^* \overleftarrow{\times} B \in \mathrm{var}_l(G). \ \square$$

LEMMA 17. *Let G satisfy the hypothesis of Lemma 16. Then*

$$var_l(G) = var_l(\textit{l-gp}(g_1, \ldots, g_s, G_{(1)}^{<G>})).$$

Proof. By Lemma 16, $G = G_{(k)}^{<G>} \nabla \mathrm{gp}(g_1, \ldots, g_n)$. Let the identity $w(x_1, \ldots, x_n) = e$ be violated in G on elements $y_i, y_{i,0}$, $1 \leq i \leq n)$. Hence $w(y_1 y_{1,0}, \ldots, y_n y_{n,0}) \neq e$ where $y_i \in \mathrm{gp}(g_1, \ldots, g_s), y_{i,0} \in G_{(k)}^{<G>}$, $1 \leq i \leq n$. Without loss of generality, we can assert that for some positive integers $1 \leq t_k \leq t_{k-1} \leq \ldots \leq t_2 \leq t_1 = s$,

$$\{g_1, \ldots, g_s\} \setminus N_G(G_{(i)}) = \{g_1, \ldots, g_{t_i}\}.$$

For each $\sigma = (\sigma_1, \ldots, \sigma_{t_i}) \in M_{t_i}$ and $x \in G_{(i)}^{<G>}$ let $\varrho_i(\tilde{x}_\sigma)^{g_1^{\sigma_1} \cdots g_{t_i}^{\sigma_{t_i}}}$ be the projection of x onto convex l-subgroup $G_{(i)}^{g_1^{\sigma_1}, \ldots g_{t_i}^{\sigma_i}}$. Let

$$X_k = \{b_k\} \cup \{\{\varrho_k(\tilde{y}_{i,0})_\sigma \mid \sigma \in M_{t_k},\ 1 \leq i \leq n\} \setminus G_{(k-1)}^{<G>}\},$$

where $b_k > e$ is some element of $G_{(k)} \setminus G_{(k-1)}^{<G>}$. By Lemma 12,

$$1-\mathrm{gp}(X_k, G_{(k-1)}^{<G>} \cap G_{(k)}) = (G_{(k-1)}^{<G>} \cap G_{(k)}) \overleftarrow{\times} B_k,$$

where $B_k \neq E$ is a finitely generated Abelian o-group such that

$$N_G(B_k) = C_G(B_k).$$

The elements $y_{i,0}$, $i \in \{1, 2, \ldots, n\}$ can be represented in the form $y_{i,0} = d_i y_{i,1}$ where $d_i \in B_k^{<G>}$, $y_{i,1} \in G_{(k-1)}^{<G>}$. Let

$$X_{k-1} = \{b_{k-1}\} \cup \{\{\varrho_{k-1}(\tilde{y}_{i,1})_\sigma \mid 1 \leq i \leq n, \ \sigma \in M_{t_{k-1}}\} \setminus G_{(k-2)}^{<G>}\}$$

where $b_{k-1} \in G_{(k-1)} \setminus G_{(k-2)}^{<G>}$, $b_{k-1} > e$. Again, by Lemma 12,

$$1-\mathrm{gp}(X_{k-1}, G_{(k-2)}^{<G>} \cap G_{(k-1)}) = (G_{(k-2)}^{<G>} \cap G_{(k-1)}) \overleftarrow{\times} B_{k-1},$$

where B_{k-1} is a non-trivial finitely generated Abelian o-group such that

$$N_G(B_{k-1}) = C_G(B_{k-1}), \ N_G(B_k) \supseteq N_G(B_{k-1}),$$

$$1-\mathrm{gp}(B_k, B_{k-1}^{<G>} \cap B) = (B_{k-1}^{<G>} \cap G_{(k)}) \overleftarrow{\times} B_k.$$

By similar arguments, $w(x_1, \ldots, x_n) = e$ is violated in l-subgroup

$$D = 1-\mathrm{gp}(B_k^{<G>}, \ldots, B_1^{<G>}) \nabla \mathrm{gp}(g_1, \ldots, g_s)$$

where $B_k, \ldots, B_1$ are non-trivial finitely generated Abelian o-groups such that

$$N_G(B_i) = C_G(B_i), N_G(B_1) \subseteq N_G(B_2) \subseteq \ldots \subseteq N_G(B_k),$$

$$1-\mathrm{gp}(B_i, B_{i-1}^{<G>} \cap G_{(i)}) = (B_{i-1}^{<G>} \cap G_{(i)}) \nabla B_i, \ i \in \{1, 2, \ldots, k(G)\},$$

$$B_0 = E.$$

We claim that $\mathrm{var}_l(D) = \mathrm{var}_l(A_i)$, where

$$A_i = 1-\mathrm{gp}(B_k^{<G>}, \ldots, B_{i+1}^{<G>}, B_{i-1}^{<G>} \ldots B_1^{<G>}) \nabla \mathrm{gp}(g_1, \ldots g_s),$$

$k = k(G)$, $i \in \{2, 3, \ldots, k(G)\}$. Let $g = \prod_{\sigma \in M_\varrho} (u_{1,i-1}^p \cdots u_{n_{i-1},i-1}^p)^{u_\sigma}$, where $\varrho = t_{i-1} - t_i$, $u_\sigma = g_{t_i+1}^{\sigma_\varrho}$ and $e < u_{1,i-1}, \ldots, e < u_{n_{i-1},i-1}$ is a basis of B_{i-1}. In the quotient group $\overline{\prod_{\alpha \in N}} A_\alpha / \prod_{\alpha \in N} A_\alpha$, $A_\alpha \cong A$ let us consider an l-subgroup generated by images $\{\bar{a}, \bar{a}_x \mid x \in A_i\}$ of the elements $\{a, a_x \mid x \in A_i\}$ where $a(\alpha) = g^\alpha$, $a_x(\alpha) = x$, $x \in A_i$, $\alpha \in N$. It is easy to see that

$$A^* \cong 1 - \mathrm{gp}(B_k^{<G>}, \ldots, B_{i+1}^{<G>}, B^{<G>}, B_{i-1}^{<G>}, \ldots, B_1^{<G>}) \nabla \mathrm{gp}(g_1, \ldots, g_s)$$

where B is an infinite cyclic o-group $(\bar{a})$ generated by the element $\bar{a}$, $B \leq B_i$. By Theorem 14.1.2 and Corollary 1 of Theorem 14.3.2, B_i is l-isomorphic to some o-subgroup of the reduced product $\prod_{\beta \in J} Z_\beta / \mathcal{D}$ for some index set J and a filter $\mathcal{D}$ on J, where $Z_\beta \cong \mathcal{Z}$. (Theorem 14.1.2 is independent of the results of this section and we can use it.) Let $a_1 > e, \ldots, a_{n_i} > e$, $(a_1, \ldots, a_{n_i} \in \overline{\prod_{\beta \in J}} Z_\beta)$ be preimages of $u_{1,i} > e, \ldots, u_{n_i,i} > e$ correspondingly. In the reduced product $\overline{\prod_{\beta \in J}} A_\beta^* / \mathcal{D}$, $(A_\beta^* \cong A^*)$ let us consider an l-subgroup A^{**} generated by the elements $\{\bar{b}_1, \ldots, \bar{b}_{n_i}, \bar{a}_x \mid x \in A_i\}$ where $\{b_1, \ldots, b_{n_i}, a_x \mid x \in A_i\} \subseteq \overline{\prod_{\beta \in J}} A_\beta^*$ and $b_j(\alpha) = \bar{a}^{a_j(\alpha)}$, $a_x(\alpha) = x$, $\alpha \in J$, $x \in A_i \subseteq A^*$ and $\bar{b}_1, \ldots, \bar{b}_{n_i}, \bar{a}_x$ are images of $b_1, \ldots, b_{n_i}, a_x$ in $\overline{\prod_{\beta \in J}} A_\beta^* / \mathcal{D}$. Thus, $A^{**} \cong D$ and $\mathrm{var}_l(G) = \mathrm{var}_l(A_i)$. Hence, the identity $w(x_1, \ldots, x_n) = e$ is violated in the l-group $B_1^{<G>} \nabla \mathrm{gp}(g_1, \ldots, g_s)$. □

Let

$$A(n,p) = \mathrm{gp}(u_1, u_2, \ldots, u_n, a \mid$$
$$[u_i, u_j] = [a, u_i] = [(u_1^{\sigma_1} \cdots u_n^{\sigma_n})^{-1} a (u_1^{\sigma_1} \cdots u_n^{\sigma_n}), a] = e),$$

where $(\sigma_1, \ldots, \sigma_n) \in M_n$, $i, j \in \{1, 2, \ldots, n\}$). We order $A(n,p)$ by the rule: $u_1 \gg u_2 \gg \ldots \gg u_n > e$, and

$$a \wedge (u_1^{\sigma_1} \ldots u_n^{\sigma_n})^{-1} a (u_1^{\sigma_1} \ldots u_n^{\sigma_n}) = e \text{ if } \sum_{i=1}^n \sigma_i \neq 0,$$

where $(\sigma_1, \ldots, \sigma_n) \in M_n$. The easy verification shows that $A(n,p)$ is an l-group and $A(n,p) \in \mathcal{L}_p$.

LEMMA 18. *Let G be a finitely generated l-group, $G \in \mathcal{M}$ and $o(G) = p^n$. Then* $\mathrm{var}_l(G) = \mathrm{var}_l(A(n,p))$.

Proof. By Lemmas 13, 16, 17, it follows that $\mathrm{var}_l(G) = \mathrm{var}_l(H)$ where $H = G_{(1)}^{<G>} \nabla \mathrm{gp}(g_1, \ldots, g_n)$ and $\mathrm{gp}(g_1, \ldots, g_n)$ is a finitely generated Abelian o-group with basis $g_1, \ldots, g_n$. Let A be a finitely generated Abelian group with basis $d_1, \ldots, d_n$ totally ordered by the rule $d_1 \gg d_2 \gg \ldots \gg d_n > e$. By Lemma 15, $H \overleftarrow{\times} A \in \mathrm{var}_l(H)$. Then the l-subgroup $l - \mathrm{gp}(d_1 g_1, \ldots, d_n g_n, b)$ where $e < b \in G_{(1)}$ is l-isomorphic to $A(n,p)$. Thus $\mathrm{var}_l(A(n,p)) \subseteq \mathrm{var}_l(G)$.

Conversely, let B be a finitely generated Abelian o-group with the basis $y_1, \ldots, y_n$ and B be l-isomorphic to $\mathrm{gp}(g_1, \ldots, g_n)$ under the l-isomorphism extending the map $y_i \to g_i$, $i \in \{1, 2, \ldots, n\}$. Since $A(n,p) \in M$, then

$$A(n,p) \overleftarrow{\times} B \in \mathrm{var}_l(A(n,p)).$$

If the identity $w(x_1, \ldots, x_n) = e$ is violated in H, then this identity is violated in the l-group $G_1^{*<G>} \nabla \mathrm{gp}(g_1, \ldots, g_n)$ for some finitely generated Abelian o-group G_1^*. So, without loss of generality, we can assume that $G_{(1)}$ is finitely generated. Then $G_{(1)} \subseteq \bar{\prod}_{\beta \in J} Z_\beta / \mathcal{D}$. Arguments similar to the proof of Lemma 17 show that the reduced product $\bar{\prod}_{\beta \in J} H_\beta / \mathcal{D}$ where $H_\beta \cong A(n,p) \overleftarrow{\times}$ for all $\beta \in J$ contains an l-subgroup which is l-isomorphic to H.

Now we define words in a free l-group on countably many generators in the following way:

$$w_1(x, y, z, y_1) = \bigvee_{\sigma_1 \in M_1} (y_1^{\sigma_1})^{-1} |[|[x, y]| \wedge |z|, y_1]| (y_1^{\sigma_1}),$$

$$\ldots$$

$$w_n(x, y, z, y_1, \ldots, y_n) =$$

$$= \bigvee_{\sigma \in M_n} (y_1^{\sigma_1} \cdots y_n^{\sigma_n})^{-1} |[w_{n-1}(x, y, z, y_1, \ldots, y_{n-1}), y_n]| (y_1^{\sigma_1} \cdots y_n^{\sigma_n}).$$

LEMMA 19. $\mathrm{var}_l(A(n,p)) \neq \mathrm{var}_l(A(n+1,p))$.

Proof. The direct verification shows that

$$A(n,p) \in \mathrm{var}_l([w_n(x,y,z,y_1,\ldots,y_n),t]=e).$$

On the other hand, the identity $[w_n(x,y,z,y_1,\ldots,y_n),t]=e$ is violated in $A(n+1,p)$ for $z=x=a,\quad y=y_1=u_1,\quad y_2=u_2,\ldots,\quad y_n=u_n,\quad t=u_{n+1}$ since

$$[w_n(a,u_1,a,u_1,u_2,\ldots,u_n),u_{n+1}]=$$
$$=[\bigvee_{\sigma\in M_n}(u_1^{\sigma_1}\cdots u_n^{\sigma_n})^{-1}(a^{2^{n-1}})(u_1^{\sigma_1}\cdots u_n^{\sigma_n}),u_{n+1}]\neq e.\Box$$

As each l-variety is uniquely determined by its finitely generated finitely irreducible l-groups, then the following result is true.

THEOREM 1. (Gurchenkov [3], Holland, Reily and Mekler [1].) *Let $\mathcal{X}\neq\mathcal{E},\mathcal{A}$ be a proper subvariety of l-variety $\mathcal{L}_p$. Then $\mathcal{X}=\mathrm{var}_l(A(n,p))$ for some natural number n.* □

COROLLARY 1. *Each subvariety of $\mathcal{L}_l$ is finitely based.*□

COROLLARY 2. *The basic rank of the l-variety $\mathcal{L}_p$ is infinite.* □

The similar description of the subvarieties of $\mathcal{L}_n\wedge\mathcal{A}^2$ where n is not necessarily a prime was obtained by S.A.Gurchenkov [6] , Ch.Holland and N.Reily [1].

12.4 Solvable l-Varieties

Let n be a positive integer and $\mathcal{L}_n$ be an l-variety defined by the identity $[x^n,y^n]=e$. The main aim of this Section is to show that if a positive integer n is a product of k prime numbers, then $\mathcal{L}_n\subseteq\mathcal{A}^{k+1}$ (and so it is solvable of the lenght $k+1$). This result was obtained by S.A.Gurchenkov [6] and N.Reily [2]. We start with some auxiliary results.

LEMMA 1. (Reily [3], Darnel [1].) *If C is a convex l-subgroup of the l-group G, then the normalizer $N_G(C)$ of C is an l-subgroup.*

Proof. It suffices to show that if $x, y \in N_G(G)$ then $x \vee y \in N$. This follows if, for all $e \leq g \in C$, we can prove that

$$h = (x \vee y)^{-1} g (x \vee y) \in C.$$

But for $e \leq g$, $g \in C$, $e \leq h = (x^{-1} \wedge y^{-1}) g (x \vee y) = (x^{-1}gx \wedge y^{-1}gx) \vee (x^{-1}gy \wedge y^{-1}gy) \leq x^{-1}gx \wedge y^{-1}gy \in C$ so that, by convexity of C, $h \in C$, as required.□

LEMMA 2. (Martinez [3].) $\mathcal{R} \cap \mathcal{L}_n = \mathcal{A}$.

Proof. Indeed, in each orderable group the equality $[x^n, y^n] = e$ is equivalent to $[x, y] = e$, therefore the result holds.□

LEMMA 3. (Reily [3].) *Let G be an l-group such that* $\mathrm{var}_l(G) \wedge \mathcal{R} = \mathcal{A}$ *and let n be a positive integer. Then the following statements are equivalent.*

(1) $G \in \mathcal{L}_n$,

(2) *For all convex l-subgroups P of G and all $x \in G$ is valid* $x^{-n} P x^n = P$.

(3) *There exists a family $\mathcal{P}r$ of prime subgroups such that* $\bigcap_{P \in \mathcal{P}r} P = E$ *and for all* $P \in \mathcal{P}r, x \in G, x^{-n} P x^n = P$.

Proof. (1) $\Rightarrow$ (2). Let $y \in P$ and $y > e$. Then $e < y < y^n \in P$ and $e < x^{-n} y x^n < x^{-n} y^n x^n = y^n$. Thus, $x^{-n} y x^n \in P$. Since each convex l-subgroup P is uniquely determined by its positive elements, then $x^{-n} P x^n = P$.

(2) $\Rightarrow$ (3). If we take $\mathcal{P}r$ to be the family of all minimal prime subgroups, then $\bigcap\limits_{P \in \mathcal{P}r} P = E$.

(3) $\Rightarrow$ (1). Let $P \in \mathcal{P}r$ and $x, y \in G$. Then $x^n, y^n \in N_G(P)$. By Lemma 1, $N_G(P)$ is an l-subgroup of G. Let H denote the l-subgroup of $N_G(P)$ generated by P, x^n and y^n. Then P is a normal prime subgroup of H so that H/P is totally ordered. But then $H/P \in \mathrm{var}_l(G) \cap \mathcal{R} = \mathcal{A}$, so that H/P must be Abelian. Thus $[x^n, y^n] = e$. Hence $G \in \mathcal{L}_n$. □

Let us recall that a prime subgroup P of l-group G is a *representing prime subgroup* if $\bigcap\limits_{g \in G} \{P^g\} = E$.

COROLLARY 1. *Let G be an l-group such that* $\mathrm{var}_l(G) \cap \mathcal{R} = \mathcal{A}$. *Let n be a positive integer and P be a representing prime subgroup. Then $G \in \mathcal{L}_n$ if and only if $x^{-n} P x^n = P$, for all $x \in G$.*

Proof. If $G \in \mathcal{L}_n$, then $x^{-n}Px^n = P$, for all $x \in G$, by Lemma 3. Conversely, let $x^{-n}Px^n = P$ for all $x \in G$. Let $\mathcal{P}r = \{P^g \mid g \in G\}$. Then $\bigcap_{P \in \mathcal{P}r} P = E$, since P is a representing prime subgroup. Further, for any $Q = P^g$, $x \in G$,

$$x^{-n}Qx^n = x^{-n}g^{-1}Pgx^n = g^{-1}gx^{-n}g^{-1}Pgx^ng^{-1}g =$$
$$= g^{-1}(gxg^{-1})^{-n}P(gxg^{-1})^ng = g^{-1}Pg = Q,$$

so by Lemma 3, $G \in \mathcal{L}_n$. □

COROLLARY 2. *For any positive integers m and n is valid $\mathcal{L}_m \cap \mathcal{L}_n = \mathcal{L}_d$ where d is the greatest common divisor of m and n.*

Proof. Clearly, $\mathcal{L}_d \subseteq \mathcal{L}_m \cap \mathcal{L}_n$. Let G be any subdirectly irreducible l-group in $\mathcal{L}_m \cap \mathcal{L}_n$. Since $G \in \mathcal{L}_m$, we have $x^{-m}Px^m = P$ and since $G \in \mathcal{L}_n, x^{-n}Px^n = P$, for any $x \in G$. Since there exist integers a, b such that $d = am + bn$ it follows that $x^{-d}Px = P$. By Corollary 1, $G \in \mathcal{L}_d$.□

For any l-group G, any $x, y \in G$ and integers n, p and r, let

$$d_0(n, p^r, x, y) = x \wedge (x^{-1}y^nx \wedge x^{-2}y^nx^2 \wedge \ldots \wedge x^{-p^r+1}y^nx^{p^r-1});$$
$$d_i(n, p^r, x, y) = x^{-i}d_0(n, p^r, x, y)x^i, \ (0 \leq i \leq p^r - 1);$$
$$e_i(n, p^r, x, y) = \bigvee_{0 \leq j \leq p^r-1}^{j \neq i} (d_i \wedge d_j), \ (0 \leq i \leq p^r - 1;$$
$$a_i(n, p^r, x, y) = d_i(n, p^r, x, y) \cdot e_i(n, p^r, x, y)^{-1}.$$

For any integers m, n, let (m, n) denote the greatest common divisor of m and m.

LEMMA 4. (Reily [3].) *Let $G \in \mathcal{L}_n$, where $n = p^rm$, for some prime number p, and $(p, m) = 1$. Let $y, z \in G$ and $x = z^m$. Let $a_i = a_i(n, p^r, x, y)$. Then*

$$x^{-1}a_ix = a_{i+1}, \quad 1 \leq < p^r - 1;$$
$$x^{-1}a_{p^r-1}x = a_0;$$
$$a_i \wedge a_j = e \qquad \textit{for } i \neq j.$$

Proof. The direct verification. □

Of course, for arbitrary chosen elements in an l-group $G \in \mathcal{L}_n$, there is no guarantee that $a_i(n, p^r, x, y)$ will not be the identity.

For any prime subgroup P of l-group G and any $g \in G, g$ induces a permutation g^* on the set $\mathcal{P}r$ of all conjugates of P by conjugation $g^* : P^a \to P^{ag}$. The set of all such actions form a permutation group on $\mathcal{P}r$. For any $g \in G$, we denote by $O(P, g)$ the orbit of g^* containing P.

LEMMA 5. (Reily [3].) *Let $G \in \mathcal{L}_n$, where $n = p^r m$ for some prime $p, r \geq 1$ and $(p, m) = 1$. Let P be a prime subgroup of G and $z \in G$ be such that $|O(P, z)| = p^r$. Let $x = z^m$. Then there exists a $y \in G$ such that $a_i = a_i(n, p^r, x, y) > e$ for $i = 0, 1, \ldots, p^r - 1, a_0 \notin P$ and the elements $a_i, i = 0, 1, \ldots, p^r - 1$, are distinct.*

Proof. Since $(p, m) = 1$, we have $|O(P, x)| = p^r$. Let $P_i = x^{-i}Px^i$, for $i = 0, 1, \ldots, p^r - 1$. Let y be such that $e < y$ and

$$y \in P_0^* = P_0 \setminus \bigcup_{i=1}^{p^r - 1} P_i.$$

Then $y^n \in P_0^*$ and $x^{-i}y^n x^i \in P_i^* = P_i \setminus \bigcup_{i \neq j} P_j$ In particular, $x^{-i}y^n x^i \in P_i \setminus P$, $i \neq 0$, while $x \notin P$, since $P^x \neq P$. Hence, $d_0(n, p^r, x, y) \in (\bigcap_{i \neq 0} P_i) \setminus P$. Similarly, $d_i(n, p^r, x, y) \in (\bigcap_{i \neq j} P_j) \setminus P_i$. It follows that $e_i = e_i(n, p^r, x, y) \in \bigcap_j P_j \subseteq P_i$. Now it is clear that $e_i \leq d_i$ for all i. Since $e_i \in P_i$ but $d_i \notin P_i$, it follows that $e_i < d_i$ and, therefore, $a_i > e$, as required. □

LEMMA 6. *Let $G \in \mathcal{L}_n$ where $n = p^r m$ for some prime p, $r \geq 1$ and $(p, m) = 1$. Let $x \in G$ and H be a convex l-subgroup of G such that $H \cap x^{-i}Hx^i = E$ for all i, $1 \leq i \leq p^r - 1$ and $H = x^{-p^r}Hx^{p^r}$. Then $H \in \mathcal{L}_m$.*

Proof. Let us assume that $H \notin \mathcal{L}_m$. Since $H \in \mathcal{L}_n$, there exists the smallest integer s, say, such that $H \in \mathcal{L}_s$. By Lemma 3, and since $G \in \mathcal{L}_n, x^{-n}Px^n = P$, for all minimal prime subgroups P in H and elements $x \in G$, whereas by Lemma 3, s must be the smallest positive integer such that convex $x^{-s}Px^s = P$, for all minimal primes P in H and all elements $x \in H$. It follows that s divides n. Since $H \notin \mathcal{L}_m$, then s does not divide m . Hence $s = p^k l$, for some integers $k, l \geq 1$. By Lemma 3, there must exist a prime subgroup P in H and $z \in H$ such that $|O(P,z)| = p^k t$, where $1 \leq t \leq l$. Replacing z by z^t, if necessary, we can suppose that $a_i = a_i(n, p^k, u, v) > e$, for $0 \leq i \leq p^k - 1$.

Now, since $(p, m) = 1$, we can replace x by x^m without altering the hypothesis and so can assert that $x = g^m$, for some $g \in G$. Next, since $u \in H$ it follows from our assumption that u commutes with $x^{-i}a_0x^i$ for $1 \leq i \leq p^r - 1$. Therefore,

$$a_0^n = (xu)^{-n}a_0^n(xu)^n = (xu)^{-n+1}u^{-1}x^{-1}a_0^n xu(xu)^{n-1} =$$
$$(xu)^{-n+1}x^{-1}a_0^n x(xu)^{n-1} = \dots$$
$$= \dots (xu)^{-n+p^r}u^{-1}x^{-p^r}a_0^n x^{p^r}u(xu)^{n-p^r} =$$
$$(xu)^{-n+p^r}u^{-1}a_0^n u(xu)^{n-p^r}(\text{ since } x^{p^r} = g^n) = \dots = u^{-m}a_0^n u^m = a_j^n$$

where $j \equiv m \pmod{p^r}$, which is a contradiction , since $(p, m) = 1$. Hence $H \in \mathcal{L}_m$, as required. □

THEOREM 1. (Gurchenkov [6], Reily [3].) *If the positive integer n can be written as a product of k prime numbers (not necessarily distinct), then $\mathcal{L}_n \subseteq \mathcal{A}^{k+1}$*

Proof. Use induction on k. If $k = 0$, then $n = 1$, and, clearly, $\mathcal{L}_1 = \mathcal{A}$. So assume that the claim holds for any integer that is a product of $k-1$ of fewer prime numbers and let n be a product of k primes. Let $n = p_1^{r_1} \cdots p_l^{r_l}$, where p_i, $i = 1, 2, \dots, l$ are distinct prime numbers and $r_i \geq 1$. Let $n_i = n/p_i$, $i = 1, 2, \dots, l$. Let $\mathcal{U} = \bigvee^{\mathbf{L}}_{1 \leq i \leq l} \mathcal{L}_{n_i}$ and $G \in \mathcal{L}_n$. By the inductive assumption, $\mathcal{U} \subseteq \mathcal{A}^k$, so that if $G \in \mathcal{U}$, then the claim holds. So suppose that $G \notin \mathcal{U}$. Since $G \in \mathcal{L}_{n_1}$, there must exist, by Lemma 3, a prime subgroup P of G and an element z of G such that $|O(P,z)| = p_1^{r_1}m$, for some integer $m \geq 1$. Replacing z by z^m, if necessary, we can assume that $|O(P,z)| = p_1^{r_1}$. By Lemma 5, there exist $x, y \in G$ such that $a_0(n, p_1^{r_1}, x, y) > e$. Let H denote the convex l-subgroup of G generated by a_0, then H is non-trivial and by Lemma 6, $H \in \mathcal{L}_m$. By the induction hypothesis, $H \in \mathcal{A}^k$ and, therefore, H has a non-trivial convex Abelian l-subgroup. Thus $\mathcal{A}(H) \neq E$ and $\mathcal{A}(G) \neq E$.

Let $K = G/\mathcal{A}(G)$. If $K \in \mathcal{U} = \bigvee^{\mathbf{L}}_{1 \leq i \leq l} \mathcal{L}_{n_i}$, then $K \in \mathcal{A}_l^k$ by the inductive hypothesis, so that $G \in \mathcal{A}^{k+1}$, as required. So suppose that $K \notin \mathcal{U}$. Then some subdirectly irreducible homomorphic image of K does not lie in $\mathcal{U}$ and so we can assume that K is subdirectly irreducible.

Then K must have a representing prime subgroup Q. Since for all i $K \notin \mathcal{L}_{n_i}$, by Corollary 1 of Lemma 3 and by the above arguments, that for each i there exists an element $z_i \in K$ such that $|O(Q, z_i)| = p_i^{r_i}$. Then, by Lemma 5, there exist $g_i, y_i' \in K$ such that for $x_i' = (g_i')^{m_i}$, where $m_i = n/p_i^{r_i}, b_i = a_0(n, p_i^{r_i}, x_i', y_i') > e$ and $b_i \notin Q$.

Let g_i and y_i be preimages of g_i' and y_i' under the natural l-homomorphism of G onto $K = G/\mathcal{A}(G)$, and let $x_i = g_i^{m_i}$. Then $a_{0i} = a_0(n, p_i^{r_i}, x_i, y_i)$ is a preimage of b_i and so $a_{0i} > e$ and $a_{0i} \notin P$, where P is the preimage of Q. By Lemma 6, the convex l-subgroup H_i of G generated by a_{0i} must lie in $\mathcal{L}_{m_i}$. Since $a_{0i} \notin P$, for all $i, a = \bigwedge_{i=1}^{l} a_{0i} \notin P$. However, if H denotes the convex l-subgroup of G generated by a,then $H \subseteq H_i$, for all i. Therefore, $H \in \overset{\mathbf{L}}{\bigwedge_{1 \le i \le l}} \mathcal{L}_{m_i}$. But the greatest common divisor of the m_i is 1. Hence $\overset{\mathbf{L}}{\bigcap_{1 \le i \le l}} \mathcal{L}_{m_i} = \mathcal{A}$ by Corollary 2 of Lemma 3. Thus, $H \in \mathcal{A}$, yet $a \in H \setminus P \subseteq G \setminus \mathcal{A}(G)$, which is a contradiction. Therefore, $K \in \mathcal{U}$. □

The value of k in Theorem 1 is the best possible, since examples constructed by J.Smith [2] and N.Reily and R.Wroblewski [1] show that if n is a product of k primes, then $\mathcal{L}_n \nsubseteq \mathcal{A}^k$.

EXAMPLE 1. (Smith [2]), Reily and Wroblewski [1].) For any l-group H and any positive integer n, let (for details, see Chapter 11)

$$H\mathrm{Wr}^{(n)}(\mathbf{Z}, \mathbf{Z}) = \{(\{h_z\}, g) \in H\mathrm{Wr}(\mathbf{Z}, \mathbf{Z}) \mid h_{z_1} = h_{z_2} \text{ if } z_1 \equiv z_2 \pmod{n},\ z_1, z_2 \in \mathbf{Z}\}.$$

It is straightforward to verify that $H\mathrm{Wr}^{(n)}(\mathbf{Z}, \mathbf{Z})$ is an l-subgroup of $H\mathrm{Wr}(\mathbf{Z}, \mathbf{Z})$. Now the construction of $H\mathrm{Wr}^{(n)}(\mathbf{Z}, \mathbf{Z})$ can be generalized. For any finite sequence $(n_1, \ldots, n_k)$ of positive integers

$$H\mathrm{Wr}^{(n_1, \ldots, n_i, n_{i+1})}(\mathbf{Z}, \mathbf{Z}) = H\mathrm{Wr}^{(n_1, \ldots, n_i)}(\mathbf{Z}, \mathbf{Z})\mathrm{Wr}^{(n_i+1)}(\mathbf{Z}, \mathbf{Z}).$$

We write $G(n_1, \ldots, n_k) = \mathbf{Z}\mathrm{Wr}^{(n_1, \ldots, n_k)}(\mathbf{Z}, \mathbf{Z})$. The direct verification shows that $G(n_1, \ldots, n_k) \in \mathcal{L}_{n_1 n_2 \cdots n_k}$. If $p_1, \ldots, p_k$ are prime numbers, then, clearly, $G(p_1, \ldots, p_k) \in \mathcal{A}^{k+1} \setminus \mathcal{A}^k$ and $G(p_1, \ldots, p_k) \in \mathcal{L}_n$ where $n = p_1 \ldots p_k$. □

12.5 Covers in the Lattice of l-Varieties

Let $\mathcal{U}, \mathcal{V}$ be l-varieties and $\mathcal{U} \subseteq \mathcal{V}$. If there is no l-variety $\mathcal{X}$ such that $\mathcal{U} \subseteq \mathcal{X} \subseteq \mathcal{V}$ and $\mathcal{U} \neq \mathcal{X}, \mathcal{V} \neq \mathcal{X}$, then $\mathcal{V}$ *covers* $\mathcal{U}$ in the lattice of l-varieties **L**. From Weinberg's result (Theorem 10.3.1) it follows that the l-variety $\mathcal{A}$ of Abelian l-groups covers the trivial l-variety $\mathcal{E}$. Holland's theorem (Theorem 9.2.2) shows that the l-variety of all l-groups $\mathcal{L}$ covers the l-variety $\mathcal{N}$ of normal-valued l-groups in the lattice of l-varieties **L**. Moreover, S.A.Gurchenkov [4] has proved that every proper l-variety $\mathcal{X}$ has a cover in L.

1^0. Let $\mathcal{L}_n$ be an l-variety defined by the identity $[x^n, y^n] = e$. Let $\mathcal{M} = \mathcal{A}^m \overset{\mathbf{L}}{\bigwedge} \mathcal{L}_n$.

LEMMA 1. *Let $G \in \mathcal{A}^s \setminus \mathcal{A}^{s-1}$. Then G has a solvable tower of ideals*

$$E = G_{s+1} < G_s < \ldots < G_2 < G_1 = G \tag{1}$$

such that each convex l-subgroup $H \in \mathcal{A}^{s-i+1}$ is contained in G_i.

Proof. Let $G_s = \mathcal{A}(G),\ G_{s-1} = \mathcal{A}^2, \ldots, G_1 = \mathcal{A}^s$. It is clear that

$$E = G_{s+1} < \mathcal{A}(G) < \mathcal{A}^2(G) < \ldots \mathcal{A}^s(G) = G.$$

Now the proof follows immediatly from Theorem 9.5.1.□

For each $G \in \mathcal{M}$ let $m(G)$ denote the lenght of the tower (1).

LEMMA 2. *Let $G \in \mathcal{M} \setminus \mathcal{A}$. Then G contains a prime subgroup A which is not ideal in G.*

Proof. Indeed, if each prime subgroup of G is an ideal in G, then by Theorem 9.3.5, G is representable. Since $\mathcal{R} \cap \mathcal{M} = \mathcal{A}$, it follows that G is Abelian group, which contradicts the choice of G.□

LEMMA 3. *Let $G \in \mathcal{M}$, A be a non-invariant convex l-subgroup of G and the positive integer $n = p_1 \cdots p_k$ be the product of different prime numbers $p_1, \ldots, p_k$. Then $m(G) \leq k+1$ and there are $x, y \in G$, $p \in \{p_1, \ldots, p_k\}$ such that $e < x$, $x = z^{\frac{n}{p}}$, $x \in G_1 \setminus G_2$, $A = x^{-p} A x^p$, $A \neq x^{-i} A x^i$, $i = 1, \ldots p-1$.*

Proof. Let $z \in G$ and $z^{-1} A z \neq A$. Let q be the least positive integer for which $z^{-q} A z^q = A$. Without loss of generality, we can assume that

$z > e$ and $z \in G_1 \setminus G_2$. We claim that q divides n. Suppose, on the contrary, that q does not divide n and $d = (q, n)$ is the greatest common divisor of q and n. Since there exist integers a, b that $d = aq + bn$, it follows that $z^{-d}Az^d = z^{-(aq+bn)}Az^{aq+bn} = A$, a contradiction to the choice of q. So we can assume that $q = p_1 \cdots p_i (i \leq k)$. Let $p_1 = \min\{p_1, \dots p_i\}$. If $z^{-\frac{n}{p}}Az^{\frac{n}{p}} = A$, then $(\frac{n}{p}, q) = p_2 \cdots p_i$ and by similar arguments $A = z^{-p_2 \cdots p_i}Az^{p_2 \cdots p_i}$, a contradiction to choice of q. Hence, the result holds for $x = z^{\frac{n}{p}}$. The inequality $m \leq k+1$ follows from Theorem 1 of Section 5. □

Let us recall that for any l-group $G \in \mathcal{M}$, any $x, y \in G$ and integers n, p

$$\begin{aligned} d_0 = \quad & d_0(n,p,x,y) = x \wedge (\bigwedge_{i=1}^{p-1} x^{-i}y^n x^i); \\ & d_i = d_i(n,p,x,y) = x^{-i}d_0x^i, \qquad (i = 1, 2, \dots, p-1); \\ & e_i = e_i(n,p,x,y) = \bigwedge_{j \neq i}(d_j \wedge d_i), \quad (i = 0, 1, \dots, p-1); \\ & a_i = a_i(n,p,x,y) = d_i e_i^{-1}, \qquad (i = 0, 1, \dots, p-1). \end{aligned}$$

LEMMA 4. *Let $G \in \mathcal{M}$ and $y, x \in G$, $x = z^{\frac{n}{p}} > e$, $y > e$. Then:*

$$\begin{aligned} c_i = c_j, \qquad & \textit{for } i, j \in \{0, 1, \dots p-1\}; \\ a_i \wedge a_j = e, \qquad & \textit{for } i, j \in \{0, 1, \dots, p-1\},\ i \neq j; \\ x^{-1}a_i x = a_{i+1}, \qquad & \textit{for } i \in \{0, 1, \dots, p-1\},\ x^{-1}a_{p-1}x = a_0. \end{aligned}$$

The proof is straightforward.□

LEMMA 5.(Scrimger [1].) *Let $G \in \mathcal{M} \setminus \mathcal{A}$. Then there are $x, y, z \in G$ and a prime divisor p of n such that $e < x = z^{\frac{n}{p}}, y > e$ and $a_i(x,y) > e (i = 0, 1, \dots, p-1)$.*

Proof. By Lemma 3, there are prime subgroups A, $x, y \in G$ such that $e < x = z^{\frac{n}{p}}$, $A = x^{-p}Ax^p$, $i = 0, 1, \dots, p-1$. Now arguments coincide with the proof of Lemma 12.4.5 .□

LEMMA 6. *Let $G \in \mathcal{M} \setminus \mathcal{A}$. Then there are $x, y, z \in G$ and a prime divisor p of n such that :*

$$x = z^{\frac{n}{p}} \in G_1 \setminus G_2,\ x > e;$$
$$e < a_i(x,y) \in G_2 \setminus G_3,\ (i = 0, 1, \dots, p-1),\ x \gg a_i(x,y).$$

Proof. We use induction on $m(G)$. If $m(G) = 2$, then by Lemma 3, $e < x = z^{\frac{n}{p}} \in G_1 \setminus G_2$ and by Lemma 5,

$$e < a_i(n,p,x,y) = \bigwedge_{j \neq i} (e \vee x^{-i} d_0 x^i d_0^{-1} d_0 x^{-j} d_0^{-1} x^j) \in G_2 \setminus E.$$

So suppose that the result holds for all $G \in \mathcal{M} \setminus \mathcal{A}$ with $m(G) \leq q$ and let $m(G) = q + 1$. Therefore, in the l-group $\bar{G} = G/G_{m(G)}$ there exist $\bar{x}, \bar{y}, \bar{z} > e$ and a prime divisor p of n such that $\bar{z}^{\frac{n}{p}} = \bar{x} \in \bar{G}_1 \setminus \bar{G}_2, e < a_i(\bar{x}, \bar{y}) \in \bar{G}_2 \setminus \bar{G}_3, i = 0, 1, \ldots, p-1$. Let $\bar{x}, \bar{z}$ be positive preimages of $\bar{x}, \bar{z}$ such that $x = z^{\frac{n}{p}}$ and y be a positive preimage of $\bar{y}$. It is clear that $a_i(x,y) \in G_2 \setminus G_3$ and $z \in G_1 \setminus G_2$.

We claim that $x \gg a_i(x,y)$. Let H be an l-subgroup of G generated by $a_i(x,y), i = 0, 1, \ldots, p-1$ and $x > e$. Let A_i be a convex l-subgroup of G generated by the elements $a_i(x,y), i \in \{0, 1, \ldots, p-1\}$ and $x > e$. Then $A_i \cap A_j = E$ for all $i, j \in \{0, 1, \ldots, p-1\}$, $i \neq j$, and $A_i^x = A_{i+1}$ for all $i \in \{0, 1, \ldots, p-2\}$ and $A_{p-1}^x = A_0$. Thus the l-subgroup $A = A_0 A_1 \cdots A_{p-1}$ is an ideal of H. Assume, on the contrary, that there is $z \in A$ for which $x \wedge z \neq z$. Let $x' = x(x \wedge z)^{-1}, z' = z(x \wedge z)^{-1}$. Then $x' \wedge z' = e$ and $z' \in A$. It is clear that $A_i^x = A_i^{x'}$ for all $i \in \{0, 1, \ldots, p-1\}$ and there is an element $u \in A_{i_0}$ for some $i_0 \in \{0, 1, \ldots, p-1\}$ such that $e < u \leq z'$. Hence $x' \wedge u = e$ and $x'^{-1} u x' = u \in x'^{-1} A_{i_0} x' \cap A_{i_0} = x^{-1} A_{i_0} x \cap A_{i_0} = E$, a contradiction the choice of the element u. Thus, H is a lexicographic extension of A by the infinite cyclic o-group (x), i.e., $x \gg a_i(x,y) > e, \quad i \in \{0, 1, \ldots, p-1\}$.□

PROPOSITION 1. *Let $G \in \mathcal{M} \setminus \mathcal{A}$. Then there exist b_i, $e < b_i$, $b_i \in G_i \setminus G_{i+1}$ and prime number $\bar{p}_i \in \{p_1, \ldots, p_k\}, i = 1, \ldots m(G)$ such that $[b_j, b_i^s] \wedge [b_j, b_i^t] = b_j^{-1}$ if $s \neq t$, $b_i \gg b_j$, $s, t \in \{0, 1, \ldots \bar{p}_i - 1\}$ and $i < j$, $i, j \in \{0, 1, \ldots, m(G)\}$.*

Proof. Use induction on $m(G)$. If $m(G) = 2$, then by Lemma 6, $e < b_1 = x \in G_1 \setminus G_2$, $e < b_2 = a_0(x,y) \in G_2 \setminus E$, $b_1 \gg b_2$ and by Lemma 4, $e = b_1^{-s} b_2 b_1^s \wedge b_1^{-t} b_2 b^t$. Then $b_2^{-1} = [b_2, b_1^s] \wedge [b_2, b_1^t]$ if $s \neq t$, $s, t \in \{0, 1, \ldots, p-1\}$ and $p \in \{p_1, \ldots, p_k\}$. So suppose that the result holds for all $G \in \mathcal{M} \setminus \mathcal{A}$ with $m(G) \leq q$. Let $G \in \mathcal{M} \setminus \mathcal{A}$ and $m(G) = q+1$. Let $B = (a_0(x,y))_l$ be a convex l-subgroup of G generated by the element $a_0(x,y)$. We claim that $m(B) \leq m(G) - 1$. Assume, on the contrary, that $m(B) < m(G) - 1$; then by Lemma 1, the convex l-subgroup B is contained in G_3 and $a_0(x,y) \in G_2 \setminus G_3$. By the inductive assumption, there are $b_{i+1} \in B_i \setminus B_j$ (where B_i are ideals of B defined in Lemma 1 and prime numbers $\bar{p}_{i+1} \in \{p_1, \ldots, p_k\}$, $i = 1, 2, \ldots, m(G)-1$ such that $b_i \gg b_j$ and $[b_j, b_i^s] \wedge [b_j, b_i^t] = b_j^{-1}$ if $s \neq t$, $s,t \in \{0, 1, \ldots, \bar{p}_i - 1\}$, $i < j$, $i,j \in \{2, 3, \ldots, m(G)\}$. By Lemma 1, $b_i \in G_i \setminus G_{i+1}$ for all $i = 2, 3, \ldots, m(G)$. Since $b_2 \in B = (a_0(x,y))_l$, then $a_0(x,y)^\lambda > b_2 > e$ for some positive integer λ and $b_1 \gg a_0$. Therefore, $b_1 \in G_1 \setminus G_2$ and $b_1 \gg b_2 \gg \ldots \gg b_{m(G)} > e$. For all $s,t \in \{0, 1, \ldots, \bar{p}_1 - 1\}$, $s \neq t$ the following inequalities hold in G

$$b_1^{-s}((a_0(x,y))^\lambda)b_1^s \wedge b_1^{-t}((a_0(x,y))^\lambda)b_1^t \geq b_1^{-s}b_jb_1^s \wedge b_1^{-t}b_jb_1^t,\ 2 \leq j \leq m(G).$$

Since $b_1^{-s}((a_0(x,y))^\lambda)b_1^t \wedge b_1^{-t}(a_0(x,y))^\lambda)b^t = e$, then $b_1^{-s}b_jb_1 \wedge b_1^{-t}b_jb_1^t = e$, $2 \leq j \leq m(G)$ and $[b_j, b_i^s] \wedge [b_j, b_i^t] = b_j^{-1}$ for all $s,t \in \{0, 1, \ldots \bar{p}_i - 1\}$, $s \neq t$. □

Let J_k be the set of positive integers which are products of k different prime numbers and if $\alpha = p_1 \cdots p_k$, $\beta = q_1 \cdots q_k \in J$, $\alpha < \beta$, then $\max\{p_1, \ldots, p_k\} < \min\{q_1, \ldots, q_k\}$. For each $\alpha \in J_k$ let $\mathcal{M}_\alpha$ denote a cover of $\mathcal{A}^k$ in the lattice of l-varieties $\mathbf{L}$ which is contained in the l-variety $\mathcal{A}^k \vee ((\mathcal{A}^{k+1}) \wedge \mathcal{L}_\alpha)$. Since by Theorem 12.2.2, the l-variety $\mathcal{A}^k$ is finitely based and $\mathcal{A}^k \neq (\mathcal{A}^{k+1} \wedge \mathcal{L}_\alpha)$, then the existence of this cover is obvious.

THEOREM 1. (Gurchenkov [4].) $\bigvee\limits_{\alpha \in J_k}{}^{\mathbf{L}} \mathcal{M}_\alpha = \mathcal{A}^{k+1}$.

Proof. Let $\mathbf{Z}$ be the additive o-group of integers, $(\mathbf{Z}, \mathbf{Z})$ be the right regular representation of o-group Z, $\mathbf{Z}\mathrm{wr}(\mathbf{Z}, \mathbf{Z})$ be the wreath product of $\mathbf{Z}$ and $(\mathbf{Z}, \mathbf{Z})$ (see Chapter 11) and $\mathrm{wr}^{k+1}(\mathbf{Z}, \mathbf{Z}) = \mathrm{wr}^k(\mathbf{Z}, \mathbf{Z})\mathrm{wr}(\mathbf{Z}, \mathbf{Z})$. We claim that $\mathrm{wr}^{k+1}(\mathbf{Z}, \mathbf{Z}) \in \bigvee\limits_{\alpha \in J_k}{}^{\mathbf{L}} \mathcal{M}_\alpha$. For each $\alpha \in J_k$ let $G_\alpha \in \mathcal{M}_\alpha$ be a solvable subdirectly irreducible l-group and the solvability lenght of G_α be equal to $(k+1)$. Then $G_\alpha \in \mathcal{A}^{k+1} \overset{\mathbf{L}}{\bigwedge} \mathcal{L}_\alpha$. By Proposition 1, there exist elements $b_{1,\alpha} \gg b_{2,\alpha} \gg \ldots \gg b_{k+1,\alpha} > e$ in G_α and prime numbers $p_{1,\alpha}, \ldots, p_{k,\alpha} \in \{p_1, \ldots, p_k\}$ such that $[b_{j,\alpha}, b_{i,\alpha}^s] \wedge [b_{j,\alpha}, b_{i,\alpha}^t] = b_{j,\alpha}^{-1}$ if $s \neq t$, $s,t \in \{0, 1, \ldots, p_{i,\alpha} - 1\}$ and $i < j$, $i,j \in \{1, 2, \ldots, k+1\}$.

Now let us consider an l-subgroup of the l-group $\prod\limits_{\alpha \in J_k} G_\alpha / \overline{\prod\limits_{\alpha \in J_k}} G_\alpha$ generated by the images $b_1, \ldots, b_{k+1}$ of the elements $\hat{b}_1, \ldots, \hat{b}_{k+1} \in$

$\prod_{\alpha \in J_k} G_\alpha$, where $\hat{b}_i(\alpha) = b_{i,\alpha} \in G_\alpha, \alpha \in J_k$; $i = 1, 2, \ldots k+1$. It is easy to see that $[b_j, b_i^s] \wedge [b_j, b_i^t] = b_j^{-1}, b_i \gg b_j > e$ for $i < j$, $i, j \in \{1, 2, \ldots, k+1\}$ and $s \neq t$. Therefore, $\text{l-gp}(b_1, \ldots, b_{k+1}) \cong \text{wr}^{k+1}(Z, Z)$. By Theorem 11.3.2, $\overset{\mathbf{L}}{\bigvee_{\alpha \in J_k}} \mathcal{M}_\alpha \supseteq \mathcal{A}^{k+1}$. □

THEOREM 2. (Gurchenkov [4].) *Every proper l-variety $\mathcal{X}$ has a cover in the lattice of l-varieties* **L**.

Proof. If $\mathcal{X} = \mathcal{E}$, then $\mathcal{A}$ covers $\mathcal{E}$ by Theorem 10.3.1. If $\mathcal{X} = \mathcal{N}$, then the l-variety $\mathcal{L}$ of all l-groups covers $\mathcal{N}$ by Theorem 9.2.2. So, we may assume that $\mathcal{X} \neq \mathcal{N}, \mathcal{E}$. Since $\mathcal{N} = \overset{\mathbf{L}}{\bigvee_{1 \leq k < \infty}} \mathcal{A}^k$ (Theorem 11.2.2), then there exists a positive integer k such that $\mathcal{A}^k \subseteq \mathcal{X}$ and $\mathcal{A}^{k+1} \nsubseteq \mathcal{X}$. If $\mathcal{M}_\alpha \nsubseteq \mathcal{X}$ for some $\alpha \in J_k$, then the l-variety $\mathcal{M}_\alpha \overset{\mathbf{L}}{\bigvee} \mathcal{X}$ covers $\mathcal{X}$ in **L**. Suppose, on the contrary, that $\mathcal{M}_\alpha \subseteq \mathcal{X}$ for all $\alpha \in J_k$, then by Theorem 1, $\mathcal{A}^{k+1} \subseteq \mathcal{X}$ contradiction the choice of the integer k.□

COROLLARY 1. *Let $\mathcal{X}$ be an l-variety and $\mathcal{X} = \mathcal{E}, \mathcal{N}, \mathcal{L}$. Then $\mathcal{X}$ has infinitely many covers in the lattice of l-varieties* **L**. □

COROLLARY 2. *Each solvable l-variety $\mathcal{X}$ has a solvable cover in the lattice of l-varieties* **L**. □

2⁰. Let $\mathbf{L}_\mathcal{R}$ be the lattice of all subvarieties of the l-variety $\mathcal{R}$ of representable l-groups. Intuition might have led us to expect that representable l-varieties would be "easier" to deal with and one might have expected to be able to describe a significant part of the bottom of the lattice. However, this is not true. Here we will prove that the lattice $\mathbf{L}_\mathcal{R}$ has no covering condition, i.e., there exists a representable l-variety $\mathcal{V}$ for which there is no l-variety of representable l-groups covering $\mathcal{V}$.

Let **R** be an additive o-group of real numbers, (β) be an infinite cyclic o-subgroup of the multiplicative group of reals numbers generated by the positive real number β, $(\beta \neq 1)$. Now for $\beta \in \mathbf{R}$, $0 < \beta, \beta \neq 1$, define T_β to be the group of ordered pairs (r, a), with $a \in \mathbf{R}$ and $r \in (\beta)$ With multiplication $(r, a)(r', a') = (rr', r'a + a')$. Then T_β is a splitting extension of **R** by infinite cyclic group (β) with the conjugation given by $\beta^{-1} r \beta = \beta r$, where $r \in R$, and β is a generator of (β). We order T_β

lexicographically by calling an element (r, a) positive if $r > 1$ in o-group (β) or if $r = 1$ and $a \geq 0$, making T_β an o-group. These o-groups T_β were defined in different wey in Section 4 of Chapter 11.

LEMMA 7. *Let* $\mathcal{U}_\beta = \mathrm{var}_l T_\beta$ *where* $\beta > 0, \beta \neq 1$. *Then there is an l-variety* $\mathcal{U}'_\beta$ *such that:* $\mathcal{U}'_\beta \subseteq \mathcal{U}_\beta$; $\quad \mathcal{U}'_\beta \neq \mathcal{U}_\beta, \mathcal{A}$.

Proof. Let us assume that $\beta > 1$. Then there exist positive integers k, n such that $\beta < n < \beta^k$. Let $T'_\beta = \{(r, a) \mid r \in (\beta^k),\ a \in R\}$ be a subgroup of T_β totally ordered by induced order. Let $\mathcal{U}'_\beta = \mathrm{var}_l T'_\beta$. Since T_β is non-abelian, then $\mathcal{U}'_\beta \neq \mathcal{A}$ and by definition, $\mathcal{U}'_\beta \subseteq \mathcal{U}_\beta$. We claim that the identity

$$(2) \qquad (|x| \vee |y|)^{-1}|[x, y]|(|x| \vee |y|) \wedge |[x, y]|^n = |[x, y]|^n$$

holds in o-group T'_β and is violated in o-group T_β. Let $x, y \in T'_\beta$. We can assume that $(|x| \vee |y|) = (\beta^{kp}, a)$, and $|[x, y]| = (1, a)$ where $p > 0, a > 0$. Then

$$(|x| \vee |y|)^{-1}|[x, y]|(|x| \vee |y|) = (1, \beta^{kp}a),$$
$$(1, \beta^{kp}a) > (1, \beta^k a) > (1, na) = (1, a)^n,$$

by the choice of the natural numbers n and k. Therefore,

$$(|x| \vee |y|)^{-1}|[x, y]|(|x| \vee |y|) \wedge |[x, y]|^n =$$
$$= (1, \beta^{kp}a) \wedge (1, a)^n = (1, a)^n = [x, y]^n$$

and identity (2) holds in o-group T'_β and in $\mathcal{U}'_\beta$. Identity (2) is violated in o-group T_β. In fact, let $x = (\beta, 0)$ and $y = (1, a), a > 0$. Then $[x, y] = (1, (1 - \beta)a)$ where $1 - \beta \neq 0$ in R and $|[x, y]| = (1, (\beta - 1)a)$. The direct calculations show that

$$(|x| \vee |y|)^{-1}|[x, y]|(|x| \vee |y|) = (\beta, 0)^{-1}(1, (\beta - 1)a)(\beta, 0) =$$
$$= (1, \beta(\beta - 1)a) < (1, n(\beta - 1)a) = (1, (\beta - 1))^n = |[x, y]|^n.$$

Therefore, identity (2) does not hold in $T_\beta (\beta > 1)$. The case $0 < \beta < 1$ is similar.□

COROLLARY 1. $T_\beta \in \mathcal{U}_\beta \setminus \mathcal{U}'_\beta$. □

Let $\mathcal{V}_0$ be an l-variety defined by the infinite basis of identities:

$$\text{a)}\quad \begin{gathered} |(|[x,y]|^2 \vee y^{-1}|[x,y]|y)|[x,y]|^{-2}|\wedge \\ \wedge|(|[x,y]|^2 \vee x^{-1}|[x,y]|x)|[x,y]|^{-2}|\wedge \\ \wedge|((|x|\vee|y|)^{-1}|[x,y]|(|x|\vee|y|)\wedge|[x,y]|^n)|[x,y]|^{-n}|\wedge \\ \wedge|((|x|\vee|y|)|[x,y]|(|x|\vee|y|)^{-1}\wedge|[x,y]|^m)|[x,y]|^{-m}| = e, \end{gathered}$$

$$\text{b)}\quad (x \wedge y^{-1}x^{-1}y) \vee e = e,$$

where m, n are positive integers and $m, n \geq 2$.

By definition, $\mathcal{V}_0 \in \mathbf{L}_{\mathcal{R}}$ and $\mathcal{V}_0 \supseteq \mathcal{W}a$. Therefore, $\mathcal{V}_0 \neq \mathcal{A}$.

LEMMA 8. *Let β be a positive real number, $\beta \neq 1$. Then $T_\beta \notin \mathcal{V}_0$ and $\mathcal{V}_0 \not\supseteq \mathcal{U}_l =$* $\text{var}_l T_\beta$.

Proof. Let $\beta > 1$. Then there exists a natural number k such that $\beta > 2$. The direct verification shows that the identities a) with $m = 2$ and $n > \beta^k$ are violated in T_β. In fact, let $x = (\beta^k, a)$, $a \geq 0$ in R and $y = (\beta^k, 0)$. Then

$$|[x,y]| = (1, (\beta^k - 1)a) = (1, b) \neq (1, 0), |[x,y]|^2 = (1, 2b),$$
$$y^{-1}|[x,y]|y = (1, \beta^k b), x^{-1}|[x,y]|x = (1, \beta^k b)$$

and

$$\begin{gathered} |(|[x,y]|^2 \vee y^{-1}|[x,y]|y)|[x,y]|^{-2}| = \\ = |(|[x,y]|^2 \vee x^{-1}|[x,y]|x)|[x,y]|^{-2}| = |(1, (\beta^k - 2)b)| > e. \end{gathered}$$

Similar calculations show that

$$\begin{gathered} |((|x|\vee|y|)^{-1}|[x,y]|(|x|\vee|y|)\wedge|[x,y]|^n)|[x,y]|^{-n}| = \\ = |((1,\beta^k b) \wedge (1, nb))(1, nb)^{-1}| = |(1, (\beta^k - n)b)| > e \end{gathered}$$

and

$$\begin{gathered} |((|x|\vee|y|)|[x,y]|(|x|\vee|y|)^{-1}\wedge|[x,y]|^2)|[x,y]|^{-2}| = \\ = (1, (2 - \beta^{-k})b) > e. \end{gathered}$$

Since T_β is an o-group, then identity a) with k such that $\beta^k > 2, n > \beta^k$ and $m = 2$ is violated in T_β. Then $T_\beta \notin \mathcal{V}_0$ and $\mathcal{U}_\beta = \text{var}_l T_\beta \not\subseteq \mathcal{V}_0$.

The case $\beta < 1$ is analogous.

COROLLARY 1. $\mathcal{V}_0 \neq \mathcal{R}$.

THEOREM 3. (Medvedev [8].) *The l-variety $\mathcal{V}_0$ has no covers in the lattice $\mathbf{L}_\mathcal{R}$.*

Proof. Assume, on the contrary, that there is an l-variety $\bar{\mathcal{V}}_0$ such that $\bar{\mathcal{V}}_0 \subseteq \mathcal{R}$ and $\bar{\mathcal{V}}_0$ covers $\mathcal{V}_0$ in the lattice of l-varieties L. Then the l-variety $\bar{\mathcal{V}}_0$ is uniquely determined by its o-groups. Therefore, there is o-group $G \in \bar{\mathcal{V}}_0 \setminus \mathcal{V}_0$. Thus, there are positive integers $m, n \geq 2$ and $x, y \in G$ such that

$$(3)\qquad |(|[x,y]|^2 \vee y^{-1}|[x,y]|y)|[x,y]|^{-2}| > e,$$
$$|(|[x,y]|^2 \vee x^{-1}|[x,y]|x)|[x,y]|^{-2}| > e;$$
$$(4)\qquad |((|x| \vee |y|)^{-1}|[x,y]|(|x| \vee |y|) \wedge |[x,y]|^n)|[x,y]|^{-n}| > e$$
$$(5)\qquad |((|x| \vee |y|)|[x,y]|(|x| \vee |y|)^{-1}|[x,y]|^{-m})|[x,y]|^{-m}| > e.$$

From (3) it follows that

$$(6)\qquad y^{-1}|[x,y]|y > |[x,y]|^2,\ x^{-1}|[x,y]|x > |[x,y]|^2.$$

From (4) and (5) we have

$$(7)\qquad (|x| \vee |y|)^{-1}|[x,y]|(|x| \vee |y|) < |[x,y]|^n,$$
$$(8)\qquad (|x| \vee |y|)|[x,y]|(|x| \vee |y|)^{-1} < |[x,y]|^m.$$

Let $|x < |y|$, Then (7) and (8) can be rewritten in the form

$$(9)\qquad |y|^{-1}|[x,y]||y| < |[x,y]|^n,$$
$$(10)\qquad |y||[x,y]||y|^{-1} < |[x,y]|^m.$$

Hence,

$$(11)\qquad |[x,y]| < |y|^{-1}|[x,y]|^m|y| = (|y|^{-1}|[x,y]||y|)^m < |[x,y]|^{mn}.$$

Hence, the elements $|[x,y]|$ and $|y|^{-1}|[x,y]||y|$ are Archimedean equivalent. Thus, the jump $G_\alpha \prec \bar{G}_\alpha$ in the system of convex subgroups of G defined by the element $|[x,y]|$ is invariant under conjugation by $(|y|)$. Let $K = \mathrm{gp}(y, [x,y])$ be a subgroup of G generated by the elements x and $[x,y]$. Let $\bar{G}_\alpha \cap K = \bar{K}_\alpha$ and $G_\alpha \cap K = K_\alpha$. Then $K_\alpha, \bar{K}_\alpha$ are normal convex subgroups of K . Let $T = K/K_\alpha$ be the quotient of K by K_α totally ordered under induced order. Then $\bar{K}_\alpha/K_\alpha = A$ is a normal convex Archimedean subgroup in T and $B/A = (\bar{y})$ is an infinite cyclic o-group. Thus, T is a splitting extension A by infinite cyclic o-group (y) ordered lexicographically by the rule: $g = y^p a \geq e$ iff $p > 0$ or $p = 0$ and $a \geq e$ in Archimedean o-group A. By Hölder's theorem (Theorem 3.2.1), A is order isomorphic to a subgroup of additive o-group of real numbers $\mathbf{R}$. Let φ be this order isomorphism. Since each order automorphism of an Archimedean o-group $(A)\varphi \subseteq \mathbf{R}$ is the multiplication on some positive real number β, $(b^{-1}ab)\varphi = \beta(a)\varphi$ for all $a \in A, b = y$. Since K is non-abelian, $\beta \neq 1$, $\beta > 0$. Now the direct verification shows that the mapping $f : y^p a \to (\beta^p, (a)\varphi)$ is an order isomorphism of o-group K into T_β and $A_\beta = (A)\varphi$ is a dense subgroup of $\mathbf{R}$. Thus, $(K)f \in \mathrm{var}_l G \subseteq \bar{\mathcal{V}}_0$ and $(K)f \leq T_\beta$. It is clear that the completion of $(K)f$ in T_β coincides with T_β. Therefore, by Bernau's theorem (Theorem 9.5.2), $T_\beta \in \mathrm{var}_l G \subseteq \bar{\mathcal{V}}_0$ and $\mathcal{U}_\beta = \mathrm{var}_l T_\beta \subseteq \bar{\mathcal{V}}_0$. By Lemma 7, there is an l-variety U'_β such that $\mathcal{A} \neq \mathcal{U}'_\beta \subseteq \mathcal{U}_\beta, \mathcal{U}'_\beta \neq \mathcal{U}_\beta$. Moreover, $\mathcal{U}'_\beta = \mathrm{var}_l T'_\beta$ and $\mathcal{U}'_\beta \not\subseteq \mathcal{V}_0$. Hence $\bar{\mathcal{V}}_0 \supseteq \mathcal{U}_\beta \vee \mathcal{V}_0 \supseteq \mathcal{V}_0, \bar{\mathcal{V}}_0 \supseteq \mathcal{U}'_\beta \vee \mathcal{V}_0 \supseteq \mathcal{V}_0$ and $\mathcal{U}'_\beta \vee \mathcal{V}_0 \neq \mathcal{V}_0$. By our assumption, $\bar{\mathcal{V}}_0 = \mathcal{U}_\beta \vee \mathcal{V}_0 = \mathcal{U}'_\beta \vee \mathcal{V}_0$. Therefore, $T_\beta \in \mathcal{U}'_\beta \vee \mathcal{V}_0$. Since the o-group T_β is finitely subdirectly irreducible, then $T_\beta \in \mathcal{U}'_\beta$ or $T_\beta \in \mathcal{V}_0$. These inclusions contradict Corollary 1 of Lemma 7 and Lemma 8. □

12.6 Covers for l-Variety of Abelian l-Groups

1⁰. In this Section we consider l-varieties that cover $\mathcal{A}$. First we will consider solvable non-representable covers. For any prime number p let G_p be a group having presentation

$$G_p = \mathrm{gp}(a_{1p}, \dots, a_{pp}, b_p \mid b_p^{-1} a_{ip} b_p = a_{jp} \ (j \equiv i+1 \pmod p)),$$
$$[a_{ip}, a_{jp}] = e \text{ for all } i, j \in \{0, 1, \dots, p-1\})$$

and be lattice-ordered by the rule: $x = b_p^n a_{1p}^{k_1} \cdots a_{pp}^{k_p} \geq e$ iff $n > 0$ or $n = 0$ and $k_1 \geq 0, \dots, k_p \geq 0$. This l-group is called a *Scrimger l-group* . Let $\mathcal{S}_p$ be the l-variety generated by the G_p. It is clear that $\mathcal{S}_p \subseteq \mathcal{L}_p$.

THEOREM 1. (Scrimger [1].) *For each prime number p the l-variety $\mathcal{S}_p$ covers $\mathcal{A}$ in the lattice of l-varieties* $\mathbf{L}$.

Proof. We claim that any non-abelian l-group contains an l-subgroup which is l-isomorphic to G_p. Let $H \in \mathcal{S}_p \setminus \mathcal{A}$. It is clear that a subgroup H^p of H generated by $\{x^p \mid x \in H\}$ is a normal Abelian subgroup of H. Therefore, there are $a, b \in H$ such that $a \in H^p$ and $a^b = b^{-1}ab \neq a$.

Let $a_1 = a$ and $a_{k+1} = b^{-1}a_k b$. Then $a_{p+1} = a_1$ and elements $a_1, \ldots, a_p$ mutually commute. Let $\bar{a} = a_1 \vee a_2 \vee \ldots \vee a_p$. Then $\bar{a} \neq a_1$ and $b^{-1}\bar{a}b = \bar{a}$. Now we fix indeces $i_1 < i_2 < \ldots < i_k$ for which the element $c = a_{i_1} \vee a_{i_2} \vee \ldots \vee a_{i_k}$ is maximal of the elements of this form and $c \neq \bar{a}$. Then $\bar{a}'_1 = \bar{a}c^{-1} > e$ and the elements $a'_i = b^{-i+1}a'_1 b^{i-1}$, $1 \leq i \leq p$ are strictly positive and mutually orthogonal. In fact, for $i < j$ we have

$$\begin{aligned} a'_i \wedge a'_j &= b^{-i+1}\bar{a}c^{-1}b^{i-1} \wedge b^{-j+1}\bar{a}c^{-1}b^{j-1} = \\ &= \bar{a}(b^{-i+1}c^{-1}b^{i-1} \wedge b^{-j+1}c^{-1}b^{j-1}) = \\ &= \bar{a}(b^{-i+1}(a_{i_1} \vee \ldots \vee a_{i_k})^{-1}b^{i-1} \wedge b^{-j+1}(a_{i_1} \vee \ldots \vee a_{i_k})^{-1}b^{j-1}) = \\ &= \bar{a}b^{-i+1}(a_{i_1}^{-1} \wedge \ldots \wedge a_{i_k}^{-1} \wedge b^{i-j}a_{i_1}^{-1}b^{j-i} \wedge \ldots \wedge b^{i-j}a_{i_k}^{-1}b^{j-i})b^{i-1} = \\ &= \bar{a}b^{-i+1}(a_{i_1} \vee \ldots \vee a_{i_k} \vee a_{i'_1} \vee \ldots \vee a_{i'_k})^{-1}b^{i-1}. \end{aligned}$$

Since p is a prime number, then the subsets of indeces $\{i_1, \ldots, i_k\}$ and $\{i'_1, \ldots, i'_k\}$ are distinct. Thus, $a_{i_1} \vee \ldots \vee a_{i_k} \vee a_{i'_1} \vee \ldots \vee a_{i'_k} = \bar{a}$ and $a'_i \wedge a'_j = \bar{a}b^{-i+1}\bar{a}^{-1}b^{i-1} = e$. Let G be an l-subgroup of H generated by the elements $b, a'_1, \ldots, a'_p$ and A_i be a convex l-subgroup of G generated by the element $a_i : A_i = \{x \in G \mid |x| \leq a_i^n$ for some integer $n\}$.

By above arguments, all convex l-subgroups A_i are mutually orthogonal and the convex l-subgroup $A = A_1 \cdot A_2 \cdots A_p$ is an ideal of G. It is clear that $A_i^x = A_i$ for all $x \in A$ and $A_i^b = A_{i+1}$ for $i = 1, 2, \ldots, p-1$ and $A_{p-1}^b = A_1$. Thus, G is a splitting extension of A by the infinite cyclic o-group (b). We claim that G is ordered lexicographically, i.e., $b \gg x$ for each $x \in A$, $x \geq e$. Assume, on the contrary, that $b \wedge x \neq x$ for some $x \in A, x > e$. Let $b_1 = b(b \wedge x)^{-1}$ and $x_1 = x(b \wedge x)^{-1}$. Then $b_1 \wedge x_1 = e$ and $x_1 \in A, x_1 > e$. It is clear that $A_i^b = A_i^{b_1}$ for all $i = 1, 2, \ldots, p$. On the other hand, there is an element y such that $y \in A_i$ for some $i \in \{1, 2, \ldots, p\}$ and $e < y < x_1$. Then $b_1 \wedge y = e$ and $b_1^{-1}yb_1 = y \in b^{-1}A_{i_0}b \cap A_{i_0} = E$, a contradiction to the choice of the element y . Therefore, G is a lexicographic extension of A by the infinite cyclic o-group (b). Now it is evident that G is generated by the elements $b, a'_1, \ldots, a'_p$ and G is l-isomorphic to G_p. □

2^0. Now we consider representable solvalle covers of $\mathcal{A}$. Let

$$N_2 = \mathrm{gp}(a, b, c \mid c = [a, b], [a, c] = [a, b] = e).$$

Any element of N_2 can be written uniquely in the form $a^m b^n c^k$ where m, n, k are integers. We totally order N_2 lexicographically by the rule:

$g = a^m b^n c^k \geq e$, $g \in N_2$ iff $m > 0$ or $m = 0$, $n > 0$ or $m = n = 0$, $k \geq 0$. The group N_2 is totally ordered under this order and is denoted as N_0. Let $\mathcal{M}_0 = \mathrm{var}_l(N_0)$.

Let W^+ denote $\mathbf{Z}\mathrm{wr}(\mathbf{Z},\mathbf{Z})$ totally ordered as follows: $(f,k) > e$ iff $k > 0$ or $k = 0$ and $f(r) > 0$ where r is the greatest element of the support of f. Here $f \in \prod_{i\in\mathbf{Z}} Z_i$, Z_i is o-isomorphic to $\mathbf{Z}$. Let $\mathcal{M}^+ = \mathrm{var}_l(W^+)$. It is clear that $(f,0)^{(\varphi,k)} \gg (f,0)$ for all $(\varphi,k) \in \mathbf{Z}\mathrm{wr}(\mathbf{Z},\mathbf{Z})$ such that $k > 0$.

Let W^- denote $\mathbf{Z}\mathrm{wr}(\mathbf{Z},\mathbf{Z})$ totally ordered as follows: $(f,k) > e$ iff $k > 0$ or $k = 0$ and $f(r) > 0$ where r is the smallest element of the support of f. Let $\mathcal{M}^- = \mathrm{var}_l(W^-)$. It is clear that $(f,0)^{(\varphi,k)} \ll (f,0)$ for all $(\varphi,k) \in Z\mathrm{wr}(Z,Z)$ such that $k > 0$.

THEOREM 2. (Medvedev [7].) *The solvable representable covers of $\mathcal{A}$ are precisely $\mathcal{M}_0, \mathcal{M}^+, \mathcal{M}^-$.*

Proof. The direct verification shows that in the l-variety $\mathcal{M}^+$ are valid the identities

$$|[x,y]|^{(|x|\vee|y|)} \wedge |[x,y]|^n = |[x,y]|^n$$

for all positive integers n and in the l-variety $\mathcal{M}^-$ are valid the identities

$$|[x,y]|^{(|x|\vee|y|)} \vee |[x,y]|^n = |[x,y]|^{(|x|\vee|y|)}$$

for all positive integers n. The l-variety $\mathcal{M}_0$ is nilpotent and, therefore, by Theorem 9.4.1, is weakly abelian. From these observations it is easy to see that $\mathcal{M}^+ \overset{\mathbf{L}}{\wedge} \mathcal{M}^- = \mathcal{M}^+ \overset{\mathbf{L}}{\wedge} \mathcal{M}_0 = \mathcal{M}^- \overset{\mathbf{L}}{\wedge} \mathcal{M}_0 = \mathcal{A}$.

Let $\mathcal{X}$ be any solvable representable l-variety different from $\mathcal{A}$. Then there exists a solvable non-abelian o-group $G \in \mathcal{X} \setminus \mathcal{A}$. Replacing G by a member of its derived series, if necessary, we can assume that G is metabelian. Let G be a weakly abelian non-abelian o-group. Then there are $a, b \in G$ for which $[a,b] \neq e$. Let H be a value of $[a,b]$ in G. Then H is normal in G and in G/H the o-subgroup $\mathrm{gp}(aH, bH)$ is isomorphic (as a group) to N_2. Thus, $\mathcal{X}$ contains an o-group $N_2 = \mathrm{gp}(a',b',c' \mid [a',b'] = c',\ [a',c'] = [b',c'] = e)$. Since o-group N_2 is weakly abelian, then the subgroup (c') is convex in each total order on N_2. Without loss of generality, we can suppose that $a', b' > e$ in o-group N_2. Taking $[a',b']$ or $[b',a']$, as appropriate, we can assume that

$[a', b'] > e$. Hence we may assert that the total order P on N_2 coincides with one of the following orders:

P_1 corresponds to the system of convex subgroups

$$E \subset (c') \subset (c', b') \subset N_2 \text{ and } e < c' \ll b' \ll a';$$

P_2 corresponds to the system of convex subgroups

$$E \subset (c') \subset (c', a') \subset N_2 \text{ and } e < c' \ll b' \ll a';$$

P_3 corresponds to the system of convex subgroups

$$E \subset N_2, \ e < c' \ll a', b',$$

and $N_2/(c')$ is an Archimedean o-group.

It is clear that o-group (N_2, P_1) is o-isomorphic to the o-group N_0. Therefore, $\mathcal{M}_0 \subseteq \mathcal{X}$. Let $(N_2, P_i) \in \mathcal{X} (i = 2, 3)$. Let $(N_2, \ P_i)_n = (N_2, P_i)$ for all positive integers n,

$$\bar{\prod_{n \in \mathbf{N}}} (N_2, P_i)_n = \bar{N}, \quad \hat{N} = \prod_{n \in \mathbf{N}} (N_2, P_i)_n.$$

Let $\bar{a}, \bar{b} \in \bar{N}$ be defined by the rule: $\bar{a}(n) = b'^n a'$, $\bar{b}(n) = b'$ for all $n \in \mathbf{N}$. Then in $\bar{N}/\hat{N}$ the subgroup $\mathrm{gp}(\bar{a} \cdot \hat{N}, \ \bar{b} \cdot \hat{N})$ is o-group order isomorphic to the o-group N_0. Thus $\mathcal{M}_0 \subseteq \mathcal{X}$. So we can claim that the l-variety $\mathcal{X}$ does not contain weakly abelian non-abelian o-group. Let H be any metabelian o-group from $\mathcal{X}$. Since $H \notin \mathcal{W}a, \mathcal{A}$, then there are $e < x, y \in H$ such that $[x, y] \neq e$. Let H_α be a value of $[x, y] = b$. It is clear that the jump $H_\alpha \prec \bar{H}_\alpha$ in the chain of convex subgroups of o-group H is not central under conjugation of x or y. Let the jump $H_\alpha \prec \bar{H}_\alpha$ be not central under conjugation of x. Thus there arise three different cases: 1) $x^{-1} H_\alpha x = H_\alpha$; 2) $x^{-1} H_\alpha x \supset H_\alpha$; 3) $x^{-1} H_\alpha x \subset H_\alpha$.

Case 1). Let $H_0 = \mathrm{gp}(xH_\alpha, bH_\alpha)$ be a subgroup of totally ordered group $\mathrm{gp}(x, b, H_\alpha)/H_\alpha$. Then H_0 is the splitting lexicographic extension of normal Archimedean subgroup A (generated as a convex normal subgroup by one element bH_α) by the infinite cyclic o-group (xH_α). Let $\bar{H} = \bar{\prod_{i \in \mathbf{N}}} H_i$ and $R = \prod_{i \in \mathbf{N}} H_i$ be the Cartesian and direct products of o-groups H_i, $(i \in \mathbf{N})$, where $H_i = H_0$, correspondingly, and $\bar{x}, \bar{b} \in \bar{H}$ be defined by the rule: $\bar{x}(i) = x^i$, $\bar{b}(i) = b$, $i \in \mathbf{N}$. Then in the quotient group $\bar{H}/R$ the elements $\bar{b}R$ and $\bar{b}^{\bar{x}}R$ are linearly independent in the l-group $G = \mathrm{gp}(\bar{x}, \bar{b}, R)/R$ and G is an o-group. It is easy to see that o-group G is order-isomorphic to W^+ if $\bar{b}R \ll \bar{b}^{\bar{x}}R$. If $\bar{b}R \gg \bar{b}^{\bar{x}}R$, then G is order isomorphic to W^-. Hence $\mathcal{X}$ contains $\mathcal{M}^+$ or $\mathcal{M}^-$.

Case 2). Then $e < b \ll b^x$ and b is a commutator. Then we have $\ldots \ll b^{x^{-1}} \ll b \ll b^x \ll \ldots$. Moreover, $A = \mathrm{gp}(b^{x^k} \mid k \in \mathbf{Z})$ is Abelian group and $b \gg a$ for all $a \in A$. The direct verification shows that $G = \mathrm{gp}(b, x)$ is order isomorphic to W^+. Then $\mathcal{M}^+ \subseteq \mathcal{X}$.

Case 3) is analogous to case 2) and the arguments similar to that of Case 2) give $\mathcal{M}^- \in \mathcal{X}$. □

3^0. There are also non-solvable representable l-varieties that cover $\mathcal{A}$. The first examples of these covers were provided by J.Bergman [1] and V.M.Kopytov [9]. Here we will prove the existence of non-solvable representable covers. This approach is due to W.Ch.Holland and N.Ya.Medvedev.

In the book of R.Botto-Mura and A.Rhemtulla [2] the construction of a total order on a free product of orderable groups is presented (this order is attributed to Philip Hall). We will describe this construction in more details.

Let F be the free group and $\{a_\gamma | \gamma \in \Gamma\}, |\Gamma| \geq 2$ be the set of free generators of F. Let the set Γ be totally ordered by the relation $>_\Gamma$. Let F be a totally ordered group under each of an infinite sequence of relations: $>_1, >_2, >_3, \ldots$. Let F_i denote the ordered group $(F, <_i)$. We will use the given sequence of orders to produce a new order on F. In the remainder of this Section, we will define an index set

$$\Lambda \subseteq \Gamma \times F_1 \times F_2 \times \ldots$$

(restricted product), and order the index set lexicographically. For each $\lambda \in \Lambda$, we will define an element $a_\lambda \in F$ and we will show that every non-identity element of F has a unique expression as

$$a_{\lambda_1}^{m_1} \cdot a_{\lambda_2}^{m_2} \cdots a_{\lambda_n}^{m_n},$$

where $0 \neq m_i \in \mathbf{Z}$, $\lambda_i \in \Lambda$, and $\lambda_1 > \lambda_2 > \ldots > \lambda_n$ in the lexicographic order on Λ. We then order F lexicographically by Λ; that is, we call such an element positive iff $m_1 > 0$ in $\mathbf{Z}$.

We need some more notations. For $\lambda \in \Gamma \times \prod F$, we write

$$\lambda = (\lambda(0), \lambda(1), \ldots).$$

For $i \geq 1$, $\lambda \mid i \in \Gamma \times \prod F$, such that

$$\lambda \mid i(r) = \begin{cases} \lambda(r) & \text{if } r < i, \\ e & \text{otherwise ,} \end{cases}$$

and $a_\lambda = a_\lambda(0)^{\lambda(1)\lambda(2)\cdots}$, the upper indexes denote conjugation.

We define, inductively, the elements of Λ and certain sets U_λ, $\lambda \in \Lambda$. For each $\lambda(0) \in \Gamma$, we declare that $\lambda = (\lambda(0)) = (\lambda(0), e, e, \ldots) \in \Lambda$. We can omit the extra parentheses and write $\lambda = \lambda(0)$. Note that $a_\lambda = a_{\lambda(0)}$ is one of the free generators of F. Define $U_\lambda = \mathrm{gp}(a_\alpha | \alpha \in \Gamma, \alpha >_\Gamma \lambda(0))$ the subgroup of F generated by the indicated elements. Then U_λ is a free subgroup of F.

We declare that $\lambda = (\lambda(0), \lambda(1), e, e, \ldots) \in \Lambda$ iff $\lambda(1) \in U_{(\lambda(0))}$. Define

$$U_\lambda = \mathrm{gp}(a_{\lambda(0)}^\alpha | \alpha \in U_{(\lambda(0))}, \alpha >_1 \lambda(1)).$$

We continue inductively. Having defined $\lambda \in \Lambda$ and U_λ for all λ of length $\leq n$, we assert that $\lambda = (\lambda(0), \lambda(1), \ldots, \lambda(n)) \in \Lambda$ iff $\lambda \mid n \in \Lambda$ and $\lambda(n) \in U_{\lambda|n}$. We then define $U_\lambda = \mathrm{gp}(a_{\lambda|n}^\alpha | \alpha \in U_{\lambda|n}, \alpha >_n \lambda(n))$, a free subgroup of F.

This completes the definition of Λ and of U_λ for each $\lambda \in \Lambda$. We now order Λ lexicographically. That is, for $\lambda, \beta \in \Lambda$ let k be the position of the first difference between λ and β; we say that $\beta > \lambda$ if either $k = 0$ and $\beta(0) >_\Gamma \lambda(0)$, or $k \geq 1$ and $\beta(k) >_k \alpha(k)$ (observing that $\alpha_k, \beta_k \in U_{\lambda|k}$).

The order we will define on F can be simply described. It is generated by all relations of the following sort: $\{a_\lambda \gg a_\mu | \lambda, \mu \in \Lambda$, and $\lambda > \mu$, where $a_\lambda \gg a_\mu$ means $a_\lambda > a_\mu^n$ for all $n \in \mathbf{Z}$. Showing that these relations do, indeed, generate a total order of F takes the rest of this Section.

Next, we consider the effect of conjugation of a_λ by a_β where $\beta > \lambda$ in the order of Λ. Still denoting by k the position of the first difference, note that $a_\lambda^{a_\beta} = a_{\lambda(0)}^\sigma$, where

$$\begin{aligned} \sigma = \quad & \lambda(1) \cdots \lambda(k)\lambda(k+1)\lambda(k+2)\lambda(k+3) \cdots \lambda(n)a_\beta = \\ & \lambda(1) \cdots \lambda(k)\lambda(k+1)a_\beta\lambda(k+2)^{a_\beta}\lambda(k+3)^{a_\beta} \cdots \lambda(n)^{a_\beta}, \end{aligned}$$

where we have placed a_β so that the result will be a legitimate index, as we indicate below. Thus, we define $\lambda\beta =$

$$= (\lambda(0), \lambda(1), \ldots, \lambda(k), \lambda(k+1)a_\beta, \lambda(k+2)^{a_\beta}, \lambda(k+2)^{a_\beta}, \lambda(k+3)^{a_\beta}, \ldots),$$

where the superscript denotes conjugation. We will soon prove that $\lambda\beta \in \Lambda$.

We now prove that $\lambda\beta \in \Lambda$ and $a_\lambda^{a_\beta} = a_{\lambda\beta}$. Since $\lambda \in \Lambda$, we have that $\lambda(i) \in U_{\lambda|i}$ for each $1 \leq i \leq k$. Next, we have that $\lambda(k+1) \in U_{\lambda|k+1}$. In order to show that $\lambda(k+1)(a_\beta) \in U_{\lambda|k+1}$, it will suffice to show that $a_\beta \in U_{\lambda|k+1}$. We follow by induction on the lenght of β. We have $a_\beta = a_{\lambda(0)}^{\lambda(1)\ldots\lambda(k-1)\beta(k)\beta(k+1)\ldots}$. If $g \in U_{\lambda|k}$ and $g >_k \lambda(k)$, then $a_{\lambda(0)}^{\lambda(1)\ldots\lambda g} \in U_{\lambda|k+1}$.

This shows both that $a_{\lambda(0)}^{\lambda(1)\ldots\lambda(k-1)\beta(k)} \in U_{\lambda|k+1}$ and that $U_{\beta|k+1} \subseteq U_{\lambda|k+1}$. Now for any $i \geq 0$, suppose we have that $a_{\lambda(0)}^{\lambda(1)\ldots\lambda(k-1)\beta(k)\ldots\beta(k+i)} \in U_{\lambda|k+1}$ and that $U_{\beta|k+i+1} \in U_{\lambda|k+1}$. Then $\beta(k+i+1) \in U_{\beta|k+i+1} \subseteq U_{\lambda|k+1}$ and so $a_{\lambda(0)}^{\lambda(1)\ldots\lambda(k-1)\beta(k)\ldots\beta(k+i)\beta(k+i+1)} \in U_{\lambda|k+1}$. Moreover,

$$\begin{aligned} U_{\beta|k+i+2} = \operatorname{gp}\{(&a_{\lambda(0)}^{\lambda(1)\ldots\lambda(k-1)\beta(k)\ldots\beta(k+i)})^\alpha | \\ &\alpha \in U_{\beta|k+i+1}, \alpha >_{k+i+1} \beta(k+i+1)\} \end{aligned}$$

Since $a_{\lambda(0)}^{\lambda(1)\ldots\lambda(k-1)\beta(k)\ldots\beta(k+i)} \in U_{\lambda|k+1}$, then if $\gamma \in U_{\beta|k+i+1} \subseteq U_{\lambda|k+1}$, it follows that

$$(a_{\lambda(0)}^{\lambda(1)\ldots\lambda(k-1)\beta(k)\ldots\beta(k+i)})^\gamma \in U_{\lambda|k+1}$$

and so, that

$$U_{\beta|k+i+2} \subseteq U_{\lambda|k+1}.$$

Therefore, by induction,

$$a_{\lambda(0)}^{\lambda(1)\ldots\lambda(k-1)\beta(k)\ldots\beta(k+i)} \in U_{\lambda|k+1}$$

for all $i \geq 0$, and so $a_\beta \in U_{\lambda|k+1}$. Thus, we finally have that $\lambda(k+1)a_\beta \in U_{\lambda|k+1}$. Next we show that

$$\lambda(k+2)^{a_\beta} \in U_{(\alpha_0,\ldots,\alpha_k,(\alpha_{k+1})a_\beta)} = U_{(\lambda\beta)|k+2}.$$

More generally, we show that

$$(U_{\lambda|k+2})^{a_\beta} = U_{(\lambda\beta)|k+2}.$$

To see this, recall that

$$\begin{aligned}(U_{\lambda|k+2})^{a_\beta} &= \mathrm{gp}((a_{\lambda|k+1})^{\alpha}|\alpha \in U_{\lambda|k+1}, \alpha >_{k+1} \lambda(k+1))^{a_\beta} = \\ &= \mathrm{gp}((a_{\lambda|k+1})^{\alpha a_\beta}|\alpha a_\beta \in U_{\lambda|k+1}, \alpha a_\beta >_{k+1} \lambda(k+1)a_\beta) = \\ &= \mathrm{gp}((a_{\lambda|k+1})^{\alpha a_\beta} \in U_{\lambda a_\beta} \mid k+1, \alpha a_\beta >_{k+1} \lambda(k+1)a_\beta) = \\ &\qquad U_{(\lambda\beta)|k+2},\end{aligned}$$

where we used the fact proved above that $a_\beta \in U_{\lambda|k+1}$, and that the order $>_{k+1}$ is at least a right order. Since $\lambda(k+2) \in U_{\lambda|k+2}$, it follows that $\lambda(k+2)^{a_\beta} \in U_{(\lambda\beta)|k+2}$.

Next, we show that $(U_{\lambda|k+3})^{a_\beta} = U_{(\lambda\beta)|k+3}$.

$$(U_{\lambda|k+3})^{a_\beta} = \mathrm{gp}((a_{\lambda|k+2})^{\alpha}|\alpha \in U_{\lambda|k+2}, \alpha >_{k+2} \lambda(k+2))^{a_\beta} =$$

$$= \mathrm{gp}((a_{\lambda|k+2})^{\alpha a_\beta}|\alpha \in U_{\lambda|k+2}, \alpha >_{k+2} \lambda(k+2)) =$$

$$\mathrm{gp}(a_{\lambda(0)}^{\lambda(1)\ldots\lambda(k)(\lambda(k+1)a_\beta)\alpha^{a_\beta}}|\alpha^{a_\beta} \in U_{(\lambda\beta)|k+2}, \alpha^{a_\beta} >_{k+2} \lambda(k+2)^{a_\beta}) =$$

$$\mathrm{gp}((a_{\lambda\beta|k+2})^{\gamma}|\gamma \in U_{(\lambda\beta)|k+2}, \gamma >_{k+2} \lambda\beta(k+2)) =$$
$$= U_{(\lambda\beta)|k+3},$$

using our previously proved fact about conjugates of $U_{\lambda|k+2}$, and this time, using the fact that $>_{k+2}$ is a two-sided order, not just a right order. As before, since $\lambda(k+3) \in U_{\lambda|k+3}$, it follows that $\lambda(k+3)^{a_\beta} \in U_{(\lambda\beta)|k+3}$.

We can now assume that for some $i \geq 0$,

$$(U_{\lambda|k+3+i})^{a_\beta} = U_{\lambda\beta|k+3+i}.$$

Then $(U_{\lambda|k+4+i})^{a_\beta} =$

$$= \mathrm{gp}((a_{\lambda|k+3+i})^{\alpha}|\alpha \in U_{\lambda|k+3+i}, \alpha >_{k+3+i} \lambda(k+3+i))^{a_\beta} =$$

$$= \mathrm{gp}((a_{\lambda|k+3+i})^{\alpha a_\beta}, \alpha >_{k+3+i} \lambda(k+3+i)) =$$

$$= \mathrm{gp}((a_{\alpha_0})^{\alpha_1 \ldots \alpha_k (\alpha_{k+1}) a_\beta a_{k+2}^{a_\beta} \ldots \alpha_{k+2}^{a_\beta} \ldots \alpha_{k+2+i}^{a_\beta} \alpha^{a_\beta}} |$$

$$\alpha^{a^\beta} \in U_{\lambda\beta|k+3+i}, \alpha^{a_\beta} >_{k+3+i} \lambda(k+3+i)^{a_\beta}) =$$

$$= \mathrm{gp}((a_{\lambda\beta|k+3+i})^{\gamma} | \gamma >_{k+3+i} \lambda\beta(k+3+i)) = U_{\lambda\beta|k+4+i},$$

using the induction assumption and the fact that $>_{k+3+i}$ is a two-sided order. As before, it follows that

$$\alpha_{k+4+i}^{a_\beta} \in U_{\lambda\beta|k+4+i}.$$

We have now proved

LEMMA 1. *If $\beta > \lambda$ in Λ, then letting*

$$\lambda\beta = (\lambda(0), \lambda(1), \ldots, \lambda(k), \lambda(k+1)a_\beta, \lambda(k+2)^{a_\beta}, \lambda(k+3)^{a_\beta}, \ldots)$$

we have $\lambda\beta \in \Lambda, a_\lambda^{a_\beta} = a_{\lambda\beta}$, and $(U_\lambda)^{a_\beta} = U_{\lambda\beta}$. □

Next, we wish to show that each non-identity element of F has a unique expression as

$$a_{\lambda_1}^{m_1} \cdot a_{\lambda_2}^{m_2} \cdots a_{\lambda_n}^{m_n}, \text{ where } m_i \in \mathbf{Z},\ \lambda_i \in \Lambda, \text{ and } \lambda_1 > \lambda_2 > \ldots > \lambda_n.$$

Given arbitrary $\prod a_{\delta_i}^{\pm 1} = w \neq e$ in $F(\delta_i \in \Gamma)$, we proceed in steps to rearrange its factors in the manner indicated by the example at the beginning of this Section. Before we start, we observe that we have a product of the form $\prod a_{\lambda_i}^{\pm 1}$ with $\lambda_i \in \Lambda$. But $\lambda's$ may not be arranged in strictly decreasing order. We begin to do so. In the first step, we find the right generator having the smallest subscript, and move it to the far right, conjugating it by each generator it passes through; that is, whenever we find two adjacent factors $\ldots a_{\delta_i}^{\pm 1} a_{\delta_j}^{\pm 1} \ldots$ with $\delta_j >_\Gamma \delta_i$, we replace it with $\ldots a_{\delta_j}^{\pm 1} \cdot (a_{\delta_j}^{\mp 1})(a_{\delta_i}^{\pm 1})(a_{\delta_j}^{\pm 1}) \ldots$

We then find the next occurence and move it to the right, and so on. Having dealt with each generator of smallest subscript, we proceed to treat those generators of next smallest subscript in the same way, and so on. Eventually, we will have a product of factors of the form $\prod(a_{\varrho_i}^{\sigma_i})^{m_i}$ where $0 \neq m_i \in \mathbf{Z}, \varrho_i \in \Gamma, \sigma_i \in U_{(\varrho_i)}$, and $\varrho_1 \geq_\Gamma \varrho_2 \geq_\Gamma \ldots$. Each of the resulting maximal contiguous blocks on which the subscripts are equal has length (in terms of its present factors) less than the original w unless no rearranging was necessary. Again, we have the product of the form $\prod a_{\lambda_i}^{m_i}$ where $\lambda_i \in \Lambda$, and now the zeroth coordinates of $\lambda's$ are in (non-strictly) decreasing order. In the next step, independently on each of the maximal contiguous blocks with equal subscripts , we rearrange as before, this time using the superscripts σ_i and the order $>_1$. The result, for any fixed $\varrho_i = \lambda(0)$, say, is a product of the form $\prod(a_{\lambda(0)}^{\sigma_i \tau_i})^{m_i}$, where $\tau_i \in U_{(\lambda(0),\sigma_i)}$ and $\sigma_1 \geq_1 \sigma_2 \geq_1 \sigma_3 \geq_1 \ldots$. The maximal contiguous blocks with constant $\lambda(0)$ and σ_i are each shorter than the containing block with only constant $\lambda(0)$. Again, we have a product of the form $\prod a_{\lambda_i}^{m_i}$, $\lambda_i \in \Lambda$, and now both the zeroth and first coordinates of $\lambda's$ are in decreasing order. The process can be repeated until after a finite number of steps we have the desired form $\prod a_{\lambda_i}^{m_i}, \lambda \in \Lambda$, with $\lambda_1 > \lambda_2 > \ldots$ in the lexicographic order of Λ. The following lemma shows that the product is unique.

LEMMA 2. *If* $w = \prod a_{\lambda_i}^{m_i} = \prod a_{\mu_i}^{k_i}$ *with* $\mu_1 > \mu_2 > \ldots > \mu_r$, *and* $\lambda_1 > \lambda_{(2)} > \ldots \lambda_n$, *then* $n = r$, *and for each* $i, \lambda_i = \mu_i$ *and* $m_i = k_i$.

Proof. If $\lambda_n = (\lambda(0), \lambda(1), \lambda(2), \ldots)$ and $\mu_r = (\mu(0), \mu(1), \mu(2), \ldots)$, then each of $\lambda(0)$ and $\mu(0)$ is the smallest member of Γ appearing in the expression of w as a word in the free group F. Hence they are equal. Now w is uniquely expressible as $w = w'w^*$ where $w' \in U_{\lambda(0)}$ and $w^* \in \mathrm{gp}(a_{\lambda(0)}^{\alpha} | \alpha \in U_{\lambda(0)})$. Then w' is simply the word w with all factors having subscript $\lambda(0)$ removed. The factorization $w'w^*$ corresponds to factorization $\prod_{i=1}^{t} a_{\lambda_i}^{m_i} \times \prod_{i=t+1}^{n} a_{\lambda_i}^{m_i}$ and $\prod_{i=1}^{s} a_{\mu_i}^{k_i} \times \prod_{i=s+1}^{r} a_{\mu_i}^{k_i}$, respectively. Since each of the corresponding factors is shorter (as a member of its respective free group) than w, we have by induction that the expressions are unique.

LEMMA 3. *If* $\beta > \alpha > \delta$ *in* Λ, *then* $\beta > \alpha\beta > \delta\beta$.

Proof. It is clear from the definition, that $\beta > \alpha\beta$. For the other part, let the first difference of β and α occur at k, with $\beta(k) > \alpha(k)$, the first difference of α and δ at l, with $\alpha(l) > \delta(l)$, and the first difference of β and δ at m, with $\beta(m) > \delta(m)$, each with the appropriate order for the position. Then we see that $m = \min\{k, l\}$. If $m = l \leq k$, then the first difference of $\alpha\beta$ and $\delta\beta$ is $(\alpha\beta)(l) = \alpha(l) > \delta(l) = \delta(m) = (\delta\beta)(l)$. If $m = k < l$, then the first difference of $\alpha\beta$ and $\delta\beta$ is in position l, and either $l = k+1$, in which case, $(\alpha\beta)(l) = (\alpha(k+1))a_\beta > (\delta(k+1))a_\beta = (\delta\beta)(l)$, or else $l > k+1$, in which case $(\alpha\beta)(l) = (\alpha(l))^{a_\beta} > (\delta(l))^{a_\beta} = (\delta\beta)(l)$, which finishes the proof of the lemma. □

Now order F lexicographically by the forms $\prod a_{\lambda_i}^{m_i} = a_{\lambda_1}^{m_1} a_{\lambda_2}^{m_2} \cdots a_{\lambda_n}^{m_n}$. That is, the non-identity element $w = \prod a_{\lambda_i}^{m_i} = a_{\lambda_1}^{m_1} a_{\lambda_2}^{m_2} \cdots a_{\lambda_n}^{m_n}$ is positive iff m_1 is a positive integer. If w is not positive, then $m_1 < 0$ and $w^{-1} = a_{\lambda_n}^{-m_n} \cdots a_{\lambda_1}^{-m_1}$, which has the canonical form $a_{\lambda_1}^{-m_1} u$ for some u. Hence w^{-1} is positive. If $w = a_{\lambda_1}^{m_1} a_{\lambda_2}^{m_2} \cdots a_{\lambda_n}^{m_n}$ and $v = a_{\mu_1}^{p_1} a_{\mu_2}^{p_2} \cdots a_{\mu_q}^{p_q}$ are both positive, then the canonical form of wv is :

$$a_{\lambda_1}^{m_1} \cdots \text{ if } \lambda_1 > \mu_1; \; a_{\mu_1}^{p_1} \cdots \text{ if} \mu_1 > \lambda_1; \text{ or } a_{\mu_1}^{p_1+m_1} \text{ if } \lambda_1 = \mu_1.$$

In any case, since m_1 and p_1 are both positive integers, wv is positive in F.

Finally, we show that the positive elements of F are a normal set. Let $w = a_{\lambda_1}^{m_1} a_{\lambda_2}^{m_2} \cdots a_{\lambda_n}^{m_n}$ in standard form be positive, that is, $m_1 > 0$. It is enough to show that for each $\beta > \lambda_1, w^{a_\beta^{\pm 1}}$ is positive. But

$$w^{a_\beta} = a_{\lambda_1\beta}^{m_1} a_{\lambda_2\beta}^{m_2} \cdots a_{\lambda_n\beta}^{m_n},$$

which, by Lemma 3, is in the standard form, so w^{a_β} is also positive. Similar arguments show that if $m_1 < 0$, then w^{a_β} is negative. Therefore with this order, F is a totally ordered group. □

At this point, we note that with the resulting order, (F, P) has the property that if $e < x \ll y$, then either $x \ll x^y$ or $x^y \ll x$. It follows that F satisfies the identity:

$$\bigwedge |(((|x| \vee |y|)^{-\varepsilon} |[x,y]| ((|x| \vee |y|)^{\varepsilon})^{-\beta} |[x,y]| \cdot$$
$$\cdot ((|x| \vee |y|)^{-\varepsilon} |[x,y]| ((|x| \vee |y|)^{\varepsilon})^{\beta} \wedge |[x,y]|^2) |[x,y]|^{-2} = e,$$

where $\varepsilon, \beta = \pm 1$. Now the direct verification shows that this identity is violated on o-groups W^+, W^- and N_0. Since $\mathcal{A}$ is finitely based and $\mathcal{A} \subseteq \mathrm{var}_l(F, P)$, then there is an l-variety X such that $\mathcal{A} \subseteq \mathcal{X} \subseteq \mathrm{var}_l(F, P)$ and $\mathcal{X}$ covers $\mathcal{A}$ in the lattice of l-varieties L.

So we have proved the following theorem:

THEOREM 3. (Bergman [1], Kopytov [9].) There are non-solvable representable covers of $\mathcal{A}$ in the lattice of l-varieties L. □

4⁰. Here we classify covers of $\mathcal{A}$ in the lattice of l-varieties **L**. Let us recall that for each $n \geq 2$ the Scrimger group has the presentation

$$G_n = \mathrm{gp}(a_{1n}, \ldots, a_{nn}, b_n \mid b_n^{-1} a_{in} b_n = a_{jn}, j \equiv (i+1) \pmod{n}, \\ [a_{in}, a_{jn}] = e \text{ for all } 1 \leq i, j \leq n)$$

and $x = b^l a_{1n}^{k_1} \cdots a_{nn}^{k_n} \geq e$ iff $l > 0$ or $l = 0$ and $k_1 \geq 0, \ldots, k_n \geq 0$.

THEOREM 4. (Gurchenkov and Kopytov [1], D.A. Van-Ree [1].) *For each non-abelian quasivariety of l-groups $\mathcal{X}$ one of the following is true:*

(1) *$\mathcal{X}$ contains a non-abelian representable l-group;*
(2) *$\mathcal{X}$ contains the metabelian l-group* $\mathrm{wr}(\mathbf{Z}, \mathbf{Z})$*;*
(3) *$\mathcal{X}$ contains the metabelian l-group G_n for some positive integer n.*

Proof. Let G be a non-abelian l-group and $G \in \mathcal{X}$. If G is representable, then the result holds. So we can assume that G is not representable. By Theorem 9.3.1, there are $x_1, y_1 \in G$ such that $x_1 > e, y_1 > e$ and $x_1 \wedge y_1^{-1} x_1 y_1 = e$. There are two different cases.

Case 1). For each positive integer $n \in \mathbf{N}$ there are elements $x_n, y_n \in G$ such that $x_n > e, y_n > e$ and $x_n \wedge y_n^{-i} x_n y_n^i = e$ for all $i = 1, 2, \ldots, n$.

Case 2). There is a positive integer $n, n \geq 2$ such that $x \wedge y^{-n} x y^n = e$ for all $x > e, y > e, x, y \in G$ for which $x \wedge y^{-i} x y^i = e$ for all $i = 1, 2, \ldots n-1$.

Let us consider Case 1). Let $\bar{\prod}_{n \in \mathbf{N}} G_n$ be the Cartesian product of l-groups G_n where $G_n = G$ for all positive integers n. Let $\mathcal{F}$ be the filter of cofinite subsets of positive integers $\mathbf{N}$. Let H be the l-subgroup of the reduced product $\bar{\prod}_{n \in \mathbf{N}} G_n / \mathcal{F}$ generated by $\hat{x}\mathcal{F}$, $\hat{y}\mathcal{F}$ where $\hat{x}, \hat{y} \in \bar{\prod}_{n \in \mathbf{N}} G_n$ and $\hat{x}(n) = x_n$, $\hat{y}(n) = y_n$ for all $n \in \mathbf{N}$. Then the direct verification shows that $(\hat{x} \wedge \hat{y}^{-m} \hat{x} \hat{y}^m)(n) = e$ for all $n \in \mathbf{N}, n > m$. Hence $(\hat{x} \wedge \hat{y}^{-m} \hat{x} \hat{y}^m)\mathcal{F} = e\mathcal{F}$ and $\hat{y}\mathcal{F} \gg \hat{x}\mathcal{F}$ in the reduced product

$\overline{\prod\limits_{n \in \mathbf{N}}} G_n/\mathcal{F}$. Thus, H is l-isomorphic to $\mathrm{wr}^2(\mathbf{Z}, \mathbf{Z})$.

Case 2). Let $x, y \in G, x > e, y > e$ such that $x \wedge y^{-k}xy^k = e$ for all $k = 1, 2, \ldots, n$ and n is the greatest positive integer with this property. By our assumption, $n \geq 2$. Let $X_i (i \in \mathbf{Z})$ be the convex l-subgroup of G generated by the element $y^{-i}xy^i$.

LEMMA 4. $X_1 \cdot X_2 \cdots X_{n-1} \cap X_{nm} = E$ *for all* $n \in \mathbf{N}$.

Proof. We use induction on m. If $m = 0$, then the result holds by the choice of the convex l-subgroups $H_i, 0 \leq i \leq n-1$. Let us suppose that the result holds for $m-1$ and $X_{nm} \cap X_1X_2 \cdots X_{n-1} \neq E$. Then there is an element $u, u > e$ such that $y^{-n}uy^n \in X_{nm} \cap X_1X_2 \cdots X_{n-1}$. Then $u \wedge y^{-n}uy^n = e$ since $X_{n(m-1)} \cap X_1X_2 \cdots X_{n-1} = E$. We claim that $u \wedge y^{-i}uy^i = e$ for all $i = 1, 2, \ldots, n-1$. In fact, $X_0 \cap X_i = E, i = 1, 2, \ldots, n-1$, then

$$y^{-n(m-1)}X_0y^{n(m-1)} \cap y^{-n(m-1)}X_iy^{n(m-1)} = E$$

for all $i = 1, 2, \ldots, n-1$. Therefore,

$$X_{n(m-1)} \cap y^{-i}X_{n(m-1)}y^i = E, \quad i = 1, 2, \ldots, n-1.$$

Then $u \wedge y^{-i}uy^i = e$ for all $i = 1, 2, \ldots, n$, a contradiction to the choice of the integer n. □

Let $X(i), i \in \{0, 1, \ldots, n-1\}$ be the l-subgroup of G generated by all l-subgroups $X_{nm+i}, m \in \mathbf{Z}$. By Theorem 3.1.1. the l-subgroups $X(i)$ are convex and $X(i) = y^{-i}X(0)y^i, X(i) \cap X(j) = E$ for all $i, j \in \{0, 1, \ldots, n-1\}$, $i \neq j$.

LEMMA 5. *For each polar P of the convex l-subgroup $X(0)$ holds* $y^{-n}Py^n = P$.

Proof. Assume, on the contrary, that there is a polar $P \subseteq X(0), P \neq E$ such that $y^{-n}Py^n \neq P$. Then there is a convex l-subgroup A of $X(0)$such that $P = A^{\perp}$ and, by our assumption, $E \neq y^{-n}Py^n \cap A$. Thus, there is $u \in P, u > e$ for which $y^{-n}uy^n \in y^{-n}Py^n \cap A$. Then $u \wedge y^{-n}uy^n = e$. Since $y^{-i}uy^i \in X(i)$ and $X(i) \cap X(j) = E$ for all $i, j \in \{0, 1, \ldots, n-1\}, i \neq j$ it follows that $u \wedge y^{-k}uy^k = e$ for all $k = 1, 2, \ldots, n$, a contradiction the choice of n. □

LEMMA 6. *The l-subgroup $X(0)$ is representable.*

Proof. Suppose, on the contrary, that $X(0)$ is not representable. Then there are elements $a, b \in X(0), a > e, b > e$ and $a \wedge b^{-1}ab = e$. Now consider the elements $yb, a \in X(0)$. We claim that $a \wedge (yb)^{-k}a(yb)^k = e$ for $k = 1, 2, \ldots, n$. In fact,

$$(yb)^k = y^k(y^{-k+1}by^{k-1}) \cdots (y^{-1}by)b.$$

Since $y^{-k}ay^k \in X(k), y^{-i}ay^i \in X(i)(k, i \in \{1, 2, \ldots, n\})$, then

$$(yb)^{-k}a(yb)^k = y^{-k}ay^k.$$

Hence, $a \wedge (yb)^{-k}a(yb)^k = a \wedge y^{-k}ay^k$ for $k = 1, 2, \ldots, n-1$ and

$$\begin{aligned} a \wedge (yb)^{-n}a(yb)^n &= a \wedge (yb)^{-1}(yb)^{-(n-1)}a(yb)^{n-1}(yb) = \\ &= a \wedge (yb)^{-1}y^{-(n-1)}ay^{n-1}yb = a \wedge b^{-1}y^{-n}ay^n b. \end{aligned}$$

By Lemma 5,

$$(bab^{-1})^{\perp} = y^{-n}(bab^{-1})^{\perp}y^n$$

and $y^{-n}(bab^{-1})^{\perp}y^n = (y^{-n}bab^{-1}y^n)^{\perp}$. Hence,

$$a \wedge b^{-1}y^{-n}ay^n b = b^{-1}(bab^{-1} \wedge y^{-n}ay^n)b = e$$

and $a \wedge (yb)^{-k}a(yb)^k = e$ for all $k = 1, 2, \ldots, n$ and we have a contradiction to the choice of integer n. □

If $X(0)$ is non-abelian, then the result holds, if $X(0)$ is Abelian group, then the l-gp$(X(0), y)$ is a non-abelian metabelian l-group which contains as l-subgroup the l-group G_n. □

COROLLARY 1. (Kopytov and Gurchenkov [1].) *If $\mathcal{X}$ is a cover of $\mathcal{A}_l$ in the lattice of l-varieties L, then one of the following is true:*

(4) $\mathcal{X} = \mathcal{S}_p = \mathrm{var}_l(G_p)$ *for some prime p, where G_p is a Scrimger l-group;*

(5) $\mathcal{X}$ *is a non-solvable representable cover of $\mathcal{A}$;*

(6) $\mathcal{X}$ *is one of the l-varieties $\mathcal{M}^+, \mathcal{M}^-$ or $\mathcal{M}_0$.*

The proof follows from Theorem 4 and Theorem 2. □

12.7 Independent Axiomatization of l-Varieties

In the theory of varieties (quasivarieties) of algebraic systems there is a well-known connection between the independent basis of identities (implications) of variety (quasivariety) and the covers in the lattice of varieties (quasivarieties) of this variety (quasivariety). We formulate this connection for l-varieties and l-quasivarieties.

PROPOSITION 1. ([Gorbunov [1].) *Let $\mathcal{V} \subseteq \mathcal{W} \subseteq \mathcal{L}$ be l-varieties (l-quasivarieties) and $\mathcal{W}$ be finitely based. Let l-variety $\mathcal{V}$ have an infinite independent basis of identities (implications). Then there is an infinite set of l-varieties (l-quasivarieties) $\mathcal{U}_i$, $i = 1, 2, \ldots, n, \ldots$ such that each $\mathcal{U}_i$ covers $\mathcal{V}$ in the lattice of l-varieties (l-quasivarieties) and $\mathcal{U}_i \subseteq \mathcal{W}$, $i = 1, 2, \ldots, n, \ldots$.*

Proof. Let $\Phi = \{\varphi_j \mid j \in \mathbf{N}\}$ be an infinite independent basis of identities (implications) of l-variety (l-quasivariety) $\mathcal{V}$. Let

$$\Phi_k = \Phi \setminus \{\varphi_k\}, \ k \in \mathbf{N}$$

and $\mathcal{V}_k$ be l-variety (l-quasivariety) defined by the identities Φ_k, $k \in \mathbf{N}$. It is clear that $\mathcal{V}_{k_1} \cap \mathcal{V}_{k_2} = \mathcal{V}$ for all $k_1, k_2 \in \mathbf{N}$, $k_1 \neq k_2$. From the Compactness theorem (Burris and Sankappanavar [1] , page 212), it follows that there exists a finite subset $K \subseteq \mathbf{N}$ such that $\mathcal{V}_k \subseteq \mathcal{W}$ for all $k \in \mathbf{N} \setminus K$. We claim that each l-variety $\mathcal{V}_k$ ($k \in \mathbf{N} \setminus K$) contains a cover of $\mathcal{V}$. Assume, on the contrary, that $\mathcal{V}_k$ has no cover of $\mathcal{V}$. Then there exists a descending chain of l-varieties (l-quasivarieties) $\mathcal{V} \subseteq \ldots \subseteq \mathcal{X}_\alpha \subseteq \ldots \subseteq \mathcal{V}_k \subseteq \mathcal{W}$ for which $\mathcal{W} = \bigcap_{\alpha \in I} \mathcal{X}_\alpha$. By our assumption, $\mathcal{V}$ is defined in the l-variety (l-quasivariety) $\mathcal{V}_k$ by one identity (implication) φ_k. By Compactness theorem, the identity φ_k follows from the identities defining the l-variety (l-quasivariety) $\mathcal{X}_{\alpha_0}$. Thus, $\mathcal{V} = \mathcal{X}_{\alpha_0}$, contradicts our assumption .□

Let $\mathcal{V}_0$ be the l-variety defined in Section 6. Then $\mathcal{V}_0 \subseteq \mathcal{R}$ and $\mathcal{R}$ is defined by one identity $(x \wedge y^{-1}x^{-1}y) \vee e = e$.

THEOREM 1. (Medvedev [8].) *The l-variety $\mathcal{V}_0$ does not have any independent bases of identities.*

Proof. If the l-variety $\mathcal{V}_0$ has an infinite bases of identities, then in the l-variety $\mathcal{V}_0$ has infinitely many different covers contained in the l-variety $\mathcal{R}$. This is impossible by Theorem 3 of Section 6.□

Now there are a lot of l-varieties without independent basis of identities (see N.Ya.Medvedev [5], S.A.Gurchenkov [7]). Moreover, in [13] the example of an l-variety without an independent basis of universal formulae is given.

Chapter 13

Ordered permutation groups and l-varieties

13.1 l-Varieties Generated by $A(\Omega)$

Let Ω be a totally ordered set and $A(\Omega)$ be the l-group of all order permutations of Ω. One should ask which l-varieties can be generated by the l-groups $A(\Omega)$. Such l-varieties are now completely known. More precisely, $\mathrm{var}_l(A(\Omega))$ is either $\mathcal{L}$, $\mathcal{N}$ or $\mathcal{A}^n$ for some positive integer n. This remarkable result is due to W.Ch.Holland [12].

Let us recall that an *orbital* of an order permutation l-group $A(\Omega)$ is the convexification of the orbit $\alpha A(\Omega)$, $\alpha \in \Omega$. Note that if Γ is an orbital of $A(\Omega)$, then the restriction of $A(\Omega)$ to Γ is $A(\Gamma)$ and $A(\Omega)$ is the Cartesian product of its restrictions to orbitals. So we can assume that a totally ordered set Ω is the convexification of orbit $\Omega_\alpha = \alpha A(\Omega)$. Let $A_\alpha(\Omega)$ be the restriction of $A(\Omega)$ to the orbit Ω_α. For any convex $A_\alpha(\Omega)$-congruence ζ on Ω_α and arbitrary class Δ of the convex $A_\alpha(\Omega)$-congruence on Ω_α let $\mathrm{Conv}(\Delta)$ be the convex closure of Δ in the totally ordered set Ω. Thus, for each $x \in A_\alpha(\Omega)$ holds $\Delta g = \Delta$ or $\Delta g \cap \Delta = \emptyset$. Hence, $\mathrm{Conv}(\Delta)g = \mathrm{Conv}(\Delta)$ or $\mathrm{Conv}(\Delta)g \cap \mathrm{Conv}(\Delta) = \emptyset$.

LEMMA 1. *Let $A(\Omega)$ be an l-group of all order permutations of the totally ordered set Ω, $A(\Omega)$ have just one orbital, and $\alpha \in \Omega$. Let there be covering pairs*

$$(\zeta_1, \zeta^1),\ (\zeta_2, \zeta^2), \ldots, (\zeta_n, \zeta^n)$$

of the convex $A_\alpha(\Omega)$-congruences on $\Omega_\alpha = \alpha A(\Omega)$, such that $\zeta_{i+1} \supseteq \zeta^i$ $(i = 1,\dots,n-1)$ and let all primitive (ζ_i,ζ^i)-components $K_i = (\widehat{A_\alpha(\Omega)}_{(\Delta^i)}, \Delta^i/\zeta_i)$ be nontrivial for all $i = 1,2,\dots,n$. Then $A(\Omega)$ contains an l-subgroup l-isomorphic to $\mathrm{Wr}^n(\mathbf{Z},\mathbf{Z})$.

Proof. We use induction on n. Since K_1 is a non-trivial ordered permutation l-group, then there is $g \in A_\alpha(\Omega)$ such that:

(1) $\Delta^1 g = \Delta^1$ for some class Δ^1 of the convex $A_\alpha(\Omega)$ -congruence ζ^1;

(2) $\delta g \neq \delta$ for some $\delta \in \Delta_1 \subseteq \Delta^1$.

Then $\mathrm{Conv}(\Delta^1)\cdot g = \mathrm{Conv}(\Delta^1)$ and there is $g_1 > e$, $g_1 \in A(\mathrm{Conv}(\Delta^1))$. Thus, the element g_1 generates an infinite cyclic o-group (g_1) which is o-isomorphic to $\mathbf{Z} = \mathrm{Wr}^1(\mathbf{Z},\mathbf{Z})$. Now, let us assume that $A(\mathrm{Conv}(\Delta^i))$ contains an l-subgroup H which is l-isomorphic to $\mathrm{Wr}^i(\mathbf{Z},\mathbf{Z})$. We can identify $A(\mathrm{Conv}(\Delta^i))$ with the subgroup of $A(\Omega)$ fixing each point from the complement of $\mathrm{Conv}(\Delta^i)$. Let $\mathrm{Conv}(\Delta^{i+1})$ be the convex closure of class Δ^i of the convex $A_\alpha(\Omega)$-congruence ζ^{i+1}. Since the primitive $(\zeta_{i+1},\zeta^{i+1})$-component $K_{i+1} = (\widehat{A_\alpha(\Omega)}_{(\Delta^{i+1})}, \Delta^{i+1}/\zeta_{i+1})$ is non-trivial then there is a positive element $g > e$, $g \in A(\Omega)$ such that $\Delta_{i+1}g \neq \Delta_{i+1}$ and $\Delta^{i+1}g = \Delta^{i+1}$. Therefore, $(\mathrm{Conv}(\Delta_{i+1}))g \neq \mathrm{Conv}(\Delta_{i+1})$ and $(\mathrm{Conv}(\Delta_{i+1}))g \neq \mathrm{Conv}(\Delta_{i+1})$ and $(\mathrm{Conv}(\Delta^{i+1}))g = \mathrm{Conv}(\Delta^{i+1})$. Thus there is an element $g' \in A(\mathrm{Conv}(\Delta^{i+1}))$ such that g' coincides with the restriction g to $\mathrm{Conv}(\Delta^{i+1})$. Therefore, the set $\{\mathrm{Conv}(\Delta_{i+1})g^j | j \in \mathbf{Z}\}$ is a pairwise disjoint. Let $H_j = g'^{-j}Hg'^j$. Clearly $A(\mathrm{Conv}(\Delta^{i+1}))$ contains the Cartesian product $\overline{\prod_{j\in\mathbf{Z}}} H_j$, and conjugation by g permutes the indexes of this product. Therefore, the l-subgroup of $A(\mathrm{Conv}(\Delta^{i+1}))$ generated by g' and $\overline{\prod_{j\in\mathbf{Z}}} H_j$ is l-isomorphic to

$$H\mathrm{Wr}(\mathbf{Z},\mathbf{Z}) \cong \mathrm{Wr}^i(\mathbf{Z},\mathbf{Z})\mathrm{Wr}(\mathbf{Z},\mathbf{Z}) \cong \mathrm{Wr}^{i+1}(\mathbf{Z},\mathbf{Z})$$

by the results of Section 2 of Chapter 11.□

THEOREM 1. (Holland [12].) *Let $A(\Omega)$ be an l-group of all order permutations of the totally ordered set Ω. Then* $\mathrm{var}_l(A(\Omega))$ *is either $\mathcal{L}$, $\mathcal{N}$ or $\mathcal{A}^n$ for some positive integer n.*

Proof. Since the listed l-varieties form a complete tower and since

$A(\Omega)$ belongs to the l-variety $\mathcal{X}$ iff the restriction of $A(\Omega)$ to each of its orbitals belongs to $\mathcal{X}$, it suffices to prove the theorem assuming $A(\Omega)$ has just one orbital.

If one of the l-group $A_\alpha(\Omega)$ contains a non-Archimedean primitive (ζ_i, ζ^i)-component, then by Theorem 9.2.2,

$$\mathrm{var}_l(A_\alpha(\Omega)) = \mathcal{L} = \mathrm{var}_l(A(\Omega)).$$

So we can assert that all primitive (ζ_i, ζ^i)-components of the l-group $A_\alpha(\Omega)$ are Archimedean. Then by Theorem 11.2.1, $A_\alpha(\Omega) \in \mathcal{N}$ and $A(\Omega) \in \mathcal{N}$. If the number of non-trivial primitive (ζ_i, ζ^i)-components of $A_\alpha(\Omega)$ is unbounded, then, by Lemma 1, the l-group $A(\Omega)$ contains l-subgroups which are l-isomorphic to $\mathrm{Wr}^n(Z, Z)$ for all positive integers n. By Theorem 11.2.2, $\mathrm{var}_l(A_\alpha(\Omega)) \supseteq \bigvee_{n \in \mathbf{N}} \mathcal{A}^n = \mathcal{N}$.

Now suppose that there is only the finite number $n(\alpha)$ of non-trivial convex $A_\alpha(\Omega)$-congruences. Then, by Theorem 11.1.2, $A_\alpha(\Omega) \subseteq \mathrm{Wr}^{n(\alpha)+1}(\mathbf{R}, \mathbf{R})$ and $A_\alpha(\Omega) \supseteq \mathrm{Wr}^{n(\alpha)+1}(\mathbf{Z}, \mathbf{Z})$. If the set of $\{n(\alpha) : \alpha \in \Omega\}$ is unbounded, then $\{\mathrm{var}_l(A_\alpha(\Omega) : \alpha \in \Omega)\} = \mathrm{var}_l(A(\Omega)) = \mathcal{N}$. If $n_0 = \max\{n(\alpha) : \alpha \in \Omega\}$, then $\mathrm{var}_l\{A_\alpha(\Omega) : \alpha \in \Omega\} = \mathrm{var}_l(A(\Omega)) = \mathcal{A}^{n(\alpha)+1}$.

The direct verification shows that if $\Omega = \underbrace{\mathbf{Z} \overleftarrow{\times} \ldots \overleftarrow{\times} \mathbf{Z}}_{n}$, then $A(\Omega) = \mathrm{Wr}^n(\mathbf{Z}, \mathbf{Z})$ and $\mathrm{var}_l(A(\Omega)) = \mathcal{A}^n$. If $\Omega = \mathbf{R}$ where $\mathbf{R}$ is the totally ordered set of reals, then $A(\mathbf{R})$ is o-2-transitive and, by Theorem 9.2.2, $\mathrm{var}_l(A(\Omega)) = \mathcal{L}$. In the case $\mathcal{N}$, let us consider totally ordered sets

$$\Omega(1) = \mathbf{Z} \overleftarrow{\bigcup} \{1\}, \ldots, \ \Omega(n) = \underbrace{\mathbf{Z} \overleftarrow{\times} \ldots \overleftarrow{\times} \mathbf{Z}}_{n} \overleftarrow{\bigcup} \{1\} \overleftarrow{\bigcup} \cdots \overleftarrow{\bigcup} \{n\}, \ldots$$

Let $\Omega = \Omega(1) \overleftarrow{\bigcup} \Omega(2) \overleftarrow{\bigcup} \ldots \overleftarrow{\bigcup} \Omega(n) \overleftarrow{\bigcup} \ldots$. Then

$$A(\Omega) \cong \bar{\prod_{n \in \mathbf{N}}} A(\underbrace{\mathbf{Z} \overleftarrow{\times} \ldots \overleftarrow{\times} \mathbf{Z}}_{n}) \cong \bar{\prod_{n \in \mathbf{N}}} \mathrm{Wr}^n(\mathbf{Z}, \mathbf{Z}). \square$$

13.2 l-Varieties Generated by Right-Ordered Groups

1⁰. Let G be a right-ordered group and $R : G \to A(G, \leq)$ be the right regular representation of the right-ordered group $(G, \leq)$. Then $R(G)$ is a partially ordered subgroup of the l-group $A(G, \leq)$ of all order permutations of the totally ordered set $(G, \leq)$. Let G^* be the l-subgroup of $A(G, \leq)$ generated by the subgroup $(G)R$. Some properties of G^* are inherited from G. For example, if a right-ordered group G is Conradian, then G^* is a normal-valued l-group by Theorem 9.2.4.

S.V.Varaksin [1], M.Darnel and A.Glass [1] investigated the relations between the group identities of G and the l-group identities of G^*.

THEOREM 2. (Varaksin [1].) *Let $(G, \leq)$ be a right-ordered group and there be a finite subnormal series*

$$G = G_n \geq \ldots \geq G_1 \geq G_0 = e$$

of convex subgroups, all factors G_k/G_{k-1} be o-groups $(k = 1, \ldots, n-1)$ and $\mathrm{var}_l(G_k/G_{k-1}) = \mathcal{X}_k$. *Then* $\mathrm{var}_l G^* \subseteq \mathcal{X}_1 \cdot \mathcal{X}_2 \cdots \mathcal{X}_n$.

The heart of this theorem is contained in the following technical lemma.

LEMMA 2. *Let G be a right-ordered group, N be a normal convex subgroup of G such that G/N is an o-group. Then for each $g \in G$ one of the following two properties hold*

(1) *if $(N)g \neq N$, then there is an element $f \in G$ such that $(Nx)g = Nxf$ for all $x \in G$;*

(2) *if $(N)g = N$, then g can be represented in the form $g = \bigvee_{i\in I}\bigwedge_{j\in J} R(f_{ij})$, where $f_{ij} \in N$.*

Proof. We use induction on the length of the form $\bigvee_{i\in I}\bigwedge_{j\in J} R(f_{ij})$, where $f_{ij} \in G$, I and J are finite index sets.

If $g \in R(G)$ and $(N)g \neq N$, then $g \notin R(N)$ and $f = R^{-1}(g)$. If $(N)g = N$, then $g \in R(N)$ and $g = R(h)$.

Let $g = g_1 \vee g_2$, where $g_1, g_2 \in G^*$ and for g_1, g_2 the result holds. If $(N)g_1 \neq N$ or $(N)g_2 \neq N$, then $(N)g_1 = Nf_1$ or $(N)g_2 = Nf_2$ for some $f_1, f_2 \in G$. There are four different cases :

a) $(N)g_1 \neq N, (N)g_2 \neq N$ and $Nf_1 > Nf_2$;
b) $(N)g_1 \neq N, (N)g_2 \neq N$ and $Nf_1 = Nf_2$;
c) $(N)g_1 = N, (N)g_2 = N$;
d) $(N)g_1 = N, (N)g_2 \neq N$ and $Nf_2 < N$.

Case a). In this case, for each $x \in G$, $(Nx)g_1 = Nxf_1 > Nxf_2 = (Nx)g_2$. Then $(Nx)g = (Nx)g_1$ and $(Nx)g = Nxf$ for all $x \in G$ and $f = f_1$.

Case b). In this case, for $f = f_1$, $(Nx)g = \max\{Nxf_1, Nxf_2\} = Nxf_1 = Nxf_2$ for all $x \in G$.

Case c). By the inductive assumption,

$$g_1 = \bigvee_{i \in I} \bigwedge_{j \in J} R(m_{ij}), g_2 = \bigvee_{k \in K} \bigwedge_{l \in L} R(n_{kl})R$$

where $m_{ij}, n_{kl} \in N$ and the index sets I, J, K, L are finite sets of indeces. Then $g = \bigvee_{r \in I_1} \bigwedge_{s \in J_1} R(h_{rs})$.

Case d). By inductive assumption, $g_1 = \bigvee_{i \in I} \bigwedge_{j \in J} R(h_{ij})$. Thus

$$(Nx)g_2 = Nxf_2 < Nx = (Nx)g_1.$$

Then $(x)g =$

$$= \max\{(x)g_1, (x)g_2\} = \max\{(\max_i \min_j(xh_{ij}), xf_2)\} =$$
$$= \max_i \min_j\{xh_{ij}\} = \bigvee_{i \in I} \bigwedge_{j \in J} (x)R(h_{ij})$$

and $(x)g = (x)g_1$ for all $x \in G$. Hence

$$g = g_1 = \bigvee_{i \in I} \bigwedge_{j \in J} R(h_{ij}). \square$$

Now the proof of Theorem uses induction on the lenght n of the finite subnormal series of convex subgroups of G . If $n = 1$, then G is an o-group. Let us assume that for each right-ordered H with the finite subnormal series of convex subgroups of H of the length $< m$ the result holds. Let G have the finite subnormal series

$$G = G_m > G_{m-1} > \ldots > G_1 > G_0 = E$$

of convex subgroups of G. So, by our assumption, the l-group G^*_{m-1} lies in the l-variety $\mathcal{X}_1 \cdot \mathcal{X}_2 \cdots \mathcal{X}_{m-1}$.

Let D_{m-1} be an l-subgroup of $G^*_m = G^*$ generated by the elements $(g)R$ where $g \in G_{m-1}$. Since elements of D_{m-1} map each right coset $G_{m-1}g$, $g \in G$, onto itself, then D_{m-1} is embedded into the Cartesian product $\bar{\prod\limits_{g\in G}} A(G_{m-1}g, \leq)$ (Proposition 4.3.2) of l-groups of order permutations $A(G_{m-1}g, \leq)$ of the totally ordered sets $(G_{m-1}g, \leq)$. Let $\mathrm{Pr}_{A(Gm-1g,\leq)}$ be the projection of the l-group D_{m-1} into $A(G_{m-1}g, \leq)$ and $\mathrm{Pr}_{A(Gm-1g,\leq)} D_{m-1} = L_{m-1}$ be the image of D_{m-1} in $A(G_{m-1}g, \leq)$. We claim that L_{m-1} is l-isomorphic to G^*_{m-1}. Let φ be the map of L_{m-1} into G^*_{m-1} defined by the rule:

$$(xg)f = \max_{i\in I} \min_{j\in J}(ygf_{ij}),$$

where $x, f \in G_{m-1}$. Then $y((f)\varphi) = \max\limits_{i\in I} \min\limits_{j\in J}(ygf_{ij}g^{-1})$, $g \in G_{m-1}$. Since G_{m-1} is normal in G, then $(y)((f)\varphi) \in G$. It is clear that φ maps the identity element of L_{m-1} onto the identity element of G^*_{m-1}. Let $f, h \in G^*_{m-1}$ and

$$(xg)h = \max_{i\in I}, \min_{j\in J}\{xgf_{ij}\}, \quad xgh = \max_{k\in K} \min_{l\in L}\{xgh_{kl}\},$$

where $f_{ij}, h_{kl} \in G_{m-1}$. Therefore, for their product fh, $(xg)(fh) = \max\limits_{i,k} \min\limits_{j,l}\{xgf_{ij}h_{kl}\}$. Thus

$$(y)((f)\varphi(h)\varphi) = \max_{i,k} \min_{j,l}\{ygl_{ij}h_{kl}g^{-1}\} = (y)((fh)\varphi).$$

Now we claim that φ preserves the lattice operations. Indeed,

$$\begin{aligned}(y)((f \vee h)\varphi) &= \max\{\max_i \min_j(ygf_{ij}g^{-1}), \max_k \min_l(ygh_{kl}g^1)\} = \\ &= (y)((f)\varphi \vee (h)\varphi).\end{aligned}$$

Hence, φ is an l-homomorphism of L_{m-1} into G^*_{m-1}. It is clear that φ maps L_{m-1} onto G^*_{m-1} since each $h = \bigvee\limits_{i\in I} \bigwedge\limits_{j\in J} \in G^*_{m-1}$ is the image of the element $\bigvee\limits_{i\in I} \bigwedge\limits_{j\in J} R(g^{-1})R(h_{ij})R(g) \in D_{m-1}$ under φ .

Let $f \in L_{m-1}$. If $(x)((f)\varphi) = \max_I \min_J \{xgf_{ij}g^{-1}\} = x$ for all $x \in G_{m-1}$, then $(yg)f = \max_I \min_J \{ygf_{ij}\} = yg$ for all $y \in G_{m-1}$. If $(f)\varphi$ is the identity order permutation of the totally ordered set $G_{m-1}g$, then f is the identity order permutation of the totally ordered set G_{m-1}. Therefore, φ is an l-isomorphism of L_{m-1} onto G^*_{m-1} and D_{m-1} is l-isomorphic to an l-subgroup of the Cartesian product of l-isomorphic copies of G^*_{m-1}. By our assumption,

$$D_{m-1} \in \mathcal{X}_1 \cdot \mathcal{X}_2 \cdots \mathcal{X}_{m-1}.$$

By Lemma 2, G_m/D is l-isomorphic to G_m/G_{m-1}. Thus, $G^*_m \in \mathcal{X}_1 \cdot \mathcal{X}_2 \cdots \mathcal{X}_m$.□

The following examples show that there are right-ordered groups G_1, G_2 for which var$G^*_1 = \mathcal{X}_1 \cdot \mathcal{X}_2 \cdots \mathcal{X}_n$ and var$_l G^*_2 \neq \mathcal{X}_1 \cdot \mathcal{X}_2 \cdots \mathcal{X}_n$.

EXAMPLE 1. Let $N_2 = \mathrm{gp}(a, b, c | [a, b] = c, [a, c] = [b, c] = e)$. Let N_2 be right-ordered by the rule:

$$g = b^k c^l a^t \geq e \text{ in } N_2$$

if and only if $k > 0$ or $k = 0$ and $l = 0$ or $k = l = 0$ and $t \geq 0$. Then the series $E < (a) < (a, c) < N_2$ is a subnormal series of convex subgroups of N_2 and all factors of this series are infinite cyclic o-groups. The interested reader with the use of the results of Ghapter 4 can without difficulty prove that var$_l N^*_2 = \mathcal{A}^2$. □

EXAMPLE 2. Let $G_n = \mathrm{gp}(a_{1n}, a_{2n}, \ldots, a_{nn}, b_n | b_n^{-1} a_{in} b_n = a_{jn}, j \equiv (i+1) \pmod n, [a_{in}, a_{jm}] = e$ for all $1 \leq i, j \leq n)$ be a Scrimger group. Now define the right order on G_n by the rule:

$$x = b_n^l a_{1n}^{k_1} \ldots a_{nn}^{k_n} \geq e$$

iff $l > 0$ or $l = 0$ and $k_1 > 0$ or $l = k_1 = 0$ and $k_2 > 0$ or, ..., or $l = k_1 = \ldots = k_{n-1} = 0$ and $k_n \geq 0$. Then the direct verification shows that var$_l G^*_n \neq \mathcal{A}^2$. □

13.3 l-Varieties Generated by Simple l-Groups

In this section we will consider l-varieties which are generated by simple l-groups.

LEMMA 1. *Let G be a simple non-representable l-group. Then for each positive integer n and each non-trivial convex l-subgroup P of G there is a convex l-subgroup $N \leq P$, $N \neq E$ and $x \in P, x > e$ such that $x^{-i}Nx^{i} \cap N = E$ for all $i \in \{1, 2, \ldots, n\}$.*

Proof. Since G is not representable and G is simple, then the representable radical $R(G) = E$. Therefore, the representable radical of each convex l-subgroup P is trivial, too. Then there are $y, x_1 > e, y, x_1 \in P$ such that $y \wedge x_1^{-1}yx_1 = e$. Let N_1 be the convex l-subgroup of P generated by y. It is clear that $N_1 \wedge x_1^{-1}N_1x_1 = E$.

Let us assume that there are the convex l-subgroup N_k and the element $x_k > e$ such that $x_k^{-i}N_kx_k^i \cap N_k = E$ for all $i \in \{1, 2, \ldots, k\}$. Let $L = x_k^{-k-1}N_kxk + 1^k \cap N_k$. If $L = E$, then we set $N_{k+1} = N_k$ and $x_{k+1} = x_k$. Let $L \neq E$, then there is a convex l-subgroup M_0 and the element $r > e$ such that $r^{-1}M_0r \cap M_0 = E$. Let $M = x_k^{-k-1}M_0x^{k+1} \cap M_0$ and two different cases: a) $M = E$; b) $M \neq E$ arise.

In case a) we have $x_k^{-i}M_0x_k^i = E$ for all $i \in \{1, 2, \ldots, k+1\}$. Then we set $N_{k+1} = M_0$ and $x_{k+1} = x_k$.

Case b). Let $K = x_k^{k+1}Mx_k^{-k-1}$. Then $x_k^{-i}Kx_k^i \subseteq x_k^{-i}M_0x_k^i$. The element r is orthogonal to the l-subgroups $x_k^{-i}M_0x_k^i$ for all $i \in \{1, 2, \ldots, k\}$. Therefore, $(x_kr)^{-i}K(x_kr)^i = x_k^{-i}Kx_k$ if $i \in \{1, 2, \ldots, k\}$ and

$$(x_kr)^{-k-1}K(x_kr)^{k+1} = r^{-1}x_k^{-k-1}Kx_k^{k+1}r \subseteq$$
$$\subseteq r^{-1}x_k^{-k-1}x_k^{k+1}Mx_k^{-k-1}x_k^{k+1}r \subseteq r^{-1}M_0r.$$

As $r^{-1}M_0r \cap M_0 = E$, then $(x_kr)^{-k-1}K(x_kr)^{k+1} \cap K = E$ and

$$(x_kr)^{-i}K(x_kr)^i \cap K = E \text{ for all } i \in \{1, 2, \ldots, k+1\}.$$

In this case we can set $N_{k+1} = K$ and $x_{k+1} = x_kr$. □

COROLLARY 1. *Let G be a simple non-representable l-group. Then for each positive integers m, n there are elements $g_1, \ldots, g_m \in G$ such that:*

a) $g_m \gg g_{m-1} \gg \ldots \gg g_1 > e$;
b) $g_i^{-k} g_j g_i^k \wedge g_j = e$ *for all* $1 \le j < i \le m, \quad -n \le k \le n$.

LEMMA 2. *Let G be a simple non-representable l-group. Then for each* $m \in \mathbf{N}$ *the l-variety* $\mathrm{var}_l(G)$ *contains the l-variety* $\mathcal{A}^m$.

Proof. Let $G_i = G$ for all $i \in \mathbf{N}$. Now let us consider an l-subgroup of the quotient l-group

$$\bar{\prod_{t\in\mathbf{N}}} G_t / \prod_{t\in\mathbf{N}} G_t$$

generated by the images $g_1, g_2, \ldots, g_m$ of $\hat{g}_1, \hat{g}_2, \ldots, \hat{g}_m \in \bar{\prod}_{t\in\mathbf{N}} G_t$, where $\hat{g}_l(t) = g_{i,t} \in G_t$ and

$$g_{m,t} \gg g_{m-1,t} \gg \ldots \gg g_{1,t} > e,$$

$g_{i,t}^{-k} g_{j,t} g_{i,t} \wedge g_{j,t} = e$, for $k, \quad -n \le k \le n$. The arguments similar to the proof of Theorem 12.6.1 show that l-gp$(g_1, g_2, \ldots g_m) \cong \mathrm{wr}^m(\mathbf{Z}, \mathbf{Z})$. □

THEOREM 1. (Varaksin [2].) *Let G be a simple non-representable l-group. Then the l-variety* $\mathrm{var}_l(G)$ *generated by G is* $\mathcal{N}$ *or* $\mathcal{L}$.

Proof. By Lemma 2,

$$\mathrm{var}_l(G) \supseteq \mathcal{A}^m$$

for all $m \in N$. Then, by Theorem 11.2.2, $\mathrm{var}_l(G) = \mathcal{N}$ or $\mathrm{var}_l(G) = \mathcal{L}$.□

COROLLARY 1.*A non-abelian solvable l-group G is not a simple l-group.*

Proof. If G is a solvable representable l-group, then the result holds by Theorem 6.4.1. If G is non-representable, then the result holds by Theorem 1. □

In [2] S.V.Varaksin using the approach of P.Hall (see the book of Kokorin and Kopytov [2]) constructed the examples of simple normal-valued l-groups which generate the l-variety of all normal-valued l-groups $\mathcal{N}$. □

Chapter 14

Quasivarieties of lattice-ordered groups

14.1 l-Quasivarieties

1^0. An implication of the signature $l = \{\cdot\ ,\ ^{-1},\ e\ , \vee, \wedge\}$ is a formula φ of the predicate calculus of the form

$$(\forall x_1)\dots(\forall x_n)(w_1(x_1,\dots,x_n) = e\&\dots\&w_k(x_1,\dots,x_n) = e \Rightarrow \\ \Rightarrow w_{k+1}(x_1,\dots,x_n) = e),$$

where $w_1(x_1,\dots,x_n),\dots,w_{k+1}(x_1,\dots,x_n)$ are l-group words. An l-group G satisfies the implication φ if whenever

$$w_1(g_1,\dots,g_n) = e,\dots,w_k(g_1,\dots,g_n) = e,$$

then $w_{k+1}(g_1,\dots,g_n) = e$

A *quasivariety* of l-groups (or, simply, l-quasivariety) is a class X of l-groups which satisfy a given set Φ of implications of the signature $l = \{\cdot,^{-1},e,\vee,\wedge\}$. Set Φ is called a basis of the implications of the l-quasivariety $\mathcal{X}$. From standard results in universal algebra (Burris and Sankappanavar [1], Theorem 2.25.) we have the following characterization of l-quasivarieties:

THEOREM 1. *Let $\mathcal{X}$ be a class of l-groups. Then the following statements are equivalent:*

(1) *$\mathcal{X}$ is a quasivariety of l-groups,*
(2) *$\mathcal{X}$ is closed under l-isomorphisms, l-subgroups and reduced products, and contains a trivial l-group,*
(3) *$\mathcal{X}$ is closed under l-isomorphisms, Cartesian products and ultraproducts, and contains a trivial l-group.*□

It is clear that the intersection of l-quasivarieties is an l-quasivariety. Hence, given any class $\mathcal{K}$ of l-groups, there is the least l-quasivariety that contains this class $\mathcal{K}$. This l-quasivariety is called the l-quasivariety generated by $\mathcal{K}$ and denoted by $q_l(\mathcal{K})$.

For any class $\mathcal{K}$ of l-groups let $P(\mathcal{K})$, $S(\mathcal{K})$ and $P_U(\mathcal{K})$ denote the classes of l-groups which are Cartesian products, l-subgroups and ultraproducts, respectively, of elements of $\mathcal{K}$.

From Theorem 1 the following usefull observation follows

THEOREM 2. *If $\mathcal{K}$ is any class of l-groups, then the l-quasivariety $q_l(\mathcal{K})$ generated by $\mathcal{K}$ is $SPP_U(\mathcal{K})$.* □

The following result can be proved by direct verification.

PROPOSITION 1. *The conjunction of a finite number of implications of the signature $l = \{\, \cdot \,, \,^{-1} \,, \, e \,, \vee, \wedge\}$ is equivalent to a simple implication in the class of l-groups $\mathcal{L}$.* □

COROLLARY 1.*If an l-quasivariety $\mathcal{X}$ has a finite basis of implications, then $\mathcal{X}$ can be defined by one implication.* □

It is easy to see that a lot of properties of l-groups which were proved in Section 2 of Chapter 2 can be formulated as implications of the signature $l = \{\cdot \,, \,^{-1} \,, \, e \,, \vee, \wedge\}$. Each l-variety is an l-quasivariety, but there are l-quasivarieties which are not l-varieties. One such l-quasivariety is the class of all orderable l-groups. For the proof of this fact we observe the following. Let D_l^0 be the l-group of all bounded piecewise linear order permutations of the real numbers R (Section 7 of Chapter 4).

Then (D_l^0, R) is o-2-transitive and orderable. Thus (D_l^0, R) generates the l-variety $\mathcal{L}$ of all l-groups, this would imply that all l-groups are orderable. But it is not true since, for example, the l-group $A(R)$ of all order permutations of real numbers is not orderable. So the set of all l-quasivarieties Λ properly contais the set of all l-varieties $\mathcal{L}$. The set of all l-quasivarieties Λ is a lattice where, for any l-varieties $\mathcal{X}, \mathcal{Y}$, the greatest lower bound, or meet, of $\mathcal{X}$ and $\mathcal{Y}$ is their set-theoretical intersection $\mathcal{X} \overset{\Lambda}{\bigwedge} \mathcal{Y} = \mathcal{X} \cap \mathcal{Y}$ and their least upper bound, or join, is the intersection of all l-quasivarieties containing both $\mathcal{X}$ and $\mathcal{Y}$:

$$\mathcal{X} \overset{\Lambda}{\bigvee} \mathcal{Y} = \bigcap \{\mathcal{W} : X \cup \mathcal{Y} \subseteq \mathcal{W}\}.$$

It follows from these definitions that the lattice of all l-varieties L is a sublattice of the lattice of all l-quasivarieties Λ.

The theory of l-quasivarieties is more rich and complicated than that of l-varieties.

2⁰. It follows from Theorems 1 and 2 that each l-quasivariety $\mathcal{X}$ has a free l-group $F_{\mathcal{X}}$. This free l-group $F_{\mathcal{X}}$ is coinsided with the free l-groups $F_{\text{var}_l(\mathcal{X})}$ of the l-variety $\text{var}_l \mathcal{X}$ generated by l-quasivariety $\mathcal{X}$. Thus the descriptions of free l-groups of l-quasivarieties were done in Chapter 10.

14.2 The Lattice of l-Quasivarieties

1⁰. The following result establishes the difference between the lattice of l-varieties **L** and the lattice of l-quasivarieties Λ.

THEOREM 1. (Gurchenkov [private communication].) *The lattice of l-quasivarieties Λ is not modular and therefore it is not distributive.*

Proof. Consider the group

$$G = \text{gp}(a, b \mid [a^{b^j},\ a^{b^i}] = [a, b, b, b] = e,\ (i, j \in \mathbf{Z})).$$

It is clear that G is metabelian and torsion-free nilpotent of the class ≤ 3. Now we define on G two total orders. The total order P_1 is uniquely defined by the inequalities $b \gg a \gg [a, b] \gg [a, b, b] > e$. Let G_1 be a group G totally ordered by P_1. The total order P_2 is uniquely defined by the inequalities: $b \gg a \gg [a, b] \gg [a, b, b]^{-1} > e$. The group G totally ordered by P_2 is denoted by G_2. Let G_3 be the l-subgroup of the direct product $G_1 \times G_2$ of o-groups G_1 and G_2 generated by the elements $\bar{b} = (b, b)$, $\bar{a} = (a, a)$. Now consider the following implications (we omit quantifiers):

$$t_1 : \quad [x, y] \geq e \Rightarrow [x, y, z] \geq e;$$

$$t_2 : \quad [x, y] \geq e \Rightarrow [x, y, y] \leq e;$$

$$t_4 : \quad [x, y] \geq e \Rightarrow [x, y, y] = e;$$

$$t_{3,1} : \quad [x, y, y] \geq e \Rightarrow [x, y, y] = e;$$

$$t_{3,2} : \quad [x, y, y] \leq e \Rightarrow [x, y, y] = e,$$

where $x = |z| \wedge |w|$, $y = |z| \vee |w|$.

The direct verification shows that the implications t_1, t_2 hold in G_1, G_2 correspondingly and $t_{3,1}, t_{3,2}$ hold in G_3. Let $\mathcal{M}$ be an l-quasiva riety of nilpotent l-groups of the class ≤ 3 defined by the implication t_4. Let

$$\mathcal{M}_1 = q_l(G_1) \overset{\wedge}{\bigvee} \mathcal{M}, \mathcal{M}_2 = q_l(G_2) \overset{\wedge}{\bigvee} \mathcal{M}, \mathcal{M}_3 = q_l(G_1) \overset{\wedge}{\bigvee} \mathcal{M} \overset{\wedge}{\bigvee} q_l(G_3).$$

It is clear that $\mathcal{M}_1, \mathcal{M}_2, \mathcal{M}_3$ are different. The direct verification shows that $\mathcal{M}_2 \neq \mathcal{M}_1 \vee \mathcal{M}_2 = \mathcal{M}_2 \vee \mathcal{M}_3 \neq \mathcal{M}_3$ and $\mathcal{M}_1 \vee \mathcal{M}_2 = \mathcal{M}$. We claim that $\mathcal{M}_3 \wedge \mathcal{M}_2 = \mathcal{M}$. Let $H \in \mathcal{M}_3 \wedge \mathcal{M}_2$ and $z, w \in H$ such that $[x, y] \geq e$. As $H \in \mathcal{M}_2$ then $[x, y, y] \leq e$. Since $H \in \mathcal{M}_3$ then H is a subdirect product of l-groups $H_1 \in M, H_2 \in q_l(G_1), H_3 \in q_l(G_3)$. Then

$$[x, y, y] = [x_1, y_1, y_1] \cdot [x_2, y_2, y_2] \cdot [x_3, y_3, y_3],$$

where $[x_i, y_i, y_i] \in H_i$ $(i = 1, 2, 3)$. Since $[x, y] \geq e$ and $[x, y, y] \leq e$ in h, then $[x_i, y_i] \geq e, [x_i, y_i, y_i] \leq e$ in H_i. Since the implications $t_{3,1}, t_{3,2}$ are valid in H_3, the inequalities are valid in H_3, and the inequalities $[x_3, y_3] \geq e$, $[x_3, y_3, y_3] \leq e$ imply $[x_3, y_3, y_3] = e$. Since the implication t_1 holds in H_2, then the inequality $[x_2, y_2] \geq e$ implies $[x_2, y_2, y_2] = e$. Finally, the validity of the implication t_4 and the inequality $[x_1, y_1] \geq e$ implies $[x_1, y_1, y_1] = e$. Hence the sublattice of the lattice of l-quasivarieties Λ generated by l-quasivarieties $\mathcal{M}_1, \mathcal{M}_2, \mathcal{M}_3$ is 5-elements non-modular lattice. □

It is also known that the sublattice of all l-quasivarieties which can be defined by implications of the signature of the group theory $\{\ \cdot\ ,\ ^{-1}\ ,\ e\ \}$ is not modular (N.Ya. Medvedev [10]).

2⁰. In Chapter 9 it was proved that the l-variety $\mathcal{N}$ of normal-valued l-groups is the greatest element in the lattice of l-varieties **L**. In the lattice of l-quasivarieties Λ this fact is not valid.

For each positive integer n, let $\mathcal{R}_n$ be the l-quasivariety defined by the implication $(x^n = y^n) \Rightarrow (x = y)$. Let $\mathcal{R}_*$ denote the l-quasivariety of all l-groups which admit unique extraction of all roots, i.e.,

$$\mathcal{R}_* = \bigcap_{n=2}^{\infty} \mathcal{R}_n.$$

It is clear that the l-quasivariety $\mathcal{O}$ of all orderable l-groups is contained in $\mathcal{R}_*$. Let us recall that G_n be a Scrimger l-group (see Section 6 of Chapter 12).

LEMMA 1. (Arora [1].) *$G_n \in \mathcal{R}_m$ if and only if $(n,m) = 1$ where (n,m) denotes the greatest common divisor of n and m .*

Proof. Assume $(n,m) \neq 1$. Then $n = \bar{n}d$ and $m = \bar{m}d$, where $\bar{n}, \bar{m}, d \in N$ and $(\bar{n}, \bar{m}) = 1$ and $d \neq 1$. Let $x = b_n a_{1n}^{k_1} \cdots a_{nn}^{k_n}, y = b_n a_{1n}^{l_1} \cdots a_{nn}^{l_n}$, be two elements of G_n defined as follows:

$$k_i \equiv i \pmod d,\ 0 \le k_i < d, \quad l_i \equiv (d-i-1) \pmod d,\ 0 \le l_i < d.$$

Then $x^d = b_n^d a_{1n}^{t_1} \cdots a_{nn}^{t_n}$ where

$$t_1 = t_2 = \cdots = t_n = 1 + 2 + \cdots + (d-1) = \frac{d(d-1)}{2},$$

$y^d = b_n^d a_{1n}^{u_1} \cdots a_{nn}^{u_n}$, where $u_1 = u_2 = \cdots = u_n = (d-2) + (d-3) + \cdots + 0 + (d-1)$. Hence $x^d = y^d$ and $x \neq y$. Thus $G_n \notin \mathcal{R}_m$.

Conversely, assume $(n,m) = 1$. Let $x, y \in G_n$ be such that $x^m = y^m$. Since $G \in \mathcal{L}_n$ we have $x^n y^n = y^n x^n$. As n and m are relatively prime, there exist integers r and s such that $nr + ms = 1$. Therefore,

$$xy^{-1} = x^{nr+ms} y^{-ms-nr} = x^{nr} x^{ms} y^{-ms} y^{-nr} =$$
$$x^{nr} y^{-nr} = y^{-nr} x^{nr} = y^{-nr} y^{-ms} x^{ms} x^{nr} = y^{-1} x.$$

Thus, $xy = yx$. Then $(xy^{-1})^m = e$. Since every l-group is torsion-free we have $xy^{-1} = e$ or $x = e$.□

COROLLARY 1. *If p and q are distinct prime numbers, then $G_p \in \mathcal{R}_q \setminus \mathcal{R}_p$ and $G_q \in \mathcal{R}_p \setminus \mathcal{R}_q$. Hence, $\mathcal{R}_p$ and $\mathcal{R}_q$ are incomparable l-quasivarieties if p and q are distinct prime numbers.*□

Let n be a positive integer ≥ 2 and $n = p_1^{n_1} \cdots p_k^{n_k}$ where $p_1, \cdots p_k$ are distinct prime divisors of n and $n_1, \cdots, n_k > 0$. Then an easy induction on k shows that $\mathcal{R}_n = \mathcal{R}_{p_1} \cap \cdots \cap \mathcal{R}_{p_k}$. By above arguments, we conclude

COROLLARY 2. *Let m and n be positive integers. Then $\mathcal{R}_n \supseteq \mathcal{R}_m$ if and only if every prime divisor of m is a prime divisor of n .*□

Let P be the set of positive prime integers and let $P' \subseteq P$ be any non-empty subset of P. We define a quasivariety (for each P') as follows: $R_{P'} = \bigcap_{p \in P'} \mathcal{R}_p$. Let P_1 and P_2 be two incomparable subsets of P; that is, $P_1 \not\supseteq P_2$ and $P_2 \not\supseteq P_1$. Then $\mathcal{R}_{P_1}$ and $\mathcal{R}_{P_2}$ are two incomparable l-quasivarieties. In fact, for $p \in P_1 \setminus P_2$ and $q \in P_2 \setminus P_1$ is valid $G_p \in \mathcal{R}_{P_2} \setminus \mathcal{R}_{P_1}$, $G_q \in \mathcal{R}_{P_1} \setminus \mathcal{R}_{P_2}$.

Since there is a continuum of pairwise incomparable subsets of P, then we have exhibited a collection of the continuum of pairwise incomparable l-quasivarieties.

Let D_l^0 be the l-group of all bounded piecewise linear order permutations of the real numbers R (see Section 7 of Chapter 4). Let $H = D_l^0 \mathrm{Wr}(Z, Z)$. For each $p \in P$, define H_p as follows:

$$H_p = \mathrm{gp}((\{g_n\}, m) \in H \mid g_i = g_j \text{ if } i \equiv j \pmod{p}).$$

Since D_l^0 is o-2-transitive, then D_l^0 is not normal-valued. Thus H_p is not normal-valued, too, for all $p \in P$.

LEMMA 2. (Arora [1].) *$H_p \in \mathcal{R}_q \setminus \mathcal{R}_p$ if p and q are different prime numbers.*

Proof. 1) Let $f \in D_l^0$ be such that $f \neq e$ in D_l^0. Let $x = (\{g_n\}, 1)$ and $y = (\{h_n, 1)$ be two elements of H_p defined by the rules: $g_n = f^{\bar{n}}$, $h_n = f^{(\overline{p-1-n})}$ where $\bar{n} \equiv n \pmod{p}$, $0 \leq \bar{n} < p$ and $\overline{p-1-n} \equiv (p-1-n) \pmod{p}$, $0 \leq \overline{p-1-n} < p$. Then $x^p = y^p$, $x \neq y$ and therefore $H_p \notin \mathcal{R}_p$.

2) Let $x = (\{g_n\}, k)$ and $y = (\{f_n\}, k')$ be two elements of H_p such that $x^q = y^q$. Then $k = k'$. If $k \equiv 0 \pmod p$, then for each $n \in \mathbf{Z}$, $g_n^q = f_n^q$ and since D_l^0 is an orderable group then, $g_n = f_n$ for every $n \in \mathbf{Z}$ and, consequently, $x = y$. Now the following cases arise: 2.1) $k \equiv 1 \pmod p$, 2.2) $k \equiv 2 \pmod p, \cdots$ 2.p-1) $k \equiv (p-1) \pmod p$.

Case 2.1) $k \equiv 1 \pmod p$. Let us further assume that $q < p$. Since $x^q = y^q$, we obtain p equations, each involving q symbols as follows:

A_0 $$g_0 g_1 \cdots g_{q-1} = f_0 f_1 \cdots f_{q-1},$$

A_1 $$g_1 g_2 \cdots g_{q-1} g_q = f_1 f_2 \cdots f_{q-1} f_q,$$

$$\cdots$$

A_{p-1} $$g_{p-1} g_0 \cdots g_{q-2} = f_{p-1} f_{p-2} \cdots f_{q-2}.$$

Multiplying equations A_i, $A_{(q+i) \pmod p}$, $\cdots$, $A_{[(p-1)q+1] \pmod p}$ in this order for $i = 0, 1, 2, \cdots, p-1$, we get a set of p equations as follows:

$A_0^1)$ $$(g_0 g_1 \cdots g_q \cdots g_{p-1})^q = (f_0 f_1 \cdots f_q \cdots f_{p-1})^q,$$

$A_1^1)$ $$(g_1 g_2 \cdots g_{p-1} g_0^q) = (f_1 f_2 \cdots f_{p-1} f_0)^q,$$

$$\cdots$$

$A_{p-1}^1)$ $$(g_{p-1} g_0 \cdots g_{p-2})^q = (f_{p-1} f_0 \cdots f_{p-2})^q.$$

Since $f_i, g_i \in D_l^0$ $(i = 0, 1, \cdots, p-1)$ and $D_l^0 \in \mathcal{O} \subseteq \mathcal{R}_q$, the following set of p equations can be obtained from $A_0^1), A_1^1), \cdots, A_{p-1}^1$:

$A_0^2)$ $$g_0 g_1 \cdots g_{p-1} = f_0 f_1 \cdots f_{p-1},$$

$A_1^2)$ $$g_1 \cdots g_{p-1} g_0 = f_1 \cdots f_{p-1} f_0,$$

$$\cdots$$

$A^2_{p-1})$ $$g_{p-1}g_0 \cdots g_{p-2} = f_{p-1}f_0 \cdots f_{p-2}.$$

Using equations $A_0), \cdots, A_{p-1}$ with equations $A^2_0), \cdots, A^2_{p-1})$ several times (if necessary), we obtain the following set of p equations each involving q_1 symbols where q_1 is the remainder when p is divided by q, so $0 < q_1 < q$.

$A^3_0)$ $$g_0g_1 \cdots g_{q_1-1} = f_0f_1 \cdots f_{q_1-1},$$

$A^3_1)$ $$g_1g_2 \cdots g_{q_1} = f_1f_2 \cdots f_{q_1},$$

$$\cdots$$

$A^3_{p-1})$ $$g_{p-1}g_0 \cdots g_{q_1-2} = f_{p-1}f_0 \cdots f_{q_1-2}.$$

Again, using equations $A^3_0), \cdots, A^3_{p-1})$ and $A^2_0), \cdots, A^2_{p-1})$ we obtain a set of p equations, namely $A^4_0), \cdots, A^4_{p-1}$, each involving q_2 symbols, where q_2 is the remainder when p is divided by $q_1, (0 < q_2 < q_1 < q)$.

Thus each time we proceed with our algorithm, we get a set of p equations each involving fewer symbols as compared to the number of symbols obtained in the previous set of equations. So, finally, after a finite number of steps we obtain a set of p equations, as follows

$A^k_0)$ $$g_0 = f_0,$$

$A^k_1)$ $$g_1 = f_1,$$

$$\cdots$$

$A^k_{p-1})$ $$g_{p-1} = f_{p-1}.$$

Thus $f_i = g_i$ for all $i = 0, 1, \cdots, p-1$. Since $x, y \in H_p$, we have $x = y$.

Now let us assume $q > p$. Since $x^q = y^q$, we have the following set of p equations each involving q symbols:

$B_0)$ $$g_0 g_1 \cdots g_{p-1} \cdots g_{q-p-1} = f_0 f_1 \cdots f_{p-1} \cdots f_{q-p-1},$$

$B_1)$ $$g_1 g_2 \cdots g_{q-p} = f_1 f_2 \cdots f_{q-p},$$

$$\cdots$$

$B_{p-1}),$ $$g_{p-1} g_0 \cdots g_{q-p-2} = f_{p-1} f_0 \cdots f_{q-p-2},$$

where all the indeces are non-negative and are residues modulo p. Multiplying equations $B_{\overline{(0+i)}}, B_{\overline{(q+i)}}, \cdots, B_{\overline{((p-1)q+1)}}$ (for $i = 0, 1, 2, \cdots, p-1$) in this order, we obtain the set of p equations

$B_0^1)$ $$(g_0 g_1 \cdots g_{p-1})^q = (f_0 f_1 \cdots f_{p-1})^q,$$

$B_1^1)$ $$(g_1 g_2 \cdots g_{p-1} g_0)^q = (f_1 f_2 \cdots f_{p-1} f_0)^q,$$

$$\cdots$$

$B_{p-1}^1)$ $$(g_{p-1} g_0 \cdots g_{p-2})^q = (f_{p-1} f_0 \cdots f_{p-2})^q.$$

Since $D_l^0 \in \mathcal{R}_q$ and f_i, g_i $(i \in \{0, 1, \cdots, p-1\}) \in D_l^0$, we obtain the following set of p equations from the set $B_0^1), B_1^1), \cdots, B_{p-1}^1$.

$B_0^2)$ $$g_0 g_1 \cdots g_{p-1} = f_0 f_1 \cdots f_{p-1},$$

$B_1^2)$ $$g_1 g_2 \cdots g_{p-1} g_0 = f_1 f_2 \cdots f_{p-1} f_0,$$

$$\cdots$$

$B_{p-1}^2)$ $$g_{p-1} g_0 \cdots g_{p-2} = f_{p-1} f_0 \cdots f_{p-2}.$$

Using equations $B_0^2), \cdots, B_{p-1}^2)$ and equations $B_0), B_1), \ldots, B_{p-1})$ we obtain the following set of p equations each involving q_1 symbols where q_1 is the remainder when q is divided by p, so $0 < q_1 < p$.

$B_0^3)$ $$g_0 g_1 \cdots g_{q_1-1} = f_0 f_1 \cdots f_{q_1-1},$$

$B_1^3)$ $$g_1 g_2 \cdots g_{q_1-1} g_0 = f_1 f_2 \cdots f_{q_1-1} f_0,$$

$$\cdots$$

$B_{p-1}^3)$ $$g_{p-1} g_0 \cdots g_{q_1-2} = f_{p-1} f_0 \cdots f_{q_1-2}.$$

With the set $B_0), \cdots, B_{p-1})$ of equations and the set $B_0^3), \cdots, B_{p-1}^3)$ we can proceed in exactly the same way we did in case 2.1, and conclude that $x = y$.

Cases 2.2), 2.3), ..., 2.p-1) can be dealt with similarly by appropriately using f_i and $g_i (i = 0, 1, \cdots, p-1)$ in each case. Hence $H_p \in \mathcal{R}_q$ whenever p and q are distinct primes numbers.□

LEMMA 3. *Let $\{H_{ij} \mid i \in I,\ 1 \leq j \leq n < \infty\}$ be a collection of l-groups. Let $\mathcal{D}$ be an ultrafilter over I. Then*

$$\bar{\prod_{i \in I}}(H_{i1} \times H_{i2} \times \ldots \times Hin)/\mathcal{D} \cong \bar{\prod_{i \in I}} H_{i1}/\mathcal{D} \times \bar{\prod_{i \in I}} H_{i2}/\mathcal{D} \times \ldots \times \bar{\prod_{i \in I}} H_{in}/\mathcal{D}$$

Proof. Let $g = (g_i) \in \bar{\prod_{i \in I}}(H_{i1} \times \ldots \times Hin)$, and $g_i = (g_{i1}, \ldots, g_{in})$ for all $i \in I$. Define $\psi : \bar{\prod_{i \in I}}(H_{i1} \times \ldots \times H_{in}) \to \bar{\prod_{i \in I}} H_{i1}/\mathcal{D} \times \ldots \times \bar{\prod_{i \in I}} /\mathcal{D}$ by the rule $(g\mathcal{D})\psi = (\bar{g}_1, \ldots, \bar{g}_n)$ where $\bar{g}_1(i) = g_{i1}, \ldots, \bar{g}_n(i) = g_{ni}$ for all $i \in I$. Clearly ψ is an l-homomorphism and is onto. Moreover, $g \in \ker\psi$ iff $\bar{g}_1 = \ldots \bar{g}_n = e$. These equalities are equivalent $\{i \in I \mid g_j(i) = e\} \in \mathcal{D}$ for all $j = 1, 2, \ldots, n$ and $\bigcap_{j=1}^{n} \{i \in I \mid g_j(i) = e\}$. The last inclusion is equivalent to $g\mathcal{D} = e$ in $\bar{\prod_{i \in I}}(H_{i1} \times \ldots \times H_{in})/\mathcal{D}$.□

LEMMA 4. *Let $\mathcal{X}_j (j \in J)$ be a finite collection of l-quasivarieties and let G be an l-group. Then $G \in \bigvee_{j \in J} \mathcal{X}_j$ if and only if there exist $M_j \in \mathcal{X}_j$ $(j \in J)$ such that G can be l-embedded in $\prod_{j \in J} M_j$.*

Proof. If G is l-embedded into $\prod_{j \in J} M_j$ with $M_j \in \mathcal{X}_j$, then $G \in \overset{\wedge}{\bigvee_{j \in J}} \mathcal{X}_j$.

Conversely, let $\mathcal{R} = \{M \in \mathcal{L} \mid M$ can be l-embedded into $\prod_{j \in J} M_j$ for some $M_j \in \mathcal{X}_j\}$. Notice that $\mathcal{X}_j \subseteq \mathcal{R}$ for each $j \in J$ and $\mathcal{R} \subseteq \overset{\wedge}{\bigvee_{j \in J}} \mathcal{X}_j$.

Thus to complete the proof, it suffices to show that $\mathcal{R}$ is an l-quasivariety. Clearly, $\mathcal{R}$ is closed under l-isomorphisms, Cartesian products, and l-subgroups. By Lemma 4, $\mathcal{R}$ is closed under ultraproducts. Hence, by Theorem 1, $\mathcal{R}$ is an l-quasivariety. □

LEMMA 5. $H_p \notin \mathcal{N} \overset{\wedge}{\bigvee} \mathcal{R}_p$.

Proof. Since D_l^0 is simple as a group (see Section 7 of Chapter 4), the only proper ideal of H_p is the set of elements $(\{g_n\}, 0)$. For reductio ad absurdum, assume, on the contrary, that that $H_p \in \mathcal{N} \overset{\wedge}{\bigvee} \mathcal{R}_p$. Then, by Lemma 4, H_p is l-embedded in $M \times K$ where $M \in \mathcal{N}$and $K \in \mathcal{R}_p$. Without loss of generality, we can suppose that H_p is an l-subgroup of $M \times K$.

Case (i). $H_p \cap M = E$. Let $\psi : H_p \to K$ be a projection onto the second coordinate; i.e., $(h_1, h_2)\psi = h_2$, where $h = (h_1, h_2) \in H_p$. It is clear that ψ is l-embedding. So H_p is l-isomorphic to an l-subgroup of K. Since $K \in \mathcal{R}_p$, we have $H_p \in \mathcal{R}_p$. This contradicts Lemma 2.

Case (ii). $H_p \cap M \neq E$. Notice $H_p \cap M$ is an ideal of H_p. But there are exactly two ideals of H_p other than E. So $H_p \cap M = H_p$ or $H_p \cap M = \{(\{g_n\}, 0) \mid g_n D_l^0,\ n \in \mathbf{Z}\}$. If $H_p \cap M = H_p$, then H_p is an l-subgroup of M which is a contradiction since D_l^0 is l-isomorphic to an l-subgroup of H_p and D_l^0 is not normal-valued whereas M is.

If $H_p \cap M = \{(\{g_n\}, 0) \mid g_n \in D_l^0,\ n \in \mathbf{Z}\}$, then G is l-isomorphic to an l-subgroup of $H_p \cap M$, again a contradiction since M is normal-valued and D_l^0 is not. □

THEOREM 2. (Arora [1].) *There exists a continuum of incomparable l-quasivarieties each properly containing the l-variety $\mathcal{N}$ of normal-valued l-groups.*

Proof. For any non-empty set of primes P_1 we define an l-quasivariety $\mathcal{K}_{P_1}$ as follows:

$$\mathcal{K}_{P_1} = \bigcap_{p \in P_1} (\mathcal{N} \overset{\wedge}{\bigvee} \mathcal{R}_p).$$

If non-empty subsets of primes P_1 and P_2 are incomparable, let $p \in P_1 \setminus P_2$ and $q \in P_2 \setminus P_1$. Then $H_p \in \mathcal{K}_{P_2}$ by Lemma 2, and, by Lemma 5, $H_p \notin \mathcal{K}_{P_1}$. Thus, $H_p \in \mathcal{K}_{P_2} \setminus \mathcal{K}_{P_1}$. Similarly, $H_q \in \mathcal{K}_{P_1} \setminus \mathcal{K}_{P_2}$.

As it was already pointed out, there exists a continuum of incomparable subsets of the set of positive primes. Thus there is a continuum of pairwise incomparable l-quasivarieties of the form $\mathcal{K}_p$.□

LEMMA 6. (Arora [1].) *Let G be a finitely subdirectly irreducible l-group. Let $\{\mathcal{X}_j \mid j \in J\}$ be a finite collection of l-quasivarieties. Then $G \in \overset{\wedge}{\bigvee}_{j \in J} \mathcal{X}_j$ if and only if $G \in \mathcal{X}_{j_0}$ for some $j_0 \in J$.*

The proof follows immediately from Lemma 4.□

LEMMA 7. (Arora [1].) $q_l(A(R)) = \mathcal{L}$

Proof. Let l-group S fail to satisfy the implication

$$(w(x_1, \ldots, x_n) = e) \Rightarrow (u(x_1, \ldots, x_n) = e)$$

so there exists a substitution in S such that $w(s_1, \ldots, s_n) = e$ and $u(s_1, \ldots, s_n) \neq e$. Let S' be the l-subgroup of S generated by $s_1, \ldots, s_n$. Then S' is countable and fails to satisfy this implication.

From Holland's Embedding Theorem (Theorem 4.1.1) it follows that each countable l-group is l-embedded in $A(\mathbf{R})$. So $A(\mathbf{R})$ fails to satisfy this implication, too.□

THEOREM 3. (Arora [1]). *The l-variety of all l-groups $\mathcal{L}$ is finitely join-irreducible in the lattice of l-quasivarieties Λ.*

Proof. Let us assume, on the contrary, that

$$\mathcal{L} = \overset{\wedge}{\bigvee_{j \in J}} \mathcal{X}_j$$

where J is a finite set and each $\mathcal{X}_j$ is a proper l-quasivariety. By Corollary 1 of Theorem 4.6.4, the l-group $A(R)$ is subdirectly irreducible and thus by Lemma 6, $A(\mathbf{R}) \in \mathcal{X}_{j_0}$ for some $j_0 \in J$. Thus $X_{j_0} = \mathcal{L}$ contradicts the choice $X_j, j \in J$. □

14.3 The Universal Theory of Abelian l-Groups

Let us recall that a first-order formula Φ is an *universal formula* if it is in prenex form and all the quantifiers are universal. An l-group G_1 is *universally equivalent* to an l-group G_2 iff the sets $Th_\forall G_1$ and $Th_\forall G_2$ of universal sentences of the signature $l = \{\cdot\ ,\ ^{-1}\ ,\ e\ , \vee, \wedge\}$ which are satisfied by G_1 and G_2 correspondingly, are the same (Burris and Sankappanavar [1], p.215). For Abelian groups we use the additive notations.

THEOREM 1. (Gurevich and Kokorin [1].) *All nontrivial totally ordered Abelian groups are universally equivalent.*

Proof. We divide the proof of the theorem into more easy steps.

Step 1. If φ is an universal sentence true in some Abelian o-group $G \neq \{0\}$, then φ holds in the additive o-group of integers $\mathbf{Z}$ (since $\mathbf{Z}$ can be embedded in every non-trivial o-group).

Step 2. Let H be an Abelian o-group. An universal sentence φ holds in H if and only if it holds in every finitely generated subgroup of H. (If φ fails in H, there are $h_1, \ldots, h_m \in H$, such that some conjugation of disjunction of atomic and negated atomic sentences holds in H. This holds, clearly, in the subgriup of H generated by $h_1, \ldots, h_m$, so φ fails in this finitely generated subgroup of H).

Step 3. If φ is an universal sentence true in $\mathbf{Z}$, then φ is true in the additive o-group of rational numbers $\mathbf{Q}$. (Since each finitely generated subgroup of $\mathbf{Q}$ is o-isomorphic to $\mathbf{Z}$).

Step 4. Let D_m be a subgroup of the additive o-group of real numbers R with m generators. Then any universal sentence holding in Q holds in D_m.

Let $y_1, \ldots, y_n \in D_m$ and $\alpha_1, \ldots, \alpha_m$ be rationally independent basis of D_m, say, $y_i = \sum_{j=1}^{m} \lambda_{ij}\alpha_i$ with $\lambda_{ij} \in \mathbf{Z}$. Let $r_j \in \mathbf{Q}$ with $|\alpha_j - r_j| < \varepsilon$ $(1 \leq j \leq m)$ where $\varepsilon \in \mathbf{R}, \varepsilon > 0$ is small. Specifically,

$$\varepsilon < \min_{y_i \neq y_{i'}} \frac{|y_i - y_{i'}|}{\sum_{j=1}^{m} |\lambda_{ij} - \lambda_{i'j}|} \text{ and } \varepsilon < \min_{y_i + y_{i'} \neq y_k} \frac{|y_i + y_{i'} - y_k|}{\sum_{j=1}^{m} |\lambda_{ij} + \lambda_{i'j} - \lambda_{kj}|}.$$

Let $x_i = \sum_{j=1}^{m} \lambda_{ij} r_j \in \mathbf{Q}$, $(1 \leq i \leq n)$. We show that the finite submodels $\{x_1, \ldots, x_n\}$ and $\{y_1, \ldots, y_n\}$ are isomorphic. So we must show

$$\begin{array}{lll} \text{a}),\text{b}) & x_i = x_{i'} & \text{iff } y_i = y_{i'}, \\ \text{c}),\text{d}) & x_i < x_{i'} & \text{iff } y_i < y_{i'}, \\ \text{e}),\text{f}) & x_i + x_{i'} = x_k \text{ iff } y_i + y_{i'} = y_k. & \end{array}$$

For b) and f) use the linear independence of $\alpha_{j'}$, e.g., f): $y_i + y_{i'} = y_k$. Then

$$\sum_{j=1}^{m} (\lambda_{ij} + \lambda_{i'j} - \lambda_{kj})\alpha = 0 \text{ and } \lambda_{ij} + \lambda_{i'j} - \lambda_{kj} = 0$$

for all j. Hence

$$x_i + x_{i'} - x_k = \sum_{j=1}^{m} (\lambda_{ij} + \lambda_{i'j} - \lambda_{kj}) r_j = 0.$$

For a), if $x_i = x_{i'}$ but $y_i \neq y_{i'}$, then

$$|y_i - y_{i'}| \leq |x_i - x_{i'}| + \sum_{j=1}^{m} |\lambda_{ij} - \lambda_{i'j}| \cdot |\alpha_j - r_j| < \varepsilon \sum_{j=1}^{m} |\lambda_{ij} - \lambda_{i'j}| < |y_i - y_{i'}|.$$

This is a contradiction.

For d),

$$x_{i'} - x_i = \sum_{j=1}^{m} (\lambda_{i'j} - \lambda_{ij}) r_j =$$

$$= \sum_{j=1}^{m} (\lambda_{i'j} - \lambda_{ij})\alpha_j + \sum_{j=1}^{m} (\lambda_{i'j} - \lambda_{ij})(r_j - \alpha_j) \geq$$

$$\geq |y_{i'} - y_i| - \varepsilon \sum_{j=1}^{m} (\lambda_{i'j} - \lambda_{ij}) > 0.$$

Other statesments are similar.

Step 5. Let K_s be a finitely generated Abelian o-group with s generators. Any universal sentence which holds in all finitely generated subgroups of R holds in K_s.

From the description of finitely generated Abelian o-groups (Section 3 of Chapter 6) it follows that $K_s \cong D_{m_1} \times \ldots \times D_{m_t}$, where each D_{m_i} is o-isomorphic to a subgroup of $\mathbf{R}$ on m_i rationally independent generators. Choose m large enough so that D_m contains isomorphic copies of $D_{m_1}, \ldots, D_{m_t}$. Let $y_1, \ldots, y_n \in K_s$, say $y_i = (\lambda_{i1}, \ldots, \lambda_{it}), 1 \leq i \leq n$. Let $x_i \in D_m$ satisfy $x_i = \sum_{j=1}^{t} \lambda_{ij} N^{j-1}$, where N is a positive integer satisfying the inequality

$$\text{A)} \qquad N > \max_{y_i \neq y_{i'}} \frac{\sum_{j=1}^{t} |\lambda_{ij} - \lambda_{i'j}|}{|\lambda_{ij_0} - \lambda_{i'j_0}|},$$

where j_0 is the greatest j with $\lambda_{ij} \neq \lambda_{i'j}$ and
B) N exceeds all solutions z of the system of equations

$$\sum_{j=1}^{t} (\lambda_{ij} - \lambda_{i'j}) z^{j-1} = 0, \quad \sum_{j=1}^{t} (\lambda_{ij} + \lambda_{i'j} - \lambda_{kj}) z^{j-1} = 0,$$

where $(1 \leq i, i', k \leq n)$

Note that since some of these coefficients are non-zero and we are dealing with $\leq n^2 + n^3$ polynomials, the numbers of roots is finite and hence bounded.

Arguments similar of the Step 4 shows that the finite substructures $\{x_1, \ldots, x_n\}$ and $\{y_1, \ldots, y_n\}$ are isomorpic. Thus D_m and $\mathbf{Q}$ are universally equivalent (see Mal'cev [7]).

Let now G and H be non-trivial Abelian o-groups. By Step 2, we can assume that G and H are finitely generated. Let φ be an universal sentence which holds in G. By Steps 1,3, φ holds in $\mathbf{Z}, \mathbf{Q}$. By Step 4, φ holds in each $D_m, m > 0$ so by Step 5, φ holds in H. □

THEOREM 2. (Khisamiev and Kokorin [1], Khisamiev [1].)

(1) *Let G and G' be Abelian l-groups of finite orthogonal ranks r and r' correspodingly.*
Then G is universally equivalent to G' if and only if $r = r'$.
(2) *Let G and G' be Abelian l-groups of infinite orthogonal rank.*
Then G and G' are universally equivalent.

Proof. (1) The direct verification shows that if l-groups $N_i (1 \leq i \leq r < \infty)$ are universally equivalent to l-groups $M_i (1 \leq i \leq r < \infty)$ respecttively, then the direct sums $\oplus \sum_{i=1}^{r} N_i$ and $\oplus \sum_{i=1}^{r} M_i$ are also universally equivalent.

Let G be an l-group of finite orthogonal rank r. Then there are r mutually orthogonal elements $0 < a_1, \ldots, a_r \in G$ and r is the greatest number with this property. It is clear that the polars $(a_i^{\perp})$, $1 \leq i \leq r$, are prime subgroups of G and $(a_1)^{\perp} \cap \ldots \cap (a_n)^{\perp} = \{e\}$. Thus

$$G \subseteq G/(a_1)^{\perp} + \ldots + G/(a_r)^{\perp}$$

and l-gp$(a_1, \ldots, a_r) = (a_1) + \ldots + (a_r) \leq G$, where (a_i) is an infinite cyclic o-group generated by a_i, $a_i > 0$, $i = 1, 2, \ldots, r$. By Theorem 1, o-groups a_i and $G/(a_i)^{\perp}$ are universally equivalent too. Thus G is universally equivalent to $(a_1) + \ldots + (a_r)$.

If the orthogonal rank r' of G' is equal to r then by above arguments G' is universally equivalent to $(a'_1) + \ldots + (a'_r)$ and consequently to G.

If $r < r'$ then l-groups $(a_1) + \ldots + (a_r)$ and $(a'_1) + \ldots + (a'_{r'})$ are not universally equivalent since the universal sentence

$$\begin{gathered}(\forall y)(\forall z)(\forall x_1)\ldots(\forall x_{r-1})(\&_{i=1}^{k-1}(x_i > 0)\&(y > 0)\&(z > 0)\& \\ \&_{i \neq j}(x_i \wedge x_j = 0)\&\&_{i=1}^{r-1}(y \wedge x_i = 0)\&\&_{i=1}^{r-1}(z \wedge x_i = 0)) \Rightarrow \\ \Rightarrow ((y \wedge z = y) \text{ or } (y \wedge z = z))\end{gathered}$$

holds in $(a_1) + \ldots + (a_r)$ and violates in $(a'_1) + \ldots + (a'_{r'})$ and G' if $x_1 = a'_1, \ldots, x_{r-1} = a'_{r-1} = a'_{r-1}, y = a'_r, z = a'_{r+1}$. (Here $\vee$ is the disjunction of sentences.)

(2) In this case $G \supseteq \sum_{i=1}^{\infty} \mathbf{Z}_i$. Since each Abelian l-group is representable then $G \subseteq \oplus \bar{\sum_{\alpha \in \Lambda}} G_\alpha$, where the index set Λ is infinite and

$G_\alpha(\alpha \in \Lambda)$ are Abelian o-groups. We claim that $\oplus \sum\limits_{\alpha\in\Lambda}^{-} G_\alpha$ and $\oplus \sum\limits_{i=1}^{\infty} \mathbf{Z}_i$ are universally equivalent.

Let $N = \{g_1, \ldots, g_n\}$ be a finite substructure of $\oplus \sum\limits_{\alpha\in\Lambda}^{-} G_\alpha$. Then $g_i = (\ldots, g_\alpha^i, \ldots) \in \oplus \sum\limits_{\alpha\in\Lambda}^{-} G_\alpha$ and $g_\alpha^i \in G_\alpha, i = 1, \ldots, n$. By Theorem 1, each o-group G_α is universally equivalent to $\mathbf{Z}$. Let $N_\alpha = \{g_1^\alpha, \ldots, g_n^\alpha\}$ be a finite substructure of G_α.

There are only finite number non-isomorphic substructures among N_α, $\alpha \in \Lambda$. Let $N_{\alpha_1}, \ldots, N_{\alpha_s}$ be all non-isomorphic substuctures. Let G_{α_s} be a finite direct sum of o-groups G_{α_i}, $i = 1, 2, \ldots, s$ and $\hat{N} = \{\hat{g}_1, \ldots, \hat{g}_n\}$ be a finite submodel of $G_{\alpha_1} + \ldots + G_{\alpha_s}$ where $\hat{g}_1 = (g_1^{\alpha_1}, \ldots, g_1^{\alpha_s}), \ldots, \hat{g}_n = (g_n^{\alpha_1}, \ldots, g_n^{\alpha_s}), g_i^{\alpha_j} \in G_{\alpha_j}, i = 1, 2, \ldots, n$.

The direct verification shows that the map $\varphi : g_i \to \hat{g}_i$ is an isomorphism N onto $\hat{N}$. By Theorem 1, there is a finite substructure M of $\oplus \sum\limits_{i=1}^{s} \mathbf{Z}_i$ which is isomorphic to N.

Thus for each finite substructure N of $\oplus \sum\limits_{\alpha\in\Lambda} G_\alpha$ there is an isomorphic substructure M of $\oplus \sum\limits_{i=1}^{\infty} \mathbf{Z}_i$ and all l-groups of an infinite orthogonal rank are universally equivalent. □

It is clear that an l-group G of a finite orthogonal rank is not universally equivalent to l-group H of an infinite orthogonal rank.

COROLLARY 1. *The l-variety $\mathcal{A}$ of Abelian l-groups is the smallest non-trivial l-quasivariety in the lattice of l-quasivarieties Λ.*

Proof. In fact each non-trivial Abelian l-group G contains an infinite cyclic o-subgroup. Then the l-quasivariety $q_l(G)$ contains all direct sums of additive o-group of integers $\mathbf{Z}$. By Theorem 2, $q_l(G)$ contains all Abelian l-groups. □

14.4 Covers of $\mathcal{A}$ in the Lattice of l-Quasivaieties

1^0. The structure of the lattice of l-quasivarieties Λ is much more complicated than the structure of the lattice of l-varieties $\mathbf{L}$. For example, the list of covers of $\mathcal{A}$ in Λ is more rich than the same in $\mathbf{L}$. As the starting point of this section, we use the results of Section 6 of Chapter 12. For definitions of l-groups G_p (p is a prime number), W^+, W^-, N_0 see Section 6 of Chapter 12.

PROPOSITION 1. *The l-variety* $\mathrm{var}_l(G_p) = S_p$ *covers* $\mathcal{A}$ *in the lattice of l-quasivarieties* Λ.

In the proof of Theorem 12.3.1 it was proved that each subdirectly irreducible finitely generated l-group of the l-variety S_p is an l-subgroup of some reduced product of isomorphic copies of G_p. It was mentioned in the proof of Theorem 12.6.1 that each non-abelian l-group $G \in S_p$ contains an l-subgroup which is l-isomorphic to G_p. Thus $\mathrm{q}_l(G_p) = S_p = \mathrm{var}_l(G_p)$ and $\mathrm{q}_l(G_p) = S_p$ for each non-abelian l-group $G \in S_p$. Hence S_p covers $\mathcal{A}$ in the lattice of l-varieties $\mathbf{L}$.□

PROPOSITION 2. *l-Quasivarieties* $\mathrm{q}_l(W^+)$ *and* $\mathrm{q}_l(W^-)$ *cover* $\mathcal{A}$ *in the lattice of l-quasivarieties* Λ.

Proof. It is clear that $\mathrm{q}_l(W^+) \subseteq \mathrm{var}_l(W^+) = \mathcal{M}^+$. Let G be any non-abelian l-group, $G \in \mathrm{q}_l(W^+)$. The direct verification shows that the following identities hold in $\mathcal{M}^+$ (and, therefore, in $\mathrm{q}_l(W^+)$):

$$(1) \qquad (|x| \vee |y|)^{-1}|[x,y]|(|x| \vee |y|) \wedge |[x,y]|, \ n \in \mathbf{N}.$$

Let us assume that $c = |[a,b]| > e$ in G. Then from the identities (1) it follows that

$$\ldots \ll c^{d^{-1}} \ll c \ll c^d \ll c^{d^2} \ll \ldots,$$

where $d = |a| \vee |b|$. Thus $[c^{d^i}, c^{d^j}] = e$ for all $i, j \in Z$ and the set $\{c^{d^i} \mid i \in \mathbf{Z}\}$ is rationally independent in $\mathrm{gp}(c,d)$. Now arguments similar to the proof of Theorem 12.6.2 show that the subgroup $\mathrm{gp}(c,d)$ is a totally ordered subgroup of G and $\mathrm{gp}(c,d)$ is l-isomorphic to W^+. Hence each non-abelian group $H \in \mathrm{q}_l(W^+)$ contains W^+ as an l-subgroup and $\mathrm{q}_l(W^+)$ covers $\mathcal{A}$ in Λ. For the l-quasivariety $\mathrm{q}_l(W^-)$ the proof is the same. □

For nilpotent l-quasivarieties the situation is differ of the previous cases. Let K_0 be a group having the presentation

$$K_0 \ = \mathrm{gp}(a,b,c,c^+,c^- \mid c = [a,b] = c^+c^-, \ e = [c^+, \ c^-] =$$

$$= [c^+, a] = [c^-, a] = [c^-, b] = [c^+, b])$$

and lattice-ordered by the rule:$a^k b^l (c^+)^m (c^-)^n \geq e$ iff $k > 0$ or $k = 0$ and $l > 0$ or $k = l = 0$ and $m \geq 0$, $n \leq 0$. We claim that each non-abelian nilpotent l-quasivariety $\mathcal{K}$ contais K_0.

Let G be a non-abelian l-group, $G \in \mathcal{K}$. Then there are $e < x, y \in G$ such that $e \neq [x, y] \in Z(G)$, where $Z(G)$ is the centre of G, $[x, y] = [x, y]^+[x, y]^-$ and there arise three different cases:

$$\begin{aligned} &1) \quad [x,y]^+ > e, [x,y]^- = e; \\ &2) \quad [x,y]^+ = e, [x,y]^- < e; \\ &3) \quad [x,y]^+ > e, [x,y]^- < e. \end{aligned}$$

Case 1). Replacing x, y by xy, y respectively, we can assume that $x > y > e$. Since the lattice-order of G is weakly-Abelian then $x > y \gg |[x, y]|$. Let $u, v \in \bar{\prod}_{i \in \mathbf{N}} G_i$, $G_i \cong G$, be such that

$$u = (x, x^2, \ldots, x^n, \ldots), v = (y, y, \ldots, y, \ldots).$$

Let $\mathcal{F}$ be the filter of cofinite subsets of positive integers $\mathbf{N}$, l-gp$(\bar{u}, \bar{v})$ be the l-subgroup of the reduced product $\bar{\prod}_{i \in \mathbf{N}} G_i / \mathcal{F}$ generated by $u\mathcal{F} = \bar{u}$ and $v\mathcal{F} = \bar{v}$. Then $\bar{u} \gg \bar{v} \gg [\bar{u}, \bar{v}] > e$ in $\bar{\prod}_{i \in \mathbf{N}} / \mathcal{F}$ and l-gp$(\bar{u}, \bar{v})$ is l-isomorphic to the nilpotent of class 2 group

$$K = \text{gp}(a_1, b_1, c_1 \mid [a_1, b_1] = c_1, \ [a_1, c_1] = [b_1, c_1] = e)$$

totally ordered by the rule: $a_1^m b_1^l c_1^k \geq e$ iff $m > 0$ or $m = 0$ and $l > 0$ or $m = l = 0$, $k \geq 0$. It is clear that o-group K belongs to the l-quasivariety $\mathcal{K}$.

Let $z, w \in \bar{\prod}_{i \in \mathbf{N}} K_i$, $K_i \cong K$, be such that

$$z = (a_1, a_1^2, \ldots, a_1^n, \ldots), \quad w = (a_1 b_1^{-1}, a_1 b_1^{-1}, \ldots, a_1 b_1^{-1}, \ldots).$$

Then in the reduced product $\prod_{i \in \mathbf{N}} K_i^- / \mathcal{F}$, where $\mathcal{F}$ is a filter of cofinite subsets of N, the elements $z\mathcal{F} = \bar{z}$, $w\mathcal{F} = \bar{w}$ generates a totally ordered subgroup $\mathrm{gp}(\bar{z},\bar{w})$ and $\bar{z} \gg \bar{w} \gg [\bar{z},\bar{w}]^{-1} > e$. The direct verification shows that the o-group $\mathrm{gp}(\bar{z},\bar{w})$ is l-isomorphic to the nilpotent group of class 2

$$\hat{K} = \mathrm{gp}(a_2, b_2, c_2 \mid [a_2, b_2] = c_2,\ [a_2, c_2] = [b_2,\ c_2] = e)$$

totally ordered by the rule: $a_2^m b_2^l c_2^k \geq e$ iff $m > 0$ or $m = 0$ and $l > 0$ or $m = l = 0,\ k < 0$. Also, o-group $\hat{K}$ belongs to $\mathcal{K}$. Let l-gp(a_1a_2, b_1b_2) be an l-subgroup of $K \times \hat{K}$ generated by the elements $a_1a_2,\ b_1b_2$. It is clear that $a_1a_2 \gg b_1b_2 \gg |[a_1a_2, b_1b_2]| > e$ and $[a_1a_2, b_1b_2] = [a_1, b_1][a_2, b_2]$. Now it is clear that l-group l-gp(a_1a_2, b_1b_2) is l-isomorphic to the l-group K_0.

Case 2) follows from Case 1) after replacing x, y by y, x correspondingly.

Case 3). Replacing x, y by $x\bar{y}, y$ respectively, if necessary, we can assume that $x > y$. Let $u, v \in \prod_{i\in\mathbf{N}} G_i,\ G_i \cong G$ be such that

$$u = (x, x^2, \ldots, x^n, \ldots), \quad v = (y, \ldots, y, \ldots).$$

Then in the reduced product $\prod_{i\in\mathbf{N}} G_i/\mathcal{F}$, where $\mathcal{F}$ is the filter of cofinite subsets of N , the elements $\bar{u} = u\mathcal{F},\ \bar{v} = v\mathcal{F}$ generate an l-subgroup which is l-isomorphic to the l-group K_0. This yields the following result.

THEOREM 1. (Medvedev [10].) *The l-quasivariety* $\mathrm{q}_l(K_0)$ *is the unique cover of* $\mathcal{A}$ *in the lattice of nilpotent l-quasivarieties.* □

Now a natural question about coinciding $\mathcal{M}_0$ and $\mathrm{q}_l(K_0)$ arises. The answer is negative since the implication $([x,y]\vee e = [x,y]) \Rightarrow ([x,y] = e)$ holds in K_0 and violated in the o-group N_0. Moreover, the representable l-quasivariety $\mathrm{q}_l(K_0)$ does not contain any non-abelian o-group.

2°. Since the l-variety $\mathcal{A}$ is defined by a single identity, by compactness theorem (Burris and Sankappanavar [1]), it follows that each cover of $\mathcal{A}$ in the lattice $\mathbf{L}$ (defined in Section 6 of Chapter 12) contains a cover of $\mathcal{A}$ in the lattice Λ. All these covers of $\mathcal{A}$ are very close to the covers of $\mathcal{A}$ in Λ . Now we will construct an infinite set of covers of $\mathcal{A}$ in the lattice Λ which cannot be defined by above arguments.

Let $G = \mathbf{Z}\mathrm{wr}(\mathbf{Z},\mathbf{Z})$ and $a = (\{a_t\},0)$, $b = (\{o_t\},1)$ where $o_t = 0 \in \mathbf{Z}$ for all $t \in \mathbf{Z}$ and $a_t = 0$ for $t \neq 0$, $a_0 = 1$. Then $\mathrm{gp}(a,b) = G$ and the lower central series

$$G = \gamma_1 G \geq \gamma_2 G \geq \dots \gamma_n \geq \dots$$

has the following properties (Kargapolov and Merzljakov [1]):

(2) $\bigcap\limits_{n \in \mathbf{N}} \gamma_n G = E$;

(3) the quotient group $\gamma_{i+1}G/\gamma_{i+2}G = ([a,\underbrace{b,\dots,b}_{i}]\gamma_{i+2}G)$ is an infinite cyclic group generated by the element $[a,\underbrace{b,\dots,b}_{i}]\gamma_{i+2}G$.

Here $\gamma_{i+1}G = [\gamma_i G, G]$ for each $i \geq 1$.

Now let for each integer $n \geq 2$ and the finite set $\{\varepsilon_0, \varepsilon_1, \dots, \varepsilon_{n-1}\}$, where $\varepsilon_0 = +1, \varepsilon_l = \pm(l = 1,2,\dots,n-1)$, $Q_n(\varepsilon_0, \varepsilon_1, \dots, \varepsilon_{n-1})$ be a total order of $G = \mathbf{Z}\mathrm{Wr}(\mathbf{Z},\mathbf{Z})$ uniquely defined by the following inequalities:

(4) $b \gg a \gg |[a,b]| \gg |[a,b,b]| \gg \dots |[a,\underbrace{b,\dots,b}_{k}]| \gg> e$;

(5) $$[a,\underbrace{b,\dots,b}_{k}]^{\varepsilon_{\bar{k}}} > e;$$

where $\bar{k} \equiv k \pmod{n}$ and $\bar{k} \in \{0,1,\dots,n-1\}$.

The direct verification shows that if $[x,y] \neq e$ in o-group

$$(G, Q(\varepsilon_0, \varepsilon_1, \dots, \varepsilon_{n-1}))$$

for some $x,y \in G$, then $|x| \vee |y| = b^s\varphi$, where $s > 0$ and φ belongs to the Abelian radical $\mathcal{A}(G)$. Let

$$v^{(i)}(x,y) = v^i(x,y,\varepsilon_0,\varepsilon_1,\dots,\varepsilon_{n-1}) =$$
$$= (u_0^{(i)}(x,y)^{\varepsilon_0} \vee e) \wedge (u_1^{(i)}(x,y)^{\varepsilon_1}) \wedge \dots \wedge (u_{n-1}^{(i)}(x,y)_{n-1}^{\varepsilon} \vee e),$$
$$w(x,y) = w(x,y,\varepsilon_0,\dots,\varepsilon_{n-1}) = \bigvee_{i=0}^{n-1} v^i(x,y,\varepsilon_0,\dots,\varepsilon_{n-1}),$$

and $\tilde{w}(x,y) = |[w(x,y), |x| \vee |y|]|$, where

$$u_0^0(x,y) = |[x,y]|, u_0^{(i)}(x,y) =$$

$$= |[u_0^{(i-1)}(x,y), |x| \vee |y|]|,\ i = 1,2,\dots,n-1;$$
$$u_j^{(i)}(x,y) = [u_{j-1}^{(i)}(x,y), |x| \vee |y|], j = 1,2,\dots,n-1.$$

Let $\varepsilon_0 = +1, \varepsilon_i = -1,\ i \in \{1,2,\dots,n-1\}$ and let (G, Q_n) be a group $G = \mathbf{Z}\mathrm{wr}(\mathbf{Z},\mathbf{Z})$ totally ordered by the order

$$Q_n = Q_n(1,-1,-1,\dots,-1).$$

LEMMA 1. *Let* $x, y \in (G, Q_n)$ *and* $[x,y] \neq e$. *Then* $w(x,y) = [a, \underbrace{b,\dots,b}_{nt-1}]^r \varphi' > e$ *where* $\varphi' \in \gamma_{nt+1}G, r \in \mathbf{Z}$, *and* $r < 0$ *in* $\mathbf{Z}$.

Proof. As was mentioned above, $|x| \vee |y| = b^s\varphi,\ \varphi \in \mathcal{A}(G)$ and $s > 0$. Then $u_0^{(0)}(x,y) = |[x,y]| = [a, \underbrace{b,\dots,b}_{nk+l}]^{m_0}\varphi_0$ for some $\varphi_0 \in \gamma_{nk+l+2}(G)$ and, $m_0 > 0$ if $l = 0$ or $m_0 < 0$ if $l \in \{1,2,\dots,n-1\}$. So $u_j^0(x,y) = [a, \underbrace{b,\dots,b}_{nk+l+j}]^{m_j}\varphi_j$, where $m_j = m_0 s^j$ and $\varphi_j \in \gamma_{nk+l+j+2}G$.

Now consider words $v^{(i)}(x,y,+1,-1,\dots,-1), i \in \{0,1,\dots,n-1\}$.

Case 1). $i = \overline{n-l}$. Then $v^{\overline{n-l}}(x,y)=$

$$= (u_0^{\overline{n-l}}(x,y) \vee e) \wedge ((u_1^{\overline{n-l}}(x,y))^{-1} \vee e) \dots \wedge ((u_{n-1}^{\overline{n-l}}(x,y))^{-1} \vee e),$$

where $u_0^{\overline{n-l}}(x,y) = [a, \underbrace{b,\dots,b}_{n(k+1)}]^{k_0}\varphi_0',\ \varphi_0' \in \gamma_{n(k+1)+2}G,\ k_0 =\mid m_0 \mid s^{\overline{n-l}}$.

Thus

$$u_j^{\overline{n-l}}(x,y) = [a, \underbrace{b,\dots,b}_{n(k+1)+j}]^{k_j}\varphi_j', \varphi_j' \in \gamma_{n(k+1)+j+2}G,$$

$k_j = k_0 s^j,\quad j \in \{0,1,\dots,n-1\}$. Hence $u_0^{\overline{n-l}}(x,y) > e$ and $u_j^{\overline{n-l}}(x,y) < e$ for each $j \in \{1,2,\dots,n-1\}$. So in (G, Q_n)

$$u_0^{\overline{n-l}}(x,y) > e, (u_1^{\overline{n-l}}(x,y))^{-1} > e, \dots, (u_{n-1}^{\overline{n-l}}(x,y))^{-1} > e.$$

Therefore,

$$v^{\overline{n-l}}(x,y) = [a, \underbrace{b,\dots,b}_{n(k+2)-1}]^{k_{n-1}}\varphi_{n-1}' > e$$

for some $\varphi'_{n-1} \in \gamma_{n(k+2)+1}G$.

Case 2). $i \neq \overline{n-l}$. We claim that all $v^{(i)}(x,y) = e$. Indeed, let

$$u_0^{(i)}(x,y) = [a, \underbrace{b,\ldots,b}_{nk+l+i}]^{m_i}\psi_0 > e,$$

where $\psi_0 \in \gamma_{nk+l+i+2}G$, $|m_i| = |m_0|s^i$ and $m_i < 0$ since $l+i \not\equiv 0 \pmod n$. Then $u_j^{(i)} = [a, \underbrace{b,\ldots,b}_{nk+l+i+j}]^{m_i s^j}\psi_j$ for some $\psi_j \in \gamma_{nk+l+i}G$. Then

$$u_j^{(i)}(x,y) < e \text{ if } (l+i+j) \equiv 0 \pmod n.$$

In the set $\{0,1,\ldots,n-1\}$ there is a unique j_0 such that $(l+i+j_0) \equiv 0 \pmod n$. Hence, $(u_{j_0}^{(i)}(x,y))^{-1} > e$. Then $(u_{j_0-1}^{(i)}(x,y))^{-1} < e$ or $(u_{j_0+1}^{(i)}(x,y))^{-1} < e$ in (G,Q_n). Thus $(u_{j_0-1}^{(i)}(x,y))^{-1} \vee e = e$ or $(u_{j_0+1}^{(i)}(x,y))^{-1} \vee e = e$. Therefore, $v^i(x,y) = e$ for each $w(x,y) = [a, \underbrace{b,\ldots,b}_{n(k+2)-1}]^{k_{n-1}}\varphi'_{n-1} > e$ for some $\varphi'_{n-1} \in \gamma_{n(k+2)+1}G$. □

It is clear that for $[x,y] \neq e$ in G

$$\tilde{w}(x,y) = |[w(x,y), |x| \vee |y|]| = [a, \underbrace{b,\ldots,b}_{nt}]^p\psi > e,$$

where $p > 0$, $\psi \in \gamma_{nt+2}G$.

LEMMA 2. *Let $x,y \in (G,Q_n)$ and $[x,y] \neq e$. Then the subgroup $H = \mathrm{gp}(\tilde{w}(x,y), |x| \vee |y|)$ of (G,Q_n) generated by the elements $\tilde{w}(x,y)$ and $|x| \vee |y|$ is order isomorphic to o-group (G,Q_n).*

Proof. By the above arguments, $\tilde{w}(x,y) > e$ in (G,Q_n). Indeed the elements $\{\tilde{w}(x,y)^{(|x|\vee|y|)^t} \mid t \in \mathbf{Z}\}$ are rationally independent in $\mathcal{A}(G,Q_n)$. Now it is readely verified that H is isomorphic to G (as groups). For order isomorphism of (G,Q_n) to H it is enough to prove that

$$[\tilde{w}(x,y), \underbrace{|x| \vee |y|,\ldots,|x| \vee |y|}_{l}] > e \text{ if } l \equiv 0 \pmod n \quad \text{and}$$

$$[\tilde{w}(x,y), \underbrace{|x| \vee |y|,\ldots,|x| \vee |y|}_{l}] < e \text{ if } l \not\equiv 0 \pmod n.$$

It is easily verified by direct verification. □

Let K be any non-abelian subgroup of the ultrapower

$$V = (G, Q_n)^I / \mathcal{F}$$

of the o-group (G, Q_n).

LEMMA 3. *Let $x, y \in V = (G, Q_n)^I/\mathcal{F}$ and $[x, y] \neq e$. Then the subgroup* $\text{gp}(\tilde{w}(x, y), |x| \vee |y|) = H$ *of V is order isomorphic to the o-group (G, Q_n).* □

LEMMA 4. *Let $x, y \in \bar{\prod\limits_{j \in J}} V_j$, where $V_j = (G, Q_n)^{I_j}/\mathcal{F}$ is the ultrapower of the o-group (G, Q_n) for each $j \in J$ and $[x, y] \neq e$. Then the subgroup* $\text{gp}(\tilde{w}(x, y), |x| \vee |y|) = H$ *of $\bar{\prod\limits_{j \in J}} V_j$ is totally ordered and is order isomorphic to (G, Q_n).* □

The proofs of Lemma 3 and Lemma 4 are similar to Lemma 2.

By Theorem 2 of Section 1, $\text{q}_l((G, Q_n)) = SPP_U(G, Q_n)$, where S, P, P_U are operators of taking l-subgroups, Cartesian products and ultraproducts correspondently. Let H be any non-abelian l-group and $H \in \text{q}_l((G, Q_n))$. Then by Lemma 4, H contains an o-subgroup generated by the elements $\tilde{w}(x, y)$ and $|x| \vee |y|$ which is order isomorphic to o-group (G, Q_n). This yields the following result.

THEOREM 2. (Isaeva and Medvedev [1].) *Each l-quasivariety $\text{q}_l((G, Q_n)), n \in \mathbf{Z}, n \geq 2$ has the following properties:*

(6) *$\text{q}_l((G, Q_l))$ covers $\mathcal{A}$ in the lattice of l-quasivarieties Λ,*

(7) *$\text{q}_l((G, Q_n))$ is contained in the l-variety $\mathcal{W}a$ of weakly abelian l-groups,*

(8) *each nilpotent l-group of $\text{q}_l((G, Q_n))$ is Abelian.* □

It is evident that $\text{q}_l((G, Q_m)) \neq \text{q}_l((G, Q_n))$ if $n \neq m$. In fact let us assume, on the contrary, that $\text{q}_l((G, Q_m)) = \text{q}_l((G, Q_n))$ for $m < n$. From the definition of the total orders Q_n and Q_m and Lemma 1 it follows that the implication $((w(x, y) = e) \Rightarrow ([x, y] = e))$ holds in (G, Q_n) and is violated in (G, Q_m) if $x = a, y = b$.

3⁰. Here we construct a new cover of $\mathcal{A}$ in the lattice Λ with the following properties:

(9) it does not contain non-abelian o-groups or nilpotent groups;
(10) it is contained in the l-variety $\mathcal{W}a$ of weakly-Abelian l-groups;
(11) it is metabelian.

Let $G = \mathbf{Z}\mathrm{wr}(\mathbf{Z}, \mathbf{Z})$ and $a = (\{a_t\}, 0)$, $b = (\{o_t\}, 1)$ where $o_t = 0 \in \mathbf{Z}$ for all $t \in \mathbf{Z}$ and $a_t = 0$ for $t \neq 0$, $a_0 = 1$. Then $\mathrm{gp}(a, b) = G$ and G has the following well-known properties:

(12) $\bigcap_{n\in\mathbf{N}} \gamma_n G = E$, where $\gamma_1 G = G$ and $\gamma_{n+1} G = [\gamma_n G, G]$;
(13) for every $n \geq 2$ the quotient group $\gamma_n G/\gamma_{n+1} G$ is an infinite cyclic group generated by the element $[a, \underbrace{b, \ldots, b}_{n-1}]\gamma_{n+1} G$.

Now let I be the set of all infinite sequences $\alpha = (\alpha_0, \alpha_1, \ldots, \alpha_n, \ldots)$, where $\alpha_n \in \{+1, -1\}$. For every sequence $\alpha \in I$, let us consider the linear order P_α on group G, defined uniquely by the following relations:

$$b \gg a^{\alpha_0} \gg [a, b]^{\alpha_1} \gg \ldots \gg [a, \underbrace{b, \ldots, b}_{n}]^{\alpha_n} \gg \ldots > e.$$

It is not difficult to verify the following properties of (G, P_α):

(14) for sequences $\alpha = (\alpha_0, \alpha_1, \ldots,)$ and $-\alpha = (-\alpha_0, -\alpha_1, \ldots,)$ the o-groups (G, P_α) and $(G, P_{-\alpha})$ are order isomorphic under an isomorphism φ such that $\varphi(b) = b$ and $\varphi(a_0^\alpha) = a^{-\alpha_0}$;
(15) the o-groups $(\bar{G}_n, \bar{P}_{\alpha(n)})$ and $(\bar{G}_n, \bar{P}_{-\alpha(n)})$ are also order isomorphic;
(17) the o-group (G, P_α) is approximable by the o-groups $(\bar{G}_n, \bar{P}^{\alpha(n)})$ because $\bigcap_{n=1}^{\infty} \gamma_n G = e$;
(18) if for some elements $x, y \in (G, P_\alpha)$ is true $|[x, y]| > e$, then $|x| \vee |y| = b^s \varphi$ for some integer $s \in \mathbf{N}$ and $\varphi \in \oplus \sum_{k\in\mathbf{Z}} (a_{b^k}) =$ fun$((b), (a))$ – basic subgroup of G.

By $\overline{\alpha(n)}$ we denote the infinite segment $(\alpha_n, \ldots, \alpha_{n+l}, \ldots)$ of sequence $\alpha = (\alpha_0, \alpha_1, \ldots, \alpha_{n+l}, \ldots)$.

LEMMA 5. *Let $\alpha = (\alpha_0, \alpha_1, \ldots, \alpha_{n-1}, \ldots, \alpha_n, \ldots)$ be a sequence from I, then the subgroup* $\mathrm{gp}(x, y)$ *of the o-group (G, P_α) generated by elements $x = b^s\varphi$, $y = [a, \underbrace{b, \ldots, b}_{n}]^t\psi > e$, where $s > e$, $t\alpha_n >$*

$0, \varphi, \psi \in$ fun$((b),(a))$ and $\psi \in \gamma_{n+2}G$ is order isomorphic to the o-group $(G, P_{\overline{\alpha(n)}})$ or $((G, P_{-\overline{\alpha(n)}}))$.

Proof. The map $b \to x$, $a \to y$ is uniquely extended to an isomorphism φ of the group G to the subgroup gp(x,y). Therefore, we only need to verify the preservation of all order relations under this isomorphism. In fact, $(a)\varphi = [a, \underbrace{b, \dots, b}_{n}]^t\psi > e$ in (G, P_α). Then

$$([a, \underbrace{b, \dots, b}_{n}]^{\alpha_{n+l}})\varphi = [[a, \underbrace{b, \dots, b}_{n}]^t\psi, b^s\varphi, \dots, b^s\varphi]^{\alpha_{n+l}} =$$
$$= [a, \underbrace{b, \dots, b}_{n+l}]^{ts^l\alpha_{n+l}}\omega_l,$$

where $t\alpha_n > 0$, $\omega_l \in \gamma_{n+l+2}G$, $s > 0$.

If $\alpha_n, t > 0$, then

$$([a, \underbrace{b, \dots, b}_{l}]^{\alpha_{n+l}})\varphi = [a, \underbrace{b, \dots, b}_{n+l}]^{ts^l\alpha_{n+l}}\omega_l \gg$$
$$\gg ([a, \underbrace{b, \dots, b}_{n+l+1}]^{\alpha_{n+l+1}})\varphi = [a, \underbrace{b, \dots, b}_{n+l+1}]^{\alpha_{n+l+1}s^{l+1}t}\omega_{l+1} > e,$$

and, therefore, the o-group gp(x,y) is order isomorphic to $(G, P_{\overline{\alpha(n)}})$.
If $\alpha_n, t < 0$, then

$$([a, \underbrace{b, \dots, b}_{l}]^{-\alpha_{n+l}})\varphi = [a, \underbrace{b, \dots, b}_{n+l}]^{-ts^l\alpha_{n+l}}\omega_l \gg$$
$$\gg ([a, \underbrace{b, \dots, b}_{l+1}]^{-\alpha_{n+l+1}})\varphi = [a, \underbrace{b, \dots, b}_{n+l+1}]^{-ts^{l+1}\alpha_{n+l+1}}\omega_{l+1} > e,$$

(where $\omega_{l+1} \in \gamma_{n+l+3}G$) and in this case gp$(x,y)$ is order isomorphic to $(G, P_{-\overline{\alpha(n)}})$.

Let us denote by $\hat{I}$ the subset of I consisting from all sequences α such that $\alpha_0 = +1$. Now consider the l-group $H = \overline{\prod_{\alpha\in\hat{I}}}(G, P_\alpha)$ and its l-subgroup $\hat{G}$ generated by elements $\hat{a}, \hat{b}$ such that $\hat{b}(\alpha) = b$, $\hat{a}(\alpha) = a$ for every sequence $\alpha \in \hat{I}$. In the l-group $\hat{G}$ we take an element $|[x,y]| \neq e$.

Because $x = w_1(\hat{a}, \hat{b})$ and $y = w_2(\hat{a}, \hat{b})$, $\hat{b} \gg \hat{a}$, then there exist l-group words $w(\hat{a}, \hat{b})$, $\hat{\psi}(\hat{a}, \hat{b}) = \hat{\psi}$ such that $||[x, y]|| = w(\hat{a}, \hat{b})$ and $||[x, y]|| = b^s\hat{\psi}$, where $s > 0$, $\hat{\psi} \in (\hat{a}^{\hat{G}})_l$ and $(\hat{a}^{\hat{G}})_l$ is the convex l-subgroup $\hat{G}$ generated by the set of elements $\hat{a}^{\hat{G}} = \{\hat{a}^{\hat{g}} \mid g \in G\}$. Therefore,

$$||[x, y]|| = w(\hat{a}, \hat{b}) = \bigvee_M \bigwedge_P w_{ij}(\hat{a}, \hat{b}),$$

where

$$w_{ij}(\hat{a}, \hat{b}) = \hat{a}^{k_{ij1}} \hat{b}^{k_{ij2}} \dots \hat{b}^{k_{ijs}}, (k_{ij1}, \dots, k_{ijs} \in \mathbf{Z}; |M|, |P| < \infty).$$

Then $||[x, y]||(\alpha) = w(\hat{a}, \hat{b})(\alpha) = \max_{i \in M} \min_{j \in P} w_{ij}(\alpha)$ (here max and min are considered in the o-group (G, P_α)). Let N_0 be the smallest natural number such that

$$(\{w_{ij}(a, b), w_{ij}(a, b) \cdot w_{i'j'}^{-1}(a, b)\} \setminus \{e\}) \cap \gamma_{N_0+1} G = e,$$

$(i, i' \in M, \quad j, j' \in P)$. Now consider an arbitrary sequence $\alpha \in \hat{I}$ such that

$$w(\hat{a}, \hat{b})(\alpha) = w_{i_0 j_0}(\hat{a}, \hat{b})(\alpha) = w_{i_0 j_0}(a, b) \neq e$$

in (G, P_α). By choice of the number N_0 and convexity of the subgroups $\gamma_n G$ in the o-groups (G, P_α) for some natural number $n \leq N_0$, the following relation is valid in the o-group $(\bar{G}_n, \bar{P}_{\alpha(n)})$:

$$\begin{aligned} w(\hat{a}, \hat{b})(\alpha)\gamma_{n+1} G &= w_{i_0 j_0}(a, b)\gamma_{n+1} G = \\ &= \max_{i \in M} \min_{j \in P} w_{ij}(a, b)\gamma_{n+1} G > \gamma_{n+1} G \end{aligned}$$

and $w_{i_0 j_0}(a, b) \in \gamma_n G$ (here max and min are considered in the o-group $(\bar{G}_n, \bar{P}_{\alpha(n)})$). It is evident that if for sequence $\beta \in \hat{I}$ equality $\alpha(n) = \beta(n)$ is fulfilled, then in the o-group $(\bar{G}_n, \bar{P}_{\beta(n)})$

$$\gamma_{n+1} G < w_{i_0 j_0}(a, b)\gamma_{n+1} G = \max_{i \in M} \min_{j \in P} w_{ij}(a, b)\gamma_{n+1} G = [a, \underbrace{b, \dots, b}_{n-1}]^t \gamma_{n+1} G$$

is true, because the o-groups $(\bar{G}_n, \bar{P}_{\alpha(n)})$, $(\bar{G}_n, \bar{P}_{\beta(n)})$ are order isomorphic under the identity isomorphisms.

It follows that the set $\hat{I}$ can be represented in the form $\hat{I} = \hat{I}_1 \cup \hat{I}_2 \cup \ldots \cup \hat{I}_p$, $(1 \leq p \leq 2^{N_0})$, where $\hat{I}_i \cup \hat{I}_j = \emptyset$, $(1 \leq i < j \leq p)$, and for arbitrary $\alpha, \beta \in \hat{I}_l$, $(1 \leq l \leq p)$,

$$w(\hat{a}, \hat{b})(\beta)\gamma_{n(l)+1}G = [a, \underbrace{b, \ldots, b}_{n(l)-1}]^{t_{n(l)-1}}\gamma_{n(l)+1}G =$$
$$= w(\hat{a}, \hat{b})(\alpha)\gamma_{n(l)+1}G, (t_{n(l)-1} \cdot \varepsilon_{n(l)-1} > 0)$$

or

$$w(\hat{a}, \hat{b})(\beta) = w(\hat{a}, \hat{b})(\alpha) = e$$

is true. Let us suppose that for some sequence $\alpha \in \hat{I}_l$

$$w(\hat{a}, \hat{b}) = [a, \underbrace{b, \ldots, b}_{n(l)-1)}]^{t_{n(l)-1}} \cdot \gamma_{n(l)+1}G > \gamma_{n(l)+1}G.$$

Therefore, for every $\beta \in \hat{I}_l$ a subgroup

$$\text{gp}(w(\hat{a}, \hat{b})(\beta), |w_1(\hat{a}, \hat{b})(\beta)| \vee |w_2(\hat{a}, \hat{b})(\beta)) =$$
$$= \text{gp}([a, \underbrace{b, \ldots, b}_{n(l)-1}]^{t_{n(l)-1}} \cdot \psi_\beta, b^s\varphi_\beta)$$

is order isomorphic to the o-group $(G, P_{\overline{\beta(n(l)-1)}})$ (or $(G, P_{\overline{-\beta(n(l)-1)}})$ if $\beta_{n(l)-1} < 0$) under a natural isomorphism, which maps

$$a \to w(\hat{a}, \hat{b})(\beta), \; b \to |w_1(\hat{a}, \hat{b})(\beta)| \vee |w_2(\hat{a}, \hat{b})(\beta)|.$$

But

$$\{\overline{\beta(n(l)-1)} \mid \beta \in \hat{I}_l\} = \hat{I} \text{ or } \{\overline{-\beta(n(l)-1)} \mid \beta \in \hat{I}_l\} = \hat{I}.$$

It follows that the l-subgroup $\bar{\prod}_{\beta \in \hat{I}_l} (G, P_\beta)$ generated by the elements $w(\hat{a}, \hat{b})_{\hat{I}_l}$, $|w_1(\hat{a}, \hat{b})_{\hat{I}_l}| \vee |w_2(\hat{a}, \hat{b})_{\hat{I}_l}|$, where $w(\hat{a}, \hat{b})_{\hat{I}_l}(\beta) = w(\hat{a}, \hat{b})(\beta)$, $|w_1(\hat{a}, \hat{b})_{\hat{I}_i}|(\beta) = |w_1(\hat{a}, \hat{b})|(\beta)$, $|w_2(\hat{a}, \hat{b})_{\hat{I}_l}|(\beta) = |w_2(\hat{a}, \hat{b})|(\beta)$ for every $\beta \in \hat{I}_l$, is l-isomorphic to the l-group $\hat{G}$ =l-gp$(\hat{a}, \hat{b})$ under an isomorphism which maps $a \to w(\hat{a}, \hat{b})_{\hat{I}_l}$, $b \to |w_1(\hat{a}, \hat{b})_{\hat{I}_l}| \vee |w_2(\hat{a}, \hat{b})_{\hat{I}_l}|$.

If $w(\hat{a},\hat{b})(\alpha) = e = w(\hat{a},\hat{b})(\beta)$ for every $\alpha, \beta \in \hat{I}_l$, then

$$(|w_1(\hat{a},\hat{b})| \vee |w_2(\hat{a},\hat{b}|)(\alpha)\mathrm{fun}((b),(a)) = b^s\mathrm{fun}((b),(a)) = (|w_1(\hat{a},\hat{b})| \vee |w_2(\hat{a},\hat{b})|)(\beta)\mathrm{fun}((b),(a))$$

and $s > 0$. Therefore, the l-subgroup $\bar{\prod}_{\beta\in\hat{I}_l}(G,P_\beta)$ generated by the elements $(|w_1(\hat{a},\hat{b})_{\hat{I}_l}| \vee |w_2(\hat{a},\hat{b})_{\hat{I}_l}|)$ and $w(\hat{a},\hat{b})_{\hat{I}_l} = e$, is an l-homomorphic image of the l-group $\hat{G}$ under an l-homomorphism which extends the map $\hat{a} \to w(\hat{a},\hat{b})_{\hat{I}_l}$, $\quad b \to |w_1(\hat{a},\hat{b})_{\hat{I}_l}| \vee |w_2(\hat{a},\hat{b})_{\hat{I}_l}|$. Then $H =$

$$= \bar{\prod_{\alpha\in\hat{I}}}(G,P_\alpha) = \bar{\prod_{\alpha\in\hat{I}_1}}(G,P_\alpha) \times \bar{\prod_{\alpha\in\hat{I}_2}}(G,P_\alpha) \times \ldots \times \bar{\prod_{\alpha\in\hat{I}_p}}(G,P_\alpha)$$

and the l-subgroup generated by the elements

$$w(\hat{a},\hat{b}) = w(\hat{a},\hat{b})_{\hat{I}_1} \cdot \ldots \cdot w(\hat{a},\hat{b})_{\hat{I}_p}, \quad |w_1(\hat{a},\hat{b})| \vee |w_2(\hat{a},\hat{b})| =$$
$$= (|w_1(\hat{a},\hat{b})_{\hat{I}_1}| \vee |w_2(\hat{a},\hat{b})_{\hat{I}_1}|) \cdot \ldots \cdot (|w_1(\hat{a},\hat{b})_{\hat{I}_p}| \vee |w_2(\hat{a},\hat{b})_{\hat{I}_p}|)$$

is l-isomorphic to the l-group $\hat{G}$. Now the proof of the next lemma follows.

LEMMA 6. *If for some elements $x, y \in \hat{G}$ $\;\|[x,y]\| \neq e$, then the l-subgroup l-gp$(\|[x,y]\|, |x| \vee |y|)$ generated in $\hat{G}$ by the elements $\|[x,y]\|$ and $|x| \vee |y|$ is l-isomorphic to the l-group $\hat{G}$ under an l-isomorphism which maps $\hat{a} \to \|[x,y]\|, \hat{b} \to |x| \vee |y|$.* □

Let $H_1 = \hat{G}^J/\mathcal{F}$ be an ultrapower of the l-group $\hat{G}$. The following statesments follow immediately from the definitions and Lemma 6.

LEMMA 7. *Let u, v be elements of $H_1 = \hat{G}^J/\mathcal{F}$ and $\|[u,v]\| \neq e$. Then the l-subgroup l-gp$(\|[u,v]\|, |u| \vee |v|)$ of the l-group H_1, generated by the elements $\|[u,v]\|$, $|u| \vee |v|$ is l-isomorphic to the l-group $\hat{G}$.* □

LEMMA 8. *Let X be a non-abelian l-subgroup of the Cartesian product $\bar{\prod}_{\gamma\in\Gamma} V_\gamma$ where $V_\gamma (\gamma \in \Gamma)$ are ultrapowers of the l-group $\hat{G}$ .Then for every pair of elements $u, v \in X$ such that $[u,v] \neq e$, l-gp$(|u| \vee |v|, \|[u,v]\|)$ is l-isomorphic $\hat{G}$.* □

By the Theorem 2 of Section 1, the l-quasivariety $q_l(\hat{G})$, generated by the l-group $\hat{G}$ coincides with the class of l-groups $SPP_U(\hat{G})$, where S, P, P_U are closure operators of forming l-subgroups, Cartesian products and ultraproducts. Therefore, if X is a non-abelian l-group from $q_l(\hat{G})$, then by Lemma 8, X contains an l-subgroup which is l-isomorphic to the l-group $\hat{G}$. It follows that every non-abelian l-quasivariety $\mathcal{Q}_l \subseteq q_l(\hat{G})$ contains the l-group $\hat{G}$ and so $\mathcal{Q}_l = q_l(\hat{G})$.

THEOREM 3. (Medvedev) *The l-quasivariety* $q_l(\hat{G})$ *has the following properties:*

(19) $q_l(\hat{G})$ *covers the l-variety of Abelian l-groups* $\mathcal{A}$ *in the lattice of l-quasivarieties* Λ;

(20) *every o-group from* $q_l(\hat{G})$ *is Abelian;*

(21) *every nilpotent l-group from* $q_l(\hat{G})$ *is Abelian;*

(22) $q_l(\hat{G})$ *is contained in the l-variety of weakly abelian l-groups* $\mathcal{W}$. □

Since the l-variety $\mathrm{var}_l(G_p)$ generated by the Scrimger's l-group G_p coinside with the l-quasivariety $q_l(G_p)$ generated by G_p, then from Theorem 12.6.4. follows the classification of the covers of $\mathcal{A}$ in the lattice of l-quasivarieties Λ.

THEOREM 4. (Gurchenkov and Kopytov [1], D. Ann Van Ree [1].) *If* $\mathcal{X}$ *is a cover of* $\mathcal{A}$ *in the lattice of l-quasivarieties* Λ, *then one of the following is true:*

(23) $\mathcal{X} = S_p = q_l(G_p)$ *for some prime* p *where* G_p *is a Scrimger l-group ;*

(24) $\mathcal{X} \subseteq q_l(\mathbf{Z}\mathrm{wr}(\mathbf{Z}, \mathbf{Z}))$;

(25) $\mathcal{X} \subseteq \mathcal{R}$.□

14.5 Covers in the Lattice of l-Quasivarieties

Here we will prove that the lattice of l-quasivarieties Λ has no covering condition. More precisely, there are l-quasivarieties which has no covers in the lattice of l-quasivarieties Λ.

Let P be the set of all prime numbers and $M \subseteq P$. Let the l-quasivariety $\mathcal{Q}_M$ is defined by the following system of implications Σ_M:

$$((|x| \vee |y|)^{-1}|[x,y]|(|x| \vee |y|) = |[x,y]|^{p^k}) \Rightarrow ([x,y] = e),$$

where $k \in N, k \geq 2, p \in M \subseteq P$. Let $G_{p^k} = \mathrm{gp}(c, d \mid c^{-1}dc = d^{p^k})$. Then the normal closure $D = (d^G) = \mathrm{gp}(d^g \mid g \in G_{p^k})$ of the element d is isomorphic to the subgroup of rational numbers $\mathbf{Q}$ which consists of the elements $\frac{a}{b}$ with $a \in \mathbf{Z}$ and $b = p^{kn},\ n \in Z$. Thus G_{p^k} is the semidirect product of D and the infinite cyclic group (c). Since D is isomorphic to the subgroup of rational numbers Q then D has only two different total orders. Let P be a total order on D such that $d > e$. Now we define the total order on G_{p^k} by the rule: $c^k d \geq e$ iff $k > 0$ or $k = 0$ and $d \in D,\ d \geq e$ in o-group (D, P).

THEOREM 1. (Medvedev [10].) *For each non-empty set of prime numbers M there is no l-quasivariety covering the l-quasivariety $\mathcal{Q}_M$ in the lattice of l-quasivarieties Λ.*

Proof. Assume, on the contrary, that an l-quasivariety $\overline{\mathcal{Q}}_M$ covers $\mathcal{Q}_M$ in the lattice Λ. Then $\mathcal{Q}_M \supseteq \overline{\mathcal{Q}}_M$ and there is an l-group $G \in \overline{\mathcal{Q}}_M \setminus \mathcal{Q}$. Since $G \notin \mathcal{Q}_M$, then at least one implication from Σ_M is violated in G. Let the implication

$$((|x| \vee |y|)^{-1}|[x,y]|(|x| \vee |y|) = |[x,y]|^{p^{k_0}}) \Rightarrow ([x,y] = e)$$

be violated in G. Then there are $a, b \in G$ such that

$$(|a| \vee |b|)^{-1}|[a,b]|(|a| \vee |b|) = |[a,b]|^{p^{k_0}} \text{ and } |[a,b]| > e.$$

Then the subgroup $A \subseteq G$ generated by the elements $d = |[a,b]|$ and $c = |a| \vee |b|$ is isomorphic (as group) to the group $G_{p^{k_0}}$. We claim that A is totally ordered and order isomorphic to o-group $G_{p^{k_0}}$. In fact, $d, c > e$ and $d \wedge c > e$. If $d \wedge c \neq d$, then $e < (c \wedge d)^{-1} \leq d$ and $(c \wedge d)^{d^{-1}} \neq d^{d^{-1}} = d$. Hence $(c \wedge d)^{d^{-1}} = cd^{p^{k_0}-1} \wedge d = (cd^{p^{k_0}-2} \wedge e)d = d$ as $d^{p^{k_0}-2} \geq d^2 > e$. Thus $c \wedge d = d$. Similarly $c \wedge d^n = d^n$ for all $n \geq 2$ and $c \gg d > e$ in A. Each element a of A can be represented in the form $a = c^k d^n$. By the above arguments, $c^k d^n > e$ iff $k > 0$, or $k = 0$ and $n > 0$. The normal closure $D = \mathrm{gp}(d^a \mid a \in A)$ is a torsion-free Abelian group of rank 1. Since the lattice order on the l-group G is isolated then any two elements of A are comparable. Thus A is an o-subgroup of G and is order isomorphic to o-group $G_{p^{k_0}}$. Let H be the subgroup of $G_{p^{k_0}}$ generated by the elements c^3 and d. It is clear that H is order isomorphic to the o-group $G_{p^{3k_0}}$. Let $q_l(G_{p^{k_0}})$ and $q_l(G_{p^{3k_0}})$ be the l-quasivarieties generated by the o-groups $G_{p^{k_0}}, G_{p^{3k_0}}$, respectively. It is clear that $\overline{\mathcal{Q}}_M \supseteq q_l(G_{p^{k_0}}) \supseteq q_l(G_{p^{3k_0}})$. Since the implication

$$(|x| \vee |y|)^{-1}|[x,y]|(|x| \vee |y|) = |[x,y]|^{p^{3k_0}} \Rightarrow [x,y] = e$$

violates in o-groups $G_{p^{k_0}}$, $G_{p^{3k_0}}$ if $x = c^3$, $y = d$, then $q_l(G_{p^{k_0}}) \not\subseteq \mathcal{Q}_M$, $\mathcal{Q}_l(G_{p^{3k_0}}) \not\subseteq \mathcal{Q}_M$. Thus

$$\overline{\mathcal{Q}}_M = \mathcal{Q}_M \vee q_l(G_{p^{k_0}}) = \mathcal{Q}_M \overset{\Lambda}{\bigvee} q_l(G_{p^{3k_0}}).$$

The implication

$$((|x| \vee |y|)^{-1}|[x,y]|(|x| \vee |y|) = |[x,y]|^{p^{k_0}}) \Rightarrow [x,y] = e$$

holds in l-quasivariety $q_l(G_{p^{3k_0}})$. Thus this implication is valid in the l-quasivariety $\mathcal{Q}_M \overset{\Lambda}{\bigvee} q_l(G_{p^{3k_0}})$. The direct verification shows that this implication violates on o-group $G_{p^{k_0}}$ if $x = c$, $y = d$. This contradicts $\overline{\mathcal{Q}}_M = \mathcal{Q}_M \overset{\Lambda}{\bigvee} q_l(G_{p^{k_0}}) = \mathcal{Q}_M \overset{\Lambda}{\bigvee} q_l(G_{p^{3k_0}})$. Therefore, $\mathcal{Q}_M$ has no any cover in the lattice Λ. □

COROLLARY 1. *There is a continuum of different l-quasivarieties each of which has no covers in the lattice* Λ. □

COROLLARY 2. *Each l-quasivariety* $\mathcal{Q}_M$ *has no independent basis of implications.*

Proof follows immedeatly from Proposition 12.7.1.□

14.6 Non-Axiomatizability of Lattice-Ordered Groups

In this section we will prove that the class of all lattice-orderable groups cannot be defined by any set of first order group sentences.

THEOREM 1. (Vinogradov [2].) *The class of all lattice-orderable groups is not axiomatizable.*

Proof. For the proof, it is enough to construct two elementary equivalent groups G_1 and G_2 such that G_1 is lattice-orderable and G_2 is not lattice-orderable. Let Q_1, Q_2 be isomorphic copies of additive group of rationals **Q** written multiplicatively: $Q_1 = \{a^q \mid q \in Q\}$, $Q_2 = \{c^q \mid q \in Q\}$. Let G be the semidirect product of Q_1 and the infinite cyclic group $B = (b)$ corresponding the automorphism : $b^{-1}ab = a^{-1}$. Let $D = (d)$ be an infinite cyclic group generated by the element d and $G_1 = G \times D$. We claim that G_1 is not lattice-orderable. Assume, on the contrary, that G_1 is lattice-ordered and $|a| = b^p a^q d^r$. Then $b^{-1}|a|b = b^{-1}b^p a^q d^r b = b^p a^{-q} d^r$. By definition of G_1 we have

$$b^{-1}|a|b = b^{-1}(a \vee a^{-1})b = a^{-1} \vee a = |a|.$$

Thus, $q = 0$. Let n be a positive integer such that $n > |p|, n > |r|$. Then $|a^{1/n}|^n = |a| = b^p d^r, |a^{1/n}| = (b^p d^r)^{1/n}$ and the equation $x^n = b^q d^r$ has the solution in G_1, a contradiction to the definition of G_1. Therefore, G_1 is not lattice-orderable.

Let $G_2 = (G \times Q_2) \times D$. Now we will define a lattice-order on G_2. In fact, the group $G \times Q_2$ is a splitting extension of the direct product $Q_1 \times Q_2$ by the infinite cyclic group (b) corresponding to the automorphism $b^{-1}ab = a^{-1}, \quad b^{-1}cb = c$. If we set $u = ac, \quad v = ac^{-1}$, then each element of $Q_1 \times Q_2$ can be uniquely represented in the form $u^r v^s$ for some rational numbers r and s. We define the lattice order on $Q_1 \times Q_2$ by the rule: $x = u^r v^s \geq e$ if and only if $r \geq 0$ and $s \geq 0$. This order is invariant under inner automorphisms of G_2. So we can order $G \times Q_2$ by the rule: $x = b^t u^r v^s \geq e$ for $x \in G \times Q_2$ if and only if $t > 0$, or $t = 0$ and $r \geq 0, \ s \geq 0$. It is readily verified that under this order $G \times Q_2$ is lattice-ordered. Now it is evident that $G \times Q_2 \times D$ is a lattice-ordered group.

From standard results of universal algebra, it follows that the Abelian group $D \times Q_2$ is elementary equivalent to $D = (d)$. Then the groups $(D \times G) = G_1$ and $(D \times Q_2) \times G$ are elementary equivalent (W.Szmelew [1]).

BIBLIOGRAPHY

ADIJAN S.I.

1. *Infinite irreducible systems of group identities of group identities (in Russian)*, Izv. Akad. Nauk SSSR, Ser. Mat. **34**(1970), 715-734.

ARORA A.K.

1. *Quasi-varieties of lattice-ordered groups,* Algebra Universalis **20**(1985), 34-50.

ARORA A.K., MCCLEARY S.H.

1. *Centralizers in free lattice-ordered groups,* Houston J. Math. **12**(1986), 455-482.

AULT J.S.

1. *Right-ordered locally nilpotent groups,* J. London Math. Soc., **2**(1972), 662-666.

BALL R.N.

1. *Full convex l-subgroups and the existence of a*-closures of lattice-ordered groups,* Pacific J. Math., **61**(1975), 7-16.

2. *The generalized orthocompletion and strongly projectable hull of a lattice-ordered group,* Can. J. Math. **34**(1982), 621-661.

BALL R.N., DROSTE M.

1. *Normal subgroups of doubly transitive automorphism groups of chains,* Trans. Amer. Math. Soc. **290**(1985), 647-664.

BERGMAN G.M.

1. *Specially ordered groups,* Comm. Algebra **12**(1984), 2315-2333.

BERNAU S.J.

1. *Orthocompletion of lattice groups,* Proc. London Math. Son., **16**(1966), 107-130.

2. *Free abelian lattice groups,* Math. Ann., **180**(1969), 48-59.

3. *Varieties of lattice groups are closed under L-completion,* Istituto Nazionale Di Alta Matematica, Symposia Mathematica **21**(1977), 349-355.

BIGARD A., KEIMEL K., WOLFENSTEIN S.

1. *Groupes et Anneaux Réticulés,* Lecture Notes in Math. **608**, Springer-Verlag, New-York, 1977.

BIGARD A., CONRAD P., WOLFENSTEIN S.

1. *Compactly generated lattice-ordered groups,* Math. Z.,**107**(1968), 201-211.

BIRKHOFF G.

1. *Lattice-ordered groups,* Ann. Math. **43**(1942), 298-331.

2. *Lattice Theory,* Amer. Math. Soc. Colloquium **25**(1960), Providence .

3. *Lattice Theory,* Amer. Math Soc. Colloquium **25**(1967), 3-rd ed., Providence.

BLEIER R.D.

1. *The SP-hull of lattice-ordered groups.* Canad. J. Math. **26** (1974), 866-878.

BLUDOV V.V.

1. *Groups ordered in unique way (in Russian),* Algebra i Logika **13**(1974), 609-634.

BLUDOV V.V., MEDVEDEV N.YA.

1. *On the completion of ordered metabelian groups (English translation),* Algebra and Logic **13**(1974), 369-373.

BOWDI A.A.

1. *On twisted products of semigroups and rings (in Russian),* Dokl. Akad. Nauk SSSR **137**(1981), 1267-1269.

BUDKIN A.I.

1. *Independent axiomatization of quasivarieties of groups (in Russian),* Mat. Zametki **31**(1980), 817-826.

BURNS R.G., HALE V.W.

1. *A note on group rings of certain torsion-free groups,* Canad. Math. Bull. **15**(1972), 441-445.

BURRIS S., SANKAPPANAVAR H.P.

1. *A Course in Universal Algebra,* Graduate Texts in Mathematics **78**, Springer-Verlag, New York-Heidelberg-Berlin, 1981.

BUSULINI B.

1. *Sulla relazione triangolare in un l-grupo,* Rend. Sem. Math. Univ. Padova **28**(1958), 68-70.

BUTTSWORTH R.N.

1. *A family of groups with a countable infinite of full orders.* Bull. Austral. Math. Soc. **4**(1974), 97-104.

BYRD R.D.

1. *Complete distributivity in lattice-ordered groups,* Pasific J. Math., **20**(1967), 423-432

BYRD R.D., LLOYD J.T.

1. *Closed subgroups and complete distributivity in lattice-ordered groups,* Math. Z. **101**(1967), 123-130.

2. *A note on lateral completion in lattice-ordered groups,* J. London Math. Soc. **1**(1969), 358-362.

3. *Kernels in lattice-ordered groups,* Proc. Amer.Math. Soc. **57**(1976), 16-18.

BYRD R.D., LLOYD J.T., MENA R.A., TELLER J.R.

1. *Retractable groups,* Acta Math. Sci. Hungar. **29**(1977), 219-233.

CHEHATA C.G.

1. *An algebraically simple ordered group,* Proc. London Math. Soc. **2**(1952), 183-197.

CLIFFORD A.H.

1. *A noncommutative ordinally simple linearly ordered groups,* Proc. Amer. Math. Soc. **2**(1951), 902-903.

COHN P.M.

1. *Group of order automorphisms of ordered sets,* Mathematika **4**(1957), 41-50.

2. *Universal algebra,* Harper & Row, London, 1965.

CONRAD P.F.

1. *Right-ordered groups,* Michigan Math.J. **6**(1959), 267-275.

2. *The structure of a lattice-ordered group with a finite number of disjoint elements,* Michigan Math. J. **7**(1960), 171-180.

3. *The lattice of all convex l-subgroups of a lattice-ordered group,* Czech. Math J. **15**(1965), 101-123.

4. *Introduction à la thèorie des groupes rèticulès,* Lect. Notes Univ. Paris 1967.

5. *Free lattice-ordered groups,* J.Algebra **16**(1970), 190-203.

6. *Free abelian l-groups and vector lattices,* Math. Ann. **190**(1971), 306-312.

7. *The hulls of representable l-groups and f-rings,* J. Austral. Math. Soc. **16**(1973), 385-415.

CONRAD P.F., MCALISTER D.

1. *The completion of a lattice-ordered group,* J.Austral. Math. Soc. **9**(1969), 182-208.

DANWOODY M.Y.

1. *On verbal subgroups of free groups,* Math. Z. **104**(1968), 91-97.

DARNEL M.R.

1. *Special-valued l-groups and abelian covers,* Order **4**(1987), 191-194.

2. *The free lattice-ordered group over a nilpotent group,* Proc. Amer. Math. Soc. **111**(1991), 301-307.

DARNEL M.R., GLASS A.M.W.

1. *Commutator relations and identities in lattice-ordered groups,* Mich. Math. J. **36**(1990), 203-211.

DLAB V.

1. *On a family of simple ordered groups,* J. Austral. Math. Soc. **8**(1968), 591-608.

DROSTE M., SHELAH S.

1. *A construction of all normal subgroup lattice of o-2-transitive automorphism groups of linearly ordered sets,* Israel J. Math. **51**(1985), 223-261.

EVERETT C.J., ULAM S.

1. *On ordered groups,* Trans Amer. Math. Soc. **57**(1945), 208-216.

FEIL T.

1. *Varieties of representable lattice-ordered groups,* Doctoral Thesis, Bowling Green State University, 1980.

2. *An uncountable tower of l-group varieties,* Algebra Universalis **14**(1981), 129-131.

3. *Varieties of representable l-groups,* in *Ordered Algebraic Structures,* Lecture Notes in Pure and Applied Math. **99**, ed. W.B.Powell and C.Tsinakis, Marcel Dekker, New York, 1985, 89-98.

FRANCHELLO J.D.

1. *Sublattices of free products of lattice-ordered groups,* Algebra Universalis **16**(1983), 101-110.

FUCHS L.

1. *Partially Ordered Algebraic Systems,* Pergamon Press, New York, 1963.

FUCHS L., SASIADA E.

1. *Note on orderable groups,* Ann. Univ. Sci. Budapest. Sect. Math. **7** (1965), 13-17.

GLASS A.M.W.

1. *An application of ultraproducts to lattice-ordered groups,* J.London Math. Soc. **4**(1972), 533-540.

2. *l-simple lattice-ordered groups,* Proc. Edinburgh Math. Soc. **19**(1974), 133-138.

3. *The word problem for lattice-ordered groups,* Proc. Edinburgh Math. Soc. **19**(1975), 217-219.

4. *Results in partially ordered groups,* Commun. Algebra **3**(1975), 749-761.

5. *Ordered permutation groups,* Bowling Green State University, Bowling Green, Ohio, 1976.

6. *Ordered permutation groups,* London Math. Soc. Lecture Notes Series No. **55**, Cambridge University Press, 1981.

GLASS A.M.W., MCCLEARY S.H.

1. *Some l-simple pathological lattice-ordered groups,* Proc. Amer. Math. Soc. **57**(1976), 221-226.

GLASS A.M.W., HOLLAND C.W., MCCLEARY S.H.

1. *The structure of l-group varieties,* Algebra Universalis **10**(1980), 1-20.

GORBUNOV V.A.

1. *Covers in the lattice of quasivarieties and independent axiomatizability (in Russian),* Algebra i Logika **16**(1977), 507-548.

GORTHAKOV YU. M.

1. *Example of G-periodic torsion-free group (in Russian),*Algebra i Logika **6**(1967), 13-24.

GURCHENKOV S.A.

1. *Minimal varieties of l-groups (in Russian),* Algebra i Logika **21**(1982),131-137.

2. *Varieties of nilpotent lattice-ordered groups (in Russian),* Algebra i Logika **21**(1982), 499-510.

3. *Varieties of l-groups with the identity* $[x^p, y^p] = e$ *have finite basis (in Russian),* Algebra i Logika **23**(1984), 27-47.

4. *On covers in the lattice of l-varieties (in Russian),* Mat. Zametki **35**(1984), 677-684.

5. *About varieties of l-groups with infinite axiomatic rank (in Russian),* Sibirsk. Mat. J. **26**(1985), 66-70.

6. *On the theory of varieties of lattice-ordered groups (in Russian),* Algebra i Logika **27**(1988), 249-273.

7. *The lattice of varieties of weakly abelian lattice-ordered groups has not covering condition (in Russian),* Mat. Zametki **47**(1990), 35-40.

8. *On three questions of the theory of l-varieties (in Russian),* Czech. Math. J. **41**(1991), 405-410.

GURCHENKOV S.A., KOPYTOV V.M.

1. *On covers of the l-variety of abelian lattice-ordered groups (English translation),* Sibirsk. Mat. J. **28**(1987), 406-408.

GUREVICH YU.S.

1. *Elementary properties of ordered abelian groups,* Transl. Amer. Math. Soc. **46**(1965), 165-192.

2. *Hereditary undecidability of the theory of lattice-ordered abelian groups (in Russian),* Algebra i Logika **6**(1967), 45-62.

GUREVICH YU.S., KOKORIN A.I.

1. *Universal equivalence of ordered abelian groups (in Russian),* Algebra i Logika **2**(1963), 37-39.

HALL M.JR.

1. *The Theory of Groups,* Macmillan , New York, 1959.

HALL PH.

1. *Finiteness conditions for solvable groups,* Proc. London Math. Soc. **4**(1954), 419-436.

2. *Nilpotent groups,* Canad. Math. Congress, Summer sem. Univ. Alberta, 1957.

3. *Wreath powers and characteristically simple groups,* Proc. Cambr. Phil. Soc. **58**(1962), 170-184.

4. *On non-strictly simple groups,* Proc. Cambr. Phil. Soc. **59**(1963), 531-553.

HIGMAN G.
1. *The units of group rings,* Proc. London Math. Soc. **46**(1940), 231-248.
HOLLAND CH. W.
1. *Extensions of ordered algebraic structures,* Doctoral Thesis, Tulane University, 1961.
2. *The lattice-ordered group of an ordered set,* Michigan Math. J. **10**(1963), 399-408.
3. *Transitive lattice-ordered groups,* Math. Z. **87**(1965),420-434.
4. *The interval topology of a certain l-group,* Czech. Math. J. **15**(1965), 311-314.
5. *A class of simple lattice-ordered groups,* Proc. Amer. Math. Soc. **16**(1965), 326-329.
6. *The characterization of generalized wreath products,* J.Algebra **13**(1969), 152-172.
7. *Ordered permutation groups:* Permutation. - Actes Coll. Univ. Renè Descartes, 1972(1974), 57-74.
8. *Outer automorphisms of ordered permutation groups,* Proc. Edinburgh Math. Soc. **19**(1975), 331-344.
9. *The largest proper variety of lattice-ordered groups,* Proc. Amer. Math. Soc. **57**(1976), 25-28.
10. *Group equations which hold in lattice-ordered groups,* Symposia Math. **21**(1977), 365-378.
11. *Varieties of l-groups are torsion classes,* Czech. Math. J. **23**(1979), 11-12.
12. *Varieties of automorphism group of orders,* Trans, Amer. Math. Soc. **288**(1985), 755-763.
HOLLAND W. CH., MARTINEZ J.
1. *Free products of lattice-ordered groups,* Algebra Universalis **2**(1972), 247-254.
HOLLAND W. CH., MCCLEARY S.H.
1. *Wreath products of ordered permutation groups,* Pacific J. Math. **31**(1969), 703-716.
2. *Solvability of the word problem in free lattice-ordered groups,* Houston J. Math. **5**(1979), 99-105.

HOLLAND W. CH., MEKLER A. H., REILY N.R.

1. *Varieties of lattice-ordered groups in which prime powers commute,* Algebra Universalis **23**(1986), 196-214.

HOLLAND W. CH., SCRIMGER E.

1. *Free Products of lattice-ordered groups,* Algebra Universalis **2**(1972), 247-254.

HOLLAND W. CH., REILY N. R.

1. *Metabelian varieties of l-groups which contain no non-abelian o-groups,* Algebra Universalis **24**(1987), 204-223.

HOLLISTER H.

1. *Contribution to the theory of partially ordered groups,* Doctoral Thesis, Univ. of Michigan, 1965.

ISAEVA O.V., MEDVEDEV N.YA.

1. *Covers in the lattice of quasivarieties of l-groups,* Sibirsk. Mat. J. **33**(1992), 102-107.

IWASAWA K.

1. *On the structure of conditionally complete lattice groups,* Jap. J. Math. **18**(1943), 177-183.
2. *On linearly ordered groups,* J. Math. Soc. Japan **1**(1948), 1-9.

JAFFARD P.

1. *Theorie des filets dans les groupes réticulés,* C. R. Acad. Sci. Paris **230**(1950), 1024-1025.
2. *Contribution à l'á theorie des groupes ordonnés,* J.Math. Pures Appl. **32**(1953), 203-280.

JAKUBIK J.

1. *Kompakt erzeugte Verbandsgruppen,* Math. Nachr. **30**(1965), 193-202.
2. *Higher degrees of distributivity in lattices and lattice-ordered groups,* Czech. Math. J. **18**(1968), 356-376.
3. *Lattice-ordered groups of finite breadth,* Colloq. math. **27**(1973), 13-20.
4. *Products of torsion classes of lattice-ordered groups,* Czech. Math. J. **25**(1975), 576-585.
5. *Radical mappings and radical classes of lattice-ordered groups,* Symposia Math. **21**(1977), 451-477.

JAKUBIKOVA M.

1. *On some subgroups of l-groups (in Russian),* Mat.-Fyz. Časopis Sloven. Akad. Vied **12**(1962), 97-107.

JÒNSSON B.

1. *Algebras whose congruence lattices are distributive,* Math. Scand. **21**(1967), 110-121.

KANTOROVICH L.V.

1. *Lineare halbgeordnete Raume,* Mat. Sb. **2**(1937), 121-168.

KAPPOS D.A., KEHAYOPULOS N.

1. *Some remarks on the representation of lattice-ordered groups,* Math. Balkan. **1**(1971), 142-143.

KARGAPOLOV M.I.

1. *Fully orderable groups (in Russian),* Algebra i Logika **2**(1963), 5-14.

KARGAPOLOV M.I., MERZLJAKOV YU.I.

1. *Fundumentals of the theory of groups,* Springer, Berlin, 1979.

KHISAMIEV N.G.

1. *Universal theory of lattice-ordered abelian groups (in Russian),* Algebra i Logika **5**(1966), 71-76.

KHISAMIEV N.G., KOKORIN A.I.

1. *An elementary classification of lattice-ordered abelian groups with a finite number of fibers (in Russian),* Algebra i Logika **5**(1966), 41-50.

KOKORIN A.I.

1. *On a class of lattice-ordered groups(in Russian),* Ural. Gos. Univ. Mat. Zap. **3**(1962), 37-38.

2. *Methods for the lattice-ordering of a free Abelian group with a finite number of generators (in Russian),* Ural. Gos. Univ. Mat. Zap. **4**(1963),45-48.

3. *On fully orderable groups (in Russian),* Dokl. Akad. Nauk SSSR **151**(1963), 31-33.

4. *On the theory of fully orderable groups (in Russian),* Algebra i Logika **2**(1963), 15-20.

5. *Fully orderadebility of the direct product of fully orderable groups (in Russian),* Ural. Gos. Univ. Mat. Zap. **4**(1963), 95-98.

6. *Γ-fully orderable and relatively convex subgroups of orderable groups (in Russian),* Siberian Math J. **7**(1966), 713-717.

KOKORIN A.I., KOPYTOV V.M.
1. *On some classes of orderable groups (in Russian),* Algebra i Logika **1**(1962), 21-23.
2. *Fully Ordered Groups,* Halstead Press (John Wiley and Sons), New York - Toronto, 1974.
KONTOROVICH P.G., KUTYEV K.M.
1. *On the theory of lattice-ordered groups (in Russian),* Izv. Vysš. Učebnich. Zaved.Mat. **10**(1959), 112-120.
KOPYTOV V.M.
1. *On the completion of the centre of orderable group (in Russian),* Ural. Gos. Univ. Mat. Zap. **4**(1963), 20-24.
2. *Some questions of the theory of orderable groups (in Russian),* Uspekhi Mat. Nauk **20**(1965), 305.
3. *On the theory of fully orderable groups (in Russian),* Algebra i Logika **5**(1966), 27-31.
4. *On totally ordered soluble groups (in Russian)* , Algebra i Logika **12**(1973), 655-666.
5. *Lattice-ordered locally nilpotent groups (in Russian),* Algebra i Logika **14**(1975), 407-413.
6. *Free lattice-ordered groups (English translation),* Algebra and Logic **18**(1979), 259-270.
7. *Nilpotent lattice-ordered groups (in Russian),* Sibirsk. Mat. Zh. **23**(1982),127-131.
8. *Free lattice-ordered groups (in Russian),* Sibirsk. Mat. Zh. **24** (1983),120-124.
9. *Nonabelian varieties of lattice-ordered groups in which every solvable l-group is abelian (in Russian),* Mat. Sb. **126**(168) (1985), 247-266.
KOPYTOV V.M., MEDVEDEV N.YA.
1.*On totally ordered groups whose system of convex subgroups is central (in Russian),* Mat. Z. **19**(1975), 85-90.
2. *Varieties of lattice-ordered groups (English translation),* Algebra and Logic **16**(1977), 281-285.
KRULL W.
1. *Idealtheorie,* Ergebnisse ser., Berlin, 1935.

La Grange R.H., Rhemtulla A.H.
1. *A remark on the group rings of ordered preserving permutation groups,* Canad. Math. Bull. **11**(1968), 679-680.
Levi F.W.
1. *Ordered groups,* Proc. Indian Acad. Sci. A. **16**(1942), 256-263.
Litvinova M.V.
1. *Product of finitely based varieties of lattice-oedered groups,* **29** Soviet Union scientific students' conference, Novosibirsk, 1991.
Lloyd J.T.
1. *Complete distributivity in certain infinite permutation groups,* Michigan Math. J. **14**(1967), 393-400.
Lorenz K.
1. *Über Strukturverbande von Verbandsgruppen,* Acta Math. Acad. Sci. Hung. **13**(1962), 55-67.
Lorenzen P.
1. *Abstrakte Begrunding der multiplikativen Idealtheorie,* Math. Z. **45**(1939), 533-553.
2. *Über halbgeordnete Gruppen,* Arch. Math. **2**(1949), 66-70.
3. *Über halbgeordnete Gruppen,* Math. Z., **52**(1949), 483-526.
Łoš J.
1. *On the existence of linear order in a group,* Bull. Acad. Polon. Sci. **2**(1954), 21-23.
Mal'cev A.I.
1. *On the embedding of group algebras in division algebras (in Russian)* , Dokl.Akad. Nauk SSSR **60**(1948),1499-1501.
2. *Nilpotent torsion-free groups (in Russian),* Izv. Akad. Nauk SSSR ser. Mat. **13**(1949), 201-212.
3. *On ordered groups (in Russian),* Izv. Akad. Nauk SSSR ser. Mat. **13**(1949), 473-482.
4. *On the full ordering of groups (in Russian),* Trudy Mat. Inst. Steklov **38**(1951), 173-175.
5. *Remark on partially ordered groups (in Russian),* Učen. Zap. Ivanovs. Gos. Ped. Inst. **10**(1956), 3-5.
6. *On partially ordered nilpotent groups (in Russian),* Algebra i Logika **1**(1962), 5-9.
7. *Algebraic systems,* Springer , Berlin, 1976.

MARTINEZ J.

1. *Approximation by Archimedean lattice cones,* Pacific J. Math. **36**(1971), 427-437.

2. *Free products in varieties of lattice-ordered groups,* Czech. Math. J. **22**(1972), 535-553.

3. *Varieties of lattice-ordered groups,* Math. Z. **137**(1974), 265-284.

4. *Torsion theory for lattice-ordered groups,* Czech. Math. J. **25**(1975), 284-299.

5. *Torsion theory for lattice-ordered groups (part 2); Homogeneous l-groups,* Czech. Math. J. **26**(1976), 93-100.

MATSUSITA S.

1. *Sur la puissance des ordres dans un groupe libre,* Proc. koninkl. Nederl. Akad. Wet. A. **56**(1953), 15-16.

MCCLEARY S.H.

1. *Pointwise suprema of order-preserving permutations,* Ill. J. Math. **16**(1972), 69-75.

2. *O-primitive ordered permutation groups,* Pacific J. Math. **40**(1972), 349-372.

3. *O-primitive ordered permutation groups (2),* Pacific J. Math. **49**(1973), 431-445.

4. *o-2-transitive ordered permutation groups,* Pacific J.Math. **49**(1973), 425-429.

5. *Some simple homeomorphisms groups having nonsolvable outer automorphism groups,* Commun. in Algebra **6**(1978), 483-496.

6. *Telling left from right,* Ordered algebraic structures (Curaçao, 1988), Kluwer Acad. Publ., Dordrecht, 1989, 93-97.

7. *Free lattice-ordered groups represented as o-2-transitive l-permutation groups,* Trans. Amer. Math. Soc. **290**(1985), 69-79.

8. *An even better representation for free lattice-ordered groups,* Trans. Amer. Math. Soc. **290**(1985), 81-100.

MEDVEDEV N.YA.

1. *Ordered groups with finite number of relatively many convex subgroups (in Russian),* Sibirsk. Mat. Z. **15**(1974), 445-449.

2. *On the theory of partially ordered groups (in Russian),* 12 Soviet Union scientific students' conference, Novosibirsk, 1975.

3. *Lattices of varieties of lattice-ordered groups and Lie algebras (in Russian),* Algebra i Logika **16**(1977), 40-45.

4. *Decomposition of free l-groups into l-direct product (in Russian),* Sibirsk. Mat. Z. **21**(1980), 63-69.

5. *l-varieties without an independendent basis of identities (in Russian),* Math. Slovaca **32**(1982), 417-425.

6. *On the theory of varieties of varieties of lattice-ordered groups (in Russian),* Czech. Math. J. **32**(1982), 364-372.

7. *Coverings in the lattice of l-varieties (in Russian),* Algebra i Logika **22**(1983), 53-60.

8. *The lattice of o-approximable l-varieties (in Russian),* Czech. Math. J. **34**(1984),6-17.

9. *Free products of l-groups (in Russian),* Algebra i Logika **23**(1984), 493-511.

10. *Quasivarieties of l-groups and groups (in Russian),* Sibirsk. Mat. Z. **26**(1985), 111-117.

11. *On nilpotent lattice-ordered groups (in Russian),* Mat. Zametki **45**(1989), 72-79.

12. *On the infinite distributuvity in the lattice of l-varieties (in Russian),* Sibirsk. Mat. Z. **30**(1989), 216-220.

13. *Independent axiomatization of varieties of lattice-ordered groups,* Czech. Math. J. **42**(1992), 53-57.

MURA R.T.B., RHEMTULLA A.H.

1. *Notes on orderable groups,* Edmonton: University of Alberta, 1975.

2. *Orderable groups,* Marcel Dekker , New York, 1977.

NEUMANN B.H.

1. *On ordered divisible rings,* Trans. Amer. Math. Soc. **66**(1949), 202-252.

2. *An embedding theorem for algebraic systems,* Proc. London Math. Soc. **4**(1954), 138-153.

NEUMANN H.

1. *Varieties of groups,* Springer, Berlin, 1967.

OHNISHI H.

1. *On linearisation of ordered groups,* Osaka J. Math. **2**(1950), 161-164.

2. *Linear order on a group,* Osaka J. Math. **4**(1952), 17-18.

OLSHANSKII A.YU.

1. *On the problem of a finite basis of identities in groups (in Russian),* Izv. Akad. Nauk SSSR, Ser. Mat., **34**(1970), 316-384.

PIERCE K.R.

1. *Amalgamating abelian ordered groups,* Pacific J. Math. **43**(1972), 711-723.

2. *Amalgamations of lattice-ordered groups,* Trans. Amer. Math. Soc. **172**(1972), 249-260.

PIERCE R.S.

1. *Homomorphisms of semigroups,* Ann. Math. **59**(1954), 287-291.

PODDERYUGIN V.D.

1. *Conditions of orderability of arbitrary ring (in Russian),* Uspekhi Mat. Nauk **9**(1954), 211-216.

2. *Conditions of orderability of a group(in Russian),* Izv. Akad. Nauk SSSR **21**(1957), 199-208.

POWELL W.B., TCINAKIS C.

1. *Free products in the class of abelian l-groups* , Pasific J. Math. **104**(1983), 429-442.

READ J.A.

1. *Non-overlapping lattice-ordered groups,* Doctoral Thesis, University Wisconsin, 1971.

2. *Wreath products of non-overlapping groups,* Canad. Math. Bull. **17**(1975), 713-722.

REILY N.R.

1. *Permutational products of lattice-ordered groups,* J.Austral. Math, Soc. **13**(1972), 25-34.

2. *A subsemilattice of the lattice of varieties of lattice-ordered groups,* Canadian J. Math. **33**(1981), 1309-1318.

3. *Varieties of lattice ordered groups that contain no non-abelian o-groups are solvable,* Order **3**(1986), 287-297.

4. *Varieties of lattice-ordered groups,* in *Lattice-ordered groups,* Advances and Techniques, Mathematics and its Applications, ed. A.M.W. Glass and W.Ch. Holland, Kluwer Academic Publishers, Dordrecht/ Boston/London, 1988.

REILY N.R., WROBLEWSKI R.

1. *Suprema of classes of generalized Scrimger varieties of lattice-ordered groups,* Math. Z. **176**(1981), 293-309.

RHEMTULLA A.H.

1. *Right-ordered groups,* Canad. J. Math. **24**(1972), 891-895.

2. *Residually F_p-groups for many primes p are orderable,* Proc. Amer. Math. Soc. **41**(1973), 31-33.

RIEGER L.S.

1. *On the ordered and cyclically ordered groups (1 - 3),* Věstn. kral. Čs.společ Nauk **6**(1946), 1-31; **1**(1947), 1-33; **1**(1948), 1-26.

RIESZ F.

1. *Sur quelques notions foundamentales dans la théorie générale des opérateurs linéaires,* Ann. Math. **41**(1940), 174-206.

ROBINSON D.J.S.

1. *Finiteness conditions and generalized soluble groups (Part 1, 2),* Springer, Berlin, 1972.

SCRIMGER E.B.

1. *A large class of small varieties of lattice-ordered groups,* Proc. Amer. Math. Soc. **51**(1975), 301-306.

ŠIK F.

1. *Über subdirekte Summen geordneter Gruppen,* Czech. Math. J. **10** (1960), 400-424.

2. *Über direkte Zerlegungen gerichteter Gruppen,* Math. Nachr. **25** (1963), 95-110.

3. *Archimedische kompakt erzeugte Verbandsgruppen,* Math. Nachr. **38**(1968), 323-340.

SMIRNOV D.M.

1. *Infrainvariant subgroups (in Russian),* Ucen. Zap. Ivanovs. Gos. Ped. Inst. **4**(1953), 92-96.

2. *On relatively free multioperator groups (in Russian),* Dokl. Akad. Nauk SSSR **150**(1963), 44-47.

3. *On generalized soluble groups and its group rings (in Russian),* Dokl. Akad. Nauk SSSR **155**(1964), 535-537.

4. *Ordered multioperator groups (in Russian),* Sibirsk. Mat. Z. **6**(1965), 433-458.

5. *On generalized soluble groups and its group rings (in Russian),* Math. Sbornik **67**(1965),366-383.

6. *Groups of automorphisms of group rings of right orderable groups (in Russian),* Algebra i Logika **4**(1965), 31-46.

7. *Right-ordered groups (in Russian),* Algebra i Logika **5**(1966), 41-59.

8. *One-side orders in groups with increasing central series (in Russian),* Algebra i Logika **6**(1967), 77-87.

SMITH J.E.

1. *The lattice of l-group l-varieties,* Trans. Amer. Math. Soc. **257**(1980), 347-357.

2. *A new family of l-group varieties,* Houston J.Math. **7**(1981), 551-570.

STONE M.N.

1. *The theory of representations for Boolean algebras,* Trans. Amer. Soc. **40**(1936), 37-111.

STROJNOWSKI A., WOJAS W.

1. *Some remarks on o-groups,* Demonstr. Math. **23**(1990), 799-801.

SZMIELEW W.

1. *Elementary properties of abelian groups,* Fundam. Math. **41** (1955), 203-271.

TEH H.H.

1. *Construction of orders in abelian groups,* Proc. Cambr. Phil. Soc. **57**(1961), 476-482.

TREVISAN G.

1. *Classificazione dei semplice ordenamenti di un gruppo libero commutativo con n generatori,* Rend. Semin. mat Univ. Padova **22**(1953), 143-156.

VAN REE D.A.

1. *Quasivarieties of l-metabelian lattice-ordered groups,* Doctoral Thesis, Bowling Green State University, 1991.

VARAKSIN S.V.

1. *Lattice-ordered groups constructed from right-ordered groups (in Russian),* Algebra i Logika **28**(1989), 524-533.

2. *Varieties generated by simple l-groups (in Russian),* Sibirsk. Mat. J. **31**(1990),167-180.

3. *Semilinear orders on soluble and nilpotent groups (in Russian),* Algebra i Logika **29**(1990), 631-636.

VAUGHAN-LEE M.R.

1. *Uncountably many varieties of groups,* Bull. London Math. Soc. **2**(1970), 280-286.

VEKSLER A.I.

1. *A new construction of the Dedekind completion of vector lattices and divisible l-groups (in Russian),* Sibirsk Mat. J. **10**(1969), 891-896.

2. *Archimedean free linear hulls of partially ordered groups (in Russian),* Ordered sets and lattices, **4**, 11-23, Izdat. Saratov. Univ., Saratov, 1977.

VINOGRADOV A.A.

1. *On the free product of ordered groups (in Russian),* Math. Sbornik **25**(1949), 163-168.

2. *Non-axiomatizability of lattice-ordered groups (in Russian),* Sibirsk. Mat. J. **13**(1971), 331-332.

WEINBERG E.C.

1. *Completely distributive lattice-ordered groups,* Pacific J. Math., **12**(1962), 1131-1137.

2. *Free lattice-ordered abelian groups,* Math. Ann. **151**(1963), 187-199.

3. *Free lattice-ordered abelian groups (2),* Math. Ann, **159**(1965), 217-222.

4. *Embedding in a divisible lattice-ordered group,* J. London Math. Soc. **42**(1967),504-506.

WIGOLD J.

1. *Semigroups covering of groups (2),* Mat. - fiz. čas. **12** (1962), 217-223.

WOLFENSTEIN S.

1. *Values normales dans un groupe réticulé,* Accademia Nazionale dei Lincei, **44**(1968), 337-342.

ZAITSEVA M.I.

1. *On the family of orders of abelian group,* Uspechi Mat. Nauk (in Russian) **8**(1953), 135-137.

2. *Right-ordered groups (in Russian),* Učen. Zap. Suisk. Gos. Ped. Inst. **6**(1958), 215-226.

Index

MIX
Papier aus verantwortungsvollen Quellen
Paper from responsible sources
FSC® C105338

If you have any concerns about our products,
you can contact us on
ProductSafety@springernature.com

In case Publisher is established outside the EU,
the EU authorized representative is:
Springer Nature Customer Service Center GmbH
Europaplatz 3, 69115 Heidelberg, Germany

Printed by Libri Plureos GmbH
in Hamburg, Germany